Electrochemical Energy Storage Materials

Electrochemical Energy Storage Materials

Topic Editors

Huang Zhang
Yuan Ma

MDPI

Basel • Beijing • Wuhan • Barcelona • Belgrade • Novi Sad • Cluj • Manchester

Topic Editors
Huang Zhang
School of Electrical and Electronic Engineering
Harbin University of Science and Technology
Harbin
China

Yuan Ma
School of Energy and Environment
Southeast University
Nanjing
China

Editorial Office
MDPI AG
Grosspeteranlage 5
4052 Basel, Switzerland

This is a reprint of the Topic, published open access by the journals *Energies* (ISSN 1996-1073), *Nanomaterials* (ISSN 2079-4991), *Materials* (ISSN 1996-1944), *Electrochem* (ISSN 2673-3293) and *Batteries* (ISSN 2313-0105), freely accessible at: https://www.mdpi.com/topics/EESM.

For citation purposes, cite each article independently as indicated on the article page online and as indicated below:

Lastname, A.A.; Lastname, B.B. Article Title. *Journal Name* **Year**, *Volume Number*, Page Range.

ISBN 978-3-7258-3861-5 (Hbk)
ISBN 978-3-7258-3862-2 (PDF)
https://doi.org/10.3390/books978-3-7258-3862-2

Contents

About the Editors vii

Preface ix

Huang Zhang and Yuan Ma
Topic "Electrochemical Energy Storage Materials"—An Overview
Reprinted from: *Energies* **2025**, *18*, 1279, https://doi.org/10.3390/en18051279 1

Kavitha Mulackampilly Joseph, Hunter J. Kasparian and Vesselin Shanov
Carbon Nanotube Fiber-Based Wearable Supercapacitors—A Review on Recent Advances
Reprinted from: *Energies* **2022**, *15*, 6506, https://doi.org/10.3390/en15186506 5

Ryohei Mori
Separator Materials for Lithium Sulfur Battery—A Review
Reprinted from: *Electrochem* **2023**, *4*, 485–522, https://doi.org/10.3390/electrochem4040032 . . . 47

Haiwei Hu, Yanyan Guo and Jiang Zhao
Manufacturing Shape-Controllable Flexible PEDOT/rGO Composite Electrodes for Planar Micro-Supercapacitors
Reprinted from: *Materials* **2024**, *17*, 2144, https://doi.org/10.3390/ma17092144 85

Xin Zhang, Shi Liu, Yuqi Zhao, Haicun Yang and Jinchun Li
Honeycomb-like Hierarchical Porous Carbon from Lignosulphonate by Enzymatic Hydrolysis and Alkali Activation for High-Performance Supercapacitors
Reprinted from: *Energies* **2023**, *16*, 3824, https://doi.org/10.3390/en16093824 97

Meruyert Nazhipkyzy, Mukhtar Yeleuov, Shynggyskhan T. Sultakhan, Anar B. Maltay, Aizhan A. Zhaparova, Dana D. Assylkhanova and Renata R. Nemkayeva
Electrochemical Performance of Chemically Activated Carbons from Sawdust as Supercapacitor Electrodes
Reprinted from: *Nanomaterials* **2022**, *12*, 3391, https://doi.org/10.3390/nano12193391 114

Lander Lizaso, Idoia Urdampilleta, Miguel Bengoechea, Iker Boyano, Hans-Jürgen Grande, Imanol Landa-Medrano, et al.
Waterborne $LiNi_{0.5}Mn_{1.5}O_4$ Cathode Formulation Optimization through Design of Experiments and Upscaling to 1 Ah Li-Ion Pouch Cells
Reprinted from: *Energies* **2023**, *16*, 7327, https://doi.org/10.3390/en16217327 127

Minki Jo, Soojin Sim, Juhyeong Kim, Pilgun Oh and Yoonkook Son
Micron-Sized SiO_x-Graphite Compound as Anode Materials for Commercializable Lithium-Ion Batteries
Reprinted from: *Nanomaterials* **2022**, *12*, 1956, https://doi.org/10.3390/nano12121956 145

Axel Durdel, Sven Friedrich, Lukas Hüsken and Andreas Jossen
Modeling Silicon-Dominant Anodes: Parametrization, Discussion, and Validation of a Newman-Type Model
Reprinted from: *Batteries* **2023**, *9*, 558, https://doi.org/10.3390/batteries9110558 155

Md Yusuf Ali, Tianyu Chen, Hans Orthner and Hartmut Wiggers
Spray-Flame Synthesis of NASICON-Type Rhombohedral (α) $Li_{1+x}Y_xZr_{2-x}(PO_4)_3$ [x = 0–0.2] Solid Electrolytes
Reprinted from: *Nanomaterials* **2024**, *14*, 1278, https://doi.org/10.3390/nano14151278 180

Hyewoo Noh, Daeil Kim, Wooyoung Lee, Boyun Jang, Jeong Sook Ha and Ji Haeng Yu
Surface Modification of Ga-Doped-LLZO ($Li_7La_3Zr_2O_{12}$) by the Addition of Polyacrylonitrile for the Electrochemical Stability of Composite Solid Electrolytes
Reprinted from: *Energies* **2023**, *16*, 7695, https://doi.org/10.3390/en16237695 . . . . **199**

Ruslan R. Samigullin, Zoya V. Bobyleva, Maxim V. Zakharkin, Emiliya V. Zharikova, Marina G. Rozova, Oleg A. Drozhzhin and Evgeny V. Antipov
On the Thermal Stability of Selected Electrode Materials and Electrolytes for Na-Ion Batteries
Reprinted from: *Energies* **2024**, *17*, 3970, https://doi.org/10.3390/en17163970 . . . . **218**

Junxian Li, Wenli Shu, Guangwan Zhang, Jiashen Meng, Chunhua Han, Xiujuan Wei and Xuanpeng Wang
Co/Al Co-Substituted Layered Manganese-Based Oxide Cathode for Stable and High-Rate Potassium-Ion Batteries
Reprinted from: *Materials* **2024**, *17*, 1277, https://doi.org/10.3390/ma17061277 . . . . **230**

Hao Tian, Huanlin Zhang, You Zuo, Lei Ling, Tengfei Meng, Hang Zhang, et al.
An Artificial $MnWO_4$ Cathode Electrolyte Interphase Enabling Enhanced Electrochemical Performance of δ-MnO_2 Cathode for Aqueous Zinc Ion Battery
Reprinted from: *Materials* **2023**, *16*, 3228, https://doi.org/10.3390/ma16083228 . . . . **242**

Sainan Liu, Yangyang Sun, Jing Yang, Yi Zhang and Zhenyang Cai
Highly Loaded and Binder-Free Molybdenum Trioxide Cathode Material Prepared Using Multi-Arc Ion Plating for Aqueous Zinc Ion Batteries
Reprinted from: *Materials* **2022**, *15*, 5954, https://doi.org/10.3390/ma15175954 . . . . **253**

Xiaoping Tan, Gaoli Guo, Kaidi Wang and Huang Zhang
Synthesis and Electrochemical Performance of the Orthorhombic $V_2O_5 \cdot nH_2O$ Nanorods as Cathodes for Aqueous Zinc Batteries
Reprinted from: *Nanomaterials* **2022**, *12*, 2530, https://doi.org/10.3390/nano12152530 . . . . . . **262**

About the Editors

Huang Zhang

Huang Zhang is a Full Professor at the School of Electrical and Electronic Engineering, Harbin University of Science and Technology, and holds a concurrent position as a visiting researcher at the Key Laboratory of Flexible Electronics of Zhejiang Province. He has previously served as a predoctoral research assistant at KU Leuven (Belgium). Afterward, he received his PhD in 2018 from the Karlsruhe Institute of Technology (Germany) and then worked as a junior scientist at the Helmholtz Institute Ulm (Germany) and an Associate Professor at the Institute of Flexible Electronics, Northwestern Polytechnical University. His primary research focus is on the development of key materials for secondary batteries, with an emphasis on creating green, safe, and efficient technologies for energy storage. He has been listed among the top 2% of scientists worldwide for his annual scientific impact, as compiled by Stanford University, and has received multiple scientific and technological achievements for his contributions to the field. To date, he has co-authored more than 90 papers, and led various projects funded by the National Natural Science Foundation of China, provincial-level natural science foundations, and corporate-sponsored research initiatives.

Yuan Ma

Yuan Ma has been a professor at Southeast University, China, since 2023. His research centers on all-solid-state battery applications and the development of next-generation battery materials for electrochemical energy storage. Before joining Southeast University, he completed a postdoctoral fellowship under the mentorship of Prof. Janek and Dr. Brezesinski at the Institute of Nanotechnology, Karlsruhe Institute of Technology (KIT), Germany. Earlier, in 2019, he earned his Ph.D. under the supervision of Prof. Passerini at the Helmholtz Institute Ulm/KIT.

Preface

The global shift toward sustainable energy systems faces a pivotal challenge: ensuring the reliable integration of renewable energy sources such as solar, wind, and tidal power into existing grids. Electrochemical energy storage (EES) technologies are critical to addressing this challenge, as they provide solutions to the intermittency issues inherent in renewable energy. This reprint, Electrochemical Energy Storage Materials, compiles groundbreaking research and insights from leading experts in electrochemistry and materials engineering, offering a comprehensive exploration of innovations in materials science that are reshaping the future of EES systems.

The scope of this reprint spans from atomic-scale mechanistic studies to sustainable synthesis methods and industrial-scale manufacturing breakthroughs. By connecting theoretical research with practical deployment challenges, this collection aims to advance next-generation batteries and supercapacitors, offering strategies that enhance performance, durability, and scalability. These advancements not only address technical limitations but also align with broader goals of ecological sustainability and equitable energy access.

The motivation for this work stems from the urgent need to accelerate the energy transition while balancing technological progress with planetary boundaries. As renewable energy adoption continues to grow, the limitations of current storage solutions highlight the importance of interdisciplinary collaboration and materials innovation. This reprint serves as a resource for researchers, engineers, and policymakers seeking to understand and implement cutting-edge solutions in energy storage.

Contributions to this reprint come from pioneering experts in the field, whose work reflects a shared commitment to advancing both scientific understanding and practical applications. We express our gratitude to all authors for their dedication and insights. Special thanks also go to the editorial team and reviewers, whose efforts have ensured the rigor and coherence of this collection.

This reprint is intended for a diverse audience, including academic researchers, industry professionals, and policymakers working at the intersection of energy storage, materials science, and sustainability. By fostering dialogue and collaboration across disciplines, we aim to transform energy storage from a technical barrier into a catalyst for building resilient and equitable societies.

Huang Zhang and Yuan Ma
Topic Editors

energies

MDPI

Editorial

Topic "Electrochemical Energy Storage Materials"—An Overview

Huang Zhang [1,*] and Yuan Ma [2]

1 Key Laboratory of Engineering Dielectric and Applications (Ministry of Education), School of Electrical and Electronic Engineering, Harbin University of Science and Technology, Harbin 150080, China
2 Confucius Energy Storage Lab, Key Laboratory of Energy Thermal Conversion and Control of Ministry of Education, School of Energy and Environment & Z Energy Storage Center, Southeast University, Nanjing 210096, China; yuan.ma@seu.edu.cn
* Correspondence: zhang.huang@hrbust.edu.cn

The quest for efficient and reliable electrochemical energy storage (EES) systems is at the forefront of modern energy research, as these systems play a pivotal role in addressing the intermittent nature of renewable energy sources and the growing demands of portable electronics and electric vehicles. The performance, reliability, and efficiency of EES systems are intrinsically linked to the materials used in their construction, making material innovation a critical driver for advancements in this field. This Topic, "Electrochemical Energy Storage Materials", in collaboration with the esteemed journals *Energies*, *Nanomaterials*, *Materials*, *Electrochem*, and *Batteries*, presents a curated collection of cutting-edge research articles and comprehensive reviews that highlight the latest advancements across a broad spectrum of topics, including, but not limited to, lithium-ion batteries, solid-state batteries, sodium-ion batteries, potassium-ion batteries, and zinc-ion batteries.

Son et al. [Contribution 1] investigated the performance of graphite-blended electrodes with micron-sized SiO_x anode material, a widely utilized Si-based anode material in lithium-ion batteries, aiming to generate a commercially viable and practical solution. Their results show that SiO_x blending electrodes have better cycling performance than silicon micron particle blending electrodes. Using post-mortem techniques, the degrading mechanisms were clarified, providing insights into their electrochemical behavior and helping to meet both industrial and academic demands for improved anode materials.

Zhang et al. [Contribution 2] reported the synthesis of orthorhombic $V_2O_5{\cdot}nH_2O$ nanorods as cathodes for aqueous zinc batteries. This material exhibited superior zinc storage performance with high reversible capacity and long-term cycle stability, due to enhanced material stability, reduced electrolyte breakdown, and improved charge transfer kinetics. Additionally, a full cell design using microsized zinc powder as the anode demonstrated high capacity and stable cycling, confirming the practical applicability of these materials.

Cai et al. [Contribution 3] demonstrated the fabrication of highly loaded and binder-free α-MoO_3 on carbon fiber cloth (α-MoO_3@CFC) up to 15 mg/cm^2 via multi-arc ion plating and subsequent oxidative heating treatment. The α-MoO_3@CFC cathode material possessed significantly improved electrochemical performance compared to commercial MoO_3 using conventional slurry-based electrode fabrication processes. This study offers a simple way to make high-loading, binder-free cathodes for aqueous zinc-ion batteries (ZIBs), suggesting potential for use in future flexible devices.

Shanov et al. [Contribution 4] provided a comprehensive review on the basics and latest advancements in CNT-fiber-based supercapacitors. They overviewed supercapaci-

Received: 17 January 2025
Accepted: 4 March 2025
Published: 5 March 2025

Citation: Zhang, H.; Ma, Y. Topic "Electrochemical Energy Storage Materials"—An Overview. *Energies* **2025**, *18*, 1279. https://doi.org/10.3390/en18051279

tor types based on charge storage mechanisms and electrode design and discussed various fiber-making methods with an emphasis on improving electrochemical performance, stretchability, and multifunctionality. Lastly, they prospected the current performance and scalability challenges of CNT-based supercapacitors, highlighting their future potential in wearable electronic devices.

Nazhipkyzy et al. [Contribution 5] demonstrated a new approach to using the sawdust of karagash and pine trees to produce activated carbon material for EDLC application. At a relatively high scan rate of 160 mV/s, a decent specific capacitance of 147 F/g and 114 F/g was obtained, leading to high energy densities of 26.0 and 22.1 Wh/kg based on average electrode mass. This study showed that the use of electrode composites consisting of activated carbon derived from an available biomass precursor leads to the improved performance of electrochemical capacitors.

Cai et al. [Contribution 6] created a $MnWO_4$ cathode electrolyte interphase (CEI) layer on a δ-MnO_2 cathode using cyclic voltammetry. This layer effectively prevents Mn dissolution and enhances reaction kinetics. Consequently, the δ-MnO_2 cathode with the CEI layer shows improved cycling performance, retaining 98.2% of its capacity after 2000 cycles at 10 A/g. This shows that the simple electrochemical method used to build the $MnWO_4$ CEI layer significantly advances the development of MnO_2 cathodes for aqueous zinc ion batteries.

Li et al. [Contribution 7] prepared a series of 3D hierarchical porous carbon materials (PCs) by a low-cost and template-free method using calcium lignosulfonate (CL) as the precursor. The effects of enzymatic hydrolysis and different KOH feeding ratios on the structure and electrochemical properties of enzymatic hydrolysis CL (EHCL)-derived PCs were evaluated. The optimal PCs with a honeycomb-like microscopic morphology possessed a high capacitance (147 F/g at 0.25 A/g), significant rate capability (capacitance retention of 78% at 10 A/g), and good long-term cycling stability (95.3% capacitance retention after 15,000 cycles), demonstrating a promising route to apply renewable lignin derivatives in high-performance supercapacitors.

Landa-Medrano et al. [Contribution 8] reported the use of a design of experiments (DoE) matrix to obtain a mathematical model that predicts the best formulation of cathodes with $LiNi_{0.5}Mn_{1.5}O_4$ (LNMO) as the active material. After obtaining and validating this formulation, its upscaling to a semi-industrial coating line is described. The optimization and upscaling of the LNMO electrode recipe resulted in obtaining high-energy pouch cells of 1 Ah that could be used to gain knowledge in the chemistry of this lithium-ion cathode material.

Mori [Contribution 9] summarized the recent development of functional separators for lithium–sulfur batteries. The classification and modification of the separators with their working mechanisms, as well as the resulting electrochemical performance, were described. In addition, necessary performance metrics and material characteristics for separators were prospected to further improve the electrochemical properties of lithium–sulfur batteries.

Durdel et al. [Contribution 10] introduced, for the first time, a Newman p2D model for a lithium-ion cell with a silicon-dominant anode and a nickel–cobalt–aluminum–oxide cathode. The parametrization was based on values from the electrode manufacturing process, measured values using lab cells, and data from the literature. Future work was emphasized with more in-depth studies on the material parameters for silicon to expand the data available in the literature and facilitate further simulation work.

Yu et al. [Contribution 11] designed a three-layer garnet-applied PAN-added polymer ceramic electrolyte (g-PPCE) consisting of a PVDF-HFP-based SPE-Ga-doped-LLZO with 5 wt.% of PAN by using a simple tape-casting and impregnation process. Both LFP||Li-metal and NCM||Li-metal batteries using the g-PPCE exhibited high discharge capacity and cycle stability. The preparation of a pouch cell with a g-PPCE and a cell capacity of 5 mAh demonstrated the potential for large-scale production.

Wang et al. [Contribution 12] reported a Co/Al co-substitution strategy to construct a P3-type $K_{0.45}Mn_{0.7}Co_{0.2}Al_{0.1}O_2$ cathode material for potassium-ion batteries. This co-substitution effectively suppressed the Jahn–Teller distortion and alleviated the severe phase transition during K^+ intercalation/de-intercalation processes, resulting in improved potassium storage performance. The Co/Al co-substitution strategy employed in this work provided an effective approach for designing high-performance Mn-based cathode materials in potassium-ion batteries.

Zhao et al. [Contribution 13] presented shape-controllable flexible composite electrodes based on Poly(3,4-ethylenedioxythiophene) and reduced graphene oxide (PEDOT/rGO) for planar micro-supercapacitors. The high electrical conductivity of rGO facilitated rapid electron transport on the electrode surface, and its high specific surface area contributed to more active sites available for electrochemical reactions, enhancing the electrochemical performance. This presented work highlighted the potential of PEDOT/rGO electrodes in flexible supercapacitor applications.

Wiggers et al. [Contribution 14] reported, for the first time, the bottom-up approach for α phase NASICON $Li_{1+x}Y_xZr_{2-x}(PO_4)_3$ (α-LYZP) synthesis. Phase-pure rhombohedral α-LYZP was synthesized by a spray-flame process followed by a short annealing step. The results showed that the addition of Y^{3+} as a dopant was efficient to stabilize the α phase and enhance the ionic conductivity.

Drozhzhin et al. [Contribution 15] used differential scanning calorimetry (DSC) and ex situ powder X-ray diffraction to examine the thermal stability of various sodium-ion electrolytes and electrode materials. The results revealed that liquid sodium-ion electrolytes exhibit good thermal stability, only beginning to decompose at 270–300 °C. Vanadium-based cathodes, while stable on their own, show increased decomposition energy and lower DSC peak temperatures when in contact with a specific electrolyte solution. Most significantly, the anode material's interaction with the electrolyte had the greatest thermal effect, with the decomposition heat of the soaked, charged electrode being nearly 40% higher than the combined decomposition heat of the dry electrode and electrolyte alone.

These contributions from this Topic, entitled "Electrochemical Energy Storage Materials", offer the latest and most innovative solutions that can overcome the current limitations of EES materials and devices, providing valuable insights into the challenges and opportunities associated with the development of next-generation EES materials, as well as strategies for overcoming current limitations and scaling up to commercial applications. Together with the co-Topic Editor, Prof. Dr. Yuan Ma, we wish to thank the authors that contributed their works to this Topic.

Acknowledgments: The Topic Editors would like to thank the anonymous reviewers for their valuable efforts, which helped the authors to improve their manuscripts.

Conflicts of Interest: The authors declare no conflicts of interest.

List of Contributions:

1. Jo, M.; Sim, S.; Kim, J.; Oh, P.; Son, Y. Micron-Sized SiOx-Graphite Compound as Anode Materials for Commercializable Lithium-Ion Batteries. *Nanomaterials* **2022**, *12*, 1956.
2. Tan, X.; Guo, G.; Wang, K.; Zhang, H. Synthesis and Electrochemical Performance of the Orthorhombic $V_2O_5{\cdot}nH_2O$ Nanorods as Cathodes for Aqueous Zinc Batteries. *Nanomaterials* **2022**, *12*, 2530.
3. Liu, S.; Sun, Y.; Yang, J.; Zhang, Y.; Cai, Z. Highly Loaded and Binder-Free Molybdenum Trioxide Cathode Material Prepared Using Multi-Arc Ion Plating for Aqueous Zinc Ion Batteries. *Materials* **2022**, *15*, 5954.

4. Joseph, K.M.; Kasparian, H.J.; Shanov, V. Carbon Nanotube Fiber-Based Wearable Supercapacitors—A Review on Recent Advances. *Energies* **2022**, *15*, 6506.
5. Nazhipkyzy, M.; Yeleuov, M.; Sultakhan, S.T.; Maltay, A.B.; Zhaparova, A.A.; Assylkhanova, D.D.; Nemkayeva, R.R. Electrochemical Performance of Chemically Activated Carbons from Sawdust as Supercapacitor Electrodes. *Nanomaterials* **2022**, *12*, 3391.
6. Tian, H.; Zhang, H.; Zuo, Y.; Ling, L.; Meng, T.; Zhang, H.; Sun, X.; Cai, S. An Artificial $MnWO_4$ Cathode Electrolyte Interphase Enabling Enhanced Electrochemical Performance of δ-MnO_2 Cathode for Aqueous Zinc Ion Battery. *Materials* **2023**, *16*, 3228.
7. Zhang, X.; Liu, S.; Zhao, Y.; Yang, H.; Li, J. Honeycomb-like Hierarchical Porous Carbon from Lignosulphonate by Enzymatic Hydrolysis and Alkali Activation for High-Performance Supercapacitors. *Energies* **2023**, *16*, 3824.
8. Lizaso, L.; Urdampilleta, I.; Bengoechea, M.; Boyano, I.; Grande, H.-J.; Landa-Medrano, I.; Eguia-Barrio, A.; de Meatza, I. Waterborne $LiNi_{0.5}Mn_{1.5}O_4$ Cathode Formulation Optimization through Design of Experiments and Upscaling to 1 Ah Li-Ion Pouch Cells. *Energies* **2023**, *16*, 7327.
9. Mori, R. Separator Materials for Lithium Sulfur Battery—A Review. *Electrochem* **2023**, *4*, 485–522.
10. Durdel, A.; Friedrich, S.; Hüsken, L.; Jossen, A. Modeling Silicon-Dominant Anodes: Parametrization, Discussion, and Validation of a Newman-Type Model. *Batteries* **2023**, *9*, 558.
11. Noh, H.; Kim, D.; Lee, W.; Jang, B.; Ha, J.S.; Yu, J.H. Surface Modification of Ga-Doped-LLZO ($Li_7La_3Zr_2O_{12}$) by the Addition of Polyacrylonitrile for the Electrochemical Stability of Composite Solid Electrolytes. *Energies* **2023**, *16*, 7695.
12. Li, J.; Shu, W.; Zhang, G.; Meng, J.; Han, C.; Wei, X.; Wang, X. Co/Al Co-Substituted Layered Manganese-Based Oxide Cathode for Stable and High-Rate Potassium-Ion Batteries. *Materials* **2024**, *17*, 1277.
13. Hu, H.; Guo, Y.; Zhao, J. Manufacturing Shape-Controllable Flexible PEDOT/rGO Composite Electrodes for Planar Micro-Supercapacitors. *Materials* **2024**, *17*, 2144.
14. Ali, M.Y.; Chen, T.; Orthner, H.; Wiggers, H. Spray-Flame Synthesis of NASICON-Type Rhombohedral (α) Li_1+xYxZr_2−x$(PO_4)_3$ [x = 0–0.2] Solid Electrolytes. *Nanomaterials* **2024**, *14*, 1278.
15. Samigullin, R.R.; Bobyleva, Z.V.; Zakharkin, M.V.; Zharikova, E.V.; Rozova, M.G.; Drozhzhin, O.A.; Antipov, E.V. On the Thermal Stability of Selected Electrode Materials and Electrolytes for Na-Ion Batteries. *Energies* **2024**, *17*, 3970.

energies

MDPI

Review

Carbon Nanotube Fiber-Based Wearable Supercapacitors—A Review on Recent Advances

Kavitha Mulackampilly Joseph [1], Hunter J. Kasparian [2] and Vesselin Shanov [1,2,*]

[1] Department of Mechanical and Materials Engineering, University of Cincinnati, Cincinnati, OH 45221, USA
[2] Department of Chemical and Environmental Engineering, University of Cincinnati, Cincinnati, OH 45221, USA
* Correspondence: shanovvn@ucmail.uc.edu

Abstract: As wearable electronic devices are becoming an integral part of modern life, there is a vast demand for safe and efficient energy storage devices to power them. While the research and development of microbatteries and supercapacitors (SCs) have significantly progressed, the latter has attracted much attention due to their excellent power density, longevity, and safety. Furthermore, SCs with a 1D fiber shape are preferred because of their ease of integration into today's smart garments and other wearable devices. Fiber supercapacitors based on carbon nanotubes (CNT) are promising candidates with a unique 1D structure, high electrical and thermal conductivity, outstanding flexibility, excellent mechanical strength, and low gravimetric density. This review aims to serve as a comprehensive publication presenting the fundamentals and recent developments on CNT-fiber-based SCs. The first section gives a general overview of the supercapacitor types based on the charge storage mechanisms and electrode configuration, followed by the various fiber fabrication methods. The next section explores the different strategies used to enhance the electrochemical performance of these SCs, followed by a broad study on their stretchability and multifunctionality. Finally, the review presents the current performance and scalability challenges affecting the CNT-based SCs, highlighting their prospects.

Keywords: carbon nanotube fiber; energy storage; wearable technologies; supercapacitor; fiber spinning

Citation: Joseph, K.M.; Kasparian, H.J.; Shanov, V. Carbon Nanotube Fiber-Based Wearable Supercapacitors—A Review on Recent Advances. *Energies* **2022**, *15*, 6506. https://doi.org/10.3390/en15186506

Academic Editors: Huang Zhang and Yuan Ma

Received: 1 August 2022
Accepted: 1 September 2022
Published: 6 September 2022

1. Introduction

Wearable electronics have become indispensable in modern life, with rapid advances in the devices and their applications, such as in displays, fitness trackers, body-mounted sensors, artificial intelligence (AI) hearing aids, virtual reality (VR) headsets, and smart textiles. These devices require equally flexible, wearable, and high-performance energy storage devices to maintain their performance and safety. Unfortunately, most conventional 3D energy storage devices are stiff, rigid, heavy, and bulky. Therefore, recently a rapid shift has been seen from 3D to 2D planar-type energy storage devices in the form of metal-coated plastics, cloths or fabrics coated with conductive carbon, and conductive papers, which are flexible, portable, and lightweight. However, they were proven to be uncomfortable as wearable devices as they lacked the flexibility to conform to various body types and could not comply with the breathability requirements (to allow circulation of air passage between the body and the garment) [1–5]. Therefore, fiber-based 1D energy storage devices with desirable flexibility, lighter weights, smaller volumes, and conformability are the most promising solution for mitigating the abovementioned issues [6,7].

Supercapacitors (SCs) are superior to conventional SCs in energy density and are superior to batteries in terms of power delivery. One of the most appropriate chemical elements suitable for electrochemical capacitor applications is carbon, due to its high electrical conductivity, mechanical strength, wide natural presence, low density, and controllable structure and properties. Being placed in the middle of the eight-element period in the periodic table, it exhibits moderate electronegativity, with an electronic structure ($1S^2$,$2S^2$,$2P^2$)

and four valence electrons. Various forms of carbon, such as carbon black, natural or artificial graphite, carbon nanotubes (CNTs), graphene, and activated carbon, are substantially used in commercial SCs as conductive additives or active materials. However, most of them are applied as conductive coatings and films on metallic supports consolidated with binders, thereby making the SCs often bulkier, heavier, and non-flexible.

On the other hand, CNT-fiber-shaped SCs are long in size, mechanically robust, and have a tunable structure. They are macroassemblages of CNTs in the 1D fiber form. They can integrate all of the desirable features of fiber-based SC, such as extraordinary electrical conductivity, mechanical strength, flexibility, tunable porosity and surface area, and excellent stability. These unique energy storage devices are also self-standing, binder-free, and can be easily integrated into textiles and wearable devices [5,7–9].

Wearable supercapacitors represent a hot research topic, and recently there have been excellent reviews regarding supercapacitors [2,10–12], microsupercapacitors [13], flexible supercapacitors [14,15], carbon-based supercapacitors [16–18], and fiber-based supercapacitors [1,4,7,9,19,20]. However, the reviews on carbon nanotube fiber-based SCs are scarce [5,21,22]. These publications present significant studies on fiber spinning, energy conversion, novel fiber fabrication methods, and smart applications of CNT yarn. However, a comprehensive review discussing the fundamental aspects and the recent advances of CNT-based SCs is still in demand. The initial sections of the current review focus on the SCs' fundamentals, including the types of SCs based on the charge storage, configuration, and fabrication method. The next section explores strategies that can be adopted for the greater performance of CNT electrodes and CNT fiber SCs as a whole, followed by a study on 'stretchable' and 'multifunctional' CNT-based SCs. Finally, the review analyzes the challenges of CNT-based SCs, followed by exploring the prospects of these energy storage devices.

2. Background and Fundamentals of Supercapacitors

Historically, the idea of storing electrical energy became a reality with the invention of the Leyden jar in 1746. However, the concept of using stored energy for practical purposes arose only in 1954 when Becker was granted a patent for storing electrical energy in a battery or cell. The patent claims the electrical energy storage at the interface of a porous carbon electrode and aqueous electrolyte in the form of a double layer [6].

Supercapacitors are devices that can store and release electrical energy by quickly transferring the charge in the form of electrons and ions during the charge and discharge process. The conventional form of a supercapacitor consists of two parallel plates, one positive and the other negative, with a thin insulating membrane between them called the dielectric (Figure 1a). The charge stored is in the form of potential energy between the plates. The electrolytic capacitor is the next-generation capacitor, where the thin insulator is replaced by an ionic conductive medium called an electrolyte, in which the ions are free to move. The free movement of ions in the bulk of the electrolyte provides a higher charge density. The capacitance obtained from an electrolytic capacitor is much greater (mF range) compared to a conventional dielectric capacitor (μF range). Supercapacitors (SCs) or ultracapacitors, also known as electrochemical capacitors, consist of two electrodes connected ionically through an electrolyte and separated by an ion-permeable membrane called a separator made of polymer, paper, fiber, glass, ceramic, or cardboard (Figure 1b). High specific capacitance rates (tens of farads instead of mfarads) are possible with these devices obtained from a minimal volume of 1 cm^3. Because of this quality, they are also called ultracapacitors. The key reasons for the high capacitance achieved are due to the two different charge storage mechanisms involved: (a) an electric double-layer capacitor (EDLC), utilizing a large surface area in porous carbon materials; and (b) a mechanism utilizing reversible redox reactions, as in the case of pseudocapacitive materials [3,6,23,24].

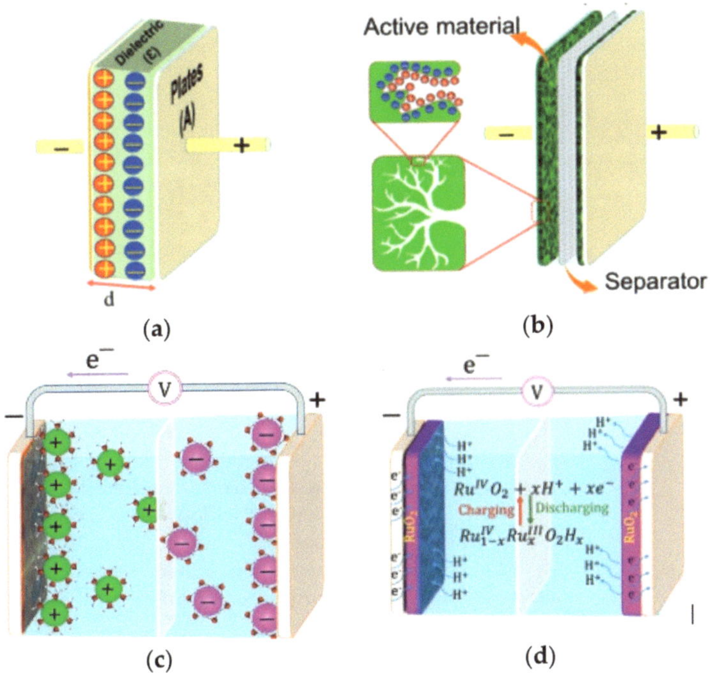

Figure 1. Schematics of (**a**) conventional capacitor, (**b**) supercapacitor, (**c**) electric double layer capacitor (EDLC), and (**d**) pseudocapacitor. Reproduced with permission from [24].

2.1. Type of SCs Based on Charge Storage

EDLCs are electrochemical double-layer capacitors (Figure 1c) that store the charge on the surface through double-layer charging. The double layer is formed by the quick adsorption–desorption of ions on the active electrode surface during charge–discharge. When power is applied to the system, a double-layer charge is formed at each electrode–electrolyte interface without any charge transfer. The process is non-Faradaic because there is no change in chemical composition or phase, and it is reversible, which renders the capacitors with high cycling stability [17].

For such capacitors, the capacitance at each electrode is given by:

$$\mathrm{Ce} = \frac{\varepsilon \mathrm{S}}{\mathrm{d}} \tag{1}$$

where ε is the permittivity of the dielectric constant, S is the surface area of the electrode–electrolyte interface, and d is the EDL thickness.

As d is usually in the nanometer range and the surface area is directly proportional to the specific capacitance, the ideal candidates for supercapacitors are nanoporous carbons with the advantage of a very large specific surface area (SSA) in the range of 1000–2500 $m^2 \cdot g^{-1}$ high electrical conductivity, and low cost. Activated carbon (AC), nanocarbons (CNTs, graphene), and nanoporous metals can provide superior capacitance via double-layer charging due to having a larger surface area [6].

In pseudocapacitors (Figure 1d), the charge storage is achieved via redox or Faradaic reactions occurring at the surface of the electrode without ion diffusion. As a result, pseudocapacitors give a higher capacitance and energy density than carbon-based EDLCs. The cause of the pseudocapacitance can be due to various mechanisms such as redox reactions, intercalation, doping, and underpotential deposition, as illustrated in Figure 2.

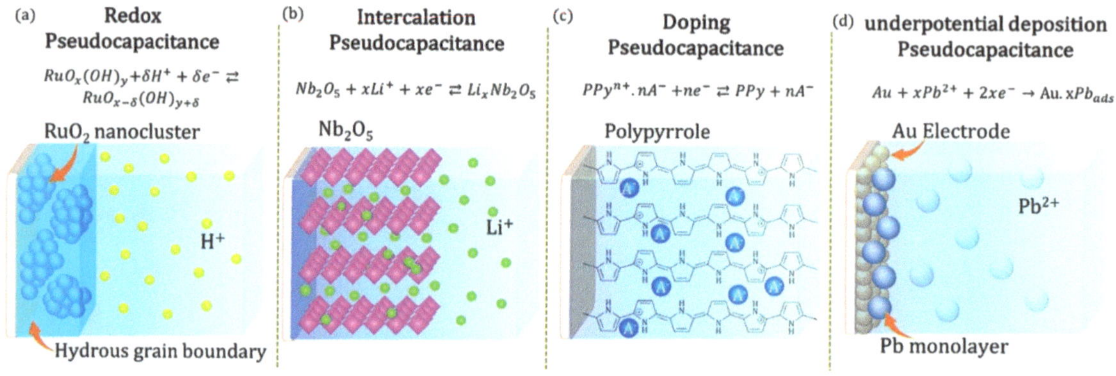

Figure 2. Scheme of different types of redox reactions giving rise to pseudocapacitance. Reproduced with permission from [24].

Metal oxides, nitrides, hydroxides, and sulfides and conducting polymers such as polyaniline, polypyrrole, and polythiophene are pseudocapacitive materials. These materials are mostly hybridized with carbon materials to improve the available capacitance and cyclic stability, because carbon-based materials can enhance the conductivity and stability [4,7,12]. Table 1 compares the different pseudocapacitor material types, showing their advantages and drawbacks.

Table 1. Comparison of different pseudocapacitor material types.

Pseudocapacitor Type	Advantages	Drawbacks
Transition metal oxides		
RuO_2	• High specific capacitance • High-rate capability • High proton conductivity • Wide potential window • Long cycle life	• High cost • Environmental hazard
MnO_2	• Outstanding theoretical specific capacitance (1370 $F{\cdot}g^{-1}$) • High oxygen evolution potential • Low cost • Abundance in nature • Environmentally friendly • Easy synthesis • Good electrochemical performance (especially in aqueous and neutral electrolytes)	• Low conductivity • Low-rate capability • Insufficient cyclic stability owing to MnO_2 dissolution • Broad band gaps • Low ionic diffusion coefficient
V_2O_5	• Can be used as a pseudocapacitive electrode in neutral and aqueous electrolytes • Easy preparation. It can be prepared via electrodeposition, sol–gel, hydrothermal, and quenching methods.	• Low specific surface area • Low solubility (limits the specific capacitance improvement)

Table 1. *Cont.*

Pseudocapacitor Type	Advantages	Drawbacks
Iron-based materials (Fe_2O_3,Fe_3O_4) & (FeC_2O_4), FeOOH	• Natural abundance • Low cost • Environmental friendliness • High theoretical capacitance	• Difficult to transfer theoretical values into practical applications due to • Relatively small surface area • Low electrical conductivity • Poor rate performance • Limited cyclic stability
Oxides/hydroxides of Ni/Co	• Simple preparation • High theoretical specific capacitance in basic electrolytes	• The strong alkaline nature probably hinders practical applications in future wearable electronics
Other oxides (InO_2, WO_3, MoO_3, SnO_2)	• Good electrochemical performance	• Poor conductivity/Slow kinetics/difficult to synthesis/Unstable
Transition metal sulfides		
(MoS_2, FeS/Cos/CuS, NiS)	• High electrical conductivity • High specific capacitance • High mechanical and thermal stability	
Transition metal nitrides		
	• High electrical conductivity (4000–5500 $S{\cdot}cm^{-1}$) • Better sustainability than metal oxides	• Synthesis demands strict operating conditions that are incompatible with wearable electronic components
Vanadium Nitride (VN)	• High theoretical specific capacitance (1340 $F{\cdot}g^{-1}$) • Remarkable electrical conductivity (1.67×10^6 Ohm·m) unlike other anodes such as carbon-based materials, MoS_2, and Fe_2N • Large potential window • Fast reversible redox reaction	• Mismatch between theoretical and experimental values of capacitance

Conductive polymers (CP) are potential candidates for future wearable electronics energy storage devices due to their inherent polymeric elastic nature, good electrical conductivity, and capability to undergo fast and reversible redox reactions. CPs become good conductors after doping and have a low equivalent series resistance. As a result, they experience quick redox reactions yielding good capacitance. They also have a wide voltage window leading to high energy density. CPs also enable the storage of charges without structural and phase changes. This allows for greater capacitance due to a larger surface area and a greater charge storage capacity.

CPs undergo redox reactions due to the presence of conjugated pi electrons. The ions travel to the backbone of the polymer during doping or oxidation and move back to the electrolyte during de-doping or reduction. Table 2 discusses the advantages and drawbacks of popular CPs used for energy storage devices [4,7,12].

Table 2. Comparison of different conductive polymers.

Type of Conducting Polymer	Advantages	Drawbacks
PANI	• High electronic conductivity • Redox and ion-exchange properties resulting in high capacitance • Ease of preparation • Excellent environmental stability • Lightweight • Low cost • High flexibility • Environmental friendliness	• Mechanical stress during charge–discharge leads to poor cycle life
PEDOT-PSS	• High conductivity • High redox active • Good chemical and electrochemical stability • Ease of preparation	• Low specific capacitance due to high molecular weight and low doping level.
Polypyrrole (PPY)	• Conductivity in the range of 10–500 $S{\cdot}cm^{-1}$ • Better rate capability • More flexible and denser than other CPs.	• Unable to be doped with n-doped materials, hence its application is limited to cathode of SCs

2.2. *Supercapacitor Designs*

A supercapacitor can have different designs based on the types of electrodes. A symmetric supercapacitor consists of an anode and cathode made of the same material, and carbon materials are commonly used. In contrast, an asymmetric supercapacitor can have various combinations of electrode materials, as shown in Figure 3 [24].

A hybrid capacitor is a combination of two electrode types—one a capacitive type and the second a capacitive Faradaic (pseudocapacitive) or non-capacitive Faradaic type (battery type). Combining both capacitive and Faradaic kinds of electrodes helps to achieve higher energy and power densities greater than EDLC devices, while also providing better cyclic stability and affordability, the absence of which has hindered the wide usage of pseudocapacitors in the past.

2.3. *Performance Metrics of Supercapacitors*

The capacitance of a supercapacitor system (C) is given by:

$$\frac{1}{C} = \frac{1}{C_1} + \frac{1}{C_2} \tag{2}$$

In the charged state, the two electrodes with capacitance $C_1$and C_2 act like two capacitors in series. The electrode with the lower capacitance determines the capacitance of the system.

A supercapacitor's energy density or specific energy (E) is the amount of energy stored per unit mass, volume, or area of the active electrode materials or the whole device.

$$E = \frac{1}{2}CU^2 \tag{3}$$

where U is the operating voltage.

The specific power or power density (P) is the amount of power a supercapacitor can deliver per unit mass, volume, or area of the electrode's active materials or the whole device [17].

$$P = \frac{U^2}{4R_S} \tag{4}$$

where Rs is the equivalent series resistance of the supercapacitor. The performance of an energy storage device is determined by the Ragone plot, which expresses the relation between the specific energy and specific power (Figure 4).

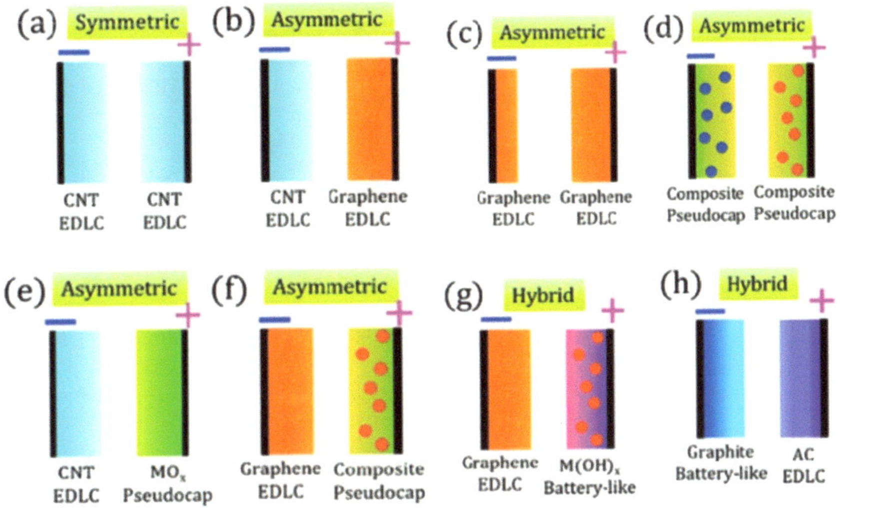

Figure 3. Schematic representation of the symmetric, asymmetric, and hybrid configurations. (**a**) Symmetric supercapacitor. As shown in the figure, asymmetric supercapacitors cover a wide range of electrode combinations, including (**b**) two different EDLC electrodes (**c**) with the same material at both electrodes but different mass loadings, (**d**) two pseudocapacitive electrodes with different compositions, (**e**) one EDLC electrode with a pseudocapacitive electrode (e.g., an activated carbon/MnO_2 aqueous supercapacitor), (**f**) one EDLC electrode with a composite electrode (that can be a composite of carbon-based materials with a metal-based species), (**g**) a hybrid configuration that is fabricated using a capacitor-type negative electrode and a battery-like positive electrode, and (**h**) a hybrid architecture (typical of a Li-ion capacitor) that is fabricated using a battery-type negative electrode and an EDLC positive electrode. Reproduced with permission from [24].

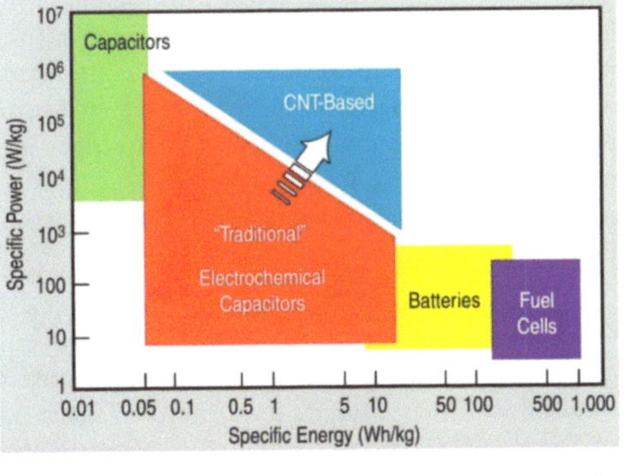

Figure 4. Ragone plot for the current generation of capacitors, supercapacitors, batteries, and fuel cells. Reprinted with permission from [25].

From the Ragone plot, it can be observed that the electrochemical capacitors have higher specific power levels than batteries and fuel cells but lower energy densities than them. Therefore, the current research is focused on designing supercapacitors with higher energy density without compromising on their excellent power density. Additionally, it can be noted from Figure 4 that carbon-nanotube-based electrochemical capacitors have superior power and energy properties over other electrochemical capacitors.

3. CNT Fiber for SCs

Carbon nanotubes (CNTs) are apt candidates with an SP^2 hybridized carbon, possessing all of the desired properties for a wearable SC. Due to the 1D nanostructure, they can be easily assembled into a 1D fiber. Furthermore, the CNTs retain high mechanical strength and flexibility and an exceptional specific surface area when assembled from the nanoscale to macroscopic scale in the fiber form. With all of these attributes, CNT assemblages in the fiber form can simultaneously be used as fiber scaffolds, current collectors, and active material electrodes for wearable SCs [5,7,8].

3.1. Fabrication of CNT Fibers

The macroscopic assemblage of CNT fibers from nanoscale carbon can be accomplished without complex treatments such as carbonization, which is an added advantage of the CNT fiber fabrication. The fabrication can be accomplished via (1) wet spinning from CNT dispersions, (2) dry spinning from CNT aerosols, (3) dry spinning from CNT arrays or forests, and (4) the twisting of CNT thin films. The CNT fiber's electrical, mechanical, and electrochemical properties depend on the fiber morphology (the structural alignment and pore structure). The fiber morphology, in turn, is determined by the assembly methods and related process parameters.

3.1.1. Wet Spinning from Dispersions

The wet spinning of CNT fibers from dispersions is a scalable process that has been practiced since 2000 [26]. It is usually accomplished by preparing a homogenous dispersion of CNTs with a surfactant or polymer and extruding it into a coagulation bath. The extruded flow becomes a solidified fiber enabled by the nanomaterials' agglomeration during the fiber's wet drawing. The latter is accomplished using a rotating spool, a rotating coagulation bath, or fiber gravity. The obtained fiber is mechanically strong but exhibits poor electrical and thermal properties due to the presence of the polymer. Polymer-free dispersion and spinning processes have also been succeeded tested, as reported by Behatbu in 2013 and Bucossi in 2015 [27,28]. The fiber morphology and fiber properties can be fine-tuned via the optimization of the wet spinning parameters, which include the following:

- The extrusion speed;
- The needle or spinneret diameter and shape;
- The quality and concentration of spinning dispersion;
- The rotation speed in the rotated spinning method;
- The nature of the coagulation bath (fast coagulation results in a non-homogenous structure, while slow coagulation imparts a homogenous structure);
- The drawing speed and ratio (a higher ratio improves the alignment, strength, and fiber conductivity);
- The post-treatment, if any (such as annealing, which is used to remove the oxides to make the fibers more conductive).

Kou et al. demonstrated a coaxial wet spinning process [29] to achieve a higher capacitance and rate capability using CNTs and reduced graphene oxide (RGO) as the spinning dope. In the coaxial spinning process, the spinneret consists of two inlets, one for the core and the other for the sheath, as shown in Figure 5a.

The core combines CNTs and RGO; the sheath is made of sodium carboxy methyl cellulose (CMC). The polyelectrolyte (CMC) acts as an ion-permeable membrane but is insulative to the electron flow. An electrode RGO+CNT@CMC is spun with a GO and CNT mixture (w/w, 1/1) of 20 mgml^{-1} and CMC 8 mgml^{-1}. SEM images of the coaxial fiber can be seen in Figure 5b–d. Two coaxial fiber electrodes (CNT + RGO @ CMC) are intertwined to form a two-ply yarn SC with a layer of PVA-H_3PO_4 gel electrolytes (Figure 5e–g). The CMC sheath layer protects the SC from short circuits. The SC yarn with a mixture of RGO + CNT outperformed the other SC with only RGO @ CMC and CNT @ CMC. The aerial capacitance C_A of CNT + RGO @ CMC exhibited superior capacitance of 177 mFcm^{-2} (158 F/cm^3) at a current density of 0.1 mA/cm^2, whereas the C_A values for RGO@ CMC and CNT@CMC were 39 mF/cm^2 and 127 mF/cm^2, respectively. The higher capacitance could be attributed to the higher surface area formed by the synergistic combination of CNT and RGO. Additionally, the superior electrochemical performance could be due to the lower internal resistance, as confirmed via electrochemical impedance spectroscopy (EIS) and shown in Figure 5i.

To assess the advantage of the coaxial spinning process, the authors fabricated a control yarn SC via the conventional wet spinning of CNT + RGO dispersions and further coated it with CMC. They also reported that the coaxial-spun SC exhibited a capacitance of 103 mF/cm^2. In contrast, the conventionally spun and post-coated SC yarn gave a much lower capacitance rate of 39 mF/cm^2 at a current density of 0.1 mA/cm^2. The main reason is that the coagulation and drying of the outer CMC layer during the spinning process are faster than for the core, which leads the sheath to tightly wrap around the core, resulting in better ion transport between the electrolyte and electrode. Additionally, the pore formation in the CNT and GO inner core during the spinning process is complicated due to the uniform shrinkage of the CMC layer. A thicker layer increases the shrinkage force, which affects the pore structure. Therefore, a thicker CMC provides higher capacitance to the SC. Finally, it can be noted that the CMC has no pseudocapacitive effect in the process, and the improved performance can be attributed to the coaxial spinning process.

Wet Spinning and Incorporation of Pseudocapacitive Materials

SCs based on nanocarbon usually exhibit limited capacitance because double-layer electrochemical charging is the only charge storage mechanism. Therefore, combining pseudocapacitive materials such as metal oxides and conducting polymers with carbon is standard practice. Furthermore, wet spinning methods are advantageous because they allow the easy incorporation of such active materials in the spinning dispersions, providing composite CNT electrodes with the required electrochemical performance and other functionalities.

I. Metal oxides

Lu et al. [30] synthesized SCs with high mechanical strength and superior electrochemical performance in the wet spinning of a liquid crystal dispersion containing single-walled CNTs and by further electrodepositing MnO_2. The "as-spun" single-wall carbon nanotubes (SWCNTs) had a tensile strength and conductivity of 225 MPa and 450 ± 15 S·cm^{-1}, respectively. These values were 246% and 50% higher than the previously reported fibers. The reason could have been the highly aligned SWCNTs and the low contact resistivity between the CNT bundles. Further, MnO_2, which has high theoretical capacitance, was electrodeposited on SWCNT fibers to form a MnO_2/SWCNT hybrid electrode. Two electrodes contained PVA/LiCl as the electrolyte to form a parallel symmetric SC. As reported by the authors, the SC exhibited an energy density of 14.1 Wh·kg^{-1} and a power density of 202 W·kg^{-1}, and showed an increased capacitance of 16% after 10,000 cycles. The impressive electrochemical performance was attributed to the fast ion transport made possible by the uniformly deposited mesoporous MnO_2 nanoflakes on the well-aligned, highly conductive CNT fiber.

The method of active material loading via post-deposition on CNT fibers has limitations for high loading rates. In contrast, the pre-deposition of the active material on a vertically aligned carbon nanotube (VACNT) array, which is then twisted into a fiber, is complicated and time-consuming. Therefore, Li et al. [31] tried using a conventional wet spinning process to fabricate a hybrid SWCNT/MnO_2 fiber, in which the aqueous dispersions of SWCNTs and MnO_2 were mixed with the help of the surfactant sodium chlorate, and were then spun, washed, and dried. The SWCNT/MnO_2 hybrid fiber had a 75% mass loading rate, conductivity of 761 S·cm^{-1}, and volumetric capacitance of 74.8 F/cm^3. Furthermore, an SC fabricated in a parallel symmetric configuration using the hybrid fiber gave an energy density of 10.4 mWh·cm^{-3} at a power density of 0.05 W·cm^{-3} and exhibited good cyclic stability (94% capacitance retention in 5000 cycles). According to the authors, the superior performance was due to the highly loaded conductive hybrid fiber facilitating quick electrical pathways thanks to the close contact between the SWCNT and MnO_2 nanosheets and the hierarchical pore structure helping the faster ion transport.

Since the conductivity of the MnO_2 used as pseudocapacitive material is low, Kim et al. [32] identified a solution incorporating platinum nanoparticles (PtNP) with MnO_2 through a one-step scalable wet spinning process to produce CNT/MnO_2/PtNP composite electrodes. CNT powder, MnO_2 powder, Pt nanoparticles, and water are mixed together and injected into a coagulation bath of PVA/water. The process is scalable; however, greater the length, the greater the resistance. The added Pt nanoparticles increased the conductivity, the PVA enhanced the mechanical strength, the CNT acted as a mechanically stable conductive core, and the MnO_2 provided the necessary pseudocapacitance. Two parallel electrodes coated with PVA-LiCl fabricated as an SC exhibited a capacitance of 53.1 mF·cm^{-2} at 10mV/s, an energy density of 1.84 μWh·cm^{-2}, and a power density of 13.3 μW·cm^{-2}. The capacitance was 49 times higher than for CNTs in a bare SC due to the presence of conductive Pt nanoparticles, which also reduced the resistance. The rate capability was retained at 60% when the scan rate varied from 10 to 100 mV/s.

II. Conducting polymers

Meng et al. [33] demonstrated the wet spinning of stable dispersions of SWCNTs and polyaniline nanowires (PANI NWs) with an outer PVA sheath. These flexible and highly conductive electrodes were combined with a PVA-H_3PO_4 electrolyte to make twisted yarn SCs. The yarn SCs with the composite SWCNT/PANI NWs exhibited 18-fold more capacitance than the SCs based on bare SWCNTs.

Zheng et al. [34] fabricated a core–sheath SC electrode with the core being hyaluronic acid (HA) and with CNTs and electrodeposited PANI as the sheath. While HA acts as a biocompatible surfactant to improve the dispersion of CNT during wet spinning, it also serves as an ion-conducting binder. PANI is deposited on the fiber via electrochemical polymerization. The authors presented data showing that the HA/CNT/PANI electrode revealed a 6-fold greater specific capacitance than the HA/CNT core. In contrast, the HA/CNT core provided high levels of electrical conductivity, mechanical robustness, and stiffness. The authors claimed the electrode's superior mechanical, electrical, and electrochemical performances were due to the in situ polymerization of PANI, which facilitates hydrogen bonding between the carboxyl groups of HA and amine groups of PANI. They also reported it as having high cyclic stability (90% capacitance retention after 3000 cycles at 100 mV/s). The high-performing, low-cost electrode was possible not only due to the synergistic interaction between the HA/CNT core and PANI sheath but also thanks to the interaction and alignments of the HA/CNT core–sheath, including the favorable pore structure and high specific surface area of the PANI.

III. Metal oxides and conducting polymers

Torres et al. studied [35] a mixture of carbon black (CB), CNT, and chitosan dispersions which were wet-spun to form a carbon skeleton, after which the MnO_2 (of two different crystalline structures, εMnO_2 and γMnO_2) was electrodeposited, followed by dip coating to form a layer of poly(3,4-ethylene dioxythiophene) (PEDOT) conducting polymer or polypyrrole. Finally, symmetric SCs were fabricated from various combinations of the above electrodes shown in Table 3, resulting in high capacitance and cyclic stability rates.

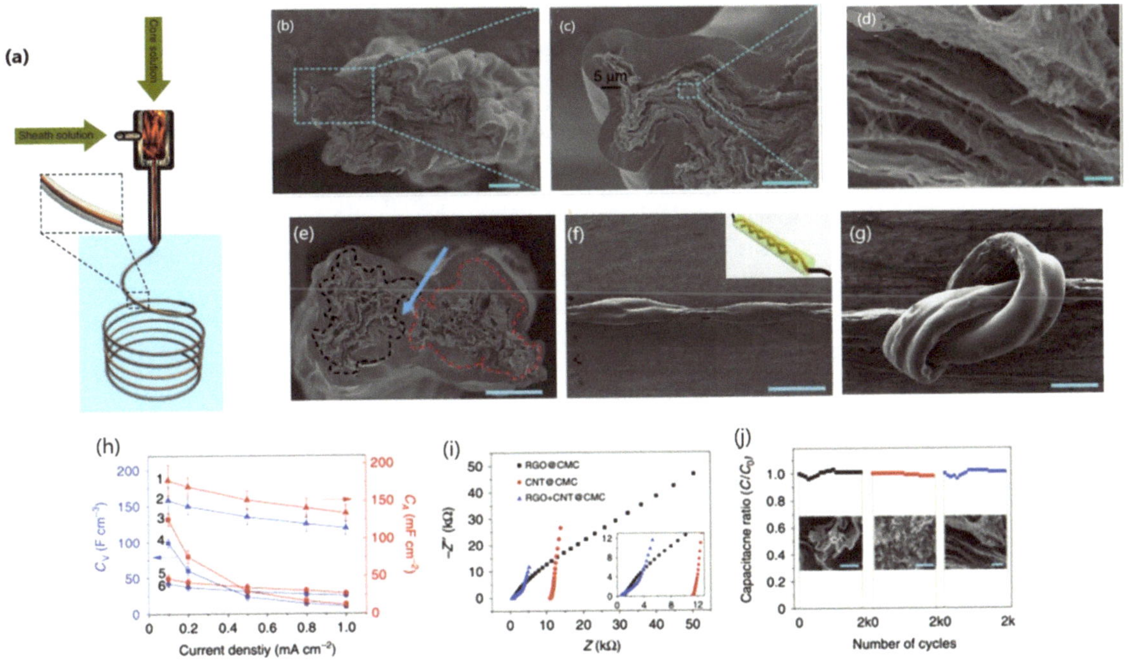

Figure 5. The coaxial wet spinning assembly process, showing fabricated yarn SCs (YSCs), their morphology, and their electrochemical performance. (**a**) Schematic of coaxial spinning. (**b–d**) SEM images of RGO + CNT@CMC. (**e–g**) SEM images of two-ply SCs in (**e**) cross-sectional view and (**f**) side view. The arrow in (**e**) is a PVA/H_3PO_4 electrolyte, and the inset of (**f**) shows the schematic illustration of the yarn SCs. (**g**) SEM image of a two-ply yarn SC knot. (**h–j**) Electrochemical performances of yarn SCs, (**h**) with the dependence of CA and CV on the charge–discharge current density for RGO@CMC (3, 4), CNT@CMC (5, 6), and RGO + CNT@CMC (1, 2) YSCs. (**i**) Nyquist plots, (**j**) showing the normalized capacitance (C/C_0, where C_0 is the initial capacitance) versus cycle number for RGO@CMC (left), CNT@CMC (middle), and RGO + CNT@CMC (right) YSCs. Scale bars: 20 μm for (**b**), 10 μm for (**c**), 0.2 μm for (**d**), 50 mm for (**e**), 500 mm for (**f**), 200 mm for (**g**), 3 mm for left, and 0.3 mm for middle and right (**j**). Reproduced with permission from [29].

Table 3. Fabrication steps to make symmetric supercapacitors. Reproduced with permission from [35].

1st Step: Wet Spinning	2nd Step: MnO_2 Electrodeposition	3rd Step: Coating of Conducting Polymer	Electrodes
CB-CNT fibers	$Mn(CH_3COO)_2$	PEDOT	$MnAc_2$-PEDOT
	$Mn(CH_3COO)_2$	PPY	$MnAc_2$-PPY
	$Mn(NO3)_2$	PEDOT	$Mn(NO_3)_2$-PEDOT
	$Mn(NO3)_2$	PPY	$Mn(NO_3)_2$-PPY

The SC with PEDOT gave a high capacitance fiber, while the one with polypyrrole revealed higher cyclic stability. Therefore, to obtain both in a single SC, the Mn $(CH_3COO)_2$/PEDOT fiber was coated with polypyrrole (PPY). As a result, the cyclic stability improved from 79% to 89%, and the capacitance obtained was maintained at 549 F/g. Thus, this study proves the opportunity for tuning the SC components from the different layers available (MnO_2, PEDOT/polystyrene sulfonate/PSS, PPY).

3.1.2. Dry Spinning from CNT Aerosols

In this process, the precursor carbon source (organic vapor or solid carbon) decomposes at a very high temperature to produce carbon atoms. These grow into CNTs with the help of catalyst nanoparticles formed in situ and floating in the vapor. With the help of van der Waals forces or pi–pi interactions, the formed CNTs self-assemble into an aerogel or network of nanobundles. These can be collected as long fibers on a spindle or as a ribbon, or on a rotating drum [36] The number of walls, diameter, and length of the CNTs can be controlled by fine-tuning the following process parameters:

- The carbon precursor;
- The catalyst concentration;
- The rate of hydrogen gas flow;
- The temperature of the pyrolysis.

The fiber morphology and mechanical and electrical properties can be optimized by controlling the following aspects:

- The CNT alignment, length, and structure;
- The post-treatment;
- The rate of the gas flow;
- The speed of the winding;
- The presence of particulate impurities and defects.

Li et al. [37] grew ultra-long and well-aligned CNT fibers by winding the CNT aerogel continuously, which was produced using the floating CVD process. A schematic of the CNT floating catalyst CVD process is shown in Figure 6.

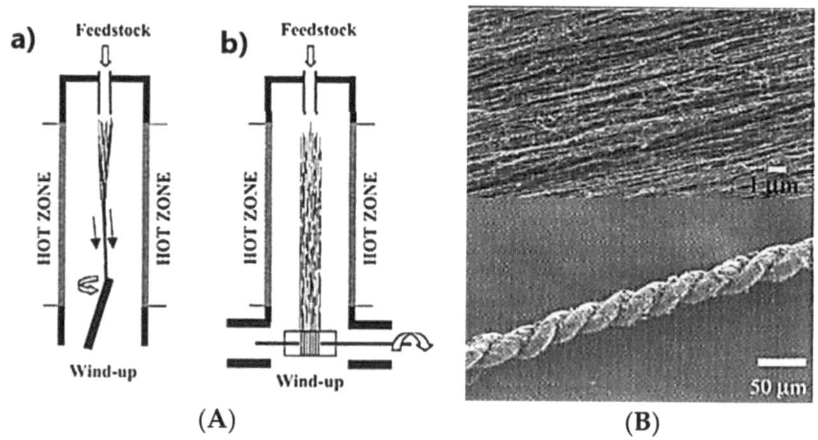

Figure 6. (**A**) (**a**) Schematic of the direct spinning process, called the floating catalyst CVD process. Liquid feedstocks such as ferrocene and thiophene are dissolved in ethanol, mixed with H_2, and then injected into a hot zone where the CNT aerogel is formed. This aerogel is captured and wound out of the hot zone by an offset rotating spindle. (**b**) Schematic of the wind-up assembly operating at a lower temperature, outside the furnace hot zone.(**B**) SEM images of well-aligned multiwall carbon nanotubes (MWCNTs) along with a twisted fiber collected from the furnace. Reprinted with permission from [37].

3.1.3. Dry Spinning from CNT Arrays

In this method, the CNT fiber is drawn primarily by twisting from a super-aligned CNT array synthesized via chemical vapor deposition (CVD) [38,39]. A schematic of the CVD process of the CNT array synthesis and the spinning process can be seen in Figure 7A,B. Our group developed and improved the synthesis process for spinnable CNT arrays and the related dry spinning and fiber post-treatment process. As a result, high-purity CNT fibers were reproducibly fabricated and characterized, and the results are published elsewhere [40,41].

The recipe for spinnable arrays has been perfected over the years regarding the gas composition, flow rates, catalyst composition, reaction time, temperature, and pressure. The fibers that are produced have a large surface area. They are further post-treated to improve the mechanical and electrical properties [41–44]. The methods used to the enhance mechanical strength and electrical conductivity practiced by our group were as follows:

- Performing fiber post-treatments such as (a) solvent densification, which decreases the cross-sectional fiber area, bringing the CNT bundles closer; and (b) post-spinning from a spun fiber (Figure 7C,D)
- Introducing optimized twisting during the spinning process;
- Annealing at a high temperature to remove the oxides and amorphous carbon and to increase the conductivity;
- Cross-linking of the CNTs fibers with appropriate polymers to increase their strength.

3.1.4. Twisting from CNT Films

This fiber fabrication method draws CNT fibers from thin CNT films and sheets via scrolling. During the scrolling, twisting, and drawing of the fibers, they are attracted to each other and self-align into a network of nanobundles with the help of van der Waal's interactions [45,46] Figure 8 illustrates the formation of biscrolling fibers with different types of twist insertions, such as Fermat-type, Archimedean-type, and dual Archemedean-type.

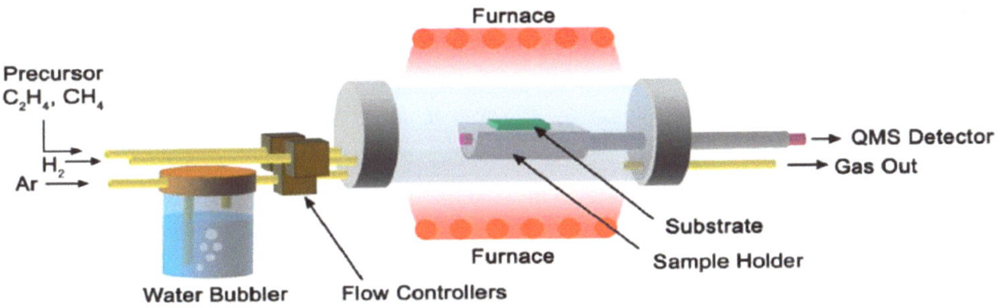

(A)

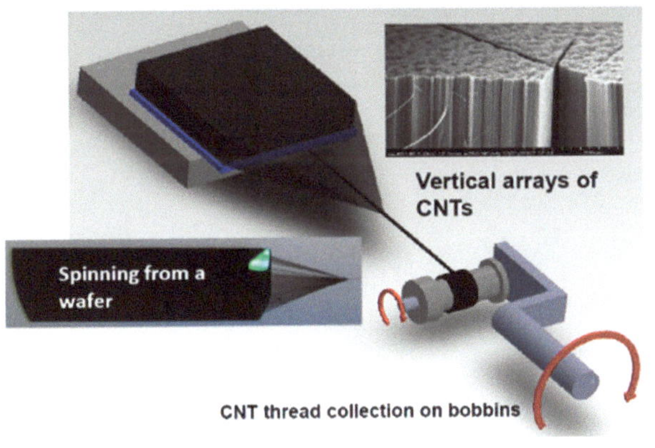

(B)

Figure 7. *Cont.*

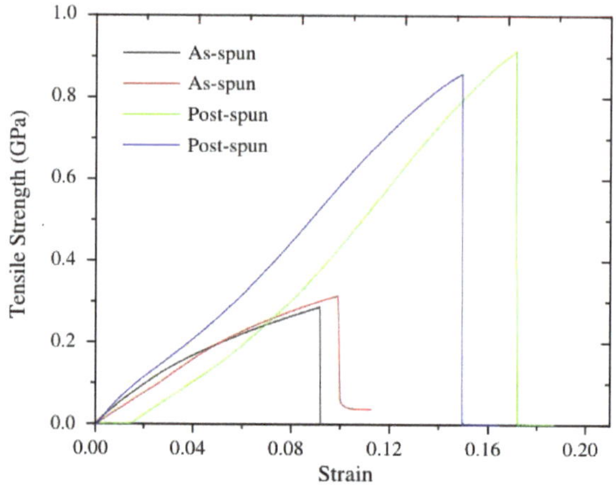

(C)

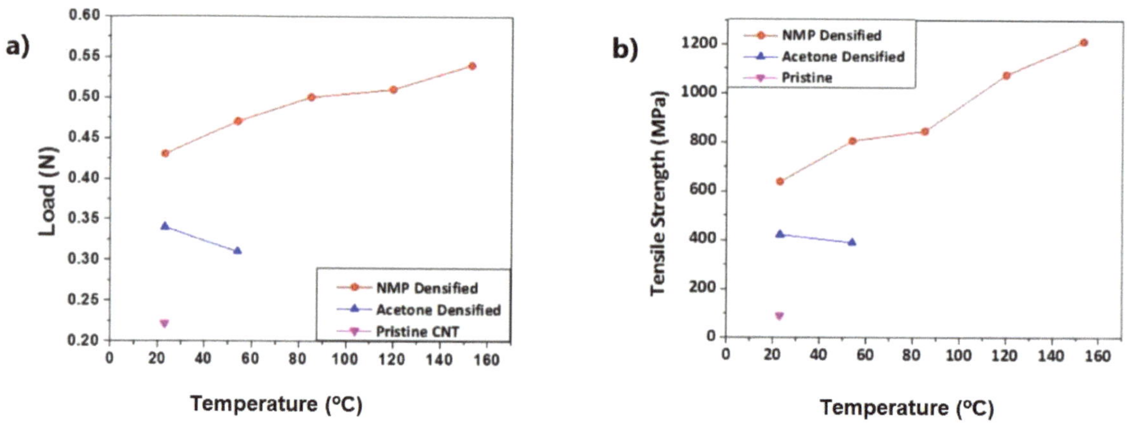

(D)

Figure 7. (**A**) Schematic of the thermal CVD process for the CNT array. (**B**) Schematic of the spinning of CNT fibers from the CNT forest array. (**C**) Schematic of as-spun fiber and post-spun fibers and comparison of the tensile strengths of "as-spun" and "post spun" CNT fibers. Reprinted with permission from [38] (**D**). Comparison of the mechanical (**a**) load-bearing and (**b**) tensile strengths of the pristine CNT fiber and the post-densified fiber with different solvents and temperatures. Reprinted with permission from [42].

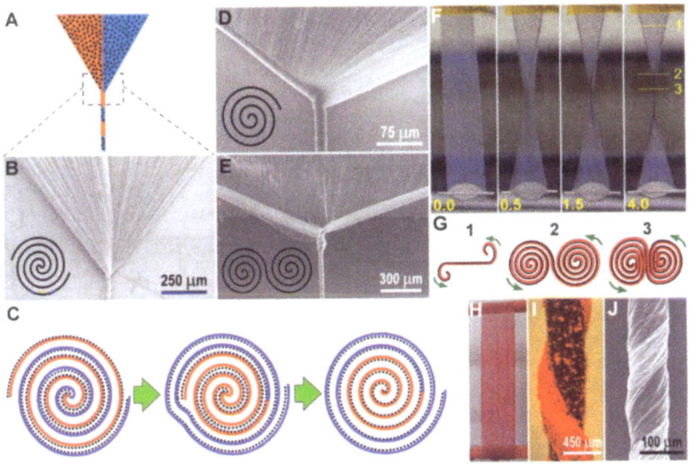

Figure 8. (**A**) Schematic of a twist induced in a CNT spinning web. The active loaded material is represented by black dots. (**B**) SEM image showing a Fermat-type twist (Fermat scroll shown in the inset). (**C**) Change from a Fermat to Archimedean scroll. (**D**) Archimedean scroll. (**E**) Dual Archimedean scroll. (**F**) Sequential pics of a liquid-state twist insertion in a multiwalled CNT(MWCNT) with filtration of the deposited boron nitride nanotubes. (**G**) Cross-sections of several twists marked in (**F**,**H**,**I**) pictures of actual MWCNTs colored with red paint, both unscrolled and scrolled. (**J**) SEM image of 70% Ti@MWCNT biscrolled yarn produced by symmetrically inserting a twist in a bilayer rectangular sheet supported in a liquid bath. Reprinted with permission from [45].

4. Strategies to Improve the CNT Fiber's Electrode Performance

As mentioned earlier, in CNTs, the only active material is carbon, so the available type of charge storage is a double layer. Double layers are insufficient for most practical applications due to their low capacitance and energy density. Hence, various strategies have been developed to improve the electrochemical performance of CNT electrodes, such as the (1) activating CNT fibers via heat treatment or surface oxidation to modify the surface properties and functionalities, (2) combining the CNT fiber with pseudocapacitive materials, and (3) co-plying with highly conductive metal wire [1,3,47,48].

4.1. Activation of CNT Fiber by Gamma Irradiation, Oxygen Plasma, or Acid Treatment

CNT fiber can undergo surface modifications via activation by gamma radiation, acids, and oxygen plasma. The treatments often lead to improvements in the electrochemical properties, which can be attributed to:

- The introduction of functional groups on the surface;
- The modification of the pore structure and the opening of new pores;
- Increasing the surface area;
- Decreasing the gravimetric density;
- Changing the surface energy by promoting hydrophilicity.

The effects mentioned above are reversed during the activation by heat treatment, with a decreased pore density and surface area, increased graphitization in the carbon microstructure, and increased gravimetric density [6].

4.1.1. Activation by Gamma Radiation

Su et al. [49] activated a CNT fiber using gamma irradiation and obtained improved mechanical, electrical, and electrochemical performances for two-ply yarn SCs. Significant improvements were seen in the mechanical strength (the breaking strength increased from 574.5 MPa to 773.4 MPa), conductivity (increased from 3596 $S{\cdot}m^{-1}$ to 5494 $S{\cdot}m^{-1}$), and gravimetric capacitance (increased from 6.2 $F \cdot g^{-1}$ to 9.2 $F \cdot g^{-1}$) after irradiation. Additionally, the XPS results showed an increase in the percentage of O_2 from 0.4% to 3.6%.

4.1.2. Activation by Oxygen Plasma Treatment

Adusei et al. [50] studied the effects of atmospheric pressure oxygen plasma functionalization on CNT fibers. The CNT fibers were subjected to plasma functionalization for various plasma parameters, such as the plasma power and oxygen flow, characterized as (a) low (60 W and 0.1 L/min), (b) medium (100 W and 0.3 L/min), and (c) high (140 W and 0.55 L/min), respectively. The secondary gas (helium) flow rate was maintained at 15 L/min for all the other plasma parameters. The fibers were passed through the plasma head in a continuous roll-to-roll process at two linear speeds (0.465 cm/s and 0.206 cm/s). The electrochemical characterization of the treated CNT fibers under various conditions revealed that the medium-oxygen plasma-treated fibers (100 W and 0.55 L/min) with a lower speed of fiber passing through the plasma head (0.206 cm/s) gave the greatest specific capacitance. The treated fibers showed an increase in surface area from 168.22 m^2/g to 208 m^2/g and a 132.8% increase in specific capacitance. The impressive capacitance improvement was due to the introduction of functional groups on the CNT fiber containing oxygen, the increased hydrophilicity, and the enhanced surface area due to pore opening.

4.1.3. Activation by Acid Treatment

The acid treatment of CNT fibers after thermal oxidation was studied by Liang et al. [51]. First, the thermal oxidation of the iron nanoparticles formed during the tube growth was conducted in air at 400 °C. Then, the acid treatment was performed by soaking the fiber in a mixture of HNO_3-H_2SO_4 (1:3 volume ratio) with stirring for 6 h, followed by rinsing with DI water and drying. As a result, the activated fiber showed an increase in oxygen level from 11.63% to 20.77%. In addition, the Brunauer, Emmett, and Teller (BET) surface area increased from 138.3 $m^2 \cdot g^{-1}$ to 236.8 $m^2 \cdot g^{-1}$. These two effects contributed to an outstanding electrochemical performance, achieving a capacitance rate of 1988 $mF \cdot cm^{-2}$ at 2 $mA \cdot cm^{-2}$ and cyclic stability of 10,000 cycles with only 3% capacitance loss.

4.2. Combining Pseudocapacitive Materials with CNT

Pseudocapacitive materials can combine redox reactions and double-layer charging to achieve higher capacitance rates and energy storage densities.

4.2.1. Deposition of Active Materials after Fiber Fabrication

Metal Oxides and Hydroxides

MnO_2 is the most attractive pseudocapacitive oxide due to its low cost, abundance, and high specific capacitance value. However, it has the limitation of poor conductivity, and this problem imposes a great challenge, especially in highly loaded composites. Usually, introducing a nanoscopic scale of MnO_2 can provide adequate electrochemical performance, but low loading often causes low energy and power densities. Therefore, Kim et al. [52] found a solution to circumvent the conductivity issue while providing high loading with the oxide. They designed a composite electrode with CNT as the core layer with electrodepositing alternating layers of MnO_2 and Ag within the resulting sheath. The highly conductive intermediate silver layer shortened the ion diffusion length in the MnO_2. In addition, it acted as a conductive bridge between MnO_2 aggregates and the Ag layer, so that during charging and discharging the flow of electrons is rapidly transferred by the Ag layer, which does not affect the rate capability of the device, irrespective of the amount of MnO_2 loading. The Ag/MnO_2 parallel symmetric two-ply sheath–core yarn SC showed a capacitance of 322.2 $mF \cdot cm^{-2}$ (208.1 $F \cdot cm^{-3}$) and an energy density of 18.3 $\mu Wh \cdot cm^{-2}$. The rate capability was increased from 21.2% to 42.6% from pristine MnO_2 sheath–core vs. Ag/MnO_2 sheath–core SC.

As discussed earlier, the MnO_2 composite not only has the disadvantage of low conductivity but also a short cycle life. Zhou et al. [53] circumvented this problem by innovatively adding black phosphorous into the CNT/MnO_2 composite. The CNT yarn was immersed in $KMnO_4$, HCl, and black phosphorous (BP) nanosheet aqueous solutions. The $KMnO_4$ was reduced to MnO_2, and the CNT/MnO_2/BP composite electrode was formed. Two polyvinyl alcohol (PVA)-KOH-coated CNT fibers were twisted into a two-ply yarn SC, which gave a promising electrochemical performance. The capacitance was reported to be 441.79 $F{\cdot}cm^{-3}$ at 0.01 $V{\cdot}s^{-1}$. The energy density reached 9.82 $mWh{\cdot}cm^{-3}$, and the power density value was 441.79 $mW{\cdot}cm^{-3}$. This study also reported a 37.9-fold increase in capacitance for the CNT-MnO_2-BP yarn SC compared to the CNT yarn and a 2.4-fold increase compared to the CNT-MnO_2 yarn SC. According to the authors, these remarkable results can be attributed to the highly conductive CNT substrate, the pseudocapacitance provided by the MnO_2, and the rapid ion diffusion enabled by the phosphoric-acid-based electrolyte formed via the reaction of black phosphorous and $KMnO_4$ during the formation of MnO_2. The rapid ion diffusion helped to increase the specific capacitance and cyclic stability.

Zhou et al. [54] fabricated a sandwiched CNT composite by electrodepositing a pseudocapacitive transition metal oxide mixture of Co_3O_4@NiO and coating a graphene layer over it. The CNTs act as a substrate with a high specific surface area for the electrodeposition of the transition metal oxide, Co_3O_4@NiO, which offers numerous active sites for redox reactions. In addition, the graphene layer decreases the equivalent series resistance (ESR) and increases the power density and cycle life. A two-ply yarn SC was fabricated by twisting two composite sandwich fibers and applying PVA-KOH as an electrolyte between them. The SC yielded a high capacitance of 263 $F{\cdot}cm^{-3}$ at 0.01 V/s, an energy density of 5.86 $mWh{\cdot}cm^{-3}$, and a power density of 263.64 $mW{\cdot}cm^{-3}$.

Le et al. [55] introduced another innovative approach to improving the electrochemical performance using a two-step strategy, namely increasing the surface area by growing ZnO nanorods on the CNT surfaces and coating them with a pseudocapacitive material like a nickel cobalt double hydroxide. As a result, the electrode exhibited an excellent capacitance of 1278 $F{\cdot}g^{-1}$ at a scan rate of 5 $mV{\cdot}s^{-1}$ and cyclic stability of 7000 cycles with a capacity retention rate of 60.5%.

Metal Sulfides

Jian et al. [56] fabricated an SC, solving the issue of the delamination of metal oxides and sulfides coated on the CNT surfaces. For this, a hybrid substrate of CNTs and reduced graphene oxide (RGO) was prepared, where the RGO served as an adhesive layer. Molybdenum sulfide was then coated onto the hybrid layer to allow the pseudocapacitance. Thus, a composite electrode of MoS_2/RGO/CNT was formed. Next, a symmetric parallel SC was fabricated using PVA-H_2SO_4 as an electrolyte. The authors reported for this SC a capacitance of 190.4 F/g at 0.2 $A{\cdot}g^{-1}$ and 85% capacitance retention after 5000 cycles.

4.2.2. Incorporation of Active Material before or during Fiber Fabrication

The deposition of active materials after the fiber formation has limitations because of the following reasons:

- Higher loading is not possible since it does not increase the capacitance, as the accessible surface area is considerably reduced;
- The resistance increases along with the ion diffusion distance, substantially reducing the power density, resulting in greater energy consumption.

As an alternative to surface deposition, embedding the active materials into the CNT structure before its formation seems to be a more practical approach. In this process, the nanosized active materials are overlaid on the CNT sheet or film before the fiber is twisted. Compared to surface cladding, this type of incorporation allows much higher loading and a closer contact of the active materials with the nanocarbon bundles, facilitating higher EDL pseudocapacitance. In addition, these materials can create more pores and channels for faster ion adsorption and diffusion, thereby increasing the capacitance, rate capability, and cyclic stability.

Shi et al. [57] successfully designed amorphous MnO_2@ MWCNT fiber electrodes for SCs with improved flexibility, energy density, and high cyclic stability. Amorphous MnO_2, which has a highly disordered structure, was incorporated into the MWCNT sheet by immersing a stacked MWCNT sheet into a mixture of a $KMnO_4/H_2SO_4$ aqueous solution, followed by twisting the sheets into a fiber with the help of an electric motor. Two such twisted yarns were used to fabricate symmetric solid-state SCs using PVA-LiCl as an electrolyte, which showed a capacitance of 8.5 $F{\cdot}cm^{-3}$ (13-fold than MWCNT fiber SC) and an energy density of 1.5 $mWh{\cdot}cm^{-3}$, and surpassed 15,000 charge–discharge cycles retaining 90% of the capacitance. Furthermore, due to the close contact between the MnO_2 and MWCNT, the synergy between them made the highly aligned MWCNT substrate with uniformly distributed MnO_2 particles highly conductive and mechanically stable. Additionally, the nano MnO_2 particles with large surface areas provided numerous accessible active sites for redox reactions, shortening the electron and ion diffusion lengths and allowing the MnO_2 to be accessible for charge storage.

5. Strategies to Improve Device Performance

As mentioned in Section 2.3, a supercapacitor device is assessed by its energy density and power density.

The energy density is given by Equation (3):

$$E = \frac{1}{2}CU^2$$

and the power density is given by Equation (4)

$$P = \frac{U^2}{4R_s}$$

where U is the working voltage, C is the capacitance, and Rs is the internal resistance.

Increasing the working voltage is more impactful than increasing the capacitance to raise the energy density. The working voltage can be increased by:

- Configuring an asymmetric type of capacitor;
- Using electrolytes with a wide working voltage;
- Modifying the electrode to widen the voltage window;
- Using redox electrolytes.

To increase the power density, the internal resistance has to be reduced by increasing the conductivity. One way to increase the conductivity is via co-plying with metals [1,58].

5.1. Metal Wire Co-Plying

CNT fiber SCs, because of their increased length, will have high internal resistance, resulting in a large potential drop under a high current density. To improve the rate capabilities at higher currents, it is often helpful to couple a highly conductive metal wire with CNTs for fast electron transfer.

Wang et al. [59] fabricated (Figure 9A) an SC with individual electrodes consisting of CNT and Pt yarn twisted together at 5000 turns/min and coated with PANI. The yarn was named Pt/CNT@ PANI. When subjected to cyclic voltammetry at a very high scan rate of 500 mV/s, the symmetric SC maintained a symmetrical shape compared with those formed at 10 mV/s. Figure 9(Bb) displays the capacitance–voltage (CV) curves at 100 mV/s for a PANI-coated CNT fiber with and without platinum. Additionally, a galvanostatic charge–discharge (GCD) test was conducted. The CV and GCD curves showed that the SC with platinum provided a greater capacitance area in the CV curve and a longer discharge time in the GCD test. Figure 9C shows the rate capability result, highlighting that a capacitance retention rate of 68% was achieved when the current density was increased from 0.8 to 40 $mA{\cdot}cm^{-2}$, revealing good stability over large currents. The electrochemical impedance spectroscopy (EIS) measurements with a frequency loop from 10^5 Hz to 10^2 Hz using an amplitude of 5 mV at an open-circuit potential (Figure 9(Bf)) showed that the SC with Pt gave an ESR of 30 ohms.

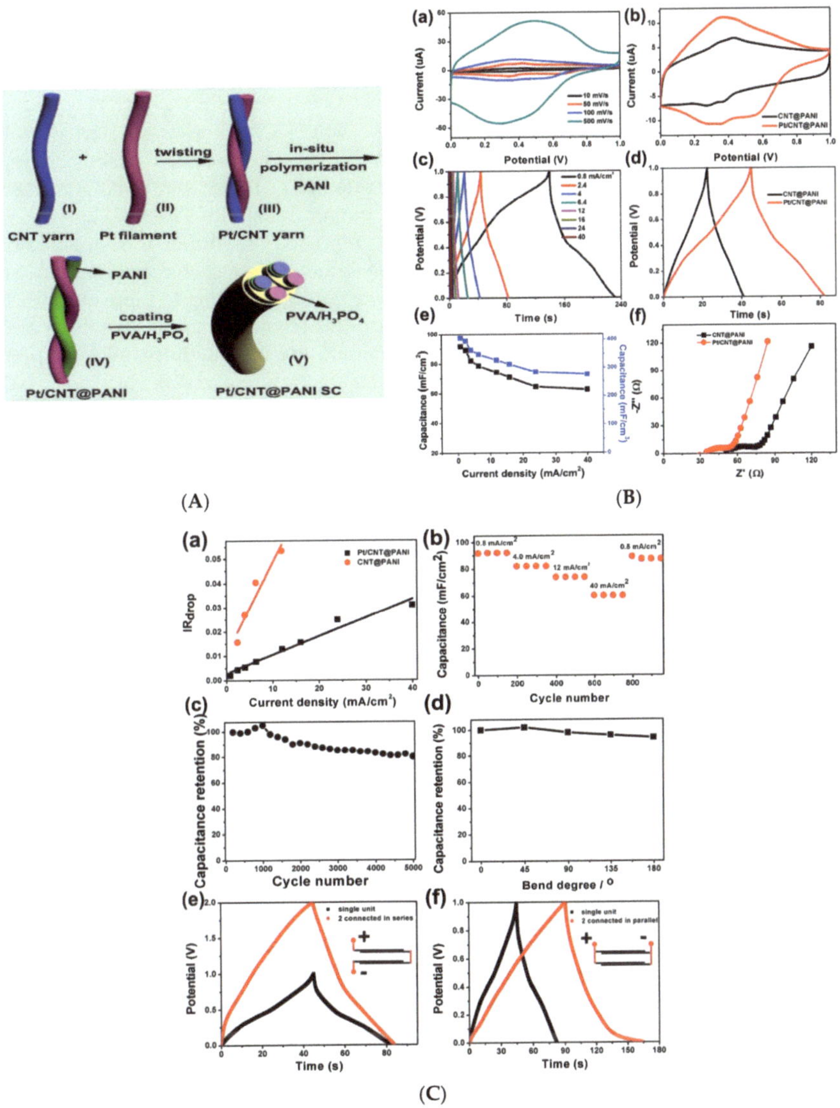

Figure 9. (**A**) Schematic of the fabrication of SCs. (**B**) The electrochemical properties of two-ply nanocomposite yarn SCs: (**a**) CV of the Pt/CNT@PANI yarn SC; (**b**) CV curves of CNT@PANI and Pt/CNT@PANI yarn SCs at 100 mV·s^{-1}; (**c**) galvanostatic charge–discharge (GCD) curves of the Pt/CNT@PANI yarn SC at different current densities; (**d**) GCD curves of CNT@PANI and Pt/CNT@PANI SCs at 2.4 mA/cm^2; (**e**) areal capacitance of the Pt/CNT@PANI yarn SC at different current densities; (**f**) EIS of CNT@PANI and Pt/CNT@PANI SC devices. (**C**) Characterization of the Pt/CNT@PANI nanocomposite yarn supercapacitor: (**a**) IR loss (drop of the cell internal resistance) as a function of the discharge current density; (**b**) cycle performance at varying current densities; (**c**) capacitance retention with different charging–discharging cycles and a constant current density of 4 mA/cm^2; (**d**) capacitance retention under different bending states; (**e**) GCD of two devices connected in series and (**f**) in parallel. Reproduced with permission from [59].

In contrast, without Pt, this parameter result was 40 ohms. The IR drop against the increasing current gave a smaller value and a less steep slope for the SC with Pt compared to the one without Pt. The SC with Pt also exhibited good cyclic stability with a capacitance retention rate of 80% after 5000 cycles. The PANI-coated CNT SC with Pt yielded a capacitance rate of 91.67 mF·cm^{-2}, an energy density of 12.68 μWh·cm^{-2}, and good cyclic stability.

Wu et al. [60] also twisted a number of CNT yarns with a platinum wire followed by PANI deposition. The 5 CNT yarns twisted with Pt gave the best results. Notably, Pt/5-CNT@PANI showed excellent electrochemical performance, revealing a specific capacitance rate of 217.7 F·g^{-1} at 0.2 A/g current density, an energy density of 30.22 Wh/kg, and a power density of 91.88 W/kg.

Wang et al. [61] reported on copper wire intertwined with CNT to be used as a current collector. The spinel-type nickel–cobalt oxide was hydrothermally deposited on the obtained composite yarn. According to these authors, the SC was fabricated from two individual CNT/copper intertwined yarns on which $NiCo_2O_4$ (NCO) nanowires were grown. The Cu enhances the conductivity, and the porous CNT allows quick charge transport between the electrolyte and the electrode. NCO was the only electroactive material used. The obtained SC exhibited a capacitance of 277.3 mF·cm^{-2} and an energy density of 35.76 μWh·cm^{-2} at a power density of 0.154 mW·cm^{-2}. In addition, this device showed excellent cyclic stability, with 89% capacitance retention after 5000 cycles.

5.2. Fabrication of an Asymmetric-Type Configuration

Symmetric SCs consist of positive and negative electrodes of the same active material with the same weight. The working voltage of a symmetric SC depends on the electrolyte and the active materials' potential window (PW). The PW is limited when an aqueous electrolyte is used due to water electrolysis (1.23 V (thermodynamic value), resulting in hydrogen and oxygen evolution. For pseudocapacitive materials such as PANI, PPY, and MnO_2, the PW can be narrower as the PW of the electrolyte can impact it. Additionally, the specific capacitance of a two-electrode SC made of pseudocapacitive materials such as PANI and PPY is always inferior to those obtained using three electrodes, because in a two-electrode cell the positive and negative electrodes will be working in different potentials, leading to different specific capacitances. This means that the one with the lower capacitance will determine the total specific capacitance, as per the Equation (2)

$$\frac{1}{C} = \frac{1}{C_1} + \frac{1}{C_2}$$

The performance matrix of an SC, like the energy and power density, can be increased by raising the working voltage, which can be expanded by using an asymmetric configuration. This can be achieved via different active materials at the positive and negative electrodes with different PWs, or when using the same active material with varying mass ratios. The PW can be wider if the positive electrode is reversibly charged to a maximum high and the negative one to a low minimum value. An asymmetric capacitor can give a maximum capacitance when $C_1 = C_2$, or $CM_+M_+ = CM_-M_-$, where CM_+, M_+, CM_-, and M_- are the gravimetric capacitance and mass ratios of the positive and negative electrodes, respectively. Asymmetric SCs can be fabricated from a host of positive and negative electrodes and can have wider working voltages and capacitances, resulting in very high energy and power densities [3,4]. The common pseudocapacitive materials and their properties are listed in Tables 1 and 2 of Section 2.1.

5.2.1. Transition Metal Oxide and Hydroxide Based Asymmetric Capacitors

Most asymmetric SCs reported in the literature are fabricated using transition metal oxides, hydroxides, sulfides, and nitrides (Ni, Co, Fe, V, Ti, Mo) as the pseudocapacitive materials. Yao's group [62] investigated the growth of Co_3O_4 on conductive substrates such CNT fibers. They used these for a positive electrode when fabricating an SC with vanadium nitride nanowires grown on the CNTs as the negative electrode. Co_3O_4 is an inexpensive and promising positive electrode material with high theoretical capacitance. Vanadium nitride, employed as a negative electrode, has a very high theoretical capacitance of 1340 F. A PVA-KOH gel electrolyte was used, which provided a cell voltage of 1.6 V. The SC exhibited an energy density of 15.79 mWh·cm^{-3} and a power density of 3.232 W·cm^{-3}. The SC gave a capacity retention rate of 93.01% after 3000 bend cycles at 90°.

Layered transition metals hydroxides (TMOH) such as Ni $(OH)_2$ and FeO(OH) are capable of fast redox reactions within the space between the layers, thereby increasing the capacitance value.

Lu's group [63] constructed a coaxial type of SC to ward off the decreased specific capacitance effects with parallel and twisted construction types and successfully fabricated an SC with MoS_2@ Fe_2O_3/CNT paper as the negative electrode and $Ni(OH)_2$ @$NiCo_2O_4$/CNT fiber as the positive electrode with a PVA-KOH electrolyte. With a voltage window of 1.6 V, the SC gave a specific capacitance of 373 F·cm^{-2} at a current density of 2 mA·cm^{-2}. The achieved energy density was 0.13 mWh·cm^{-2}, the power density 3.2 mW·cm^{-2}, and the capacity retention rate was 83.3% after 10,000 cycles.

Various combinations of positive and negative electrodes were attempted, with the researchers claiming higher working voltages. However, the obtained energy density values are still unsatisfactory for practical applications.

Yao's group [64] successfully fabricated a ternary metal oxide (ZnNiCo) with nickel hydroxide nanowire arrays (NWAs), cladding the CNT fiber as the positive electrode (ZNCO@Ni(OH)2NWAs/CNTF/KOH-PVA) and a thin layer of carbon coated vanadium nitride on the CNT fiber as the negative electrode (VN@C NWAs/CNTS) in a coaxial fiber configuration (Figure 10A). The coaxial fiber-based asymmetrical supercapacitor (CFASC) gave an impressive specific capacitance of 94.67 F·cm^{-3}, with an energy density of 33.66 mWh·cm^{-3} and with more than 90% capacitance retention after 3000 bend cycles (Figure 10B). Other transition metal oxide (TMO)-based asymmetric capacitors are displayed in Table 4.

Yao's group [65] fabricated a fiber supercapacitor (FSC) with a ternary oxide (MoNiCo) on CNT fiber as the positive electrode and thin carbon-coated vanadium nitride nanowires on CNT fiber as the negative electrode with PVA-KOH as the gel electrolyte. The use of a ternary oxide is intended to increase the electronic conductivity and electrochemical activity due to the mixed valence metal cation. This might not be the case when mono- or bi-metal oxides such as Co_3O_4 are used, which suffer from low capacitance and electrical conductivity levels. Furthermore, the hierarchical dandelion-like ternary oxide structure on the CNTs provides a greater surface area, leading to more accessible sites on the electrode for the electrolyte, resulting in fast electron ion transport. The authors of this work reported an SC with a voltage window of 1.6 V, capacitance of 62.3 F·cm^{-3}, energy density of 22.2 mWh·cm^{-3}, and 90.2% capacitance retention after 3500 bend cycles.

Peng's group [66] fabricated an FSC using a polyimide-based polymer for high mechanical strength and MnO_2 for high pseudocapacitance. As a result, the FSC revealed a reasonably high working voltage of 2.1 V, mechanical strength of 264 MPa (tensile strength), energy density of 36.4 μWh·cm^{-2}, and power density of 0.78 mW·cm^{-2}. These results are credited to the combination of MnO_2/CNT as the positive electrode and polyimide/CNT as the negative electrode with carboxy methyl cellulose (CMC)/Na_2SO_4 as the electrolyte in a parallel configuration with a voltage window of 2.1 V.

Zhang's group [67] successfully investigated the pre-insertion of cations such as Na^+ onto MnO_2 nanosheets to enhance the capacitance (Figure 11A).

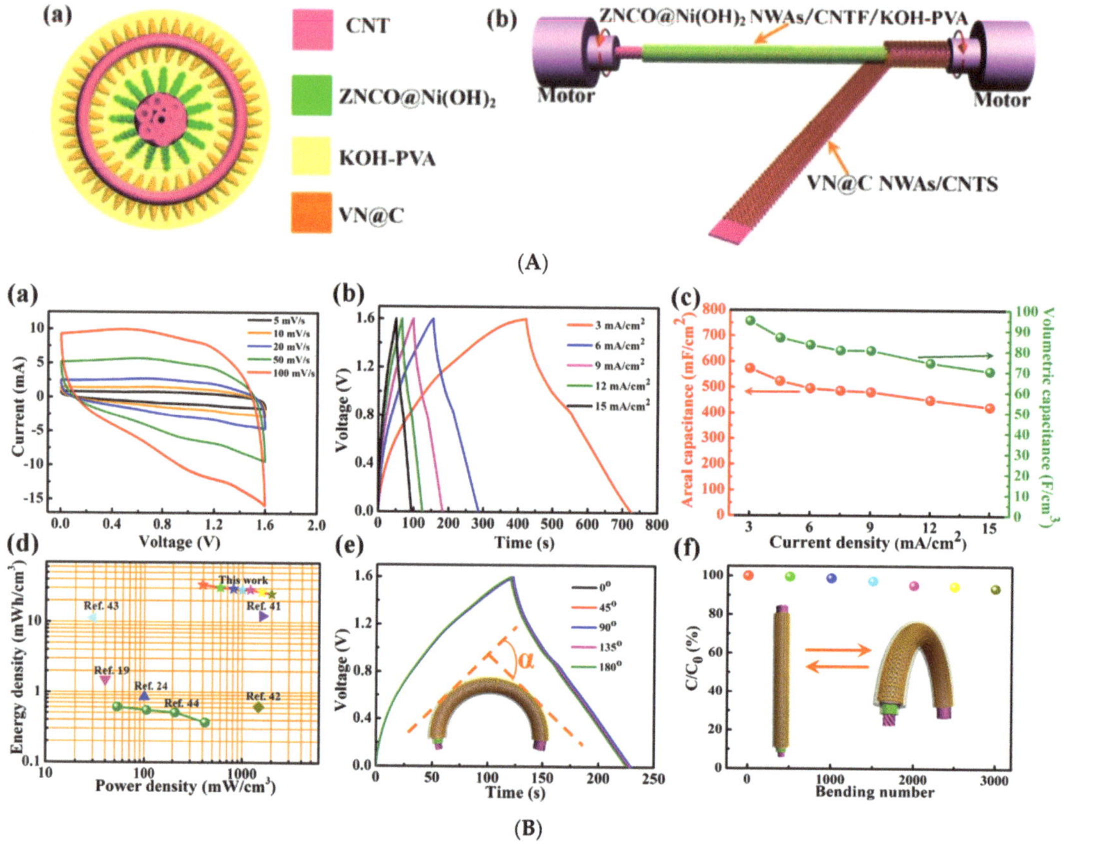

Figure 10. (**A**) (**a**) Cross sectional structure of the CFASCs (coaxial fiber shaped asymmetric supercapacitors). (**b**) Wrapping of the VN@C NWAs/CNTs around the surface of the ZNCO@Ni$(OH)_2$ NWAs/CNTF/KOH–PVA. (**B**) Electrochemical characterization and flexibility analysis of the as-assembled CFASCs: (**a**) CV curves of the assembled CFASCs measured at different scan rates between 0 and 1.6 V; (**b**) GCD curves collected at different current densities between 0 and 1.6 V; (**c**) areal and volumetric specific capacitances calculated from the GCD curves as a function of the current density; (**d**) volumetric energy and power densities of CFASCs compared with previously reported FASCs (fiber shaped asymmetric supercapacitors); (**e**) GCD curves of the as-assembled CFASCs bent at various angles and a current density of 10 mA/cm^2; (**f**) Normalized capacitances of our CFASCs bent 90° for 3000 cycles. Reproduced with permission from [64].

The low conductivity of MnO_2 results in poor ion diffusion, and the insertion of Na/K ions improves the conductivity, thereby enhancing the capacitance. An FSC with a twisted configuration, as shown in Figure 11A, showed good performance due to the effective design of the positive electrodes as Na-doped thin MnO_2 nanosheets on the CNT fiber. The adequately chosen negative electrode, which was a flowerlike thin MoS_2 nanosheet array, provided the SC with a high working voltage of 2.2 V. As reported in this publication, the FSC exhibited a high specific capacitance of 265.4 mF·cm^{-2} and energy density of 178.4 μWh·cm^{-2}. Moreover, no capacitance loss was noticed after 3500 bending cycles at 120° (Figure 11(Bf)).

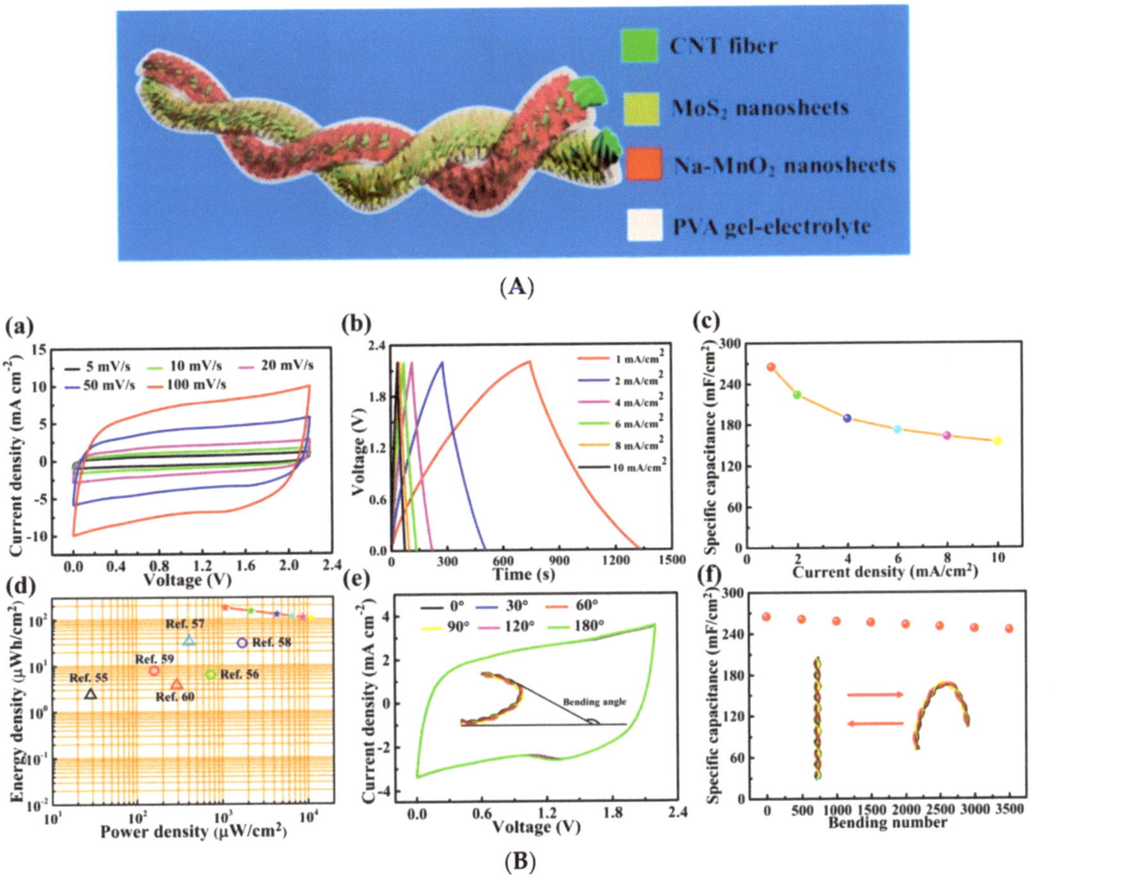

Figure 11. (**A**) Detailed illustration of the fiber shaped asymmetric supercapacitors (FASCs) (**B**) Electrochemical study of the FASCs: (**a**) FASC CV curves at various scanning rates of 5–100 mV/s; (**b**) GCD curves with different current densities of 1–10 mA/cm^2; (**c**) areal specific capacitance as a function of the current density; (**d**) FASC power density and areal energy density compared to previous FSCs; (**e**) CV curves of the FASCs under various bending angles; (**f**) capacitance retention of the FASCs exposed to bending angles of 120° for 3500 cycles. Reproduced with permission from [67].

5.2.2. Conducting Polymer-Based Asymmetric Supercapacitors

Chen's group [68] developed an easy synthesis technique wherein a mixture of CNTs plus PPY is deposited on the CNT core (positive electrode), and a mixture of RGO plus CNTs is electrodeposited on the CNT core as the negative electrode (Figure 12A). The created fiber shaped asymmetric SC (FASC) gave a higher specific capacitance and greater mechanical strength. The gel electrolyte PVA-H_3PO_4 provided a voltage of 1.6 V, which yielded a capacitance of 58.82 mF·cm^{-2}. The FSC (fiber-based SC) was fabricated as illustrated in Figure 12A, and the hierarchical construction resulted in a 3D pore structure providing a better electrochemical performance. The authors reported an energy density of 20.91 μWh·cm^{-2} and a power density of 6.4 mW·cm^{-2}. The specific capacitance retention rate was 98.6% after 200 bend cycles and 90% after 10k regular charge–discharge cycles (Figure 12C).

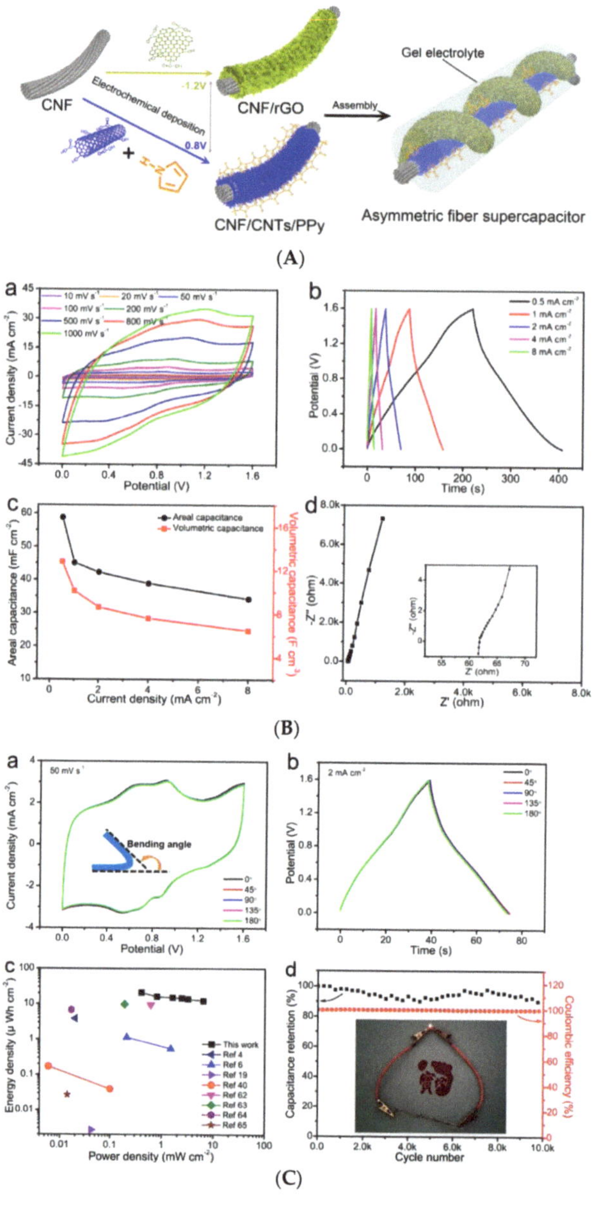

Figure 12. (**A**) Schematic of the fabrication of an FASC. (**B**) Electrochemical performances of FASC: (**a**) CV curves at various scan rates ranging from 10 to 1000 mV·s^{-1}; (**b**) GCD curves at the current densities from 0.5 to 8 mA·cm^{-2}; (**c**) areal and volumetric specific capacitances at various current densities; (**d**) Nyquist plots carried out in a frequency range from 100 kHz to10 MHz. The inset shows the enlarged plot in the high-frequency region (**C**): (**a**) CV curves for the asymmetric capacitor at a scan rate of 50 mV·s^{-1}, under different bending states, with the inset showing an illustration of an asymmetric supercapacitor at different bending angles; (**b**) GCD curves at a current density of 2 mA·cm^{-2} for the asymmetric supercapacitor under different bending states; (**c**) Ragone plots of the presently and previously reported supercapacitors; (**d**) capacitance retention during 10,000 GCD cycles. The inset shows a LED lit up at 2.4 V by two asymmetric supercapacitors in series. Reprinted with permission from [68].

Table 4 summarizes the current asymmetric SCs reported recently with details about their configuration, electrodes, electrolyte, working voltage, and electrochemical performance.

5.3. Use of Electrolytes with a Wide Potential Window

Electrolytes with a wide potential window can increase the working voltage to improve the device's performance. The most commonly used electrolytes in SCs are aqueous solutions based on sodium sulfate (Na_2SO_4), potassium hydroxide (KOH), and sulfuric acid (H_2SO_4). These have the following advantages:

- High ionic conductivity;
- Environmental benignity;
- Non-flammability;
- Small ionic radius.

However, their PW is limited to 1.23 V due to the earlier mentioned reasons. Another option is to use organic electrolytes with a greater potential window of 2.5–3 V. Nevertheless, organic solvents are flammable, volatile, toxic, and unsafe. Additionally, the power density could be poor due to the low conductivity of the organic electrolytes. Another group of non-volatile electrolytes is ionic electrolytes, which are less flammable with a PW greater than 3. Therefore, it is the safest group among the above three. Gel polymer electrolytes represent a recent group of electrolytes that are used widely for solid-state SCs. They are made by mixing a polymer with an ionic or organic electrolyte and can provide a wide voltage window that ensures good performance [3]. For example, Shanov's group [50] compared the specific capacitance of an oxygen-plasma-functionalized CNT fiber SC in three electrolytes. The SC with the PVDF-EMIMBF_4 electrolyte gave 25% and 65% more capacitance than the PVDF-EMIMTFSI and PVA-H_2SO_4 electrolytes, respectively.

Table 4. Summary of the asymmetric supercapacitors and their characteristics recently reported in the literature.

Electrodes							
Positive	**Negative**	**Electrolyte, Cell Voltage**	**Capacitance**	**Energy Density (Ed)**	**Power Density (Pd)**	**Capacity Retention/No. of Cycles**	**Ref**
NCO/VG@ CNTF	VG@CNTF	PVA-KOH (1.8 V)	188 F·g^{-1} @ 5 A·g^{-1}	65 Wh·kg^{-1}	100 W·kg^{-1}	93.1% after 5k bending cycles	[69]
PANI-CNTF	FeC_2O_4/ FeOOH-CNTF	PVA-KOH (2.1 V) PVDF-EMIMBF4 (2.8 V)	-	0.05–4.07 μWh·cm^{-2} 0.17–3.06 μWh·cm^{-2}	0.18–0.92 μW·cm^{-2} 0.26–0.97 mW·cm^{-2}	96.76% of Ed retention after 4k bending cycles	[70]
PEDOT: PSS	CNTF	Chitosan (1.4 V)	21.4 F·g^{-1}	5.83 Wh·kg^{-1}	1399 W·kg^{-1}	90% after 200 bending cycles	[71]
$NiCo_2S_4$@CNTF	VN@CNTF	PVA-LiCl (1.6 V)	86.2 F·cm^{-3} @ 0.1 mA·cm^{-3}	30.64 mWh·cm^{-3}	80 mW·cm^{-3}	91.94% after 5000 bending cycles	[72]
CNTF/Porous CNT/PPY	CNTF/RGO	PVA-H_3PO_4 (1.6 V)	58.82 mF. cm^{-2} (12.8 F·cm^{-3})	20.91 μWh·cm^{-2}	6.4 mW·cm^{-2}	98.6% after 200 bend cycles, 90% after 10k cycles	[68]
GCF@NC@NCO	P-GCF	PVA-KOH (1.55 V)	33.6 mF·cm^{-2}@ 0.61 mA·cm^{-2}	11.2 μWh·cm^{-2}	472.1 μW·cm^{-2}	93% after 10k cycles	[73]
N doped CNT on CNT/LTO	N doped CNT on CNT	LiPF6 (in 1:1 EC/DEC) (2.5 V)	-	0.296 mWh·cm^{-2}	0.172 mW·cm^{-2}	100% after 4k cycles	[74]
CNT@PPY	CNT@$CuCo_2O_4$	PVA-KOH (1.6 V)	59.55 mF·cm^{-2}	20 μWh·cm^{-2}	5.115 mW·cm^{-2}	80.1% after 8k cycles	[75]
VO_2@PPY/CNT	VN@NC/CNT	PVA-LiCl (1.8 V)	60.6 F·cm^{-3}	27.3 mWh·cm^{-3}	225 mW·cm^{-3}	88.9% after 4k bending cycles	[76]
$Ni(OH)_2$@ $NiCo_2O_4$/ CNTF	MoS_2@Fe_2O_3/ CNTpaper	PVA-KOH (1.6 V)	373 F·cm^{-2} @2 mA·cm^{-2}	0.13 mWh·cm^{-2}	3.2 mW·cm^{-2}	80.3% after 2k cycles@ 20mAcm^{-2}	[63]

Table 4. *Cont.*

Electrodes							
Positive	**Negative**	**Electrolyte, Cell Voltage**	**Capacitance**	**Energy Density (Ed)**	**Power Density (Pd)**	**Capacity Retention/No. of Cycles**	**Ref**
ZNCO@Ni(OH)$_2$ NWAs/CNTF	VN@C NWAs/CNTF	PVA-KOH (1.6 V)	94.67 F·cm^{-3} @3 mA·cm^{-2}	33.66 mWh·cm^{-3}	396 mW·cm^{-3}	93.6% after 3k bending cycles at 90°	[64]
Co$_3$O$_4$ NWAs/CNTF	VN (NWA)/CNTF	PVA-KOH (1.6 V)	44.4 F·cm^{-3} @0.4 A·cm^{-3}	15.79 mWh·cm^{-3}	3.232 W·cm^{-3}	93.12% after 4k cycles 93.01% after 3k bending cycles	[62]
MNCO/CNTF	VN@C(NWA)/ CNTF	PVA-KOH (1.6 V)	62.3 F·cm^{-3} @1 mA·cm^{-2}	22.2 mWh·cm^{-3}	213.3 mW·cm^{-3}	90.2%after 3.5k bending cycles	[65]
ZNCO NWA/CNTF	VN/CNTF	PVA-KOH (1.6 V)	50.0 F·cm^{-3} @0.1 A·cm^{-3}	17.78 mWh·cm^{-3}	80.0 mW·cm^{-3}	91.0% after 3K bending cycles at 90°	[77]
Ti@TiO$_2$	CNTF	PVA-H$_3$PO$_4$ (1.4 V)	-	11.7 Wh·kg^{-1}	2060.0 W·kg^{-1}	91.0% after 1.2k bending cycles	[78]
CNT/MnO$_2$	CF/PPY	LiCl-PVA (1.6 V)	66.27 mF·cm^{-2}	23.56 μWh·cm^{-2}	-	83% after 5k cycles	[79]
MnO$_2$/CNT	PI/CNT	CMC/Na$_2$SO$_4$ (2.1 V)	59.2 mF·cm^{-2} @0.74 mA·cm^{-2}	36.4 μWh·cm^{-2} 30.2 μWh·cm^{-2}	0.78 mW·cm^{-2} 15.6 mW·cm^{-2}	96.3% after 2k cycles	[66]
Na-MnO$_2$@CNTF	MoS$_2$@CNTF	PVA/Na$_2$SO$_4$ (2.2 V)	265.4 mF·cm^{-2} @1 mA·cm^{-2}	178.4 μWh·cm^{-2}	1100.9 μW·cm^{-2}	90% after 5k cycles	[67]

6. Stretchability of Wearable Supercapacitors

One of the most critical and emerging requirements of wearable fiber-based energy storage devices is the stretchability. In addition to other criteria for these devices such as the safety, bendability, low gravimetric density, and durability, an important one factor is the undeterred electrochemical properties during the deformations caused by the user's movements. An evaluation metric for the feasibility of such devices is their ability to withstand the frequency of human motion, including small- and large-scale motions falling in the frequency range of 0.1 and 10 Hz. At a minimum, such devices need to tolerate strains of 100% [80]. This section summarizes the advancements in developing CNT fiber supercapacitors by prioritizing the mechanical properties that can withstand high torsional strains and bending, while maintaining high capacitance retention rates.

Several variations in the manufacturing process impact the CNT fiber supercapacitor's mechanical and electrochemical properties. As already mentioned, adding pseudocapacitive materials such as metal oxides or organic polymers can improve the individual fibers' electrochemical properties. The electrodeposition of MnO$_2$, PANI, RuO$_2$, and Au have all been investigated. The results of their mechanical testing are presented in Table 5, which is displayed later. The tradeoff between using metal oxides and organic polymers is the material cost and the added weight to the fibers. Metals at optimal concentration contribute a substantial portion to the total weight of the fibers. The varying configurations all utilized different pseudocapacitive materials and were adopted to test the mechanical properties of the fibers in different ways, making it difficult to draw direct comparisons from the information available in the public domain.

6.1. Physical Configurations

6.1.1. Buckled CNT Sheets with an Elastomer Core

Buckling CNT sheets around an elastomer core are manufactured for creating individual stretchable electrodes utilized in supercapacitors. Their fabrication method includes stretching an elastomer core and wrapping a single or several CNT sheets around the elastomer. When the elastomer is relaxed, the CNT sheet buckles, as depicted in Figure 13, which enables a high surface area to be in contact with the fiber [81].

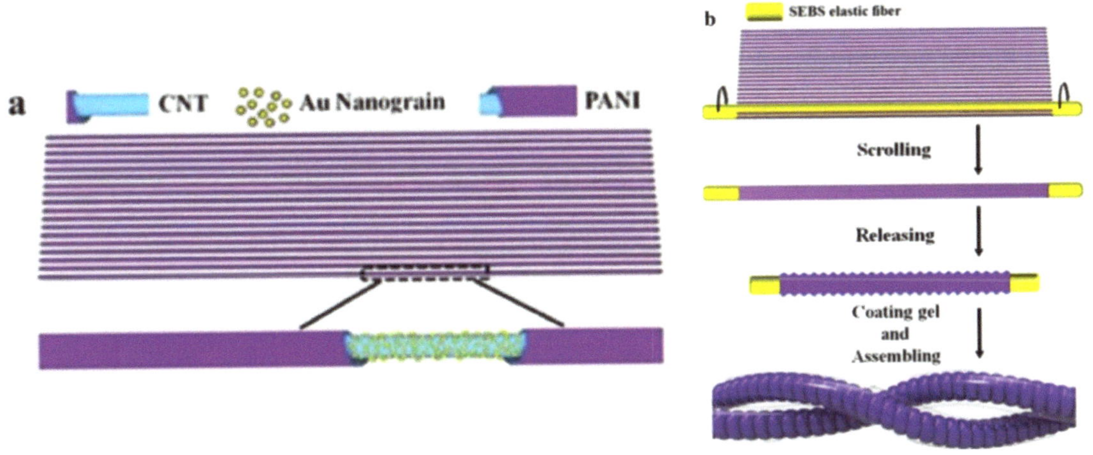

Figure 13. Schematic of the fabrication process of a stretchable linear elastomer-based FSC. (**a**) fabrication of the PANI@Au$_x$@CNT electrode, where 'x' represents the deposition time of Au nanoparticles. PANI was electrodeposited onto Au$_x$@CNT sheets. (**b**) Stretching of elastic fiber, wrapping of the electrodes, and twisting to fabricate the SC. Reprinted with permission from [81].

The buckling pattern is one of the topologically effective patterns used to save space for an otherwise stretchable and lengthy fiber capacitor, which helps to meet one of the most desirable requirements in wearable devices. The densification is often used to adhere the CNT sheets to the elastomer. The latter does not improve the electrochemical properties of the devices and adds additional weight and the benefit of enhanced mechanical properties. Many elastomers have been utilized across the industry, including Eco Flex 0040 (silicone-based), polystyrene, styrene-ethylene/butylene-styrene (SEBS), and others [81–84]. The strain tolerances of such devices have ranged from 100% to 200%, with various changes in capacitance retention. In a strained, bent, or twisted state, a decrease in the capacitance retention of the fibers is observed due to the loss of a buckled contact, reducing the available pathway for electron diffusion. Additionally, the coating of a PVA-based gel electrolyte seals the two interfaces without any issues of delamination caused by various deformations [83].

6.1.2. Coaxial CNT Layers with an Elastomer Core

Another method utilizing an elastomer core includes layering two super shearable CNT(ShCNT) sheets around a polydimethylsiloxane (PDMS) elastomer with PVA acting as the gel electrolyte and sealant to prevent a shortage, as demonstrated in Figure 14 [85]. This method resulted in a CNT supercapacitor in one fiber able to withstand 20,000 rad/m of twisting without significantly decreasing its capacitance retention. Minimal cracks and damage were observed to the CNT alignment in the film after twisting. The super shearable CNT film releases stress through the rearrangement of CNT fibrils instead of permanent deformation. Further information regarding the mechanical properties of the CNT fiber is displayed in Table 5.

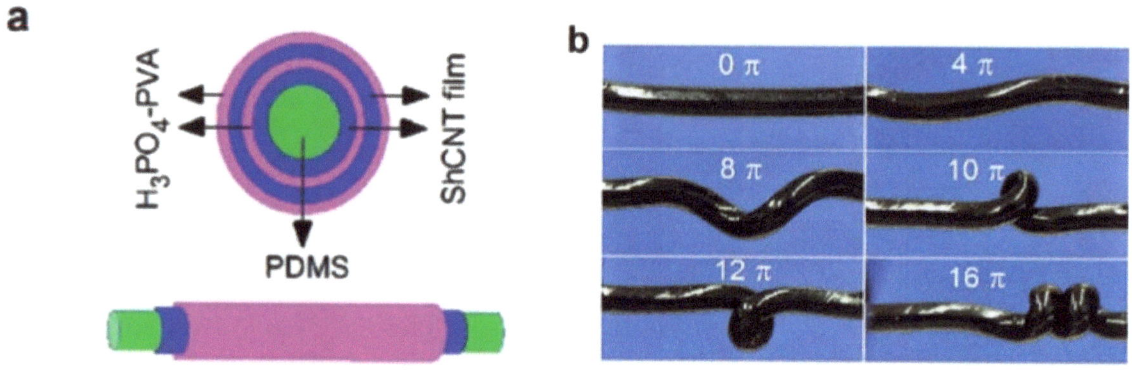

Figure 14. (**a**) The structure of a torsionable supercapacitor in cross-sectional and side views. (**b**) Digital photos showing the supercapacitor at different torsion levels. Reprinted with permission from [85].

6.1.3. Helical CNT Fibers with an Elastomer Core

An alternative method of manufacturing supercapacitors involves helically winding CNT fibers in parallel around an elastomer core. This method utilizes dry or wet spun CNT fibers, followed by doping with pseudocapacitive materials, which are then sealed with an electrolyte gel, typically a PVA solution. Symmetrical or asymmetrical fibers employing different pseudocapacitive materials are then wound helically around an elastomer core, followed by a final coat of PVA gel electrolytes to adhere to the components. The process is depicted in Figure 15 [86]. The SC in this figure is super elastic, sustaining strains as high as 850%. Varying methods produce tightly wound CNT coils with good surface contacts between the electrodes. When introduced to stress, the packing of the coils loosens, enabling movement without permanent deformation to the CNT electrodes [86–89]. The data regarding the performance of the devices under strain are summarized in Table 5.

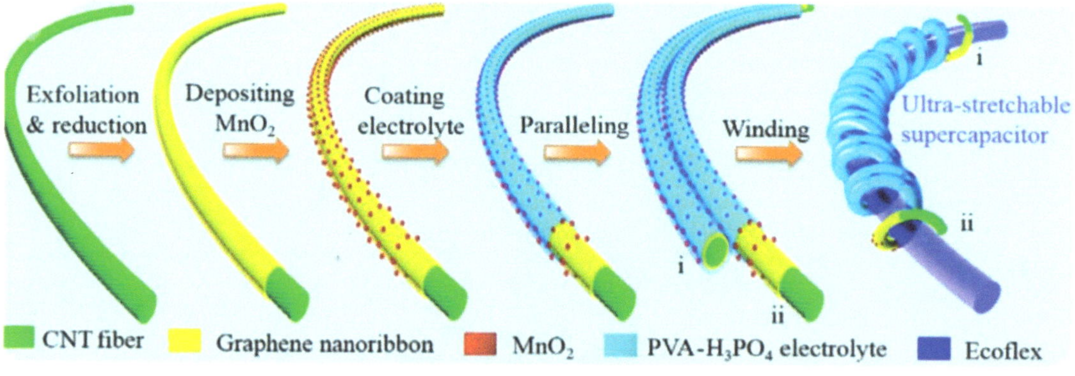

Figure 15. Schematic illustration showing the fabrication of an ultra-stretchable wire-shaped supercapacitor [86].

A modified manufacturing method for helical CNT fibers wrapped around an elastomer core was developed, demonstrating improved mechanical properties. The fibers were produced from a solution of CNTs and graphene sheets. The graphene sheet helped to create a porous structure that enhanced the ion diffusion during the electrochemical performance and contributed to its improved mechanical properties. A PVA gel electrolyte coat was applied to the individual fibers. The symmetrical electrodes were twisted together in a spring-like configuration around a temporary metal rod. The two fibers were cast in SEBS solution to provide a protective coating and to maintain a flexible spring-like design, as depicted in Figure 16a,b. Strain rates of up 800% without permanent deformation were demonstrated [90].

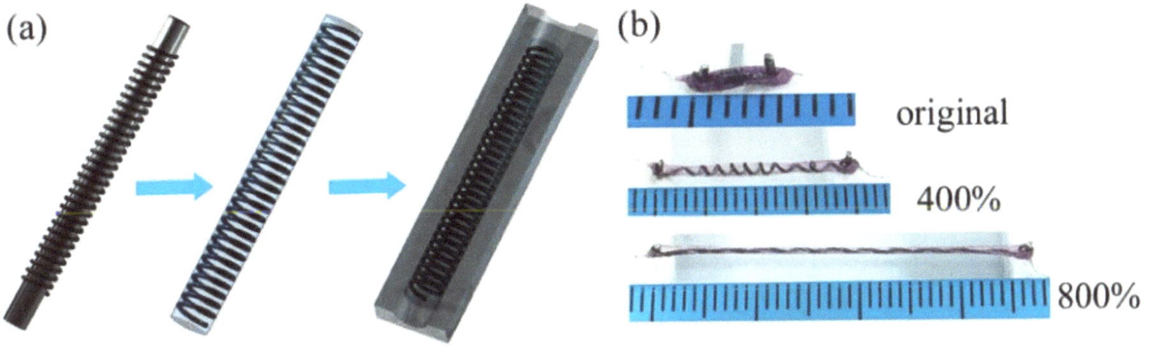

Figure 16. (**a**) Schematic illustration of the fabrication of an FSC in which the yarn is coiled on a rod and cast using SEBS polymer to fix the coil structure. The rod is removed, and the device is placed in a mold to cover the coil yarn with rubber. (**b**) Illustration of the prepared supercapacitor with different strain rates up to 800%. Reprinted with permission from [90].

6.1.4. Freestanding Helical CNT Fibers

The presence of an elastomer contributes additional weight to the CNT supercapacitors without improving the electrochemical performance. Ideally, physical configurations without elastomers are preferred but are limited due to the inherent mechanical properties of the CNT fibers. Innovative arrangements of CNT fibers have produced mechanically competitive supercapacitors [91]. Tightly coiling individual CNT fibers in a helical pattern followed by a coating of PVA allows the fibers to be elongated without damaging the connectivity. Two symmetrical CNT fibers coated with PVA can produce a freestanding helical supercapacitor (Figure 17a) with similar mechanical properties to elastomer-based supercapacitors. Figure 17b,c depict the configuration of the freestanding supercapacitor in a position of rest and elongation [92].

Two variations of the applications of this method produced CNT supercapacitors with remarkable mechanical properties. One method utilized a solution of bacteria culture (BC) cellulose structure, CNT, and poly(3,4-ethylene dioxy-thiophene) (PEDOT) to spin helically coiled fibers sealed in PVA. The PEDOT and BC support the mechanical structure while providing a highly porous surface for improved interaction with PVA [93].

The other variation utilized multiple CNT fibers spun helically, followed by a priming coat of amine groups to improve the adherence to metal–organic frameworks (MOF) grown directly on the fibers, as depicted in Figure 18. The MOF improved the mechanical integrity of the fibers while also providing a porous surface for increased diffusion of the gel electrolyte [94]. The observed mechanical properties are provided in Table 5.

6.1.5. Other Configurations

A configuration that does not fit into the general categories previously discussed is the CNT knitted coaxial supercapacitor. CNT fibers are knitted uniformly using a machine, producing a freestanding flexible structure coated in PVA. A double-walled electrode is made by sheathing one of the knitted structures inside the other to produce a single fiber for the supercapacitor (Figure 19). The drawback of such a material is the quantity of CNT fiber required to manufacture the knitted fiber. A strain of 100% was observed with minimal impact on the capacitance retention. A specific area capacitance of 511.28 mF/cm^2 was demonstrated [95].

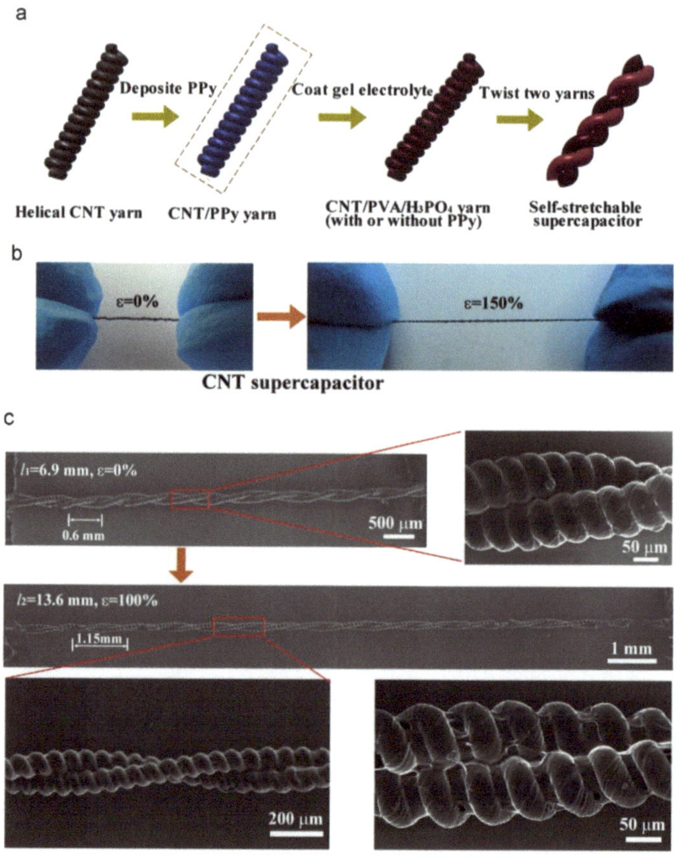

Figure 17. Fabrication of self-stretchable fiber-shaped supercapacitors. (**a**) Schematic of the freestanding helical CNT yarn SC fabrication process. Two of such yarns were twisted into a double-helix supercapacitor. (**b**) Photograph of manually stretched (1.5 times the original length) double-helix CNT yarn supercapacitor. (**c**) SEM images of a 6.9-mm-long supercapacitor in its original state (l_1 = 6.9 mm, ε = 0) and after stretching to l_2 = 13.6 mm, ε = 100%, showing uniform deformation of helical loops and a continuous electrolyte coating on the surfaces of the yarns. Reprinted with permission from [92].

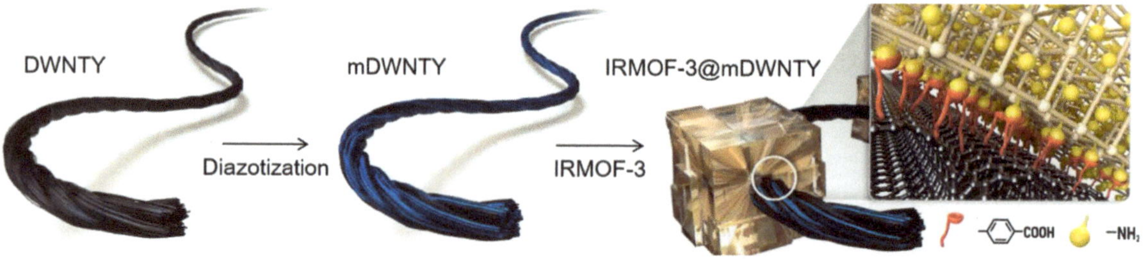

Figure 18. Schematics illustrating the synthesis process used to fabricate IRMOF3@mDWNTY (IRMOF3 = ligand-based isoreticular metal–organic framework; mDWNTY = modified double-walled nanotube yarn). Reprinted with permission from [94].

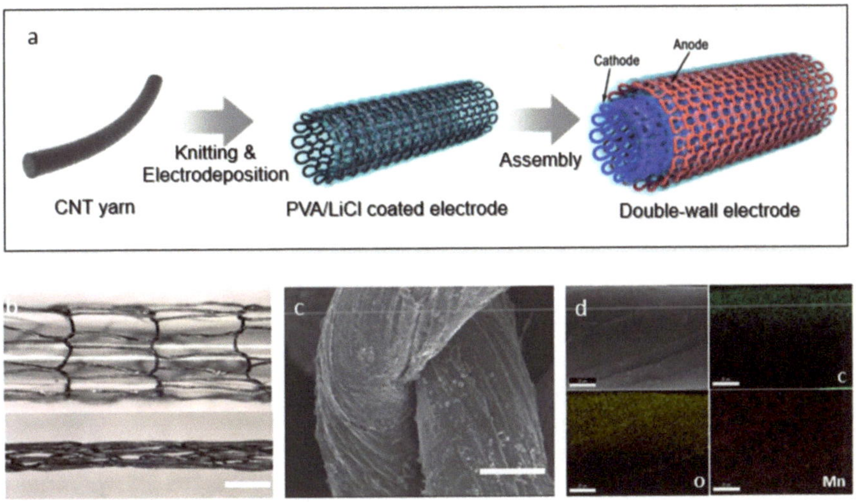

Figure 19. Schematic and images of knitted MnO_2@CNT fiber SC. (**a**) Fabrication scheme of a double-walled knitted MnO_2@CNT FSC. (**b**) Optical image of non-stretched (ε = 0) and stretched (ε = 100) knitted structures (scale bar: 1 mm). (**c**) SEM image of a knitted MnO_2@CNTfiber supercapacitor (scale bar, 500 μm, and 50 μm). (**d**) EDX mapping images of knitted MnO_2@CNTfiber supercapacitor (scale bar: 20 μm). Reprinted with permission from [95].

One alternative approach for a wearable supercapacitor configuration utilizes two CNT fibers arranged sinusoidally and encased in a poly(acrylamide) (PAAm) hydrogel. The latter is biocompatible, stretchable, and has excellent ionic conductivity when doped with LiCl salts. Figure 20 demonstrates the physical configuration of the CNT fibers in PAAm hydrogel [80]. This multifunctional composite can also be used as a strain sensor, as discussed in the next section.

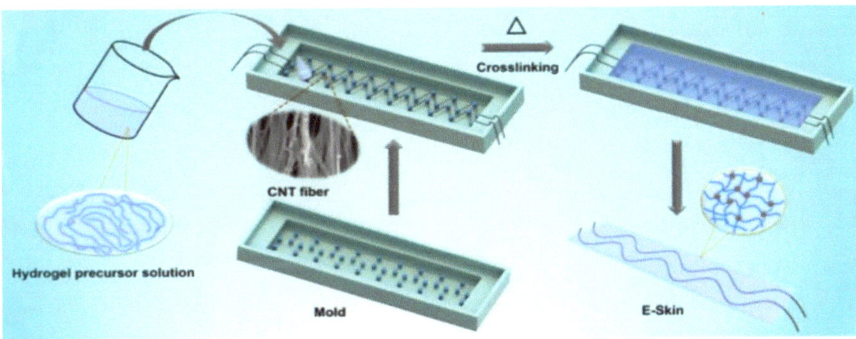

Figure 20. Schematic diagram of the CNT fiber/PAAM hydrogel (CFPH) composite preparation process. Reprinted with permission from [80].

Table 5. Summary of the stretchable supercapacitors and their characteristics, as recently reported in the literature.

Configuration	Pseudocapacitive Material/Electrolyte	Strain, Capacitance Retention	Bend, Capacitance Retention	Ultimate Strength	Capacitance	Ref
Symmetrical coiled CNT sheets, Nylon core	MnO_2/ PVA-LiCl	150%, 88%	-	- 68 MPa	5.4 mF·cm^{-1} 40.9 mF·cm^{-2}	[96]
Wire-Drawn-Die _ Free Standing Symmetrical Helical CNT-PPy composite fiber	PPY/ PVA-H_3PO_4	-	180°, 100%/ 100% after 1k cycles @90°	-	69 F·g^{-1}	[91]
Free Standing Symmetrical Helical CNT yarn	PPY/PVA-H_3PO_4	150%, 94%	-	-	19 F·g^{-1}	[92]
Asymmetrical Buckled CNT sheets Eco Flex 0040 Elastomer Core	MnO_2/PEDOT/ PVA-LiCl	200%, 97.7%	150%, 96.8%	-	2.38 mF·cm^{-1} 11.88 mF·cm^{-2}	[82]
Symmetrical Buckled CNT Sheets, SEBS Elastomer Core	Au and PANI/ PVA-H_3PO_4	400%/ 100% after 1k cycles	-	-	8.7 F·g^{-1} @1A·g^{-1}	[81]
Asymmetrical Helical Oxidized CNT Fiber, Elastomer Core	MnO_2@PEDOT:PSS@ OCNTF (positive) MoS_2@CNTF(negative) PVA-LiCl	100%, 92% after 3k cycles	-	-	278.6 mF·cm^{-2}	[87]
Symmetrical Coaxial Layered CNT Sheets, Eco flex Elastomer Core	- PVA-H_3PO_4	650%, 88%	-	-	2.42–2.68 mF·cm^{-1}	[85]
Twisted symmetrical Buckled CNT Sheets, SEBS Elastomer Core	MnO_2 & RuO_2 PVA-H_3PO_4	200%, 100%	-	-	25.0 F·g^{-1}	[83]
Symmetrical CNT Dipped Yarns/Ag plated Double-covered yarn polymer	(PVDF-HFP)/ $EMIMBF_4$	150%, 75% 5k cycles	120°, 80%, 5k cycles	46.6 MPa	4.8 F·cm^{-3} @200 mA·cm^{-3}	[88]
Symmetrical Buckled CNT Sheets Elastomer Core	PANI PVA-H_2SO_4	100%, 98%, 100 cycles	-	-	394 F·g^{-1}	[84]
Helically Coiled CNT/ Graphene Fibers, Coated in SEBS	PANI PVA-H_3PO_4	800%, 99.2% 800%, 77.3%, 5k cycles	-	-	138 F·g^{-1}	[90]
Helically Coiled Symmetrical CNT/Graphene Fibers, Eco flex Elastomer Core	MnO_2 PVA-H_3PO_4	850%, 82% after 1k cycles @ 700% strain	90°, 100%, 1k cycles	-	14.02 mF·cm^{-2}	[86]
Helically Coiled Symmetrical CNT Fibers, Spandex Elastomer Core	MnO_2 PVA-KOH	80%, 92.1%, 500 cycles	180°, 95.3%, 500 cycles	-	685 mF·cm^{-2}	[89]
Coaxial Knitted CNT Fibers	MnO_2 PVA-LiCl	100%, 98%, 500 cycles	160°, 100%	-	321.08 mF·cm^{-1} 511.28 mF·cm^{-2}	[95]
Parallel/Symmetrical Sinusoidal CNT Fibers, PAAm Hydrogel	- PAAm-LiCl	50%, 90.4%, 3k cycles	-	130 MPa	10.6 mF·cm^{-2}	[80]
Helically Coiled Symmetrical CNT Fibers, Silicone Elastomer Core	PVA-HCl	70%, 97%, 1k cycles	180°, 97%, 1k cycles	-	170 mF·cm^{-2} @100 mV·s^{-1}	[97]
Helically Coiled Symmetrical PEDOT@BacterialCellulose (BC)/CNT Fiber	PEDOT, BC PVA-H_3PO_4	1000%, 94.4%, 1k cycles	-	-	175.1 F·g^{-1}	[93]
Helically Coiled Symmetrical CNT Fibers in Metal-Organic Framework	MOF, PVA-DMSO	-	88%, 500 cycles	492.8 MPa	225.2 F·cm^{-3}	[94]

7. Multifunctional CNT Supercapacitors

Although the review is focused on CNT fiber supercapacitors, some novel applications of these devices have also been highlighted.

7.1. Strain Sensors

One application of CNT supercapacitors lies in their tolerance to deformation, undergoing twisting and stretching while returning to an original state with negligible changes in the capacitance retention. Choi et al. manufactured an asymmetrical buckled supercapacitor with a dielectric rubber core by doping both electrodes with MnO_2, where the anode was treated with poly(3,4-ethylenedioxythipene) (PEDOT). Both electrodes were exposed to PVA-LiCl gel electrolyte around an elastomer comprised of Ecoflex 0040 [82] (Figure 21A). The supercapacitor fiber was then subjected to various twists and stretching, corresponding to a predictable change in the capacitance of the wire. The increase in capacitance was due to the decrease in fiber thickness, which increased the surface area contact between the electrodes. The device reliably displayed capacitance changes of 115.7% and 26% during elongation of 200% and twisting at 1700 rad/m, respectively (Figure 21B) [82].

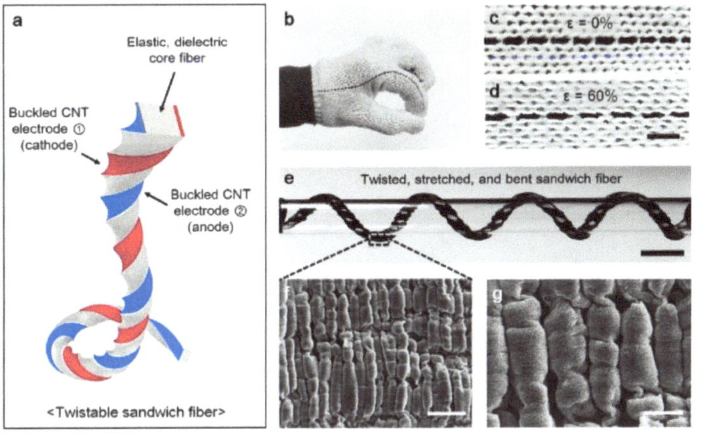

(A)

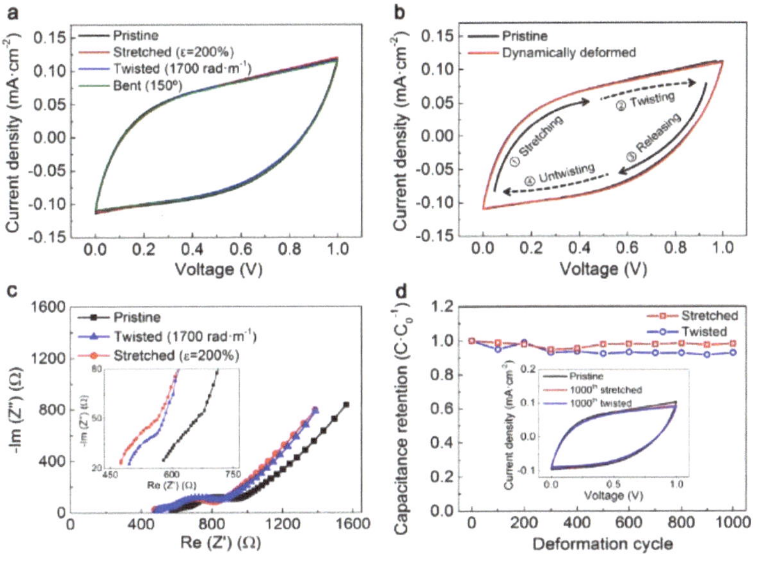

(B)

Figure 21. (**A**): (**a**) Twistable and stretchable sandwich fiber with dielectric Ecoflex 0040 rubber core and two symmetric buckled CNT electrodes. (**b**) Photograph of a 20 cm long sandwich fiber integrated into a glove. (**c**) SEM images, unstrained and (**d**) strained (60%) (scale bar: 8 mm); (**e**) photograph of the sandwich fiber wound around a 7mm diameter glass tube (scale bar: 50 mm), with SEM images showing microscopic CNT buckles at two levels of magnification, (**f**) low (scale bar: 50 μm) and (**g**) high (scale bar: 20 μm), which were formed during fiber relaxation from the fabrication strain. (**B**) Effects of severe stretching, twisting, and bending on sandwich-based fiber supercapacitors that contained MnO_2–infiltrated buckled CNT electrodes, which were coated with 15 wt.% PVA−LiCl gel electrolytes: (**a**) comparison of CV curves for pristine, stretched, twisted, and bent sandwich fiber supercapacitors; (**b**) CV curves obtained at a scan rate of 10 mV·s^{-1}, measured during the successive application of (i) 200% stretching, (ii) 1700 rad·m^{-1} twisting, (iii) release of the tensile stretching, and (iv) fiber untwisting to provide the initial state, with the CV curve before these mechanical deformations (black line) presented for comparison; (**c**) Nyquist curves (frequency rates range from 0.2 Hz to 100 kHz), with the inset showing the high-frequency region; (**d**) capacitance retention graph. The inset shows CV curves at 10 mV·s^{-1} measured before and after these 1000 cycles. Reprinted with permission from [82].

An alternative approach to strain sensing is to use CNT fibers as a piezoelectric harvester to generate a voltage upon the physical strain. This voltage generation is due to homochiral and heterochiral CNT fibers that twist in the opposite direction when strained. The output voltage of the fiber is dependent on the strain and was used to charge a CNT supercapacitor, showing promising potential for the application in a self-powered strain sensing device [97]. These devices are flexible and lightweight, enabling them to be incorporated into fabrics for real-time strain sensing applications.

7.2. Chromatic Supercapacitors

One novel utilization of CNT supercapacitors includes the application of fluorescent dye particles to produce electrodes that are visible in dark environments (Figure 22). Two methods of production exist, co-spinning and direct dipping. The co-spinning method involves dipping CNT sheets in a uniform aqueous solution of fluorescent dye particles before spinning the CNT fibers. The co-spinning process creates a uniform distribution of dye particles throughout the fiber. The direct dipping method includes dipping CNT fibers in a dye solution to anchor the particles to the fiber's surface. The electrochemical and mechanical properties of the dye-doped fibers are not diminished. Both methods can utilize a wide range of dye colors; however, it has been shown that the co-spinning process produces more distinguishable colors [98].

7.3. Self-Healing CNT Supercapacitors

During the wide range of motion, a break in fibers can inevitably occur in wearable electronics devices. For most of them, this can cause the failure of the entire device. Investigating the CNT supercapacitors with self-healing capabilities is a promising solution to this problem. One method of developing self-healing CNT supercapacitors involves using a self-healing polymer (SHP) fiber as a core wrapped in CNT sheets to produce a fiber electrode. The CNT/SHP fiber has self-healing properties that enable the reconnection of a broken fiber. The polymer forms hydrogen bonds between the two pieces separated via fracture sections when the two pieces are brought into contact.

Additionally, the CNTs interact via van der Waals forces. Although weak, the huge number of individual CNTs in the sections contribute to the adhesive forces. Pressing the two sections together enhances the hydrogen and van der Waals force for improved healing. In a reported study, a CNT fiber was subjected to breaking five times, displaying mechanical strength rates of 79.1% and 72.5% after the first and fifth break, respectively [99]. Two CNT self-healable polymers were twisted with PVA-H_2SO_4 as a gel electrolyte. The maximum capacitance observed with the self-healable SC was 140 F·g^{-1}, and 92% was restored after self-healing (Figure 23A,B). The specific capacitance was assessed with and without PANI, and the SC with PANI gave a higher capacitance rate. The approach of employing self-healing polymers for the fabrication of CNT supercapacitors paves the way for improving the durability of wearable electronics.

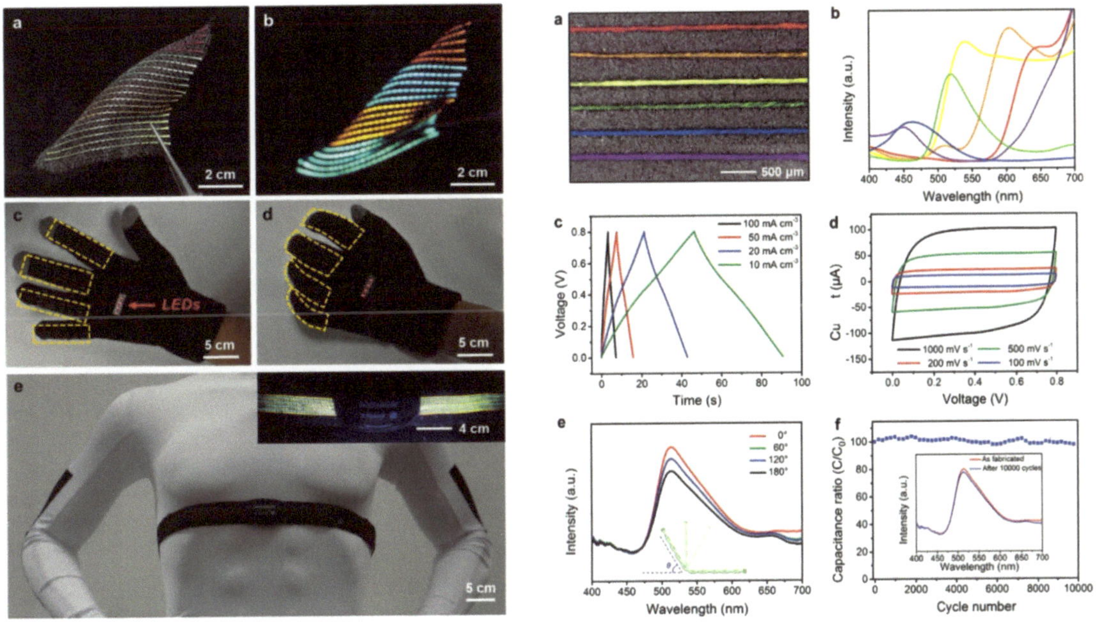

(A) (B)

Figure 22. (**A**): (**a**,**b**) Photographs of twenty fluorescent supercapacitor fibers with four colors woven into a black textile under natural and ultraviolet light, respectively. (**c**,**d**) Sixteen fluorescent supercapacitor fibers are divided into four groups and woven into a glove to power four red-light-emitting diodes. (**e**) Eighteen fluorescent supercapacitor fibers integrated into a commercial heart rate monitoring strap to extend its running time (inset: photograph of the integrated system under ultraviolet illumination). (**B**): (**a**,**b**) Optical images and spectra of various colored fiber electrodes, (**c**) GCD curves of fluorescent fiber SC based on bare MWCNT at various current densities, (**d**) CV curves at different scan rates, (**e**) spectra of green fluorescent SCs under various bending angles, and (**f**) capacitance retention rates after 10k GCD cycles. (Inset: spectrum before and after 10k GCD cycles). Reproduced with permission from [98].

7.4. Shape-Retaining CNT Supercapacitors

Wearable electronics will be exposed to various strains and movements, stressing the devices. A CNT supercapacitor that relieves stress by returning to its originally intended shape is desirable to maintain the long-term health of the device. Combining a CNT supercapacitor with a shape memory polyurethane (SMP) substrate produced a device that retained its original shape, despite experiencing deformation (Figure 24A). The electrochemical performance of the shape memory polymer fiber SC (SMPFSC) is given in Figure 24B. The desired shape can be set by heating the SMP-encased CNT fiber above its transition temperature and then cooling it in the desired configuration. Various shape-retaining polymers can be used that have different transition temperatures. This process does not harm the electrochemical properties of the CNT supercapacitor. It has also been shown that the same CNT fiber can be remolded repeatedly by heating it above the transition temperature [100]. The potential for a device to be reused in various applications while also preserving the device's health is a promising application of shape retention polymers.

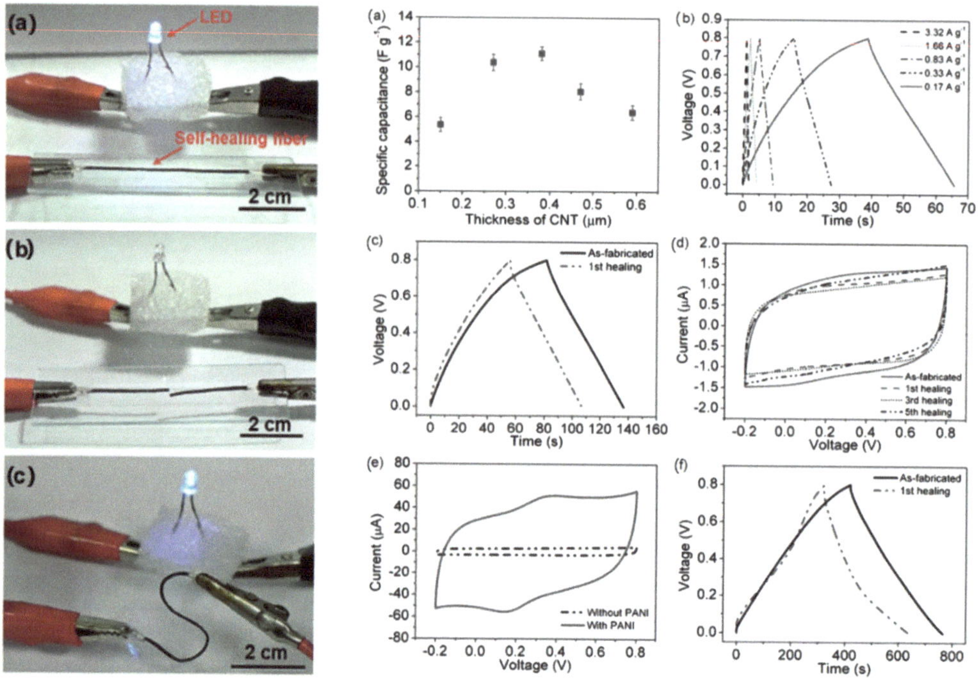

(A) (B)

Figure 23. (**A**): (**a**–**c**) Photographs of a self-healing polymeric fiber (SHP) serving as a conducting wire to power a LED before breaking, after breaking, and after healing, including being bent. The voltage was set at 3.2 V. (**B**): (**a**) Dependence of the specific capacitance on the thickness of the aligned-CNT film. (**b**) GCD curves of a wire-shaped supercapacitor with increasing current densities at a CNT thickness of 0.38 μm. (**c**) GCD curves before breaking and after healing at a current density of 0.08 A·g^{-1}. (**d**) CV curves at a scan rate of 50 mV·s^{-1} after 1, 3, and 5 healing cycles. (**e**) CV curves of wire-shaped supercapacitors based on aligned CNT/SHP fibers before and after incorporating PANI. (**f**) GCD curves for a PANI-incorporated SC before breaking and after healing at a current density of 0.17 A· g^{-1}. Reprinted with permission from [99].

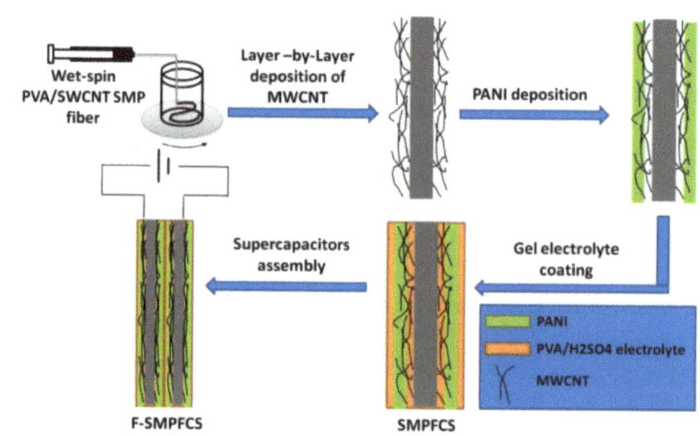

(A)

Figure 24. *Cont.*

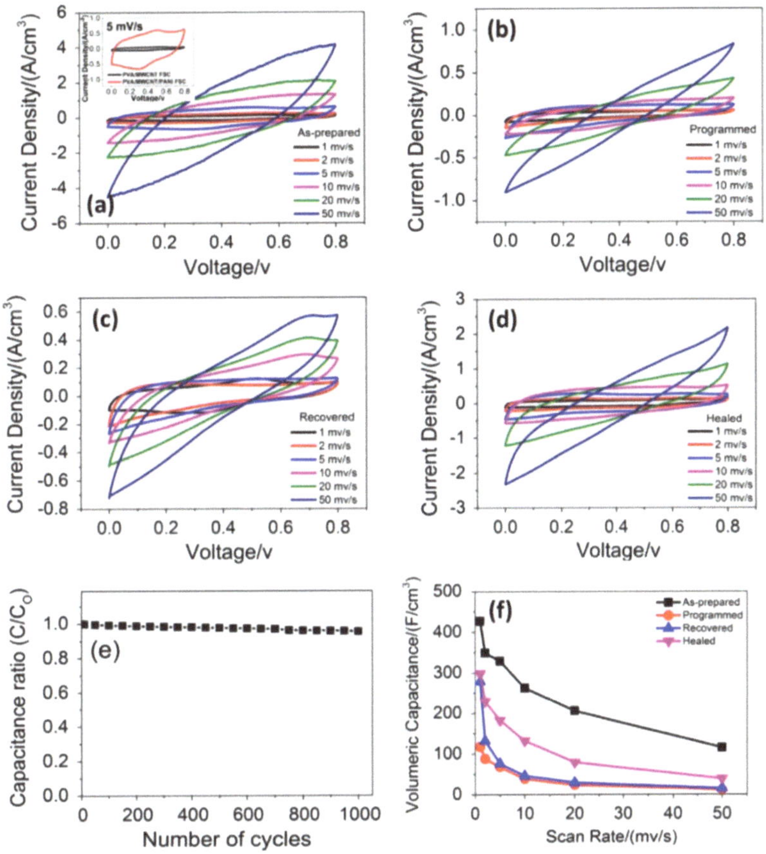

(B)

Figure 24. (**A**) Schematics of the fabrication process for a full shape memory polymer fiber SC (F–SMPFSC). (**B**) Electrochemical performance of solid-state F–SMPFSC at different shape memory cyclic stages in a two-electrode configuration. CV curves for (**a**) an as-prepared F–SMPFSC with the inset showing that the PANI layer can significantly increase the volumetric capacitance of SMPFCS, (**b**) a programmed SMPFSC with a temporary strain of 103%, (**c**) a recovered F–SMPFSC with a residual strain of 15%, (**d**) a healed F–SMPFSC obtained by immersing a recovered F–SMPFSC in a liquid H_2SO_4/PVA solution for 2 h, (**e**) the cycle life of the F–SMPFSC (scan rate: 10 mV/s), and (**f**) a comparison of the volumetric capacitance rates with the scan rates at different shape memory cyclic stages. Reprinted with permission from [100].

8. Current Challenges Related to CNT-Based Fiber Supercapacitors

Despite the rapid progress, there are challenges to overcome, both in terms of the practical applications and fundamental research on wearable CNT-based and multifunctional SCs. Some of them are listed below:

- The energy densities of the CNT SCs are low compared to other energy storage devices such as microbatteries. The CNT fiber electrodes used as scaffolds and active materials, even with a high specific surface area, result in moderate energy density. This could be due to the non-accessibility of the electrolytes to the electrode. Modifying the surface area to make it electrolyte-accessible and matching the electrode's pore diameter with the electrolyte ions' size could result in higher performance. The optimization of the fiber design and configurations to ensure the complete utilization of the electrode needs to be considered;

- Most of the current wearable SC research is conducted using PVA-KOH electrolytes, as they are mechanically robust with good electrochemical stability and low costs. However, this causes the limitation of a narrow PW, resulting in a relatively low energy density. As an alternative, electrolytes with wider PWs with an organic or ionic nature can be used. Additionally, the development of new electrolytes that are safer and ionically conductive with wide PWs need to be pursued;
- Electrolytes based on sulfuric acid, phosphoric acid, and KOH are corrosive. Most research studies do not use encapsulation, making their use unsafe. Wearable SCs employing corrosive electrolytes must be encapsulated with a polymer that does not affect the device's electrochemical performance;
- The unit cost of CNT yarn is still high, which hinders the scalability and mass production of CNT-yarn-based devices. Nevertheless, inexpensive CNT fibers are expected to appear on the market in the near future;
- As wearable electronics represent a rapidly advancing field, there is a huge market for products such as smart textiles, which necessitate scalability combined with low-cost production. Moreover, the mechanical strength aspects of the CNT fibers (flexibility, stretchability, stiffness, and good endurance life after dynamic mechanical stress cycles) must match the textile standards for the satisfactory integration with the fabric via knitting, sewing, or weaving. Therefore, more collaborations between textile or garment experts and CNT fiber researchers are necessary, which will bring to life technologies that will align the CNT fibers' physical and mechanical properties with the conventional fabric materials, thereby resulting in viable wearable electronics.

9. Conclusions

The comprehensive review conducted here confirms that the CNT fibers are valuable candidates for making wearable energy storage devices. We believe that there is a huge amount of room for developing wearable CNT-based SCs and multifunctional devices. The research on using smart CNT yarns as sensors, actuators, and energy harvesters, along with integrating CNT-based SCs with advanced electronic devices, is still in its nascent stage. Breakthroughs could be accomplished in this arena via the joint efforts and insights of researchers and industrialists, supplemented by advanced materials engineering and technology developments.

Funding: This research received no external funding.

Institutional Review Board Statement: Not applicable.

Informed Consent Statement: Not applicable.

Data Availability Statement: Not applicable.

Conflicts of Interest: The authors declare no conflict of interest.

References

1. Meng, F.; Li, Q.; Zheng, L. Flexible fiber-shaped supercapacitors: Design, fabrication, and multi-functionalities. *Energy Storage Mater.* **2017**, *8*, 85–109. [CrossRef]
2. Huang, Q.; Wang, D.; Zheng, Z. Textile-Based Electrochemical Energy Storage Devices. *Adv. Energy Mater.* **2016**, *6*, 1600783. [CrossRef]
3. Chen, S.; Qiu, L.; Cheng, H.-M. Carbon-Based Fibers for Advanced Electrochemical Energy Storage Devices. *Chem. Rev.* **2020**, *120*, 2811–2878. [CrossRef] [PubMed]
4. Sun, J.; Huang, Y.; Sea, Y.N.S.; Xue, Q.; Wang, Z.; Zhu, M.; Li, H.; Tao, X.; Zhi, C.; Hu, H. Recent progress of fiber-shaped asymmetric supercapacitors. *Mater. Today Energy* **2017**, *5*, 1–14. [CrossRef]
5. Lu, Z.; Raad, R.; Safaei, F.; Xi, J.; Liu, Z.; Foroughi, J. Carbon Nanotube Based Fiber Supercapacitor as Wearable Energy Storage. *Front. Mater.* **2019**, *6*, 138. [CrossRef]
6. Conway, B.E. *Electrochemical Supercapacitors: Scientific Fundamentals and Technical Applications*; Kluwer Academic/Plenum: New York, NY, USA, 1999.
7. Senthilkumar, S.T.; Wang, Y.; Huang, H. Advances and prospects of fiber supercapacitors. *J. Mater. Chem. A* **2015**, *3*, 20863–20879. [CrossRef]

8. Xie, P.; Yuan, W.; Liu, X.; Peng, Y.; Yin, Y.; Li, Y.; Wu, Z. Advanced carbon nanomaterials for state-of-the-art flexible supercapacitors. *Energy Storage Mater.* **2020**, *36*, 56–76. [CrossRef]
9. Pan, Z.; Yang, J.; Zhang, Y.; Gao, X.; Wang, J. Quasi-solid-state fiber-shaped aqueous energy storage devices: Recent advances and prospects. *J. Mater. Chem. A* **2020**, *8*, 6406–6433. [CrossRef]
10. Cherusseri, J.; Kumar, K.S.; Choudhary, N.; Nagaiah, N.; Jung, Y.; Roy, T.; Thomas, J. Novel mesoporous electrode materials for symmetric, asymmetric and hybrid supercapacitors. *Nanotechnology* **2019**, *30*, 202001. [CrossRef] [PubMed]
11. Joseph, N.; Shafi, P.M.; Bose, A.C. Recent Advances in 2D-MoS_2 and its Composite Nanostructures for Supercapacitor Electrode Application. *Energy Fuels* **2020**, *34*, 6558–6597. [CrossRef]
12. Forouzandeh, P.; Kumaravel, V.; Pillai, S.C. Electrode Materials for Supercapacitors: A Review of Recent Advances. *Catalysts* **2020**, *10*, 969. [CrossRef]
13. Wang, J.; Li, F.; Zhu, F.; Schmidt, O.G. Recent Progress in Micro-Supercapacitor Design, Integration, and Functionalization. *Small Methods* **2018**, *3*, 1800367. [CrossRef]
14. Dong, L.; Xu, C.; Li, Y.; Huang, Z.-H.; Kang, F.; Yang, Q.-H.; Zhao, X. Flexible electrodes and supercapacitors for wearable energy storage: A review by category. *J. Mater. Chem. A* **2016**, *4*, 4659–4685. [CrossRef]
15. Yang, P.; Mai, W. Flexible solid-state electrochemical supercapacitors. *Nano Energy* **2014**, *8*, 274–290. [CrossRef]
16. Yang, Z.; Ren, J.; Zhang, Z.; Chen, X.; Guan, G.; Qiu, L.; Zhang, Y.; Peng, H. Recent Advancement of Nanostructured Carbon for Energy Applications. *Chem. Rev.* **2015**, *115*, 5159–5223. [CrossRef] [PubMed]
17. Beguin, F.; Frackowiak, E.; Gogotsi, Y. *Carbons for Electrochemical Energy Storage and Conversion Systems*; CRC Press: Boca Raton, FL, USA, 2009. [CrossRef]
18. Frackowiak, E.; Abbas, Q.; Béguin, F. Carbon/carbon supercapacitors. *J. Energy Chem.* **2013**, *22*, 226–240. [CrossRef]
19. Chen, D.; Jiang, K.; Huang, T.; Shen, G. Recent Advances in Fiber Supercapacitors: Materials, Device Configurations, and Applications. *Adv. Mater.* **2019**, *32*, e1901806. [CrossRef] [PubMed]
20. Zhang, X.; Chen, X.; Bai, T.; You, X.-Y.; Zhao, X.; Liu, X.-Y.; Ye, M.-D. Recent advances in flexible fiber-shaped supercapacitors. *Acta Phys. Sin.* **2020**, *69*, 178201. [CrossRef]
21. Jang, Y.; Kim, S.M.; Spinks, G.M.; Kim, S.J. Carbon Nanotube Yarn for Fiber-Shaped Electrical Sensors, Actuators, and Energy Storage for Smart Systems. *Adv. Mater.* **2019**, *32*, 1902670. [CrossRef] [PubMed]
22. Yao, Y.; Li, N.; Lv, T.; Chen, T. Carbon Nanotube Fibers for Wearable Devices. In *Industrial Applications of Carbon Nanotubes*; Elsevier: Amsterdam, The Netherlands, 2016; pp. 347–379. [CrossRef]
23. Wang, Y.; Song, Y.; Xia, Y. Electrochemical capacitors: Mechanism, materials, systems, characterization and applications. *Chem. Soc. Rev.* **2016**, *45*, 5925–5950. [CrossRef] [PubMed]
24. Noori, A.; El-Kady, M.F.; Rahmanifar, M.S.; Kaner, R.B.; Mousavi, M.F. Towards establishing standard performance metrics for batteries, supercapacitors and beyond. *Chem. Soc. Rev.* **2019**, *48*, 1272–1341. [CrossRef] [PubMed]
25. Kumar, Y.; Rawal, S.; Joshi, B.; Hashmi, S.A. Background, fundamental understanding and progress in electrochemical capacitors. *J. Solid State Electrochem.* **2019**, *23*, 667–692. [CrossRef]
26. Vigolo, B.; Pénicaud, A.; Coulon, C.; Sauder, C.; Pailler, R.; Journet, C.; Bernier, P.; Poulin, P. Macroscopic Fibers and Ribbons of Oriented Carbon Nanotubes. *Science* **2000**, *290*, 1331–1334. [CrossRef] [PubMed]
27. Behabtu, N.; Young, C.C.; Tsentalovich, D.E.; Kleinerman, O.; Wang, X.; Ma, A.W.K.; Bengio, E.A.; Ter Waarbeek, R.F.; De Jong, J.J.; Hoogerwerf, R.E.; et al. Strong, Light, Multifunctional Fibers of Carbon Nanotubes with Ultrahigh Conductivity. *Science* **2013**, *339*, 182–186. [CrossRef] [PubMed]
28. Bucossi, A.R.; Cress, C.D.; Schauerman, C.M.; Rossi, J.E.; Puchades, I.; Landi, B.J. Enhanced Electrical Conductivity in Extruded Single-Wall Carbon Nanotube Wires from Modified Coagulation Parameters and Mechanical Processing. *ACS Appl. Mater. Interfaces* **2015**, *7*, 27299–27305. [CrossRef] [PubMed]
29. Kou, L.; Huang, T.; Zheng, B.; Han, Y.; Zhao, X.; Gopalsamy, K.; Sun, H.; Gao, C. Coaxial wet-spun yarn supercapacitors for high-energy density and safe wearable electronics. *Nat. Commun.* **2014**, *5*, 3754. [CrossRef] [PubMed]
30. Lu, Z.; Chao, Y.; Ge, Y.; Foroughi, J.; Zhao, Y.; Wang, C.; Long, H.; Wallace, G.G. High-performance hybrid carbon nanotube fibers for wearable energy storage. *Nanoscale* **2017**, *9*, 5063–5071. [CrossRef] [PubMed]
31. Li, G.-X.; Hou, P.-X.; Luan, J.; Li, J.-C.; Li, X.; Wang, H.; Shi, C.; Liu, C.; Cheng, H.-M. A MnO_2 nanosheet/single-wall carbon nanotube hybrid fiber for wearable solid-state supercapacitors. *Carbon* **2018**, *140*, 634–643. [CrossRef]
32. Kim, J.H.; Lee, J.M.; Park, J.W.; Kim, K.J.; Kim, S.J. MnO_2/PtNP Embedded Wet-Spun Fiber Supercapacitors. *Adv. Mater. Technol.* **2018**, *3*, 1800184. [CrossRef]
33. Meng, Q.; Wang, K.; Guo, W.; Fang, J.; Wei, Z.; She, X. Thread-like Supercapacitors Based on One-Step Spun Nanocomposite Yarns. *Small* **2014**, *10*, 3187–3193. [CrossRef] [PubMed]
34. Zheng, T.; Wang, X.; Liu, Y.; Bayaniahangar, R.; Li, H.; Lu, C.; Xu, N.; Yao, Z.; Qiao, Y.; Zhang, D.; et al. Polyaniline-decorated hyaluronic acid-carbon nanotube hybrid microfiber as a flexible supercapacitor electrode material. *Carbon* **2019**, *159*, 65–73. [CrossRef]
35. Garcia-Torres, J.; Crean, C. Multilayered Flexible Fibers with High Performance for Wearable Supercapacitor Applications. *Adv. Sustain. Syst.* **2017**, *2*, 1700143. [CrossRef]
36. Koziol, K.; Vilatela, J.; Moisala, A.; Motta, M.; Cunniff, P.; Sennett, M.; Windle, A. High-Performance Carbon Nanotube Fiber. *Science* **2007**, *318*, 1892–1895. [CrossRef]

37. Li, Y.-L.; Kinloch, I.A.; Windle, A.H. Direct Spinning of Carbon Nanotube Fibers from Chemical Vapor Deposition Synthesis. *Sci. China Technol. Sci.* **2004**, *304*, 276–278. [CrossRef]
38. Shanov, V.; Cho, W.; Malik, R.; Alvarez, N.; Haase, M.; Ruff, B.; Kienzle, N.; Ochmann, T.; Mast, D.; Schulz, M. CVD growth, characterization and applications of carbon nanostructured materials. *Surf. Coat. Technol.* **2013**, *230*, 77–86. [CrossRef]
39. Cho, W.; Schulz, M.; Shanov, V. Growth and characterization of vertically aligned centimeter long CNT arrays. *Carbon* **2014**, *72*, 264–273. [CrossRef]
40. Shanov, V.N.; Schulz, M.J. Method of Growing Carbon Nanotube and Forming a Carbon Nanotube Thread Espacenet. Spinning Patent US9796121B2, 24 October 2017.
41. Alvarez, N.T.; Miller, P.; Haase, M.; Kienzle, N.; Zhang, L.; Schulz, M.J.; Shanov, V. Carbon nanotube assembly at near-industrial natural-fiber spinning rates. *Carbon* **2015**, *86*, 350–357. [CrossRef]
42. Kanakaraj, S.N.; Alvarez, N.T.; Gbordzoe, S.; Lucas, M.S.; Maruyama, B.; Noga, R.; Hsieh, Y.-Y.; Shanov, V. Improved dry spinning process at elevated temperatures for making uniform and high strength CNT fibers. *Mater. Res. Express* **2018**, *5*, 065036. [CrossRef]
43. Li, S.; Zhang, X.; Zhao, J.; Meng, F.; Xu, G.; Yong, Z.; Jia, J.; Zhang, Z.; Li, Q. Enhancement of carbon nanotube fibres using different solvents and polymers. *Compos. Sci. Technol.* **2012**, *72*, 1402–1407. [CrossRef]
44. Tran, T.; Fan, Z.; Liu, P.; Myint, S.M.; Duong, H. Super-strong and highly conductive carbon nanotube ribbons from post-treatment methods. *Carbon* **2016**, *99*, 407–415. [CrossRef]
45. Lima, M.D.; Fang, S.; Lepró, X.; Lewis, C.; Ovalle-Robles, R.; Carretero-González, J.; Castillo-Martínez, E.; Kozlov, M.E.; Oh, J.; Rawat, N.; et al. Biscrolling Nanotube Sheets and Func-tional Guests into Yarns. *Science* **2011**, *331*, 51–55. [CrossRef] [PubMed]
46. Lee, J.A.; Shin, M.K.; Kim, S.H.; Cho, H.U.; Spinks, G.M.; Wallace, G.; Lima, M.D.; Lepro, X.; Kozlov, M.E.; Baughman, R.H.; et al. Ultrafast charge and discharge biscrolled yarn supercapacitors for textiles and microdevices. *Nat. Commun.* **2013**, *4*, 1970. [CrossRef] [PubMed]
47. Queirós, G.; Rey-Raap, N.; Pereira, C.; Pereira, M.F.R. CNT-based Materials as Electrodes for Flexible Supercapacitors. *U. Porto J. Eng.* **2021**, *7*, 151–162. [CrossRef]
48. Beidaghi, M.; Gogotsi, Y. Capacitive energy storage in micro-scale devices: Recent advances in design and fabrication of micro-supercapacitors. *Energy Environ. Sci.* **2014**, *7*, 867–884. [CrossRef]
49. Su, F.; Miao, M. Flexible, high performance Two-Ply Yarn Supercapacitors based on irradiated Carbon Nanotube Yarn and PEDOT/PSS. *Electrochim. Acta* **2014**, *127*, 433–438. [CrossRef]
50. Adusei, P.K.; Gbordzoe, S.; Kanakaraj, S.N.; Hsieh, Y.-Y.; Alvarez, N.T.; Fang, Y.; Johnson, K.; McConnell, C.; Shanov, V. Fabrication and study of supercapacitor electrodes based on oxygen plasma functionalized carbon nanotube fibers. *J. Energy Chem.* **2019**, *40*, 120–131. [CrossRef]
51. Liang, Y.; Luo, X.; Weng, W.; Hu, Z.; Zhang, Y.; Xu, W.; Bi, Z.; Zhu, M. Activated Carbon Nanotube Fiber Fabric as a High-Performance Flexible Electrode for Solid-State Supercapacitors. *ACS Appl. Mater. Interfaces* **2021**, *13*, 28433–28441. [CrossRef]
52. Kim, J.H.; Choi, C.; Lee, J.M.; De Andrade, M.J.; Baughman, R.H.; Kim, S.J. Ag/MnO_2 Composite Sheath-Core Structured Yarn Supercapacitors. *Sci. Rep.* **2018**, *8*, 1–8. [CrossRef]
53. Zhou, Q.; Tang, G.; Chen, X.; Su, F. Flexible Supercapacitors Based on CNT/MnO_2-BP Composite Yarn Synthesized by In Situ Reduction. *J. Electrochem. Soc.* **2021**, *168*, 080524. [CrossRef]
54. Zhou, Q.; Chen, X.; Su, F.; Lyu, X.; Miao, M. Sandwich-Structured Transition Metal Oxide/Graphene/Carbon Nanotube Composite Yarn Electrodes for Flexible Two-Ply Yarn Supercapacitors. *Ind. Eng. Chem. Res.* **2020**, *59*, 5752–5759. [CrossRef]
55. Le, T.S.; Truong, T.K.; Huynh, V.N.; Bae, J.; Suh, D. Synergetic design of enlarged surface area and pseudo-capacitance for fiber-shaped supercapacitor yarn. *Nano Energy* **2019**, *67*, 104198. [CrossRef]
56. Jian, X.; Li, H.; Li, H.; Li, Y.; Shang, Y. Flexible and freestanding MoS2/rGO/CNT hybrid fibers for high-capacity all-solid supercapacitors. *Carbon* **2020**, *172*, 132–137. [CrossRef]
57. Shi, P.; Li, L.; Hua, L.; Qian, Q.; Wang, P.; Zhou, J.; Sun, G.; Huang, W. Design of Amorphous Manganese Oxide@Multiwalled Carbon Nanotube Fiber for Robust Solid-State Supercapacitor. *ACS Nano* **2016**, *11*, 444–452. [CrossRef] [PubMed]
58. He, Y.; Chen, W.; Gao, C.; Zhou, J.; Li, X.; Xie, E. An overview of carbon materials for flexible electrochemical capacitors. *Nanoscale* **2013**, *5*, 8799–8820. [CrossRef]
59. Wang, Q.; Wu, Y.; Li, T.; Zhang, D.; Miao, M.; Zhang, A. High performance two-ply carbon nanocomposite yarn supercapacitors enhanced with a platinum filament and in situ polymerized polyaniline nanowires. *J. Mater. Chem. A* **2016**, *4*, 3828–3834. [CrossRef]
60. Wu, Y.; Wang, Q.; Li, T.; Zhang, D.; Miao, M. Fiber-shaped Supercapacitor and Electrocatalyst Containing of Multiple Carbon Nanotube Yarns and One Platinum Wire. *Electrochim. Acta* **2017**, *245*, 69–78. [CrossRef]
61. Wang, Q.; Zhang, D.; Wu, Y.; Li, T.; Zhang, A.; Miao, M. Fabrication of Supercapacitors from $NiCo_2O_4$Nanowire/Carbon-Nanotube Yarn for Ultraviolet Photodetectors and Portable Electronics. *Energy Technol.* **2017**, *5*, 1449–1456. [CrossRef]
62. Sun, J.; Man, P.; Zhang, Q.; He, B.; Zhou, Z.; Li, C.; Wang, X.; Guo, J.; Zhao, J.; Xie, L.; et al. Hierarchically-structured Co3O4 nanowire arrays grown on carbon nanotube fibers as novel cathodes for high-performance wearable fiber-shaped asymmetric supercapacitors. *Appl. Surf. Sci.* **2018**, *447*, 795–801. [CrossRef]
63. He, H.; Yang, X.; Wang, L.; Zhang, X.; Li, X.; Lü, W. Enhanced Energy Density of Coaxial Fiber Asymmetric Supercapacitor Based on MoS_2@Fe_2O_3/Carbon Nanotube Paper and $Ni(OH)_2$@$NiCo_2O_4$/Carbon Nanotube Fiber Electrodes. *Chem.—A Eur. J.* **2020**, *26*, 17212–17221. [CrossRef]

64. Zhang, Q.; Xu, W.; Sun, J.; Pan, Z.; Zhao, J.; Wang, X.; Zhang, J.; Man, P.; Guo, J.; Zhou, Z.; et al. Constructing Ultrahigh-Capacity Zinc–Nickel–Cobalt Oxide@Ni(OH)$_2$ Core–Shell Nanowire Arrays for High-Performance Coaxial Fiber-Shaped Asymmetric Supercapacitors. *Nano Lett.* **2017**, *17*, 7552–7560. [CrossRef]
65. Sun, J.; Zhang, Q.; Wang, X.; Zhao, J.; Guo, J.; Zhou, Z.; Zhang, J.; Man, P.; Sun, J.; Li, Q.; et al. Constructing hierarchical dandelion-like molybdenum–nickel–cobalt ternary oxide nanowire arrays on carbon nanotube fiber for high-performance wearable fiber-shaped asymmetric supercapacitors. *J. Mater. Chem. A* **2017**, *5*, 21153–21160. [CrossRef]
66. Huang, G.; Zhang, Y.; Wang, L.; Sheng, P.; Peng, H. Fiber-based MnO$_2$/carbon nanotube/polyimide asymmetric supercapacitor. *Carbon* **2017**, *125*, 595–604. [CrossRef]
67. Zong, Q.; Zhang, Q.; Mei, X.; Li, Q.; Zhou, Z.; Li, D.; Chen, M.; Shi, F.; Sun, J.; Yao, Y.; et al. Facile Synthesis of Na-Doped MnO$_2$ Nanosheets on Carbon Nanotube Fibers for Ultrahigh-Energy-Density All-Solid-State Wearable Asymmetric Supercapacitors. *ACS Appl. Mater. Interfaces* **2018**, *10*, 37233–37241. [CrossRef]
68. Liu, J.-H.; Xu, X.-Y.; Yu, J.; Hong, J.-L.; Liu, C.; Ouyang, X.; Lei, S.; Meng, X.; Tang, J.-N.; Chen, D.-Z. Facile construction of 3D porous carbon nanotubes/polypyrrole and reduced graphene oxide on carbon nanotube fiber for high-performance asymmetric supercapacitors. *Electrochim. Acta* **2019**, *314*, 9–19. [CrossRef]
69. Kim, J.; Yu, H.; Jung, J.Y.; Kim, M.J.; Jeon, D.; Jeong, H.S.; Kim, N.D. 3D Architecturing Strategy on the Utmost Carbon Nanotube Fiber for Ultra-High Performance Fiber-Shaped Supercapacitor. *Adv. Funct. Mater.* **2022**, *32*, 2113057. [CrossRef]
70. Adusei, P.K.; Johnson, K.; Kanakaraj, S.N.; Zhang, G.; Fang, Y.; Hsieh, Y.-Y.; Khosravifar, M.; Gbordzoe, S.; Nichols, M.; Shanov, V. Asymmetric Fiber Supercapacitors Based on a FeC$_2$O$_4$/FeOOH-CNT Hybrid Material. *C* **2021**, *7*, 62. [CrossRef]
71. Mirabedini, A.; Lu, Z.; Mostafavian, S.; Foroughi, J. Triaxial Carbon Nanotube/Conducting Polymer Wet-Spun Fibers Supercapacitors for Wearable Electronics. *Nanomaterials* **2020**, *11*, 3. [CrossRef] [PubMed]
72. Liu, N.; Pan, Z.; Ding, X.; Yang, J.; Xu, G.; Li, L.; Wang, Q.; Liu, M.; Zhang, Y. In-situ growth of vertically aligned nickel cobalt sulfide nanowires on carbon nanotube fibers for high capacitance all-solid-state asymmetric fiber-supercapacitors. *J. Energy Chem.* **2019**, *41*, 209–215. [CrossRef]
73. Xu, T.; Yang, D.; Liu, Y.; Zhang, S.; Zhang, M.; Zhao, T.; Li, X.; Yu, Z.-Z. Hierarchical Transition Metal Oxide Arrays Grown on Graphene-Based Fibers with Enhanced Interface by Thin Layer of Carbon toward Solid-State Asymmetric Supercapacitors. *ChemElectroChem* **2020**, *7*, 1860–1868. [CrossRef]
74. Kanakaraj, S.N.; Hsieh, Y.-Y.; Adusei, P.K.; Homan, B.; Fang, Y.; Zhang, G.; Mishra, S.; Gbordzoe, S.; Shanov, V. Nitrogen-doped CNT on CNT hybrid fiber as a current collector for high-performance Li-ion capacitors. *Carbon* **2019**, *149*, 407–418. [CrossRef]
75. Liang, X.; Wang, Q.; Ma, Y.; Zhang, D. A high performance asymmetric supercapacitor based on in situ prepared CuCo$_2$O$_4$ nanowires and PPy nanoparticles on a two-ply carbon nanotube yarn. *Dalton Trans.* **2018**, *47*, 17146–17152. [CrossRef] [PubMed]
76. Man, P.; Zhang, Q.; Sun, J.; Guo, J.; Wang, X.; Zhou, Z.; He, B.; Li, Q.; Xie, L.; Zhao, J.; et al. Hierarchically structured VO2@PPy core-shell nanowire arrays grown on carbon nanotube fibers as advanced cathodes for high-performance wearable asymmetric supercapacitors. *Carbon* **2018**, *139*, 21–28. [CrossRef]
77. Guo, J.; Zhang, Q.; Sun, J.; Li, C.; Zhao, J.; Zhou, Z.; He, B.; Wang, X.; Man, P.; Li, Q.; et al. Direct growth of vanadium nitride nanosheets on carbon nanotube fibers as novel negative electrodes for high-energy-density wearable fiber-shaped asymmetric supercapacitors. *J. Power Sources* **2018**, *382*, 122–127. [CrossRef]
78. Li, T.; Wu, Y.; Wang, Q.; Zhang, D.; Zhang, A.; Miao, M. TiO$_2$ crystalline structure and electrochemical performance in two-ply yarn CNT/TiO$_2$ asymmetric supercapacitors. *J. Mater. Sci.* **2017**, *52*, 7733–7743. [CrossRef]
79. Xu, Y.; Yan, Y.; Lu, W.; Yarlagadda, S.; Xu, G. High-Performance Flexible Asymmetric Fiber-Shaped Supercapacitor Based on CF/PPy and CNT/MnO$_2$ Composite Electrodes. *ACS Appl. Energy Mater.* **2021**, *4*, 10639–10645. [CrossRef]
80. Yi, F.-L.; Meng, F.-C.; Li, Y.-Q.; Huang, P.; Hu, N.; Liao, K.; Fu, S.-Y. Highly stretchable CNT Fiber/PAAm hydrogel composite simultaneously serving as strain sensor and supercapacitor. *Compos. Part B Eng.* **2020**, *198*, 108246. [CrossRef]
81. Zhu, W.; Zhang, Y.; Zhou, X.; Xu, J.; Liu, Z.; Yuan, N.; Ding, J. Miniaturized Stretchable and High-Rate Linear Supercapacitors. *Nanoscale Res. Lett.* **2017**, *12*, 448. [CrossRef] [PubMed]
82. Choi, C.; Lee, J.M.; Kim, S.H.; Kim, S.J.; Di, J.; Baughman, R.H. Twistable and Stretchable Sandwich Structured Fiber for Wearable Sensors and Supercapacitors. *Nano Lett.* **2016**, *16*, 7677–7684. [CrossRef] [PubMed]
83. Lee, D.W.; Lee, J.H.; Min, N.K.; Jin, J.-H. Buckling Structured Stretchable Pseudocapacitor Yarn. *Sci. Rep.* **2017**, *7*, 12005. [CrossRef] [PubMed]
84. Ren, D.; Dong, L.; Wang, J.; Ma, X.; Xu, C.; Kang, F. Facile Preparation of High-Performance Stretchable Fiber-Like Electrodes and Supercapacitors. *ChemistrySelect* **2018**, *3*, 4179–4184. [CrossRef]
85. Meng, F.; Zheng, L.; Luo, S.; Li, D.; Wang, G.; Jin, H.; Li, Q.; Zhang, Y.; Liao, K.; Cantwell, W.J. A highly torsionable fiber-shaped supercapacitor. *J. Mater. Chem. A* **2017**, *5*, 4397–4403. [CrossRef]
86. Wang, H.; Wang, C.; Jian, M.; Wang, Q.; Xia, K.; Yin, Z.; Zhang, M.; Liang, X.; Zhang, Y. Superelastic wire-shaped supercapacitor sustaining 850% tensile strain based on carbon nanotube@graphene fiber. *Nano Res.* **2017**, *11*, 2347–2356. [CrossRef]
87. Zhang, Q.; Sun, J.; Pan, Z.; Zhang, J.; Zhao, J.; Wang, X.; Zhang, C.; Yao, Y.; Lu, W.; Li, Q.; et al. Stretchable fiber-shaped asymmetric supercapacitors with ultrahigh energy density. *Nano Energy* **2017**, *39*, 219–228. [CrossRef]
88. Shi, M.; Yang, C.; Song, X.; Liu, J.; Zhao, L.; Zhang, P.; Gao, L. Stretchable wire-shaped supercapacitors with high energy density for size-adjustable wearable electronics. *Chem. Eng. J.* **2017**, *322*, 538–545. [CrossRef]

89. Zheng, X.; Zhou, X.; Xu, J.; Zou, L.; Nie, W.; Hu, X.; Dai, S.; Qiu, Y.; Yuan, N. Highly stretchable CNT/MnO_2 nanosheets fiber supercapacitors with high energy density. *J. Mater. Sci.* **2020**, *55*, 8251–8263. [CrossRef]
90. Lu, Z.; Foroughi, J.; Wang, C.; Long, H.; Wallace, G.G. Superelastic Hybrid CNT/Graphene Fibers for Wearable Energy Storage. *Adv. Energy Mater.* **2017**, *8*, 1702047. [CrossRef]
91. Guo, F.M.; Xu, R.Q.; Cui, X.; Zhang, L.; Wang, K.L.; Yao, Y.W.; Wei, J.Q. High performance of stretchable carbon nanotube–polypyrrole fiber supercapacitors under dynamic deformation and temperature variation. *J. Mater. Chem. A* **2016**, *4*, 9311–9318. [CrossRef]
92. Shang, Y.; Wang, C.; He, X.; Li, J.; Peng, Q.; Shi, E.; Wang, R.; Du, S.; Cao, A.; Li, Y. Self-stretchable, helical carbon nanotube yarn supercapacitors with stable performance under extreme deformation conditions. *Nano Energy* **2015**, *12*, 401–409. [CrossRef]
93. Liang, Q.; Wan, J.; Ji, P.; Zhang, D.; Sheng, N.; Chen, S.; Wang, H. Continuous and integrated PEDOT@Bacterial cellulose/CNT hybrid helical fiber with "reinforced cement-sand" structure for self-stretchable solid supercapacitor. *Chem. Eng. J.* **2021**, *427*, 131904. [CrossRef]
94. Na, Y.W.; Cheon, J.Y.; Kim, J.H.; Jung, Y.; Lee, K.; Park, J.S.; Park, J.Y.; Song, K.S.; Lee, S.B.; Kim, T.; et al. All-in-one flexible supercapacitor with ultrastable per-formance under extreme load. *Sci. Adv.* **2022**, *8*, eabl8631. [CrossRef] [PubMed]
95. Park, T.; Jang, Y.; Park, J.W.; Kim, H.; Kim, S.J. Quasi-solid-state highly stretchable circular knitted MnO_2@CNT supercapacitor. *RSC Adv.* **2020**, *10*, 14007–14012. [CrossRef]
96. Choi, C.; Kim, S.H.; Sim, H.J.; Lee, J.A.; Choi, A.Y.; Kim, Y.T.; Lepro, X.; Spinks, G.M.; Baughman, R.H.; Kim, S.J. Stretchable, Weavable Coiled Carbon Nanotube/MnO_2/Polymer Fiber Solid-State Supercapacitors. *Sci. Rep.* **2015**, *5*, 9387. [CrossRef] [PubMed]
97. Mun, T.J.; Kim, S.H.; Park, J.W.; Moon, J.H.; Jang, Y.; Huynh, C.; Baughman, R.H.; Kim, S.J. Wearable Energy Generating and Storing Textile Based on Carbon Nanotube Yarns. *Adv. Funct. Mater.* **2020**, *30*, 2000411. [CrossRef]
98. Liao, M.; Sun, H.; Zhang, J.; Wu, J.; Xie, S.; Fu, X.; Sun, X.; Wang, B.; Peng, H. Multicolor, Fluorescent Supercapacitor Fiber. *Small* **2017**, *14*, e1702052. [CrossRef] [PubMed]
99. Sun, H.; You, X.; Jiang, Y.; Guan, G.; Fang, X.; Deng, J.; Chen, P.; Luo, Y.; Peng, H. Self-Healable Electrically Conducting Wires for Wearable Microelectronics. *Angew. Chem. Int. Ed.* **2014**, *53*, 9526–9531. [CrossRef] [PubMed]
100. Zhong, J.; Meng, J.; Yang, Z.; Poulin, P.; Koratkar, N. Shape memory fiber supercapacitors. *Nano Energy* **2015**, *17*, 330–338. [CrossRef]

electrochem

MDPI

Review

Separator Materials for Lithium Sulfur Battery—A Review

Ryohei Mori [1,2]

[1] Green Science Alliance Co., Ltd., 2-22-11 Obana, Kawanishi City 666-0015, Hyogo Prefecture, Japan; moriryohei@fuji-pigment.co.jp; Tel.: +81-72-759-8501; Fax: +81-72-759-9008
[2] Fuji Pigment Co., Ltd., 2-23-2 Obana, Kawanishi City 666-0015, Hyogo Prefecture, Japan

Abstract: In the recent rechargeable battery industry, lithium sulfur batteries (LSBs) have demonstrated to be a promising candidate battery to serve as the next-generation secondary battery, owing to its enhanced theoretical specific energy, economy, and environmental friendliness. Its inferior cyclability, however, which is primarily due to electrode deterioration caused by the lithium polysulfide shuttle effect, is still a major problem for the real industrial usage of LSBs. The optimization of the separator and functional barrier layer is an effective strategy for remedying these issues. In this article, the current progress based on the classification and modification of functional separators is summarized. We will also describe their working mechanisms as well as the resulting LSB electrochemical properties. In addition, necessary performance for separators will also be mentioned in order to gain optimized LSB performance.

Keywords: lithium sulfur battery; rechargeable battery; separator; electrochemistry; anode; cathode

Citation: Mori, R. Separator Materials for Lithium Sulfur Battery—A Review. *Electrochem* **2023**, *4*, 485–522. https://doi.org/10.3390/electrochem4040032

Academic Editors: Huang Zhang and Yuan Ma

Received: 14 September 2023
Revised: 22 October 2023
Accepted: 2 November 2023
Published: 13 November 2023

1. Introduction

The propagation of modern human society based on the electronic industry has demanded stronger and cheaper energy storage systems. Among these, lithium ion batteries (LIBs) have been aggressively studied and developed because of their high endurance against electrochemistry variation and their stable usage for application in various types of electronic devices, including smart phones, mobile phones, EVs, etc., even though LIBs still suffer from limited capacity, high price, and safety issues, including their flammable nature. Due to these reasons, it would be difficult for LIBs to fulfill the demand for upcoming new industrial applications, such as drones, EVs, military power supplies, and stationary electrical power stations, because they always demand higher capacity as well as safer and economically friendly rechargeable batteries [1–3]. In this regard, lithium sulfur batteries (LSBs) could be one of several alternative candidate rechargeable batteries because of their high theoretical capacity and energy density (1675 mAh g^{-1} and 2600 Wh Kg^{-1}, respectively). The corresponding values for LIBs are 240–280 mAh g^{-1} and 350–400 Wh Kg^{-1}, respectively. Apart from the nature of LIBs, the electrochemical reaction of LSBs is different from that of LIBs due to the presence of a sulfur cathode, and can be expressed as:

$$S + 2Li \rightarrow Li_2S \ (E_0 = 2.20 \text{ V vs. } Li/Li^+) \quad (1)$$

As a result, LSBs can offer 5–7 times higher energy density than that of LIBs. Combined with its other advantages, i.e., sulfur being naturally abundant, cheap, and a non-toxic element, LSBs can be considered as one of the candidates for next-generation rechargeable batteries [4–7]. The typical configuration of an LSB is shown in Figure 1.

Lithium metal is the anode and a carbon–sulfur composite is the cathode [8]. As an ordinary battery, the separator and an electrolyte exist between the anode and cathode.

When the LSB is discharged, sulfur is reduced to Li_2S by a series of electrochemical reaction while forming various types of polysulfide intermediates. Figure 2 indicates an ordinary LSB profile during a charge–discharge electrochemical reaction [9].

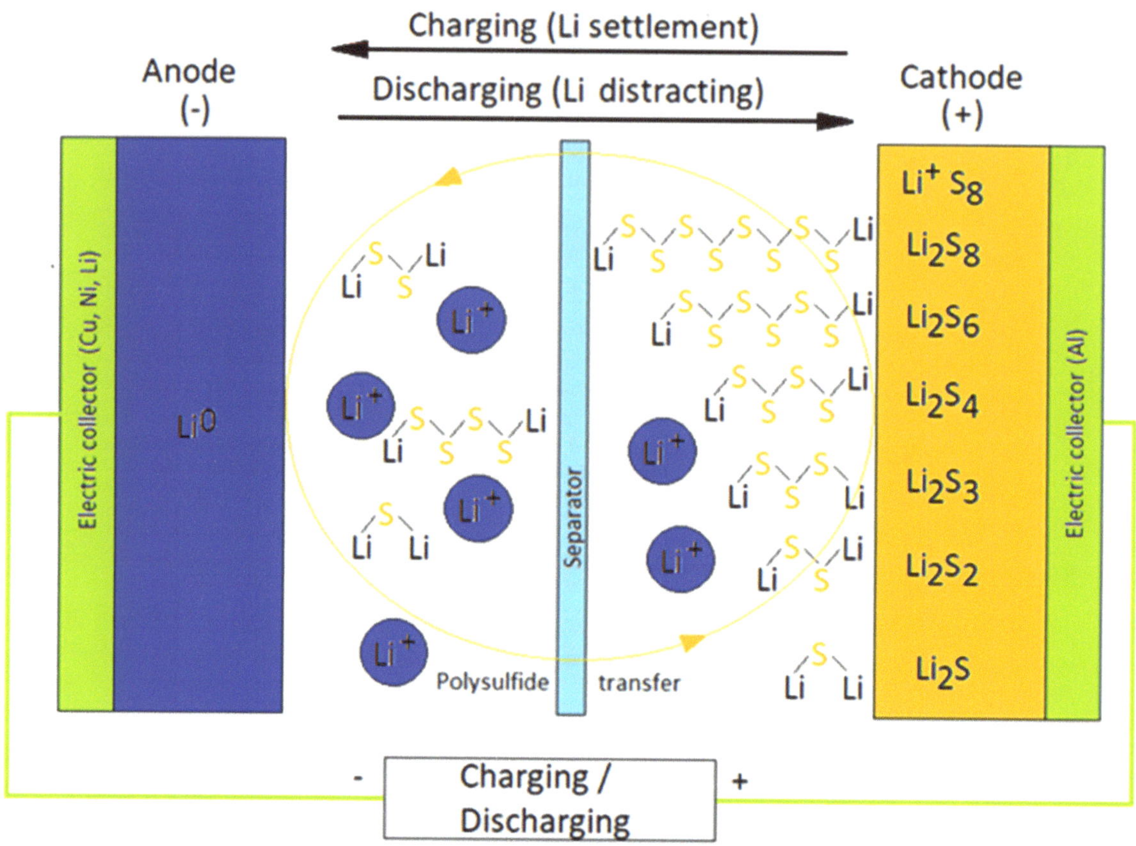

Figure 1. Structure image of LSBs with liquid electrolyte [8].

During the electrochemical reaction, sulfur is reduced to Li_2S_x (x = 2–8) via a stepwise electrochemical reaction to form each lithium sulfide compound. The lithium polysulfide (LiPS) compound with a relatively long chain (Li_2S_x (x = 4–8)) can be highly soluble. On the other hand, the solubility of Li_2S_x (x = 2–4) with short chain is lower compared to the long-chained one in the electrolyte [10]. The lithiation of sulfur occurs stepwise. At a voltage of around ~2.3 V, S_8 as the solid state is reduced to soluble $S_8{}^{2-}$ via an initial electrochemical reaction. After that, dissolved $S_8{}^{2-}$ is electrochemically reduced to $S_4{}^{2-}$ on the cathode. During this sequential chemical process, LiPS intermediates such as $S_6{}^{2-}$, $S_3{}^{2-}$, and S_3 are formed [11,12].

As LiPS concentration increases gradually throughout the electrochemical reaction, the viscosity of the electrolyte is enhanced. Gradual voltage decrease can be seen at this stage of reaction. After these events, one can observe a prolonged voltage of around 2.0–2.2 V, which is the major origin of LSB capacity. During this stage, an electrochemical reaction, which transforms soluble low-ordered LiPS to insoluble Li_2S_2 or Li_2S, takes place.

At the final stage, Li_2S_2 would become Li_2S via a reduction process. However, it should be noted that since this is the solid–solid phase reaction, the kinetics are slow due to sluggish ion diffusion. The insulating characteristics of both Li_2S_2 and Li_2S are an additional cause for a slow electrochemical reductive reaction [13].

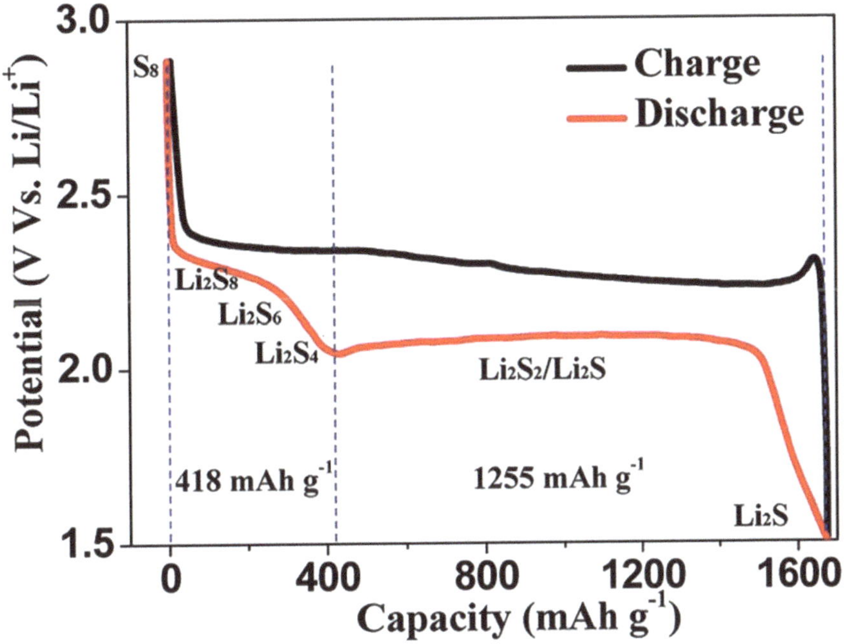

Figure 2. An ordinary charge/discharge profile curve for LSBs [9].

In spite of these advantages, LSBs still face with obstacles for practical industrial usage. For example, sulfur as a cathode active material has insulating characteristics for both the electron and the ion. LiPS as byproducts are formed in the liquid electrolyte (shuttle effect) during the discharge process, which reduce the mass and utilization of cathode sulfur. In addition, the dendrite formation of a lithium anode may cause fire once they reach the other side of the electrode through a state of battery short circuit. Furthermore, volume expansion induced by the electrochemical reaction of S to Li_2S would result in loosing stability in the cathode structure [14,15].

A tremendous amount of effort has been carried out to overcome these problematic issues by increasing the cathode materials' conductivity as well as by reducing LiPS dissolution in the electrolyte [16–19]. For example, an increase in conductivity in the cathode structure is being attempted by researchers by applying conductive carbon, carbon nanotube, graphene, as well as conductive polymer. To prevent LiPS dissolution in the electrolyte, various kinds of oxides, sulfides, nitrides, and carbides, alongside some functional materials such as quantum dot and a metal organic framework, have been replaced with carbon to reduce the shuttle effect [20–23].

Further challenges facing the cathode and electrolyte development for LSBs have been recently accelerated due to the strong influence upon the electrochemical properties. The key function of an LSB electrolyte is to transport Li^+ ions in the battery efficiently, and this requires a high level of Li^+ conductivity. The composition of electrolyte also largely influences the cathode electrode reaction as well as the behavior of LiPS byproducts [24–26].

In addition to designing cathodes and electrolytes, a separator is also a critical LSB component to effect its performance, especially in terms of suppressing the LiPS problem as separators can be reservoirs or capturing materials for polysulfide intermediates. Additionally, when separators have conductivity, for example, they can behave as the second current collector for electrons, which would result in enhancing the LSB electrochemical performance [27]. Nevertheless, it should be mentioned here that any additional separator weight due to modifications should be carefully regulated to maintain the battery energy density as a whole [28].

In general, polypropylene (PP) and polyethylene (PE) are utilized as separators, although it would be difficult for these nonpolar and hydrophobic materials to suppress the LiPS shuttle effect, because lithium polysulfide is more polar in nature [29]. As a preparation procedure, direct coating, slurry coating, or the filtration of functional materials onto a commercial separator would be enough to make such separators in order to prevent the LiPS shuttle effect. Besides LiPS suppression, such functional separators are also expected to possess catalytic activity and good mechanical strength for LSBs [30].

Fortunately, this kind of process, such as the direct coating of functional materials which ink or slurry onto the separator, can be applied to an industrial manufacturing process easily. In particular, important parameters, including thickness, weight, size, and the kind of functional materials on a separator, can be readily controlled and applied to real practical industrial processes, even though sulfur, as an active material, is essentially an electrical insulator, and so it needs to be encapsulated in a conductive carbon matrix to provide appropriate electronic contact.

Furthermore, it is necessary to provide ionic pathways through the entire cathode to ensure the conversion reaction between the sulfur and lithium ions. Thus, intimate contact and particle distribution among the carbon, sulfur, and additives is a critical factor. On top of that, the practical slurry manufacturing of actual pouch cell mass production is still challenging. In this regard, Kaskel et al. introduced the ball milling procedure for a scalable preparation method in order to achieve high sulfur utilization and loading. They clarified the carbon/sulfur agglomeration breakdown, decreasing particle size, suitable pore size volume, and distribution, which resulted in enhancing the electrical conductivity of the carbon/sulfur composite as well as achieving a higher electrochemical performance [31]. Similar studies of an improved electrochemical performance by introducing the high-energy ball milling method can be found in some other literature works [32,33]. Younes also investigated the aggregation process between carbon nanomaterials and metal oxide particles, and improved the dispersion status in aqueous solutions [34].

Since commercial separators possess a hydrophobic nature in general as explained above, it is more ideal to use hydrophilic materials to capture LiPS and prevent the shuttle effect. The thickness, weight, as well as porosity of functional layers on separators also need to be taken into consideration because they directly influence the rate of electrons and lithium ions, as well as liquid electrolyte uptake [35].

One has to also consider the modified separators: whether they are either facing the anode or cathode side. This is because the separator's role is largely influential, depending on whether they are placed toward the anode or the cathode direction. Although the main issue with LSBs is the shuttle effect induced by LiPS, lithium dendrite growth during the electrochemical process is another problematic issue. Lithium metal dendrites are formed by a side chemical process between polysulfide and lithium metal. Thus, functional interlayers on the anode side could be one of the possible solutions to prevent dendrite growth by placing the internal layer at the extra space on the anode side. This kind of interlayer may act as a second barrier layer against any leaked LiPS from the cathode side, as well as preventing dendrite physical puncture to the separator. In addition, a functional interlayer or separator could decrease the interfacial resistance of metallic lithium anode by introducing a lithium compatible material. An anode-facing functional separator may also act as the scaffold for lithium ion in order to suppress lithium dendrite growth [36].

Compared with cathode-facing separators, anode-facing separators should have a somewhat lithiophilic nature, endowing them with a high compatibility with lithium metal, in addition to their ability to functionally suppress the shuttle effect. This will control the spatial distribution of lithium deposition during redox reactions. When these properties are optimized, they can be expected to more efficiently hinder the dendritic growth of the lithium metal [37].

The remainder of this paper will explain: whether the functional layers or separators face the anode or cathode side; the material classifications of separators and their working mechanisms; and the electrochemical properties of the resulting corresponding assembled LSB. The separator materials which we have selected for this review are as follows.

2. Separators Classified by Materials

2.1. Separator Material Facing Anode Side

2.1.1. Separator Material: Metal

Nanoparticle- or nanolayer-based materials can inhibit dendritic Li growth by controlling lithium nucleation [38]. The nanoparticles in anode-facing separators behave as nucleation sites for lithium metal during electrochemical reactions as they decrease the Gibbs free energy of lithium metal electrodeposition. Initially, lithium crystal seeds would gradually grow until they cover the entire surface of the separator surface. As the process proceeds, dense lithium metal layers without dendrites are formed on the separator. Liu et al. demonstrated this phenomenon by applying lithiophilic Mg nanoparticles to an anode, whereby an LSB was prepared with this unique functional separator and which exhibited high-capacity retention (>80% after 400 cycles) [39]. Furthermore, the morphology of nanoparticles can confer the cathode structure with appropriate pores and voids, which could be a favorable lithium ion transportation channel to enhance LSB electrochemical performance. It should be noted here that Mg, Ag, and Au are also known to lower the Gibbs free energy of lithium electrodeposition.

Zuo et al. investigated the Li metal deposition phenomenon in a symmetrical cell system. As a result, lithium dendrites were seen on the surface of the pristine Li metal anode after an electrochemical discharge process, as confirmed and shown in Figure 3a. When Ag was co-deposited, Li dendrites had a tendency to fill the interspace among the Ag particles, which resulted in a lateral growth (Figure 3b). When the cross-sectional SEM image was observed, without Ag introduction, clear Li dendrites were seen after 50 cycles of the electrochemical charge–discharge process (Figure 3c). When Ag was introduced to the Li metal, in contrast, the electrode surface with a dense Ag–Li alloy layer without any Li dendrites was observed (Figure 3d). This surprising Li-dendrite-suppression effect of Ag introduction to the Li metal anode was suggested to be due to the strong affinity of Li with Ag metal. We would like to look at one example of LIB apart from LSB. With this interesting mechanism, Li ions were deposited uniformly during an electrochemical charge–discharge process. As a result, when $LiFePO_4$ | Li full cell was prepared with Ag, introducing the Li metal as anode, it exhibited a 131 mAh g^{-1} capacity, even after 300 cycles [40]. Not only was a metal material introduced, but copper was also functionalized to the lithium metal anode to prepare a 3D Cu/Li and copper mesh/Li composite anode. Yang et al. clarified that Li dendrite growth was successfully suppressed with a 3D Cu/Li-structured anode, which can operate for 600 h and demonstrated low voltage hysteresis [41,42].

2.1.2. Separator Material: Ceramic

Boron nitride was applied to optimize the commercial separators, owing to its intrinsic high levels of insulation and high thermal conductivity [43]. Boron nitride (BN)/carbon-modified composite separator was prepared with a coating method on a separator. Figure 4A shows a schematic figure and the functions of the BN/carbon separator in an LSB. During discharge, LiPS spreads out from the sulfur cathode toward the lithium anode metal. The carbon layer hinders the diffusion of the LiPS acting as the first blocking layer, and any LiPS that diffuses through the carbon layer is subsequently trapped by the BN layer acting as the second blocking layer. This BN/carbon separator was made by direct ink casting (ink composed of carbon nanopowder and BN nanopowder) onto a polypropylene separator via the slurry coating method. Owing to this simple procedure, large-scale separator fabrication (dimensions 150 mm long and 60 mm wide) could be achieved (Figure 4B). The prepared separator had sufficient mechanical strength and was able to resist physical twisting due to the strong level of adhesion between the separator, carbon, and BN nanopowders, having been bonded using a polyvinylidene fluoride (PVDF) binder (Figure 4C). Figure 4D shows the surface of a pristine commercial polypropylene separator, on which some distributed pores can be observed. After coating the separator with BN (particle size ~ 100 nm; Figure 4E) and carbon (particle size ~ 50 nm; Figure 4F) nanopowders, the pores were fully covered. BN and the carbon layer thickness were approximately 7 and 6 μm, respectively (Figure 4G). Some of the electrochemical properties of the prepared LSBs were measured. After the surface modification, lithium dendrite formation was suppressed by forming on the Li surface by an ideal Li plating/striping process (Figure 4H). Fortunately, LiPS diffusion was mitigated by an additional BN layer on the separator, thus resulting in obtaining a higher LSB electrochemical performance (Figure 4I) [43].

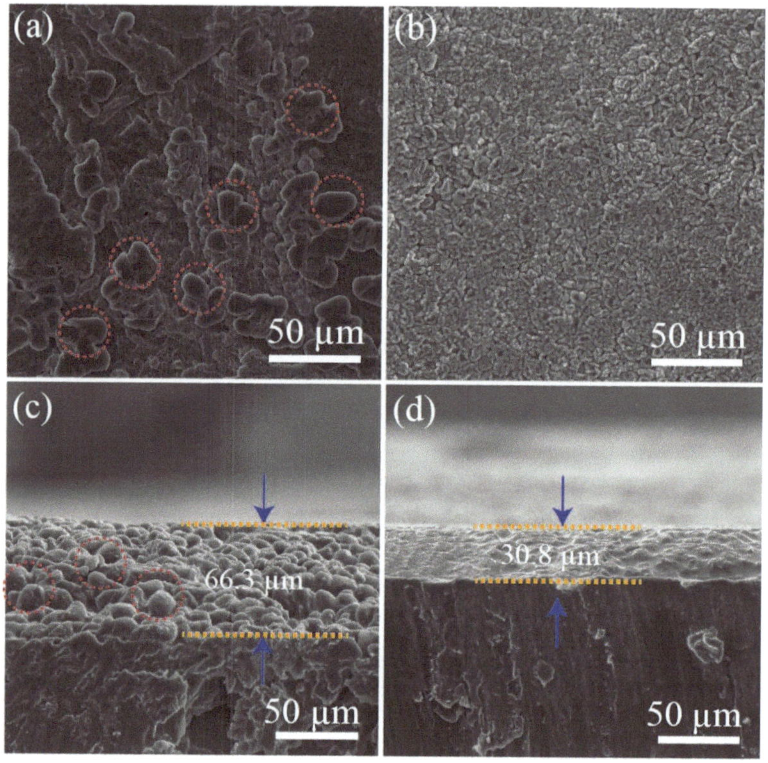

Figure 3. Surface observation: (**a**) bare lithium; (**b**) Ag co-deposited lithium with SEM. SEM observation of cross-sectional (**c**) bare lithium; (**d**) Ag co-deposited lithium [40].

In addition to controlling the lithium metal growth and nucleation, the oxide functional interlayer could also be ideal in suppressing the LiPS effect. Silica (SiO_2) is one such material effective for preventing the LiPS shuttle effect and improving LSB performance. One of its advantages is its abundance. The thermal stability of SiO_2 is also a good feature in order to confer the separator with high thermal stability [44]. However, its direct contact with the cathode should be prevented as SiO_2 is known to react with lithium metal [45]. However, it should be noted that one of the SiO_2-coated separator example will be presented in a later section. In addition, SiO_2 has high affinity with LiPS, making it suitable for suppressing the LiPS shuttle effect [46]. Therefore, SiO_2 utilization as functional LBS separators could be one of several effective methods. Li et al. attempted an SiO_2 coating of a polypropylene (PP) separator using the tetraethoxysilane (TEOS)-based sol–gel method. They have elucidated that by applying PP–SiO_2 separator, wettability and thermal stability were properly improved. As a result, rate and cyclic voltammetry performance improved, owing to suppressed LiPS shuttle effect as well as promoting lithium ion transport inside LSBs [47].

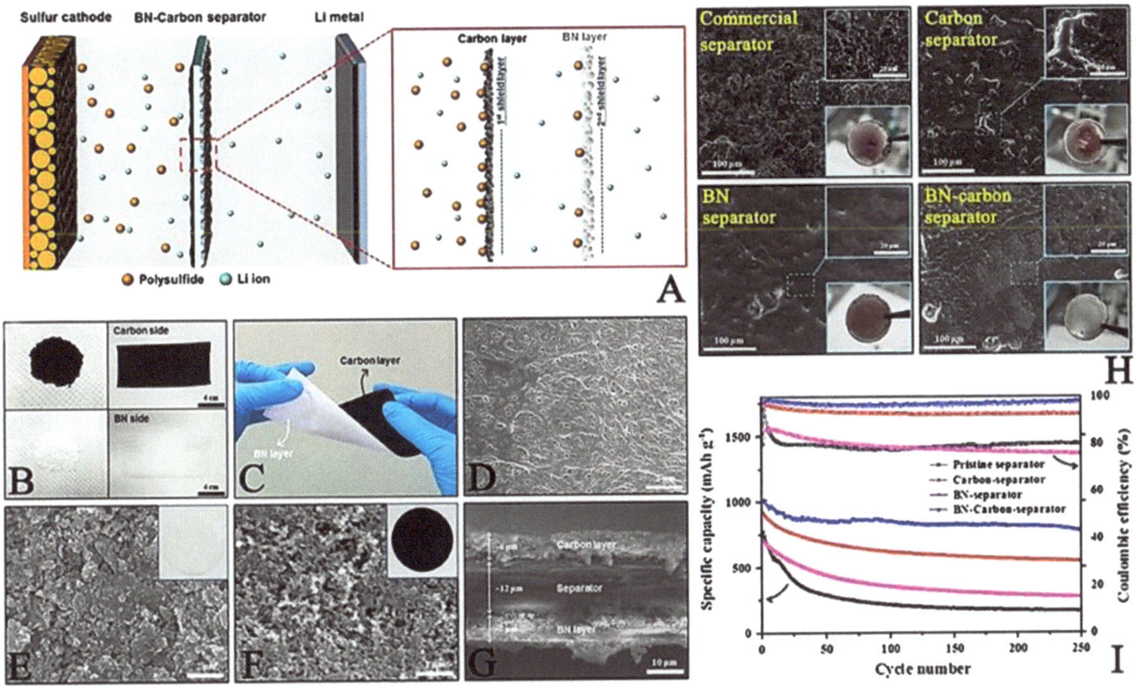

Figure 4. (**A**) Speculated image of the BN-carbon separator role. (**B**) Separator of BN coating layer. (**C**) Bent BN-coated separator image. Top view of SEM image of (**D**) intact separator. (**E**) BN side separator. (**F**) Carbon side of a BN-carbon separator. (**G**) Cross-sectional SEM view of a BN–carbon separator. (**H**) Photographs of a lithium metal anode of LSB prepared with described type of separators in each picture, after 250 cycles. (**I**) Cycle performances of various kinds of separators at 0.5 C [43].

The chemical stability and LiPS adsorption ability of alumina (Al_2O_3) are relatively high. Hou et al. elucidated that the oxygen atom with a lone electron pair of oxide materials such as Al_2O_3 and SiO_2 has strong dipole–dipole interactions with LiPS [48]. Wang et al. prepared a purely inorganic separator with Al_2O_3 nanowires. In general, a purely inorganic separator without any base organic polymer would have a brittleness problem compared to conventional organic PP separators, whereas the advantages of a ceramic separator are a higher uptake of liquid electrolyte and higher porosity. They often also exhibit improved thermal stability and enhanced conductivity compared to the organic ones. This mechanism can be applied not only for LSBs, but also LIBs. For example, when an alumina separator was applied for Li | $LiFePO_4$ (LFP), the battery electrochemical characteristics were better than a conventional PP separator-based battery [49].

Regarding LSB, He et al. prepared an Al_2O_3 thin film on a commercial separator on the anode-facing side in order to examine the lithium dendrite formation prevention effect. It was found that the Al_2O_3 layer successfully mitigated the chemical reactions between LiPS and the lithium anode. They also discovered that the porous structure of the prepared Al_2O_3 accelerated uniform lithium nucleation, which resulted in preventing lithium dendrite growth to form a smooth and dense lithium anode [50].

2.1.3. Separator Material: Solid Electrolyte

$Li_{6.4}La_3Zr1.4Ta_{0.6}O_{12}$ (LLZTO) is the lithium ion conductive solid electrolyte, and it was coated on the PP separator on the anode side. Smooth lithium deposition was achieved due to a uniformly dispersed transportation path for the lithium ion in three-dimensional structure in LLZTO. It was also clarified that lithium ion deposition was further enhanced by adding a higher content of LLZTO by localizing a higher content of anions. Owing to these improvements, the prepared LSB presented an improved electrochemical performance with better safety due to solid electrolyte stability [51].

2.1.4. Other Functional Separator Materials

As we have seen above, a modified anode-facing separator could act as a simple barrier to push back LiPS immigration from the cathode to the anode, and this would also be expected to impede Li dendrite formation. For example, there is a kind of conducting polymer that is known to capture LiPS, promoting Li ion flux and electron transportation. Li et al. prepared a conductive polypyrrole (PPy) film for a commercial anode-facing separator and succeeded in improving LSB electrochemical properties [52]. They deposited the PPy film on the separator by applying an Fe-based precursor and pyrrole monomer and succeeded in obtaining a PPy-modified separator (Figure 5a,b). A scanning electron microscope (SEM) observation of the pristine commercial separator exposed 100–200 nm sized nanopores (Figure 5c). In contrast, slightly smaller nanopores with sizes of 30–50 nm were observed on the PPy-coated separator (Figure 5d). The thickness and weight of the coated PPy layer were as small as 15–25 nm and ~0.13 mg cm^{-2}, respectively, which is advantageous for industrial use in terms of energy density per weight (Figure 5e). X-ray photoelectron spectroscopy (XPS) was carried out upon both separators. We have observed a peak at 399.8 eV, which can be ascribed to N 1s with the PPy-coated separator (Figure 5f). Nitrogen-originated peaks, including –N= (399.6 eV), –N– (400.7 eV), and N^+ (401.7 eV), are observed from PPy [53,54]. From these spectra results, we suggest that the hydrophilicity of PPy is effective in capturing and suppressing LiPS from the shuttle effect [55,56]. PPy also chemically adsorbed LiPS, owing to its special structure and functional group. As a result, Li dendrite formation was suppressed by the enhancement of the homogenous Li ionic flux and Li plating/stripping.

Preparing a physical barrier by introducing some kinds of lithium compound to a separator is an alternative way to reduce the unfavorable reactions between LiPS and lithium metal. This concept is different from using a material that would not react with any LSB component. Based on this concept, the lithium compound, LiF, was applied to a separator. Lewis acid lithium atoms in LiF could relate with 1,2-dimethoxyethane (DME) as Lewis alkali to form LiF–DME clusters. These clusters are viscous sol and form dense layers that would behave as a shield to prevent the shuttle effect of LiPS. Li et al. utilized this method to improve the electrochemical properties of LSBs [57].

Cathode-facing functional separators could also be effective as anode-facing counterparts because they can serve as the first barrier to LiPS, potentially increasing the chances of sulfur utilization as an active material. To date, polymers, metal compounds, ceramics, carbonaceous material, and their composites have been used to alter commercial separators in order to endow them with additional positive properties [58]. Separators should be mechanically strong, chemically and electrochemically stable, and lightweight in order to avoid negatively impacting the density of energy of the final LSB product. Furthermore, they should possess an enhanced ionic conductivity and the ability to suppress the shuttle effect of LiPS. The functions of a cathode-facing separator can be generally classified as physical adsorption, chemical adsorption, catalytic conversion, and dual-mechanism functions, as shown in Figure 6 [35].

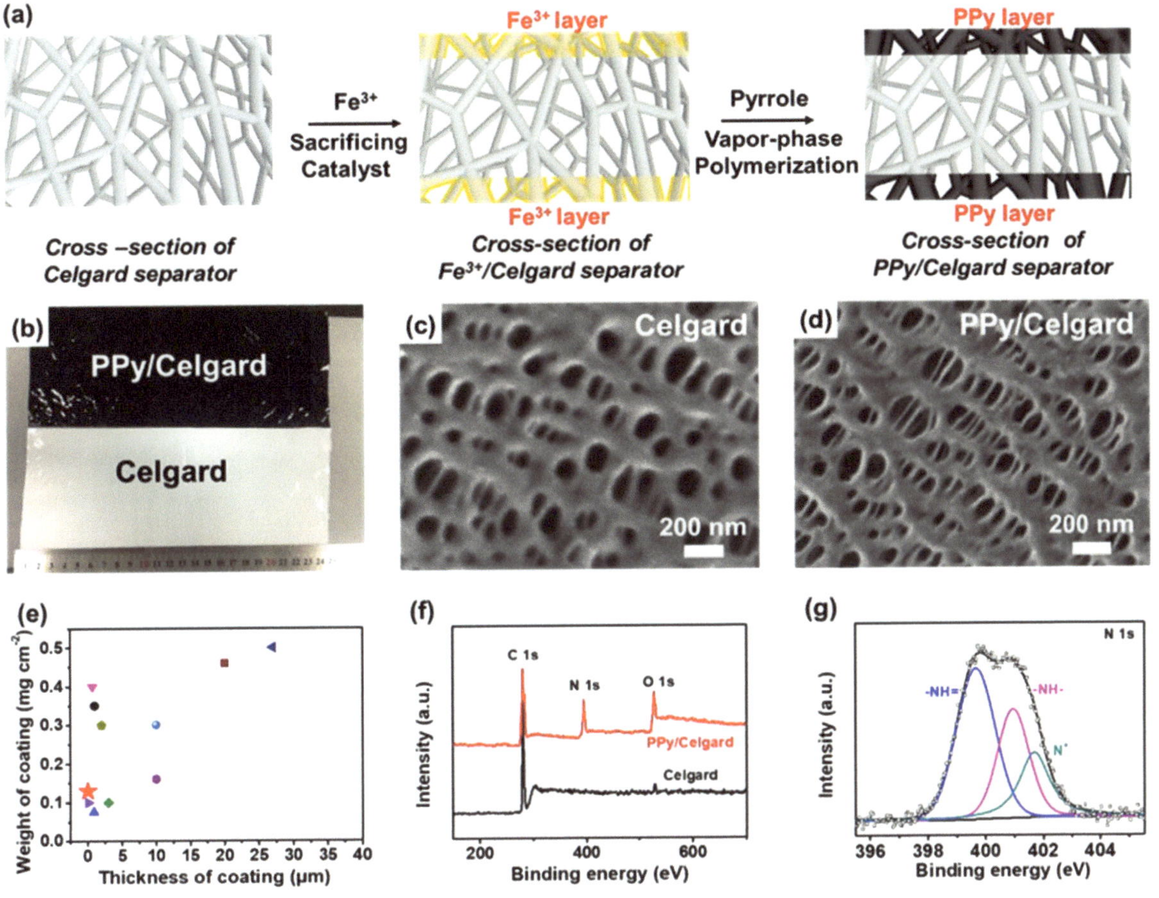

Figure 5. (**a**) The speculated image of PPy layer on a commercial separator by a vapor–phase polymerization method. (**b**) Actual image of commercial separator with and without PPy treatment. SEM image of commercial separator (**c**) without and (**d**) with PPy treatment. (**e**) Comparison and other examples of the weight and thickness with various types of coating in this work and previous studies. (**f**) XPS spectra of a separator with and without PPy treatment. The spectrum of N is originated from the PPy-treated layer. (**g**) Expanded N 1s spectrum of the separator with PPy treatment [52].

2.2. Separator Material Facing Cathode Side

2.2.1. Separator Material: Carbonaceous Materials

Commercial separators are composed primarily of PP and PE and are considered to be purely physical barriers. In light of this, a great number of research were carried out to study the influence of applying carbonaceous material to the cathode side of the separator, owing to its high electrical conductivity. These carbonaceous materials, including conductive carbon black, carbon nanotube, and graphene, are being studied with an aim to enhance LiPS suppression, cycle stability, and capacity of LSBs [59,60]. The improved electrochemical characteristics of LSBs are mainly the result of high electrical conductivity, which can confer them to behave as additional current collectors to lower the battery resistance [61]. In addition, a sulfur cathode-facing carbon-loaded separator would act as an obstacle to suppress the shuttle effect of LiPS.

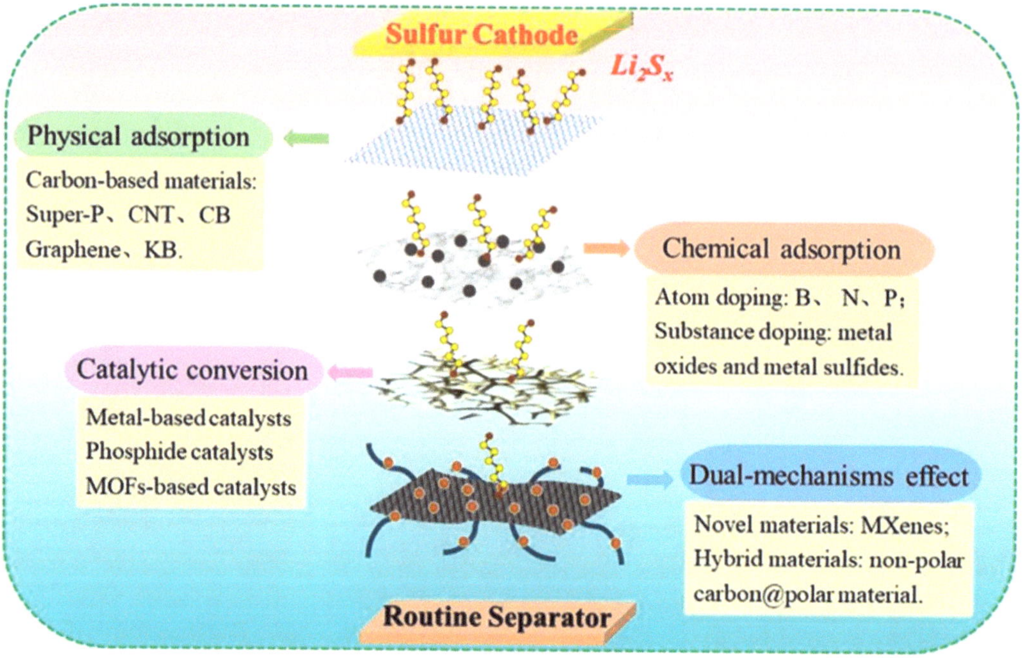

Figure 6. Classification and functional mechanism of functional separators in LSBs [35].

For example, eight separators coated with the conducted carbon of various nanoporosities were investigated by Huang et al. to see how the material characteristics contributed to LSB separator performance [62]. The LSB with the nonporous carbon-loaded separator demonstrated a 1112 mAh g^{-1} capacity at a C/10 cycling rate, and even after 200 cycles, a stable reversible capacity of 710 mAh g^{-1} was maintained [63].

Kim et al. prepared a separator treated with carbon nanotubes conjugated with hydroxyl groups. The experimental results and theoretical calculation confirmed that the hydroxyl groups on carbon nanotube surface promoted good LiPS-capturing effect on the cathode side. Furthermore, the LiPS migration decrease to the anode side conferred the lithium electrode with good stability. In addition, the high conductivity of carbon nanotube with hydroxyl group assisted in the reuse of adsorbed intermediates for further electrochemical reaction (Figure 7). Because of these efforts, the prepared LSB demonstrated 1056 mAh g^{-1} at the beginning and a capacity fading rate of 0.11% per cycle over 400 cycles at a 0.5 C rate [63].

Besides conductive carbon and carbon nanotubes, graphene can be also considered as an attractive functional coated material to improve the separator properties. In general, graphene possesses high electrical conductivity, is super lightweight, has high chemical stability and ductility, and is mechanically strong. The separator coated with graphene blocks pores in the polyolefin separator, which can prevent LiPS dissipation. Reduced graphene oxide (rGO) was obtained using Hammer's method, and it was dispersed in DMF (dimethylformamide) and combined with sodium ligno-sulfonate (SL) and hexamethylene diisocyanate (HDI). SL is rich in hydroxyl, carboxyl, and dendritic groups and it is considered an industrially cheap and available material. IR spectra elucidated a broad peak at 3400 cm^{-1}, which can be ascribed to stretching vibrations of the hydroxyl group, and 1634 cm^{-1}, 1402 cm^{-1}, and 1112 cm^{-1} peaks were originated from carbonyl C=C, –OH, and C–O bonds in GO and SL, respectively [64–66]. The -OH bond peak disappeared at the GO/SL composite material after reacting with HDI and SL; this is due to GO reduction to rGO (Figure 8B). Zeta-potential for the rGO/SL composite was 75.14 mV. This indicates that the significantly strong electrostatic repulsion force is formed against polysulfide ions, which is negatively charged, in order to suppress the LSB shuttling effect [67,68].

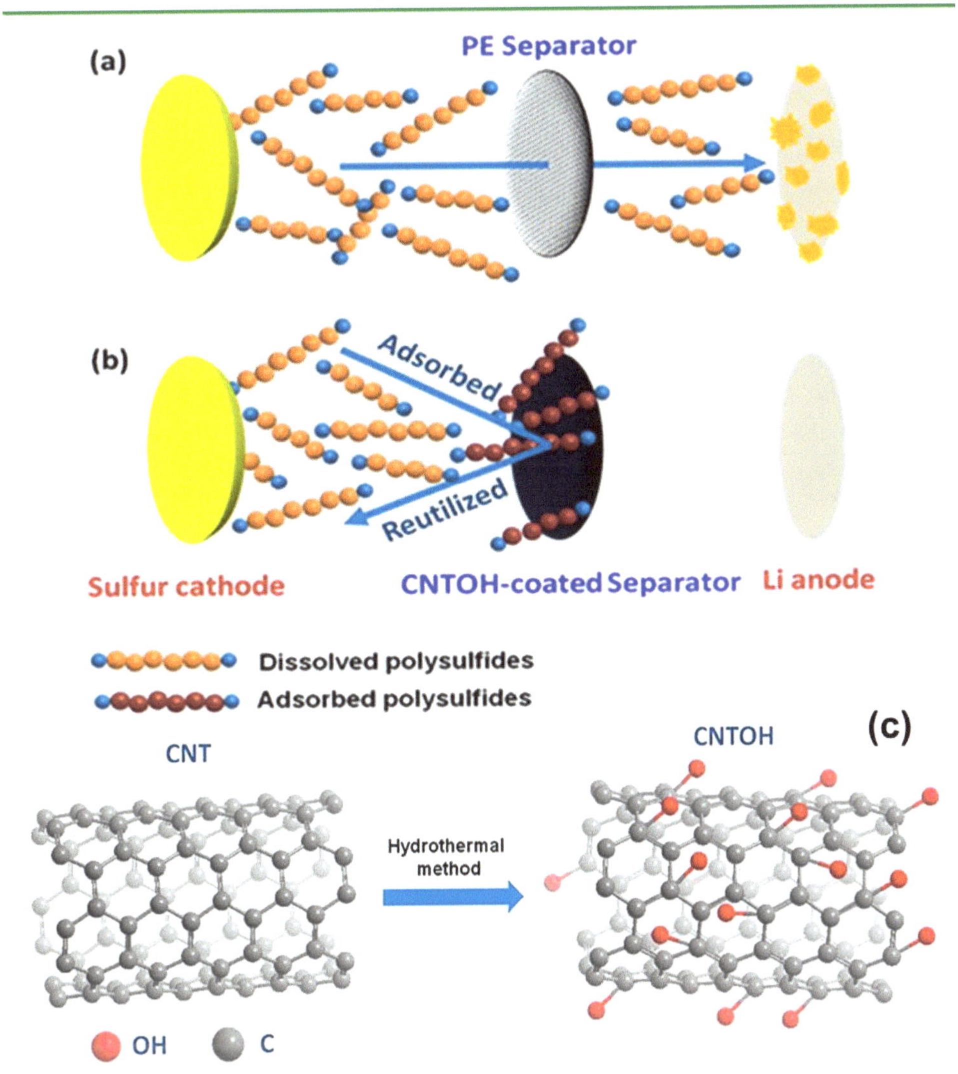

Figure 7. Speculated image of LiPS capture by the separator coated with CNTOH [63]. (**a**) LiPS transport to the anode side through separators, resulting in a decreased sulfur utilization. (**b**) In case of the separator with CNTOH application, hydroxyl functional groups on CNTOH would adsorb LiPS and utilized it to create polysulfides with short chains. (**c**) Surface treatment method of CNT via a hydrothermal procedure to form CNTOH.

This type of prepared rGO/SL composite solution was applied to a commercial PP separator using a facile vacuum filtration process. Sulfonic groups in the lignin can confer a separator with enough negative charge and reduce negatively charged LiPS diffusion. Applying this reduced graphene oxide/sodium ligno-sulfonate/separator achieved a capacity retention of 74% over 1000 cycles [69].

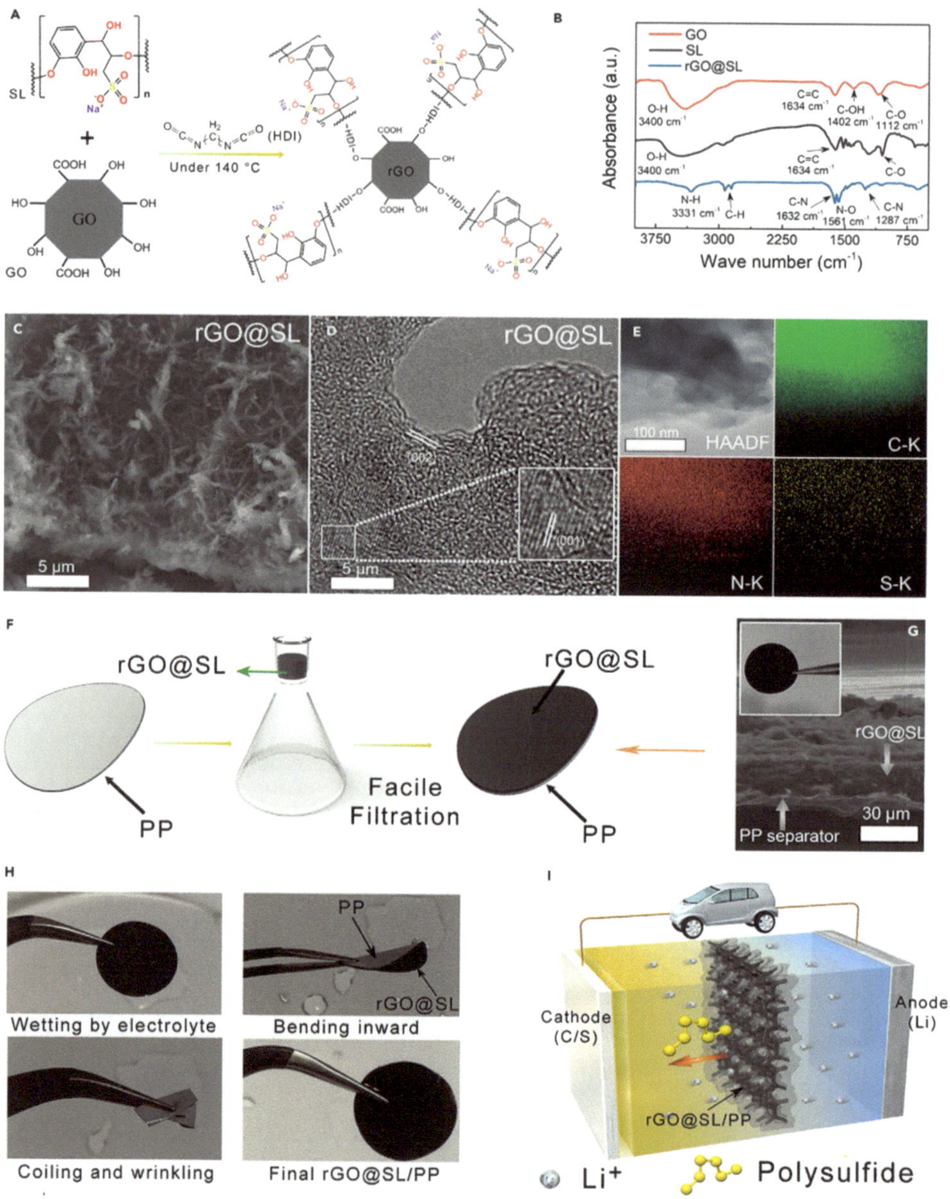

Figure 8. Synthesis procedure, microscopic observation and preparation of the rGO/SL composite and rGO/SL/PP separator [69]. (**A**) Preparation steps to make rGO/SL. (**B**) FTIR spectra of GO, SL, and rGO/SL. (**C**) Scanning electron microscopy images of the rGO/SL composite. (**D**) HRTEM observation of the rGO/SL composite. (**E**) Elemental analysis of C, N, and S for rGO/SL. (**F**) Preparation for the rGO/SL/PP separator. (**G**) Scanning electron microscopy picture of the rGO/SL/PP separator. (**H**) rGO/SL/PP separator image against mechanical stresses. (**I**) Speculated image of the rGO/SL/PP separators for suppressing LiPS shuttle effect.

In addition to simply physically blocking LiPS, chemical bonding between polar substances is an alternative procedure to suppress LiPS diffusion. To capture polar polysulfide, polar materials are superior to hydrophobic carbon materials. Carbonaceous materials doped with various kinds of elements (e.g., N, B, S, P, F, O, Cl, Co, Ni, Fe) tend to possess polarity and have greater ability to capture LiPS [70]. The composite material composed of a nitrogen-doped reduced form of graphene oxide with CoS_2 and NiS_2 was made via a hydrothermal method [71]. Figure 9a shows the adsorption–desorption isotherms and pore size distribution of the cobalt nickel sulfur/nitrogen-doped reduced graphene oxide (CNS/N-rGO) composite and the one combined with sulfur, respectively. BET-specific surface area of CNS/N-rGO was 177 m^2/g, and the pore volume was 0.416 cm^3/g. The desorption isotherm of Barrett–Joyner–Halenda (BJH) is presented in Figure 9b. An obvious strong peak at 4 nm is proof of the porous structure of the sample, which can promote lithium ion transportation and help adsorb the LiPS, thus improving the LSB electrochemical performance [72,73]. However, after combining with sulfur, the pore volume and surface area significantly reduced, suggesting that the CNS/N-rGO was deeply immersed with sulfur. Figure 9c is the XRD pattern of the CNS/N-rGO composite material. A 26° broad peak can be ascribed to the (002) facet of the graphite. One can deduce that 53.04°, 49.03°, and 32.01° peaks corresponded to the (200), (220), and (311) CoS_2 facets, respectively [74]. Furthermore, characteristic peaks at 1350 cm^{-1} (D band) and 1580 cm^{-1} (G band) as can be observed in the Raman spectra are inferred to have derived from the structural defect and the stretching mode of the C–C bond of graphene, respectively [75]. Wu et al. coated this composite material onto a separator surface and used it in an LSB [71]. Owing to its morphological advantages and LiPS affinity, the LSB exhibited 1524 mAh g^{-1} at the first discharge and 610 mAh g^{-1} at 0.1 C and 8 C, respectively. It exhibited good long-term stability, with a 700 mAh g^{-1} capacity even after 350 cycles under 1 C. It was deduced that nitrogen doping and cobalt nickel, and sulfur nanoparticles in the composite, could provide a favorable electrochemical structure and active sites for capturing soluble LiPS and promote LiPS conversion in redox reactions, thus preventing the shuttle effect [74].

Some study. applied sulfur and metal, nitrogen co-doped carbon nanoparticles as the separator material for LSBs [76,77]. They were applied onto the separator to suppress LiPS diffusion. It was found that for nitrogen, the sulfur co-doped carbon nanoparticles improved both capacity and cycle characteristics. The capacity was 841 mAh g^{-1} at 0.2 C even after 100 cycles with a 99.3% of coulombic efficiency.

Song et al. used a lightweight, functional layer composed of porous iron and nitrogen-co-doped carbon nanofiber materials to modify a PP separator for LSB. The Fe–N-co-doped carbon possesses high conductivity and LiPS adsorption capability to the separator, which was effective in preventing LiPS diffusion. Prepared separator behaved as the conductive current collector to reuse the active sulfur-based compounds as well.

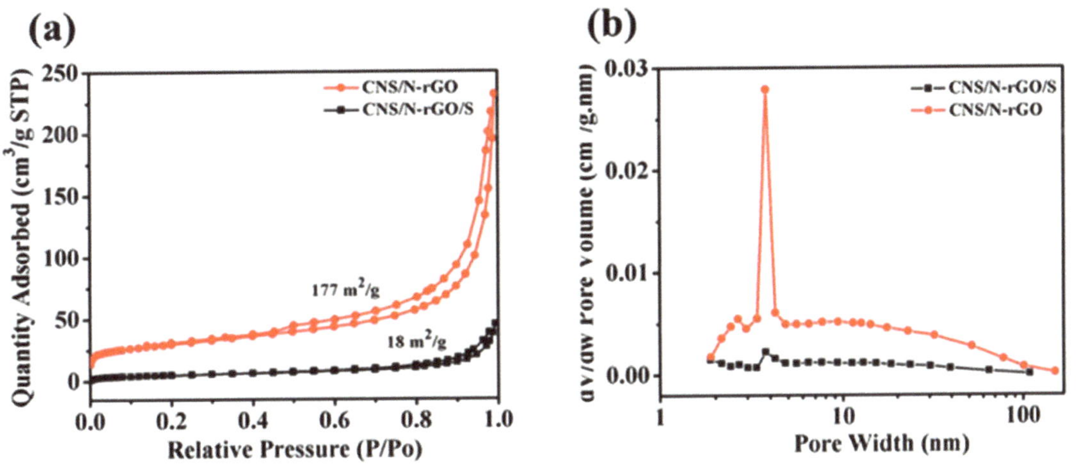

Figure 9. *Cont.*

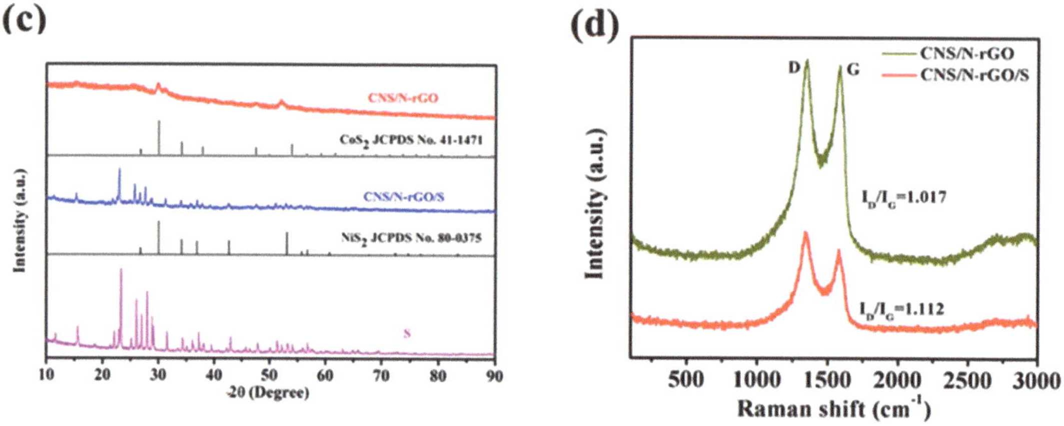

Figure 9. (**a**) The adsorption–desorption isotherms of N_2. (**b**) BJH pore size distribution curves of the CNS/N-rGO composite and CNS/N-rGO/S. (**c**) XRD pattern of the CNS/N-rGO composite. (**d**) Raman spectrum of the CNS/N-rGO composite [72].

2.2.2. Separator Material: Metal Oxide

Metal oxides were investigated for application as LSB separators as they can generate different bonds for LiPS. Metals lose their electrons in certain conditions and become metal ions, which can generate chemical bonds to capture LiPS to prevent the shuttle effect. In addition, the metal oxides in nanostructure states would be easier to combine with polymers to become functional separators. It has been shown that LiPS diffusion can be suppressed with the Al_2O_3-coated layer on the separator's cathode side as a physical block layer for LiPS. The first discharge capacity of the LSB prepared with an Al_2O_3-applied separator was 967 mAh g^{-1} at 0.2 C and deteriorated to 593.4 mAh g^{-1} after 50 cycles. The electrode resistance was reduced because of the porous structure of the Al_2O_3-coated layer that worked as an ion-conducting scaffold for capturing sulfur containing active materials [78].

Han et al. created a PP separator with a TiO_2/carbon-coated layer and applied it to an LiPS capturing barrier for LSB [76]. The highly conductive carbon with its porous structure endowed the separator with the ability to adsorb LiPS while acting as a second electrical current collector. In addition, TiO_2 on the porous carbon surface captured LiPS. Owing to these properties of the TiO_2/porous carbon composite material, 926 mAh g^{-1} capacity was observed initially at 0.1 C; also, it maintained a 75% capacity even after 150 cycles [79].

Gao et al. also prepared a separator coated with a TiO_2/surface-modified carbon nanotube composite. The LSB assembled with this type of separator—having strongly polar TiO_2 and highly conductive carbon nanotubes—demonstrated an enhanced battery performance. The capacity was initially 1104 mAh g^{-1} and was still as high as 848 mAh g^{-1} even after 200 cycles at 0.5 C, demonstrating a capacity decay of only 0.066% per cycle over 900 cycles [80].

A SiO_2 nanoparticle-modified PP separator was made by Li et al. by dipping a PP separator into a sol–gel TEOS solution and Tween-80 [78]. This kind of separator has high wettability and thermal stability, and the resulting LSB showed considerable improvement in its electrochemical performance. The capacity decay of the LSB prepared with this SiO_2–PP separator was 64% after 200 cycles at 0.2 C, which was superior than the one made with a commercial separator (45%). In addition, the capacity of LSB with SiO_2–PP separator achieved 956.3, 691.5, 621, and 567.6 mAh g^{-1} at a 0.2, 0.5, 1, and 2 C current density, respectively [81].

Increasing the ratio of sulfur utilization would increase LSB capacity. Thus, it is necessary to not only capture LiPS, but also to reactivate sulfur [82]. Certain types of separators with adsorption abilities cannot satisfy this requirement. Therefore, it is necessary to look for the material which can play a role for LiPS reutilization. In addition, the material with catalytic ability is also important.

During the LSB charging process, fast-reaction kinetics is achieved only when the energy which is higher than the activation energy of solid Li_2S_2/Li_2S conversion to polysulfide is applied [83]. Fast battery reaction rate can be accelerated when activation energy is lowered so that one needs a good catalyst to lower the activation energy. This concept differs from the LiPS adsorption materials, which we have previously described above. This kind of catalytic material can accelerate the oxidation reduction rate so that the reaction of polysulfides' conversion reaction to Li_2S_2/Li_2S can be enhanced. This is also a highly efficient way to prevent the LiPS shuttle problem besides using LiPS adsorption materials. Electrons on the catalytic material can transport to the S–S bond of LiPS to become Li_2S_2/Li_2S. In addition, the catalyst can reduce the overpotential of the sulfur species, which would enhance the sulfur electrochemical reactions [84]. Therefore, finding the appropriate catalytic material that can accelerate LiPS conversion into Li_2S_2/Li_2S is highly recommended.

In this regard, La_2O_3 was coated onto a commercial PP separator in order to adsorb LiPS and accelerate redox reactions at the same time [85]. Charge/discharge measurements demonstrated that the resulting LSB showed a 966 mAh g^{-1} capacity on the first cycle at 1 C and the capacity became 720 mAh g^{-1} after 200 cycles. It was suggested that the enhanced electrochemical properties were due to the LiPS capturing effect of La_2O_3 as well as being a chemical catalyst to promote the electrochemical reaction by lowering the activation energy [85].

2.2.3. Separator Material: Metal Sulfide

Some metal sulfides are known to adsorb LiPS, owing to their polar nature, and thus have been studied for use as a separator material. Tan et al. prepared a reduced graphene oxide/MoS_2 (rGO/MoS_2) layer on the cathode side of the commercial separator. The rGO behaves as a barrier for LiPS diffusion and an extra current collector, while MoS_2 can capture LiPS. The prepared LSB demonstrated 1122 mAh g^{-1} at 0.2 C, with a low capacity fading rate of 0.116% for 500 cycles at 1 C, and an excellent performance rate of 615 mAh g^{-1} at 2 C [86].

Another study demonstrated the potential use of edge-rich MoS_2/C hollow microspheres as LSB separators. The MoS_2/C hollow microspheres were prepared via hydrothermal synthesis with MoO_3, aniline, thiourea, and sucrose, following a carbonization process [87]. Carbonization was carried out in order to improve the graphitization degree of the carbon in MoS_2/C hollow microspheres (Figure 10). The prepared MoS_2/C hollow microspheres carry a high chemical absorption property and a high density of LiPS-capturing sites, resulting in showing excellent LiPS diffusion prevention. In addition, the phase conversion reversibility of the active sulfur species could be regulated in a stable matter with this MoS_2/C composite material particularly at high C-rates and sulfur loading. Thus, the LSB arranged with the MoS2/C separator exhibited a capacity of 935 mAh g^{-1} at an initial cycle atof 1.0 C and it was 494 mAh g^{-1} even after 1000 cycles [88].

Functional separators using two-dimensional exfoliation nanosheets could efficiently contain LiPS for long-life LSBs. Mao et al. fabricated new two-dimensional ZnS exfoliation nanosheets using the microwave-assist exfoliation method, and then combined it with graphene to modify a commercial separator. The nanosheet dispersion was mixed with graphene and filtered on the separator to form ZnS exfoliation nanosheets/graphite/Celgard separators. An LSB assembled with this type of separator exhibited an 1165.9 mAh g^{-1} as initial capacity and became 685.3 mAh g^{-1} after 100 cycles. Capacity was 58.8% compared to the initial one at 0.1 C and 0.036% of capacity decay per cycle when the cycle was carried out for 1000 times at 0.5 C [89].

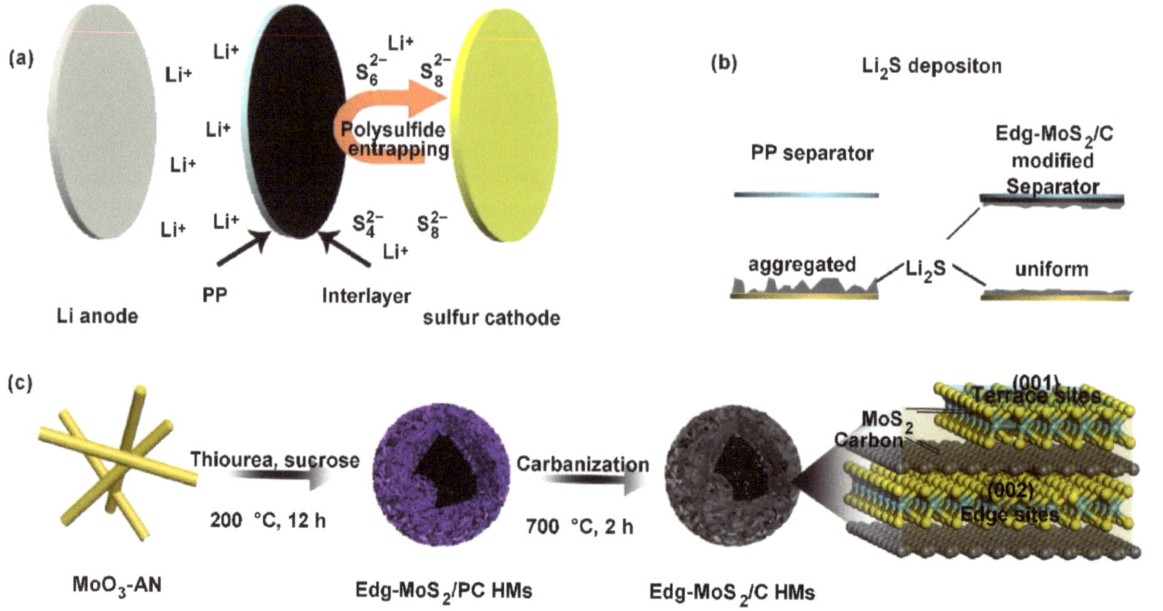

Figure 10. Schematic model of the operation and preparation of the Edge-MoS_2/C/PP separators [87]. (**a**) Image illustration of polysulfides' capture in LSB with Edge-MoS_2/C/PP separator. (**b**) Image illustration of Li_2S deposition on the separator with edge-MoS_2/C/PP and PP. (**c**) Synthesis procedure of the Edge-MoS_2/C hollow microspheres and illustration of the Edg-MoS_2/C hollow microspheres.

Yao et al. utilized the high conductivity and absorption ability of SnS, and the strong catalytic property of ZnS, to prepare a ZnS–SnS heterojunction [84]. A polydopamine-derived nitrogen-doped carbon was coated onto this ZnS–SnS composite material and was compared with its single-component counterparts (ZnS/N-doped carbon shell and a SnS/N-doped carbon shell). When coated onto the separator, the ZnS–SnS/N-doped carbon-based LSB demonstrated a high reversibility of 1149 mAh g^{-1} capacity for 300 cycles at 0.2 C, and a 661 mAh g^{-1} at 10 C, and capacity decay was 0.0126% each cycle after 2000 cycles at 4 C [90].

In addition, it was reported that a WS_2/Prussian blue (PB)–PPy composite material-modified separator could inhibit the back and forth movement of LiPS and the formation of inactive sulfur-related substances. Lithium ions, however, could still be transferred homogeneously inside the battery. When compared with the LSB with ordinary commercial PP separators, the capacity and cycle properties of this modified LSB were remarkably ameliorated and showed high capacity (1050 mAh g^{-1}) and improved capacity retention (650 mAh g^{-1} after 300 cycles) as well as a capacity retention rate that reached 62%. This study also showed that three porous layers as an ion-sieve structure indeed prevents the transfer of LiPS through the electrolyte to the anode [91].

2.2.4. Separator Material: Metal Carbide

Titanium carbide (TiC) is regarded as a representative non-oxide ceramic material with high electrical and thermal conductivity, robust hardness, and high chemical stability [92]. In a recent study, TiC is acknowledged to be utilized for energy storage purposes owing to these excellent properties. Especially for LSBs, TiCs are considered to improve the cycle stability via its high polarity to capture sulfur species and its ability to reduce the LiPS shuttle effect [93]. In addition, the high electrical conductivity of TiC is an additional good reason for it to be an ideal sulfur scaffold material.

Liu et al. synthesized TiC nanoparticles from waste polytetrafluoroethylene as a carbon source at 500 °C, that is lower than general TiC synthesis temperature. They used this TiC as a coating material for an LSB separator and successfully mitigated the shuttling problem. The visual influence was investigated by dispersing prepared TiC nanoparticles in an LiPS solution. At the TiC–Li_2S_6 bottle, the yellowish color became more transparent after two hours of adsorption, whereas the bottle without TiC remained unchanged, as presented in Figure 11A,B. These results clearly imply that waste polytetrafluoroethylene-derived TiC nanoparticles have high potential in reducing LiPS shuttle effect in LSBs [94]. The bending exam proved TiC's strong adhesion to the separator as shown in Figure 11C. One can easily visualize that the dissolved LiPS shuttle effect could be suppressed with this separator. A capacity as high as 1242 mAh g^{-1} was detected at the first cycle with the LSB prepared with the TiC-attached separator, and even after 100 cycles, a 736 mAh g^{-1} capacity was retained, which are superior results compared to an LSB with a commercial separator (respectively, 827 and 373 mAh g^{-1}). They also proved that the high electronic conductivity of the TiC resulted in a low capacity decay during rate performance [95]. Zhao et al. also fabricated a novel TiC-modified PP separator and applied it as an LBS separator. Compared to the commercial separator, the TiC-modified separator exhibited superior wettability and adsorption ability for LiPS. As a result, the LSB with a TiC-modified PP separator displayed a higher specific capacity and more stable cycling performance [96].

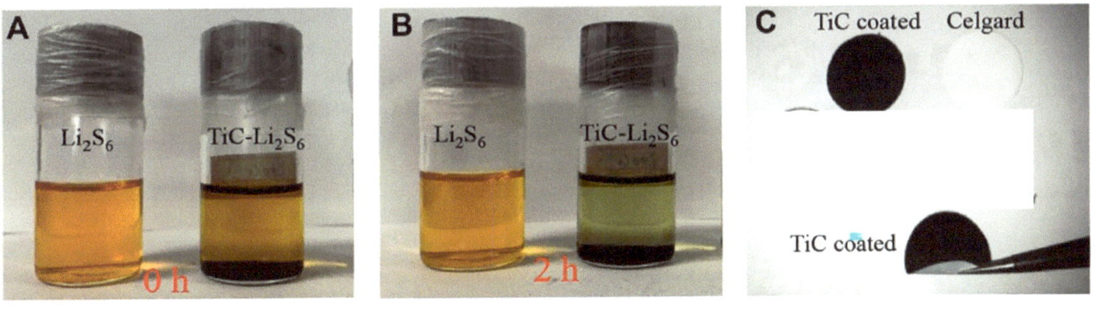

Figure 11. The pictures of LiPS capturing: (**A**) Before and (**B**) after 2 h. (**C**) Digital image of the TiC-coated separator and bent testing image [94].

Moon et al. made an LSB assembled with a separator that contained mesoporous tungsten carbide (meso-WC) and reduced graphene oxide [91]. The battery exhibited good performance: ~950 mAh g^{-1} at 1 C after 100 cycles. It was deduced that the improved characteristics were due to the shuttle effect reduction due to the high adsorption of LiPS, the effective reutilization of active materials, and the acceleration of the conversion reaction [97].

A Co_3W_3C/C composite was prepared by Zhao et al. [94] using a simple pyrolysis procedure, and it was applied on a commercial separator. The Co_3W_3C/C-modified separator was effective in reducing the shuttle effect. In addition, it worked as a catalytic membrane to promote the redox reaction of the sulfur species. For the LSB prepared with the Co_3W_3C/C-coated separator, 1345 mAh g^{-1} at 0.1 A g^{-1} was observed as initial discharge capacity. Rate performance was also good by achieving high capacity of 670 mAh g^{-1} even at 7 A g^{-1}. Cycle performance was also excellent since the decay rate was as low as 0.06% per cycle within 500 cycles at 1 A g^{-1}. These results indicate that this material utilizes Co–W metallic carbide advantage in capturing LiPS and promote the electrochemical reaction to gain high-performance LSB [98].

A nanocrystalline niobium carbide was synthesized using practical scalable autoclave technology and was used as the interlayer material for an LSB by Cai et al. [95]. The prepared nanocrystalline niobium carbide exhibited a strong ability to anchor LiPS species, which were highly effective in improving cycling performance and rate capabilities. The conductive niobium carbide interlayer not only acted as a shield to confine LiPS within the cathodic side, but also acted as the second current collector to reutilize the trapped active material and significantly enhance sulfur utilization. Because of these superior characteristics, the LSB assembled with this type of separator showed outstanding cycle stability and electrochemical performance^{-2} [99].

2.2.5. Separator Material: Nitride

Transition metal nitrides have been applied as a catalyst and as electroanalysis devices because of their high conductivity, reactivity, and stability [100]. For example, in contrast to their oxide (1.0×10^{-3} Sm^{-1}) and sulfide (9.7×10^{-2}–10^{3} Sm^{-1}) counterparts, molybdenum nitrides possess high electrical conductivity [101]. Chen et al. applied molybdenum nitride on a commercial separator and the capacity of the assembled LSB was 566 mAh g^{-1} after 500 cycles at 0.5 C [102].

Kim et al. prepared a boron nitride nanotube (BNNT)-based separator for the cathode side. Lithium stripping/plating analysis was conducted to elucidate the lithium metal's stability by observing the overpotential. When stripping/plating of lithium was conducted at 0.35 mA/cm^2, a similar overpotential profile was observed for LSBs composed of all separator types; however, when it was carried out at 1 mA/cm^2, the overpotentials of the commercial and BNNT separator were much more obvious than those of the p-BNNT (Figure 12a,b). This occurred because of dendrite formation and detachment caused by non-uniform lithium stripping/plating. In contrast, the higher ionic conductivity of the p-BNNT separator demonstrated the good stability of the lithium anode (Figure 12c–e). The p-BNNT also showed a uniform pore structure due to partial charge on the surface, which is a result of the electronegativity difference between B and N. Thus, in contrast to the commercial separator, the p-BNNT-applied separator suppresses dendrite growth on the Li metal anode, promotes ion transfer, and mitigates LiPS diffusion. Owing to these characteristics, the capacity of the p-BNNT-loaded PP separators-based LSB was 1429 mAh/g and exhibited stability until 200 cycles [103].

Chen et al. also prepared high-quality boron nitride nanosheets using a scalable exfoliation process and used them as a separator material [101]. They implemented the sonication-assisted liquid-phase exfoliation scalable process and investigated the effects of a spectrum of exfoliation factors (e.g., ultrasonic conditions and solvent and bulk material feeding) on the boron nitride nanosheet yield. A high yield of 72.5% could be achieved while maintaining the boron nitride nanosheets to a few layers and is defect free. Owing to the Lewis acid sites of the boron atoms, the boron nitride nanosheets could interact with the polysulfide anions in the liquid electrolyte and facilitate uniform lithium deposition, with the end result being an LSB with a long life [104].

Carbon nitride was used as an LSB separator material by Luo et al. [99]. Co-doped g-C_3N_4 nanosheets were synthesized via a calcination process. The capacity of the LSB prepared with the Co–g-C_3N_4-modified separator was 1121 mAh g^{-1} at 0.2 C with a high performance rate of up to 3 C. The Co–g-C_3N_4-modified separator-installed LSB also indicated an excellent stability of 640 mAh g^{-1} after 250 cycles. This amelioration was because of the restriction of LiPS diffusion owing to the high adsorption and high efficiency of LiPS catalytic conversion induced by the Co–g-C_3N_4 nanosheet-implemented separator [105].

2.2.6. Separator Material: Phosphide

Transition metal phosphides generally have good electrical performance, adequate chemical adsorption ability, and significantly high catalytic capability for LiPS, which makes them an interesting choice as a separator modifier for LSBs [106].

Chen et al. applied CoP nanospheres as a separator coating material for LSBs [101]. It was found that the conductive CoP could efficiently anchor LiPS because of its polar nature and partial surface oxidation state (as confirmed via the XPS analysis), which induce Co sites to chemically capture LiPS with Co–S bonding. Owing to these characteristics, this LSB exhibited high electrochemical characteristics and excellent rate performance (725 mAh g^{-1} at 5 C) [107].

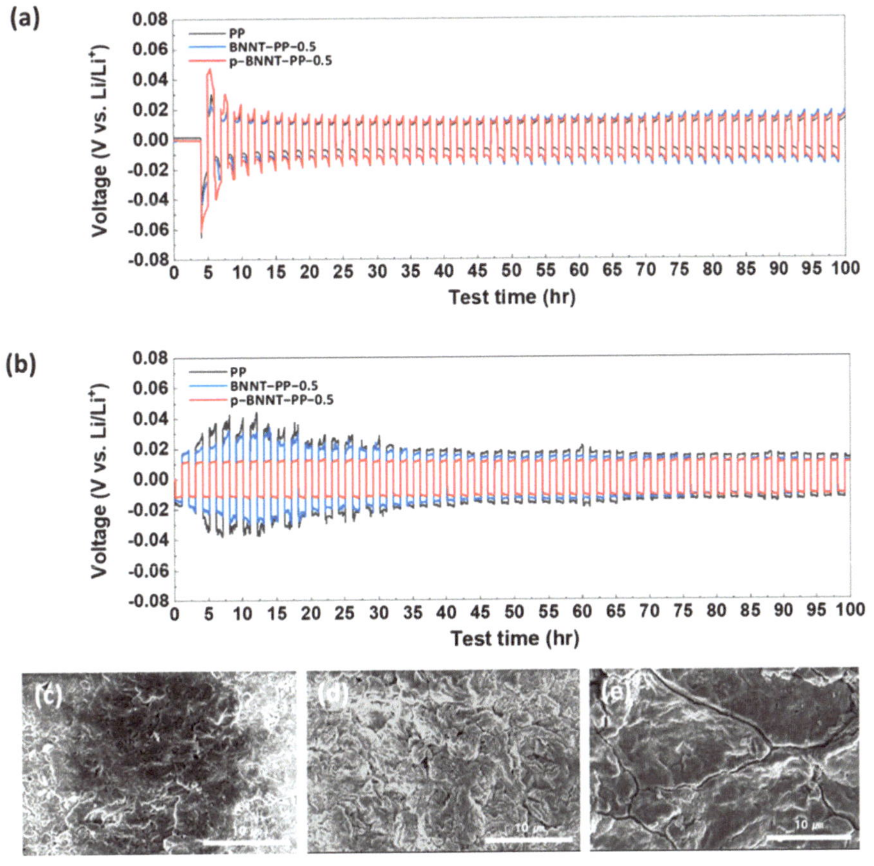

Figure 12. Lithium stripping/plating reaction with PP, BNNT, and p-BNNT-applied separator at (**a**) 0.35 mA/cm^2 and (**b**) 1 mA/cm^2. Top view of SEM observation of lithium metal after lithium stripping/plating reaction with (**c**) PPY (**d**) BNNT-PP-0.5, and (**e**) p-BNNT-PP-0.5 [103].

Lin et al. also investigated the possibility of CoP as a separator coating material [106]. They modified a PP membrane with nano-cubic CoP/C. CoP/C-deposited separator captured LiPS with high efficiency via its strong chemical affinity. The LSB arranged with the CoP/C-deposited separator presented a minimum capacity decay as low as 0.08% per cycle over 500 cycles at 1 C with a 938 mAh g^{-1} capacity as the initial cycle. Rate performance was 594 mAh g^{-1} at 4 C [108].

Another transitional metal phosphide was also studied by Zhao et al., who prepared FeP/spongy carbon composites with multiple adsorptions and catalytic sites as a modified material for an LSB separator [109]. The spongy carbon was shown to possess suitable structural stability and long ion/electron transmission channels. The addition of the FeP-endowed spongy carbon reacted with the LiPS to block shuttling and catalyze the conversion of sulfur. It was found that the FeP/spongy carbon-modified separator could reduce the flammability of the completed LSB. Benefiting from these features, the prepared LSB exhibited high cycling stability and the LSB could retain a 618 mAh g^{-1} capacity even after 150 cycles [109].

A heterostructured Ni/Ni_2P-embedded mesoporous carbon nanosphere composite (Ni/Ni_2P–MCN) was developed to accelerate polysulfide catalytic conversion in LSBs.

Figure 13 indicates cyclic voltammetry curves of an LSB prepared with four types of separators. Two cathodic peaks can be ascribed to the sulfur transition of soluble polysulfides to solid-state Li_2S_2/Li_2S, and they are stronger for the LSB prepared with Ni/Ni_2P–MCN. Thus, it can be suggested that Ni/Ni_2P–MCN can enhance the efficiency of polysulfide conversion kinetics [110]. One can also observe the anodic peak approximately at 2.4 V. The one with Ni/Ni_2P–MCN/PP indicated an anodic peak with a lower potential than the remaining three types, proving an enhanced Li_2S_2/Li_2S oxidation reaction [111,112].

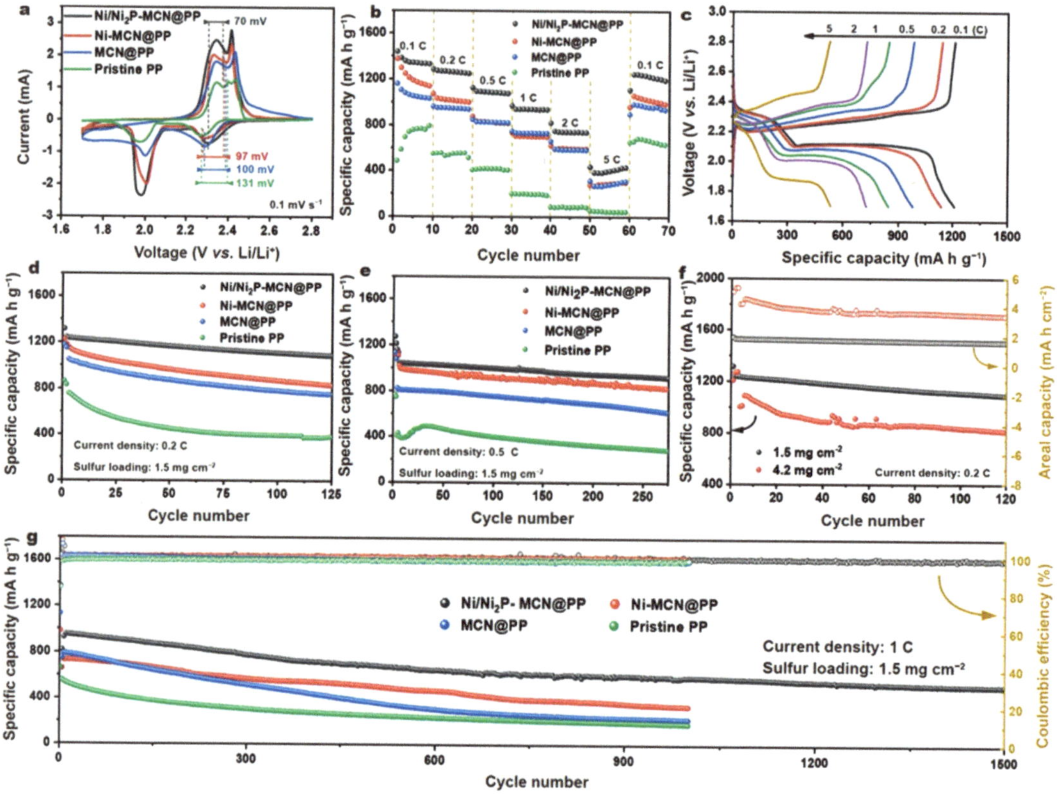

Figure 13. (**a**) Cyclic voltammetry. (**b**) Rate performances of LSB prepared with Ni/Ni_2P-MCN/PP, Ni-MCN/PP, MCN/PP, and commercial PP. (**c**) Charge–discharge profiles of LSB prepared with Ni/Ni_2P-MCN/PP at various current densities from 0.1 to 5 C. (**d**,**e**) LSB cyclic performance prepared with Ni/Ni_2P-MCN/PP, MCN/PP, and pristine PP at 0.2 and 0.5 C. (**f**) LSB cycling performance prepared with Ni/Ni_2P-MCN/PP at 0.2 C. (**g**) LSB cycling performance prepared with Ni/Ni_2P-MCN/PP, MCN/PP, and commercial PP at 1 C [110].

The Ni/Ni_2P–MCN/PP LSB also presented the strongest capacity among all LSBs (Figure 13b). Figure 13c presents the charge–discharge curves and the Ni/Ni_2P–MCN/PP LSB showed a weak degree of polarization, a typical discharge plateau, indicating its relatively high sulfur utilization. As shown in Figure 13d, the Ni/Ni_2P–MCN/PP LSB delivered an initial discharge capacity of 1315 mAh g^{-1} at 0.2 C and retained a capacity of 1086 mAh g^{-1} after 125 cycles. When the applied current was increased to 0.5 C (Figure 13e), the discharge capacity was still 916 mAh g^{-1} and a high capacity retention was maintained to keep approximately 87.9% even after 275 cycles.

The Ni/Ni_2P–MCN/PP LSB indicated an enhanced discharge capacity of 815 mAh g^{-1} even after 120 cycles (Figure 13f). An extended cycling performance test was conducted and the Ni/Ni_2P–MCN/PP LSB presented a discharge capacity of approximately 953 mAh g^{-1} at 1 C (Figure 13g). It had an excellent discharge capacity of approximately 518 mAh g^{-1} even after 1500 cycles. Thus, we can say that the Ni/Ni_2P-mesoporous carbon nanosphere composite-deposited separator not only suppressed the LiPS diffusion by their chemical capturing sites, but also possesses strong catalytic abilities for the polysulfide reactions.

2.2.7. Separator Material: Metal Organic Framework-Based Materials

Metal organic frameworks (MOFs) are the class of porous materials composed of metal ions and organic ligands. One of MOF application can be molecular sieves and selective gas separation on a molecular scale [113–117]. MOF and MOF-based composite materials have been investigated as porous host materials for LSB cathodes to retain sulfur due to their large specific surface area and catalytic effect for LiPS. A porous material with optimally regulated pore sizes can work as a sieve to separate desirable ions from an ionic solution such as LiPS, thus resulting in reducing their shuttle effect. In this regard, MOF and MOF-based materials are good candidate materials for LSB separators owing to their tunable pore size and specific surface area.

Songyan et al. prepared an MOF-based LSB separator to investigate its effectiveness in mitigating the LiPS shuttling issue [118]. The separator behaved as an ionic sieve for LSB and successfully sieved Li^+ ions while mitigating LiPS migration to lithium metal anode side.

One example is in applying MOF–graphene oxide composite as a separator material. The preparation procedure is exhibited in Figure 14a. The MOF layer was grown initially. A desirable amount of graphene oxide (GO) solution, which was obtained by filtration, was applied onto uniformly dispersed crystalline MOF particles. It was confirmed that the MOF and GO layers were strongly adhered onto the membrane. This procedure was repeated several times to assure the quality of the preparation process. The MOF nanoparticles filled the void space tightly near the grain boundaries. A self-standing separator can be obtained by peeling off this type of material from the filter. It was found that the structural scaffold of the MOF remained intact even after over 200 electrochemical cycles. (Figure 14c). The morphological observation of the MOF/GO separator is presented in the SEM image (Figure 14d,e). When this type of MOF-based separator was implemented as a separator, the resulting LSB exhibited a low decay rate, as low as 0.019% per cycle in a long duration of up to 1500 cycles. Almost no capacity fade was observed at the initial 100 cycles [118].

Dang et al. prepared a composite membrane separator composed of cerium-based UiO-67 (MOF) and a glass fiber. The prepared Ce-UiO-67 efficiently adsorbed LiPS and catalyzed its conversion, thus suppressing the associated shuttle effects. In addition, this type of MOF theoretically provides rich lithiophilic functional groups to attain rapid Li^+ ion transportation, leading to achieving stable Li plating and stripping. The resulting LSB delivered 919 mAh g^{-1} at initial cycle and a small decay rate of 0.04% per cycle for 500 cycles at 1 C [119].

Su et al. also prepared a cerium-based MOF separator and used Ce-UiO-66-NH_2 as the MOF material. Strong interaction between LiPS and the metal sites in UiO-66-NH_2, as well as being the physical barrier for LiPS, enabled this type of separator to inhibit the shuttle effect. The assembled LSB offered a high capacity of 1366 mAh g^{-1} initially with a stable cycling property at 0.2 C. In addition, capacity decay was only 0.09% per cycle at 300 cycles at 1 C [120].

A conductive carbon (Ketjenblack) and zeolite imidazole framework-8 (ZIF-8)/polypro pylene composite separator was prepared by Ma et al. [115]. ZIF-8 and Ketjenblack were mixed in a 1:4 mass ratio in ethanol for 24 h, and then ultrasonically dispersed for 20 min. After that, the homogeneous solution was coated onto a commercial separator via vacuum filtration and was followed by vacuum drying at 60 °C for 24 h. The resulting separator comprised Ketjenblack, ZIF-8, and PP with a low coating load of 0.06 mg cm^{-2}.

Ketjenblack/ZIF-8/PP can efficiently absorb LiPS due to a Lewis acid–base interaction between ZIF-8 and LiPS. This interaction could decrease the dissolution of LiPS as well as the shuttle effect, thereby enhancing the electrochemical properties of the assembled LSB. When 0.1 C of current density was applied, the assembled LSB exhibited low polarization, a capacity of 1235.6 mAh/g initially, and good capacity retention rate of 59.27% after 100 cycles [121].

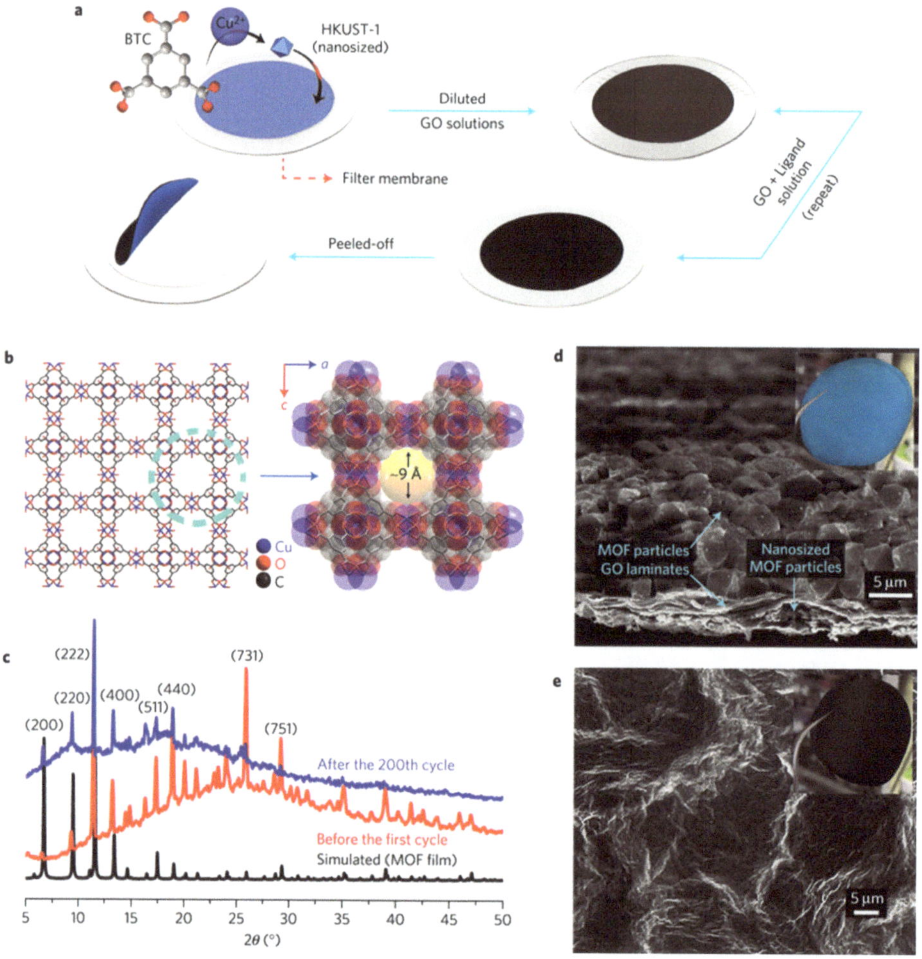

Figure 14. Preparation and material characteristics of MOF/GO separators [118]. (**a**) Illustration of preparation method to make MOF/GO separators. (**b**) Schematic image of HKUST-1. (**c**) PXRD patterns of MOF/GO separators. MOF structure remains unbroken after the discharge/charge process. (**d**) SEM photographs of the MOF/GO separator. The inset presents a digital image from the MOF side aspect. (**e**) SEM photograph of GO layer. The inset is an image from the GO side aspect.

2.2.8. Separator Material: Quantum Dot

Quantum dots (QDs) are generally super small crystalline particles with a size range of 1.0–10 nm. Recently, QDs have gained much attention for being interesting materials for electrochemical energy storage due to their large specific surface area, tunable size, short ion/electron transportation path, adjustable photoluminescence, and feasible surface functionalization [122,123]. In addition, it was reported that QDs can modify separators to suppress the shuttle effect, owing to their effective interaction with LiPS [124].

Atomic-layer deposition method was applied to prepare a TiO_2 quantum dot-modified multiwalled carbon nanotube as the deposition material for LSB separators and succeeded in preventing the LiPS shuttling effect and improving the coulomb efficiency and cycle stability (Figure 15a) [121]. It was suggested that these positive effects were achieved due to an interaction between TiO_2 quantum dots and LiPS that could adsorb soluble polysulfide compounds, leading to suppressing the shuttle effect. The interlayer also had abundant spacing and excellent conductivity because of the multiwalled carbon nanotube. The assembled LSB showed 1083 mAh g^{-1} as the initial capacity and kept a cycle capacity of 610 mAh g^{-1} after 600 cycles at a rate of 838 mAh g^{-1}. Capacity decay rate was only 0.072% per cycle (Figure 15b) [125].

A superlight PP-coated film with multiwalled carbon nanotubes/nitrogen-doped carbon quantum dots (MWCNTs/NCQDs) was synthesized by Pang et al. (Figure 15c) [126]. The weight of the MWCNT/NCQD coating per area was as low as 0.15 mg cm^{-2} (Figure 15d).

It was found to have superior capacity retention and self-discharge suppression compared to the LSB made by Chung et al. with an MWCNT-modified separator [127]. The synergistic influence of the MWCNTs and NCQDs was ultimately positive, resulting in 1331 mAh g^{-1} as an initial capacity and presenting a stable cycling performance. The capacity decay was as low as 0.05% per cycle at 0.5 C, over 1000 cycles (Figure 15e).

Yu et al. developed a Mo_2C quantum dot (MQD)-anchored nitrogen-doped graphene-deposited separator (MQD/NG) [128]. Figure 15f shows the optical and TEM pictures of the separator with MQD/NG/PP surface after 200 cycles. Polar Mo_2C QDs offer a uniform lithium deposition and good chemical adsorption of LiPS. The LSB operated more than 1600 h with dendrite-free lithium deposition at a current density of 10 mA cm^{-2}. The capacity of 1230 mAh g^{-1} was observed under stable cycle performance after 100 cycles at 0.2 C without obvious capacity decay (Figure 15g).

Liu et al. developed a zinc sulfide quantum dot/reduced graphene aerogel-modified separator [126]. The ZnS quantum dots were effective as LiPS-anchoring and catalytic sites, which could promote the redox reaction of sulfur and mitigate the shuttle effect. The 3D porous reduced graphene aerogel assisted in physically blocking the migration of LiPS. As a consequence, 1211 mAh g^{-1} at initial capacity was observed at 0.1 C and presented a stable cycling performance over 500 cycles at 1 C [129].

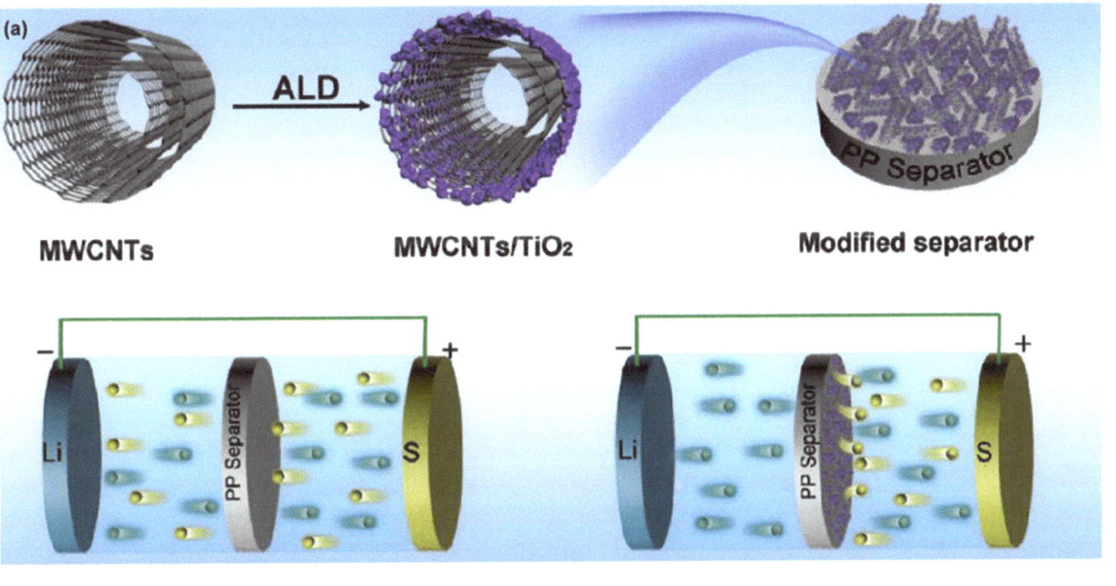

Figure 15. *Cont.*

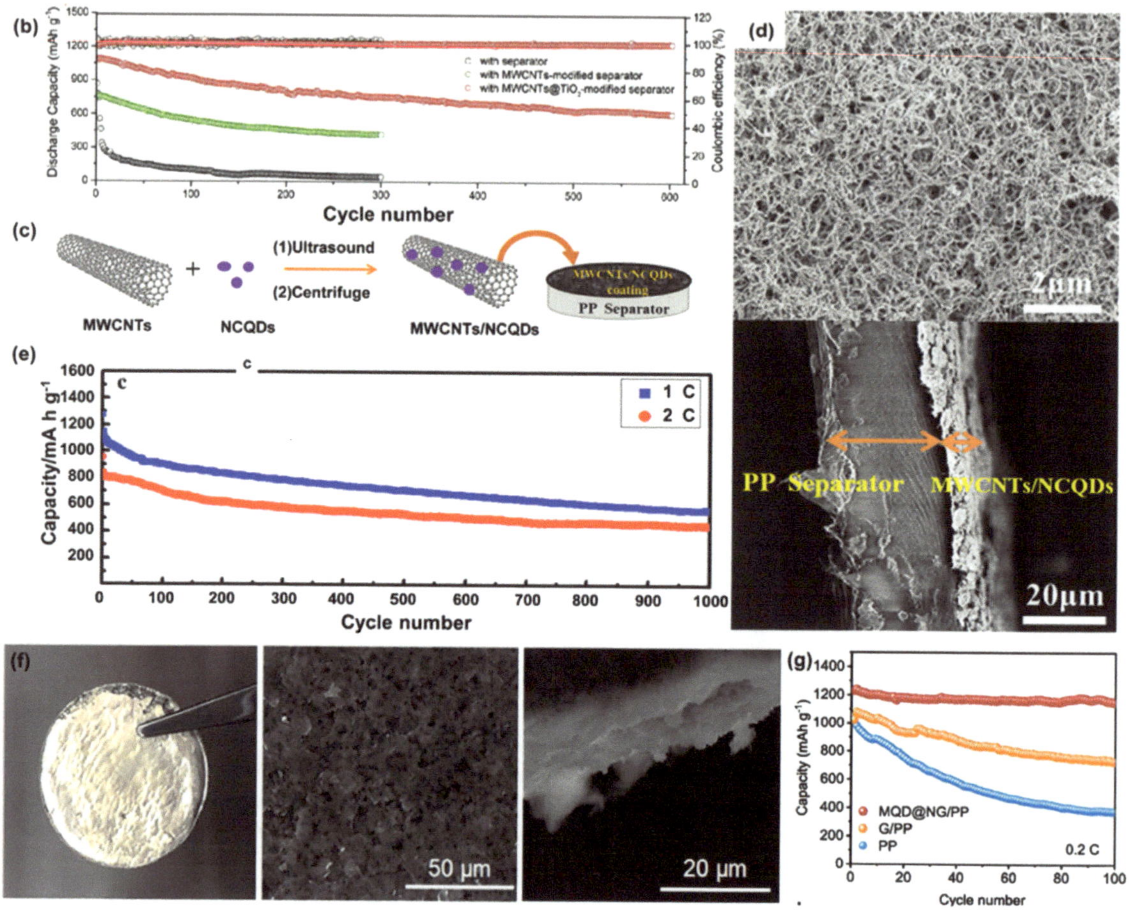

Figure 15. (**a**) Speculated image of the synthesis procedure of MWCNTs/TiO_2 quantum dots and LSB with conventional PP and applied MWCNTs/TiO_2 quantum dots. (**b**) Cycle tests of LSB prepared with PP, MWCNTs/PP, and MWCNTs/TiO_2 quantum dots/PP. (**c**) Speculated image of MWCNTs/NCQDs composite and HRTEM picture of NCQDs anchored on MWCNTs surface. (**d**) SEM image of MWCNTs/NCQDs -applied separator and cross-sectional view of MWCNTs/NCQDs-applied separator. (**e**) Cycle test of the LSB assembled with MWCNTs/NCQDs-applied separator. (**f**) Digital and SEM image of lithium metal after plating/stripping experiments carried out with MQD/NG/PP separators for 200 cycles. (**g**) Cycle performance of LSB prepared with PP, G/PP, and MQD/NG/PP [125–128].

Zhang et al. applied MoP quantum dot and nitrogen, phosphorous co-doped polypyrrole (PPy) composite material to a commercial separator for LSBs [130]. The MoP quantum dots exhibited chemisorption and catalytic conversion performance for LiPS capture and conversion. In addition, N,P co-doped PPy substrates offered flexible pathways for Li^+/electron transportation and also behaved as a physical barrier to mitigate the shuttling effect. The capacity of 739 mAh g^{-1} at 3 C was observed with a stable cycle performance (600 mAh g^{-1} at 1 C after 600 cycles, 0.052% decay per cycle). Density functional theory calculations elucidated that the MoP quantum dots possess enough adsorption energy for S_8 and Li_2S_n, which could lower the nucleation energy barrier of Li_2S that might be helpful to accelerate Li_2S reaction kinetics.

Cobalt aluminum-layered double-hydroxide quantum dots (LDH-QDs) was combined with nitrogen doped graphene (NG) as a coated material for LSB separator [125]. Experimental results and analysis, including XPS, clarified that the porous LDH-QDs/NG owned rich active hydroxyl groups and Co^{2+} sites that were effective in catching LiPS through strong chemical interactions and promote the conversion reaction kinetics. The average thickness of the prepared LDH-QDs/NG-coated separator was approximately 17 μm and had a high ionic conductivity of 2.67 mS cm^{-1} [131].

2.2.9. Separator Material: Mxenes

"MXenes" are materials known as two-dimensional transition metal carbides, nitrides, and carbonitrides and have been gaining significant attention owing to their high conductivity, rich functionality on Mxene surface, and exclusive two-dimensional morphologies, especially for energy storage devices [132,133].

Gogotsi et al. prepared Ti_3C_2 by removing Al atoms selectively from layered hexagonal ternary carbide in 2011, Ti_3AlC_2, with hydrofluoric acid (HF) treatment in water [134]. This study focused on Ti_3C_2 as a typical MXene material and prompted more researchers to investigate its physical and electrochemical properties [135,136]. In general, MXenes (such as Ti_3C_2Tx and V_2CTx) are two-dimensional layered materials, which can be obtained from transition metal carbides, nitrides, and carbonitrides (denoted as the MAX/$M_{n+1}AXn$ phase) [137]. These interesting materials can be acquired by selectively etching SP element layers from their corresponding three-dimensional MAX phases. MAX phases indicate layered ternary metal carbides, nitrides, or carbonitrides, with a formula of Mn + 1AXn (n = 1, 2, 3), in which M, A, and X express early d-block transition metals, main group SP elements, and either or both C and N atoms, respectively. Thus far, more than 70 MAX phases have been reported [138]. Typical MXene are Ti_3C_2 [139], Ti_2C [140], $(Ti_{0.5}, Nb_{0.5})_2C$, $(V_{0.5}, Cr_{0.5})_3C_2$, Ti_3CN [141], Ta_4C_3, Nb_2C, V_2C, and Nb_4C_3, etc. Additional interesting properties of Mxene is that the surfaces generally terminate with F, OH, and/or O groups. Therefore, these terminated MXene species will be regarded as $M_n + 1X_nT_x$, in which T indicates the surface groups (F, OH, and/or O) and "x" is the number of terminations [142]. These surface groups can also assist in capturing LiPS diffusion.

Glass fiber and Ti_3C_2 composite separator were prepared by Lin et al. via a vacuum filtration process [143]. As displayed in Figure 16a, a few layers of a conductive Ti_3C_2 nanosheet with an average thickness of 1–4 nm were deposited on a glass fiber separator. This type of separator can offer strong LiPS adsorbing sites due to their high porosity, and LiPSs shuttle effect was successfully reduced when compared to a glass fiber separator without Mxene. The LSB assembled with Mxene/glass fiber composite separator offered a capacity of 820 mAh g^{-1} at initial cycle and it was 721 mAh g^{-1} after 100 cycles at 0.5 A g^{-1}, that is approximately 15 times stronger than LSBs with commercial separators (Figure 16b). High performance was achieved due to its conductive property and the strong interaction of Ti_3C_2 toward LiPS.

TiO_2–MXene heterostructures were prepared by Jiao et al. via the partial oxidation of $Ti_3C_2T_x$ nanosheets [144]. It was shown that this crafted structure had a large surface area, strong ability to catch LiPS, and high conductivity and electrocatalytic activity. Figure 16c depicts how TiO_2 uniformly distributed over MXene sheets can offer numbers of effective adsorbing sites to catch LiPS, while its heterostructure interface assures the rapid diffusion of LiPS to MXene, leading to its high catalytic activity for rapid LiPS conversion. The discharge capacity of LSB assembled with this TiO_2–MXene-deposited separator was 662 mAh g^{-1} after 200 cycles at 0.5 C, which indicates 93% of capacity retention, proving its high sulfur utilization efficiency (Figure 16d).

MXene sulfation has also been studied. Yao et al. developed a flexible, conductive MXene ($Ti_3C_2T_x$) sandwich-structured layer, and the heterostructure surface of TiS_2/TiO_2 was created via vulcanization [145]. The TiO_2 nanoparticles worked as an adsorbent to catch LiPSs and the TiS_2 functioned as a catalyst to accelerate the long-chain LiPSs conversion to short-chain Li_2S_2/Li_2S, with the help of the high conductivity of $Ti_3C_2T_x$. It was found that the prepared layer protected the lithium anode by suppressing LiPS depositing on its surface. As a result, the long-term stability of a high electrochemical performance was achieved (capacity decay rate was 0.048% per cycle up to 500 cycles at 1 C).

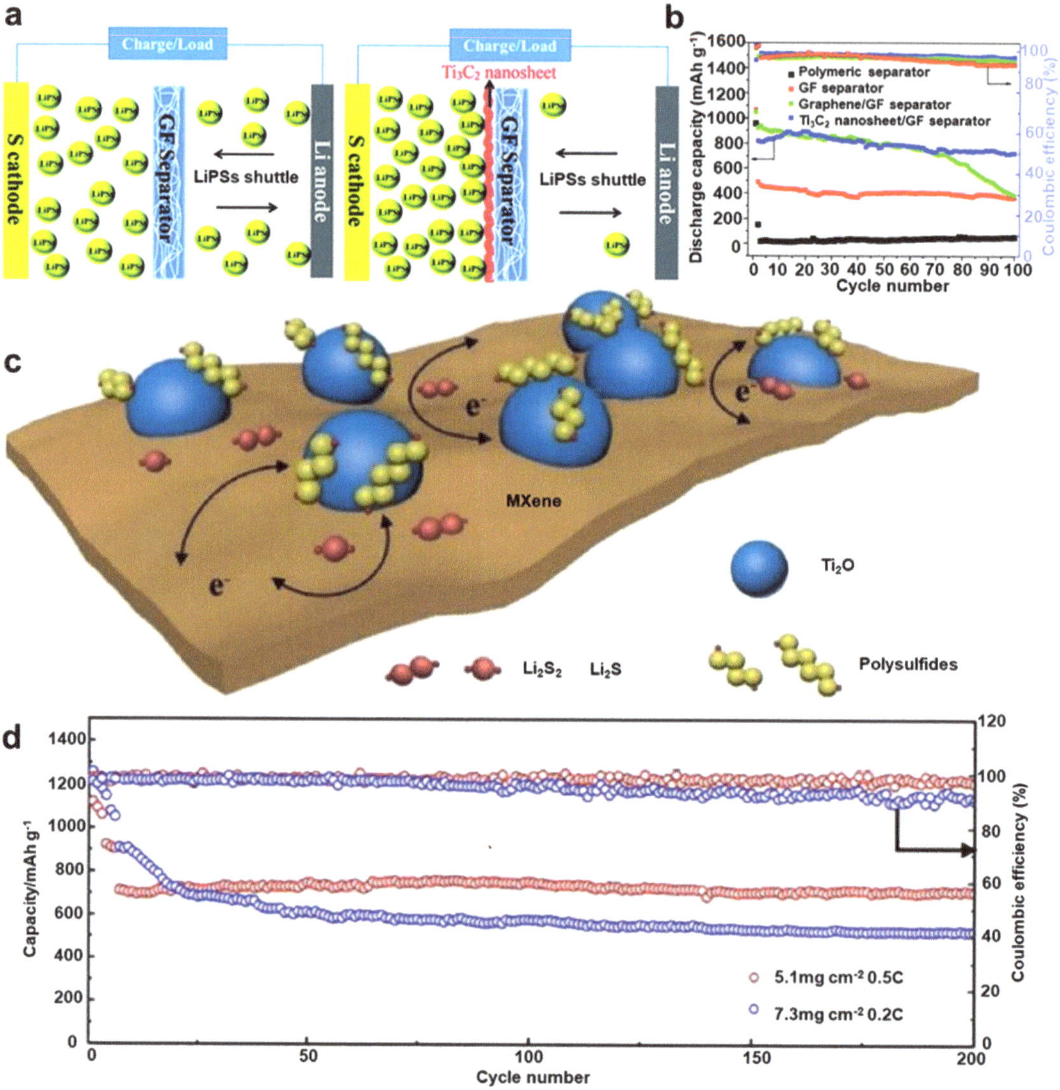

Figure 16. (**a**) Speculated image of the cell with a GF or GF/Ti_3C_2 separator. (**b**) Cycling performance and coulombic efficiency at 0.5 A g^{-1}. (**c**) Speculated image of the LiPS capture and process on TiO_2–$Ti_3C_2T_x$ structures. (**d**) Cycling performance of LSBs with a $Ti_3C_2T_x$–GN with different amounts of sulfur loadings [143,144].

Mxene has been combined with other functional materials to create even more desirable separators. Covalent organic frameworks (COFs) obtained with guanidinium salts have been gaining increasing attention as they possess strong covalent bonds and rich pore channels. It was suggested that guanidinium salts could capture LiPS, owing to their electrostatic interaction nature. Li et al. prepared ionic covalent organic nanosheets with guanidinium salts to prevent Ti_3C_2 restacking and capturing LiPS. This material was uniformly applied to a polymer separator with a simple vacuum filtration process. The charge transfer resistance of LSB assembled with this type of separator was low so as to demonstrate good performance [146].

Combining Mxene with functional polymers can be a useful way to facilitate Li^+ transport. Wang et al. prepared a laminar Nafion–MXene composite [147]. Interestingly, Nafion worked as a surfactant, or dispersant, which led to obtaining an ordered-structured MXene. When this composite was compared to each Nafion and MXene material, the shuttle effect was mitigated. LSB made with this functional separator exhibited good stability, and the capacity decay was as low as 0.03% per cycle over 1000 cycles. It was also found that the MXene promoted the reutilization of polysulfides, owing to its high electrical conduction, and an effective Li^+ transport was achieved by Nafion.

2.2.10. Other Functional Separator Materials

There are some additional functional materials that are being researched for use as LSB separators that either demonstrate the ability to capture LiPS to suppress the shuttle effect or provide catalytic sites to promote the redox reaction of sulfur-related species.

He et al. developed vanadyl phosphate ($VOPO_4$) sheets for such a purpose [148]. Specifically, two-dimensional $VOPO_4$ sheets with rich active sites were designed to adsorb LiPS through V–S bond formation. Due to the intrinsic electrical repulsion between the polysulfide anions, advanced time/space-resolved operando Raman analysis revealed that polysulfides' rich surfaces could further evolve to a "polysulfide-phobic" interface. By implementing these "polysulfide-phobic" sites in the separator, the LSB would have superior long-term cycling stability.

Pure Nafion, i.e., without Mxene, was coated onto a commercial separator by Huang et al. [149] in order to create ion selectivity. SO_3^- groups in sulfonate-ended perfluoroalkyl ether groups in separators allowed for Li^+ hopping as a positively charged species, but did not allow for the hopping of negative ions, such as polysulfide anions (S_n^{2-}), due to coulombic interactions. This cation-selective membrane worked as an effective electrostatic shield for polysulfide anions and succeeded in confining the LiPS to the cathode side.

Some nature-derived materials have also been studied. A lignin nanoparticle-coated commercial separator was prepared by Zhang et al. [150]. The lignin-coated separator had abundant electron-donating groups and was expected to result in the chemical binding of LiPS to reduce the shuttle effect. With an LSB, a cathode composed of sulfur and commercially available acetylene black (approximately 73.8 wt% sulfur content) was assembled and showed improved cycling stability compared to a commercial separator for over 500 cycles at a current density of 1 C.

Cellulose is another abundant and environmentally friendly organic resource and can be obtained from plant biomass. Nano-fibrillated cellulose (NFC) has been used to prepare porous natural separator membranes with a process learned from the paper industry. A separator composed of NFC was investigated regarding its porosity, thickness, wettability, and electrochemical stability, and its electrochemical performance as a functional separator for an LSB was evaluated. Interestingly, the mass of the commercial separator and the thinnest separator prepared in their study were very similar, although the cellulose-based separator was thicker. Top-down views of the NFC_20PO (the separator prepared by adding paraffin oil and polyethylene glycol hexadecyl ether as a surfactant) and NFC_20 (the separator prepared without any additive in water dispersion) are shown in Figure 16. The NFC_20PO (Figure 17a) separator has an open-pore structure with interconnected pores of diameters between 100 nm and 1 μm. The addition of PO and surfactant enabled the formation of pores; however, some parts remained partially closed, which implies that the preparation procedure is still not fully optimized. In contrast, the separator prepared without additives showed a much more closed, densely packed surface without any visible macro-pore structure (Figure 17b). It was suggested that the NFC fibers tend to compactly collapse from capillary action during the evaporation of water and are fixed with strong hydrogen bonds of cellulose chains, thus yielding dense-structured membranes [151]. It was also demonstrated that the electrochemical performance of the NFC separator was superior to the conventional polyolefin separator [152].

Figure 17. Morphology of the surface of separators prepared from water dispersion [152] (**a**) containing NFC (nano-fibrillated cellulose) fiber, paraffin oil, and surfactant; and (**b**) containing only NFC fibers.

It should be mentioned here that various types of materials were applied as LSB separators, as described so far, and are classified based on whether they are on the anode or cathode side of the separator. However, despite the explanation based on which side of the separator they are applied, some of them are applied on both sides of the separator. We explain this by emphasizing that the applied material acts as a functional material on a particular side. One such example, which clearly states that the MnO_2 functional layers are on both the anode and cathode sides, is the study of Tian et al., who demonstrated that MnO_2 was able to suppress LiPS migration with its chemical adsorption ability, which resulted in improving LSB electrochemical performance [153].

In addition, all of the materials applied to LSB separators in this study are summarized in Tables 1 and 2, depending upon whether they are on the anode or cathode side. One of the major roles for these separator materials are to capture LiPS in order to suppress the shuttle effect. Furthermore, the applied materials with both lithium polysulfide adsorption and catalytic activities are presented in Table 3 as multifunctional materials.

Table 1. Comparative performances of LSB with various types of materials on separator anode side.

Materials on Separator	Current Density (C, 1675 mAh g^{-1})	Cycle Number	Initial Discharge Capacity (mAh g^{-1})	Capacity Retention (%) (Fadeing Rate)	Reference
		Metal			
Mg nanoparticles		400	135–140	80	[39]
Silver	0.5	300	131	92	[40]
		Ceramic			
Boron Nitride	4	250	1018	69	[43]
SiO_2 Nanotubes	0.5	100	1266		[44]
Hollow Porous SiO_2 Nanocubes	100 mAh g^{-1}	30	919 (30th)		[45]
SiO_2	0.2	200	956.3	36	[47]
Al2O3	0.2	100	1067.7	75	[48]
		Other Functional Materials			
polypyrrole	0.5	250		(0.083% per cycle)	[52]
porphyrin-derived graphene	0.5	300		(0.099% per cycle)	[56]
lithium fluoride	0.2	200		69.30%	[57]

Table 2. Comparative performances of LSB with various types of materials on separator cathode side.

Materials on Separator	Current Density (C, 1675 mAh g^{-1})	Cycle Number	Initial Discharge Capacity (mAh g^{-1})	Capacity Retention (%) (Fadeing Rate)	Reference
		Carbonaceous Materials			
Ketjen black	0.1	100	1318		[60]
carbon	0.1	200	1112	63.8	[62]
carbon nanotube	0.5	400	1056	(0.11% per cycle)	[63]
Graphene Composite	2	1000	707	74.0%	[69]
nitrogen-doped reduced GO/Co-Ni-S composite	0.1	350	1524		[74]
nitrogen and sulfur doped carbon	2	500	841 (0.1C)	(0.089% per cycle)	[76]
Fe, nitrogen-doped carbon nanofibers and 2D graphene	0.5	500	847.9	(0.053% per cycle)	[77]
		Metal Oxide			
Al2O3	0.2	50	967	70.0%	[78]
TiO2/carbon composite	0.1	150	926	75.0%	[79]
TiO2 modified carbon nanotubes	0.5	900	1103.9	(0.066% per cycle)	[80]
SiO2	0.2	200	956.3	64.0%	[81]
CoSe2/grapheme oxide	6	500	916	50.1%	[84]
Lanthanum oxide	1	200	966	74.5%	[85]
		Metal Sulfide			
reduced Graphene Oxide@MoS2	0.5	500	1122	0.116% per cycle	[86]
MoS2/C hollow microsphere	1	1000	935	(0.053% per cycle)	[88]
ZnS nanosheet/graphene	0.1	100	1165.9	58.8%	[89]
WS2 Prussian Blue-Polypyrrole		300	1050	62.0%	[91]
		Metal Carbide			
TiC-TiO2	1	500	1218	58.6%	[93]
WC/reduced Graphene Oxide composite	1	300		83.0%	[97]
Co3W3Carbide@C	1 A g−1	500		(0.06% per cycle)	[98]
NbC	5	1500		(0.037% per cycle)	[99]
		Metal Nitride			
MoNx	0.1	500	1298	(0.063% per cycle)	[102]
BN Nanotube	0.3	200	1429		[103]
Co-doped g-C3N4	0.2	100	1121	95.0%	[105]
		Phosphide			
CoP nanosphere	1	500		(0.078% per cycle)	[107]
CoP/C	1	500	938	(0.08% per cycle)	[108]
FeP/C	1	400	526		[109]
Ni/Ni2P-Carbon	5	1500	431	(0.031% per cycle)	[110]
		Metal Organic Framework			
Ce UiO-67	1	500	919	0.04%/cycle	[119]
Ce-UiO-66-NH2	0.2	300	1366.3	0.09%/cycle	[120]
ZIF-8	0.1	199	1235.6		[121]
		Quantum Dot			
TiO2 Quantum Dot		600	1083	0.072%/cycle	[125]
MWCNT/Nitrogen Doped Carbon Quantum Dot	0.5	1000	1330.8	0.05%/cycle	[126]

Table 2. *Cont.*

Materials on Separator	Current Density (C, 1675 mAh g^{-1})	Cycle Number	Initial Discharge Capacity (mAh g^{-1})	Capacity Retention (%) (Fadeing Rate)	Reference
Mo2C quantum dot	0.2	100	1230		[128]
ZnS Quantum Dot	0.1	500	1211		[129]
MoP Quantum Dot	1	600	500	0.052%/cycle	[130]
		Mxene			
Ti3C2		100	820	0.879%/cycle	[143]
TiO2-Ti3C2Tx	0.5	200	662	0.035 %/cycle	[144]
TiS2/TiO2-Ti3C2Tx	1	500		0.048 %/cycle	[145]
		Other Functional Materials			
Lignin	1	500	487		[150]
Nano-fibrillated cellulose	1		580		[152]

Table 3. Comparative performances of LSB with various types of materials with both lithium polysulfide adsorption and catalytic activities.

Materials on Separator	Current Density (C, 1675 mAh g^{-1})	Cycle Number	Initial Discharge Capacity (mAh g^{-1})	Capacity Retention (%) (Fadeing Rate)	Reference
Lanthanum oxide	1	200	966	74.5%	[85]
Co_3W_3Carbide@C	1	500		(0.06% per cycle)	[92]
Co-doped g-C_3N_4	0.2	100	1121	95.0%	[99]
FeP/C	1	400	526		[109]
Ni/Ni_2P-Carbon	5	1500	431	(0.031% per cycle)	[110]
MOF and MOF Composite					[113]
ZnS Quantum Dot	0.1	500	1211		[126]
MoP Quantum Dot	1	600	500	0.052%/cycle	[130]
TiO_2-Mxene					[144]

Each material has advantages and disadvantages. In addition, for future study, combining these materials and creating composite materials could also be another strategy to gain even better performance.

3. Necessary Properties for LSB Separators

We have explained the actual materials for which attempts have been made to apply them to LSB separators so far. Now, we would like to further discuss the necessary properties for LSB separators.

In general, electronic resistance and ion conductivity need to be high for LSB separators. Ionic conductivity is related to separator structure, including porosity and tortuosity. Porosity is necessary to save enough electrolytes in order to keep ionic transference in an enhanced level. Tortuosity is an additional important factor. Tortuosity is the amount to explain the morphology. As tortuosity becomes smaller, conductivity will be higher. Furthermore, wettability is also a key factor to determine ionic conductivity. Obviously, a higher electrolyte permeation could serve as an easier way for lithium ion transportation. It should also be noted that the pore size and tortuosity should be uniform in order to obtain stable current density [154].

3.1. Separator Porosity, Pore Size, and Thickness

In general, a thinner separator should possess reduced ionic resistance, which would result in gaining a higher energy density. So far, separator thickness ranges within 20–50 μm, whereas that of commercial separator thickness is 20–25 μm [155]. A thinner separator tends to be more fragile, which would be the reason for short circuits and explosions. The uniformity and equality of thickness can be critical because an uneven separator can be the reason for dendrite formation, which would cause disaster during battery operation.

In addition, separator porosity can be expressed by the following equation:

$$\text{Porosity (\%)} = (1 - \rho m/\rho p) \times 100 \tag{2}$$

where ρm is the apparent density and ρ*p* is the density of the separator material [156].

In general, the porosity is calculated by the weight difference before and after separator immersion in the liquid, as shown in the equation:

$$\text{Porosity (\%)} = W - w0/\rho LV0 \tag{3}$$

where W is the intact separator weight and w0 is the weight of the separator after liquid immersion; ρL and V0 denote the liquid density and the separator volume, respectively.

Porosity needs to be optimized so that electrolytes are filled in the pores to assure ion conductivity. For example, the separator itself would be weak if porosity is too high. In contrast, if the porosity is too small, ion conductivity would not be high enough to ensure good electrochemical performance. The optimized porosity of separators for a commercial lithium ion battery is known to be approximately 40%, and that of lithium sulfur battery is still controversial, since there are not as many commercialized lithium sulfur batteries in number [157].

3.2. Tortuosity and Permeability

Tortuosity for a separator can be a factor indicating the influence of the separator shape upon the ionic conductivity [158], and can be expressed by the following equation:

$$\tau = \text{sqrt}\ (\varepsilon \times Rs/R0) \tag{4}$$

where ε is the porosity, Rs and R0 are the separator resistance before and after being soaked in liquid electrolyte, respectively.

Geometric effective transport coefficient can be a measure of the effective ionic transport of separators' morphology, and can be calculated as $\delta = \varepsilon/\tau$.

The permeability is similar to tortuosity. Geometric structure can influence the ionic conductivity by pressure difference and the permeability can be applied in this respect.

Darcy's law can be applied to define the rate of the fluid through the porous surface, which can be expressed by the following equation:

$$u = -\kappa/\eta\ \nabla P \tag{5}$$

where η is the viscosity, ∇P is the applied pressure gradient, and u is the average velocity of the fluid when penetrating the porous surface [159].

3.3. Wettability

Wettability is also one of the critical factors because it influences the electrolyte absorption. When wettability is high, one can say that the separators' resistance becomes small, which would be ideal for the LSB performance [160]. It is also known that wettability is related to the porosity, pore size, morphology, and characteristics of each separator.

3.4. Mechanical Properties and Thermal Behavior

Mechanical strength is necessary for separators to suppress the dendrite growth and short circuit. Tensile and puncture strength are the most important mechanical properties. Particularly, high tensile strength is necessary in the stretched directions for separators, whereas the puncture strength is the weight which is necessary for a needle to make a puncture on the separator [158]. Puncture strength has been applied to define how hard separators can bear a short circuit during battery assembly, manufacture, and electrochemical reactions. Sufficient puncture strength is necessary to confirm that separators do not break, or otherwise, a short circuit will occur, which would cause critical damage to the battery. The separator should also not deform or shrink because of temperature rise.

In summary, even though there are many strategies and candidate materials for separators, which may improve the LSB electrochemical performance, the abovementioned characteristics, such as separator thickness, porosity (pore size), tortuosity, electrolyte wettability, and permeability, need to be carefully considered when selecting the materials. Moreover, the mechanical strength of a separator is also important. In brief, one has to choose good separator materials and they need to be optimized to achieve high LSB performance.

The author group of this article also successfully improved the LSB electrochemical performance by optimizing the cathode materials and composition, electrolyte composition, and separator. The separator was coated with a metal organic framework-derived ink to enhance its LiPS-capturing ability [115,117,161]. Details will be described in a forthcoming paper.

4. Conclusions

In summary, it is necessary to think about whether an applied material on a separator is facing the anode or cathode side because of the different roles the orientation plays. The main roles of a separator are to adsorb LiPS to suppress the shuttle effect, boost catalytic activity to promote the sulfur-based redox reaction, function as a physical or chemical barrier to minimize LiPS migration from the cathode to the anode, and provide lithium affinity to control the lithium metal dendrite growth on the anode. For these purposes, materials such as metals, carbonaceous materials, oxides, nitrides, carbides, metal organic frameworks, quantum dots, Mxenes, biomass-derived lignins, and cellulose have all been studied. One also needs to remember that it is important to optimize necessary characteristics such as thickness, porosity, electrolyte permeability, and mechanical strength for separators to obtain high LBS performance. Each material and mechanism has its own merits and demerits and these intensive research efforts are necessary to improve the electrochemical properties of LSBs.

Funding: This research received no external funding.

Institutional Review Board Statement: Not applicable.

Informed Consent Statement: Not applicable.

Data Availability Statement: The data presented in this study are available on request from the corresponding author.

Acknowledgments: The author wishes to thank Hideki Yoshioka, Katsuji Konishi, Yoishi Inubushi, and Yoshiyuki Nakano for their helpful discussions.

Conflicts of Interest: Author Ryohei Mori was employed by the company Green Science Alliance Co., Ltd. and Fuji Pigment Co., Ltd. The author declares that the research was conducted in the absence of any commercial or financial relationships that could be construed as a potential conflict of interest.

References

1. Kim, T.; Song, W.; Son, D.Y.; Ono, L.K.; Qi, Y. Lithium-ion batteries: Outlook on present, future, and hybridized technologies. *J. Mater. Chem. A* **2019**, *7*, 2942–2964. [CrossRef]
2. Manthiram, A. A reflection on lithium-ion battery cathode chemistry. *Nat. Commun.* **2020**, *11*, 1550. [CrossRef] [PubMed]
3. Etacheri, V.; Marom, R.; Elazari, R.; Salitra, G.; Aurbach, D. Challenges in the development of advanced Li-ion batteries: A review. *Energy Environ. Sci.* **2011**, *4*, 3243–3262. [CrossRef]
4. Xu, G.; Ding, B.; Pan, J.; Nie, P.; Shen, L.; Zhang, X. High performance lithium–sulfur batteries: Advances and challenges. *J. Mater. Chem. A* **2014**, *2*, 12662–12676. [CrossRef]

5. Kang, W.; Deng, N.; Ju, J.; Li, Q.; Wu, D.; Ma, X.; Li, L.; Naebe, M.; Cheng, B. A review of recent developments in rechargeable lithium–sulfur batteries. *Nanoscale* **2016**, *8*, 16541–16588. [CrossRef]
6. Lopez, C.V.; Maladeniya, C.P.; Smith, R.C. Lithium-sulfur batteries: Advances and trends. *Electrochem* **2020**, *1*, 226–259. [CrossRef]
7. Manthiram, A.; Fu, Y.; Chung, S.H.; Zu, C.; Su, Y.S. Rechargeable lithium–sulfur batteries. *Chem. Rev.* **2014**, *114*, 11751–11787. [CrossRef]
8. Li, G.; Li, Z.; Zhang, B.; Lin, Z. Developments of electrolyte systems for lithium–sulfur batteries: A review. *Front. Energy Res.* **2015**, *3*, 1. [CrossRef]
9. Guo, J.; Liu, J. A binder-free electrode architecture design for lithium–sulfur batteries: A review. *Nanoscale Adv.* **2019**, *1*, 2104–2122. [CrossRef]
10. Yang, Y.; Zheng, G.; Cui, Y. A membrane-free lithium/polysulfide semi-liquid battery for large-scale energy storage. *Energy Environ. Sci.* **2013**, *6*, 1552. [CrossRef]
11. Zhang, S.S.; Read, J.A. A new direction for the performance improvement of rechargeable lithium/sulfur batteries. *J. Power Sources* **2012**, *200*, 77–82. [CrossRef]
12. Barchasz, C.; Molton, F.; Duboc, C.; Leprêtre, J.C.; Patoux, S.; Alloin, F. Lithium/sulfur cell discharge mechanism: An original approach for intermediate species identification. *Anal. Chem.* **2012**, *84*, 3973–3980, Corrigendum in **2020**, *92*, 13610. [CrossRef] [PubMed]
13. Yamaki, J.I.; Tobishima, S.I.; Sakurai, Y.; Saito, K.I.; Hayashi, K. Safety evaluation of rechargeable cells with lithium metal anodes and amorphous V_2O_5 cathodes. *J. Appl. Electrochem.* **1998**, *28*, 135–140. [CrossRef]
14. Cao, R.; Xu, W.; Lv, D.; Xiao, J.; Zhang, J.G. Anodes for rechargeable lithium–sulfur batteries. *Adv. Energy Mater.* **2015**, *5*, 1402273. [CrossRef]
15. Brieske, D.M.; Warnecke, A.; Sauer, D.U. Modeling the volumetric expansion of the lithium-sulfur battery considering charge and discharge profiles. *Energ. Storage Mater.* **2023**, *55*, 289–300. [CrossRef]
16. Kamisan, A.I.; Kudin, T.I.T.; Kamisan, A.S.; Omer, A.F.C.; Taib, M.F.M.; Hassan, O.H.; Ali, A.M.M.; Yahya, M.Z.A. Recent advances on graphene-based materials as cathode materials in lithium–sulfur batteries. *Int. J. Hydrogen Energy* **2022**, *47*, 8630–8657. [CrossRef]
17. Choi, Y.J.; Chung, Y.D.; Baek, C.Y.; Kim, K.W.; Ahn, H.J.; Ahn, J.H. Effects of carbon coating on the electrochemical properties of sulfur cathode for lithium/sulfur cell. *J. Power Sources* **2008**, *184*, 548–552. [CrossRef]
18. Ji, X.; Lee, K.T.; Nazar, L.F. A highly ordered nanostructured carbon–sulphur cathode for lithium–sulphur batteries. *Nat. Mater.* **2009**, *8*, 500–506. [CrossRef]
19. Liu, X.; Huang, J.Q.; Zhang, Q.; Mai, L. Nanostructured metal oxides and sulfides for lithium–sulfur batteries. *Adv. Mater.* **2017**, *29*, 1601759. [CrossRef]
20. Guo, Y.; Zhang, Y.; Zhang, Y.; Xiang, M.; Wu, H.; Liu, H.; Dou, S. Interwoven V_2O_5 nanowire/graphene nanoscroll hybrid assembled as efficient polysulfide-trapping-conversion interlayer for long-life lithium sulfur batteries. *J. Mater. Chem. A* **2018**, *6*, 19358–19370. [CrossRef]
21. Hu, N.; Lv, X.; Dai, Y.; Fan, L.; Xiong, D.; Li, X. SnO_2/reduced graphene oxide interlayer mitigating the shuttle effect of Li–S batteries. *ACS Appl. Mater. Interfaces* **2018**, *10*, 18665–18674. [CrossRef] [PubMed]
22. Cho, J.; Ryu, S.; Gong, Y.J.; Pyo, S.; Yun, H.; Kim, H.; Lee, J.; Yoo, J.; Kim, Y.S. Nitrogen-doped MoS_2 as a catalytic sulfur host for lithium–sulfur batteries. *Chem. Eng. J.* **2022**, *439*, 135568. [CrossRef]
23. Zhong, Y.; Xia, X.; Shi, F.; Zhan, J.; Tu, J.; Fan, H.J. Transition metal carbides and nitrides in energy storage and conversion. *Adv. Sci.* **2016**, *3*, 1500286. [CrossRef]
24. Gao, J.; Lowe, M.A.; Kiya, Y.; Abruna, H.D. Effects of liquid electrolytes on the charge-discharge performance of rechargeable lithium/sulfur batteries: Electrochemical and in-situ X-ray absorption spectroscopic studies. *J. Phys. Chem. C* **2011**, *115*, 25132–25137. [CrossRef]
25. Zhang, S.S. Liquid electrolyte lithium/sulfur battery: Fundamental chemistry, problems, and solutions. *J. Power Sources* **2013**, *231*, 153–162. [CrossRef]
26. Aurbach, D.; Pollak, E.; Elazari, R.; Salitra, G.; Kelley, C.S.; Affinito, J. On the surface chemical aspects of very high energy density, rechargeable Li-sulfur batteries. *J. Electrochem. Soc.* **2009**, *156*, A694. [CrossRef]
27. Fan, W.; Zhang, L.; Liu, T. Multifunctional second barrier layers for lithium–sulfur batteries. *Mater. Chem. Front.* **2018**, *2*, 235–252. [CrossRef]
28. Chang, C.H.; Chung, S.H.; Manthiram, A. Ultra-lightweight PANiNF/MWCNT-functionalized separators with synergistic suppression of polysulfide migration for Li–S batteries with pure sulfur cathodes. *J. Mater. Chem. A* **2015**, *3*, 18829–18834. [CrossRef]
29. Kong, W.; Yan, L.; Luo, Y.; Wang, D.; Jiang, K.; Li, Q.; Fan, S.; Wang, J. Ultrathin MnO_2/graphene oxide/carbon nanotube interlayer as efficient polysulfide trapping shield for high-performance Li–S batteries. *Adv. Funct. Mater.* **2017**, *27*, 1606663. [CrossRef]
30. Huang, J.Q.; Zhang, Q.; Wei, F. Multi-functional separator/interlayer system for high-stable lithium–sulfur batteries: Progress and prospects. *Energy Storage Mater.* **2015**, *1*, 127–145. [CrossRef]
31. Fiedler, M.; Cangaz, S.; Hippauf, F.; Dörfler, S.; Abendroth, T.; Althues, H.; Kastel, S. Mechanistic Insights into the Cycling Behavior of Sulfur Dry-Film Cathodes. *Adv. Sustain. Syst.* **2023**, *7*, 2200439. [CrossRef]

32. Xu, J.; Shui, J.; Wang, J.; Wang, M.; Liu, H.-K.; Dou, S.X.; Jeon, I.-Y.; Seo, J.-M.; Baek, J.-B.; Dai, L. Sulfur–Graphene Nanostructured Cathodes via Ball-Milling for High-Performance Lithium–Sulfur Batteries. *ACS Nano* **2014**, *8*, 10920–10930. [CrossRef] [PubMed]
33. Li, X.; Gao, M.; Du, W.; Ni, B.; Wu, Y.; Liu, Y.; Shang, C.; Guo, Z.; Pan, H. A mechanochemical synthesis of submicron-sized Li_2S and a mesoporous Li_2S/C hybrid for high performance lithium/sulfur battery cathodes. *J. Mater. Chem. A* **2017**, *5*, 6471–6482. [CrossRef]
34. Younes, H.; Hong, H.; Peterson, G.P. A Novel Approach to Fabricate Carbon Nanomaterials–Nanoparticle Solids through Aqueous Solutions and Their Applications. *Nanomanuf. Metrol.* **2021**, *4*, 226–236. [CrossRef]
35. Li, J.; Xiao, Z.; Chen, A.; Zhang, W.; Zhu, D.; Jin, Y.; Mao, Q.; Wang, G.; He, J.; Xia, Y. Functionally modified polyolefin-based separators for lithium–sulfur batteries: Progress and prospects. *Front. Energy Res.* **2020**, *8*, 593640. [CrossRef]
36. Yan, J.; Liu, X.; Li, B. Capacity fade analysis of sulfur cathodes in lithium–sulfur batteries. *Adv. Sci.* **2016**, *3*, 1600101. [CrossRef]
37. Peng, H.J.; Huang, J.Q.; Cheng, X.B.; Zhang, Q. Review on high loading and high-energy lithium–sulfur batteries. *Adv. Energy Mater.* **2017**, *7*, 1700260. [CrossRef]
38. Deng, N.; Liu, Y.; Li, Q.; Yan, J.; Lei, W.; Wang, G.; Wang, L.; Liang, Y.; Kang, W.; Cheng, B. Functional mechanism analysis and customized structure design of interlayers for high performance Li–S battery. *Energy Storage Mater.* **2019**, *23*, 314–349. [CrossRef]
39. Liu, Y.; Xiong, S.; Wang, J.; Jiao, X.; Li, S.; Zhang, C.; Song, Z.; Song, J. Dendrite-free lithium metal anode enabled by separator engineering via uniform loading of lithiophilic nucleation sites. *Energy Storage Mater.* **2019**, *19*, 24–30. [CrossRef]
40. Zuo, Z.; Zhuang, L.; Xu, J.; Shi, Y.; Su, C.; Lian, P.; Tian, B. Lithiophilic silver coating on lithium metal surface for inhibiting lithium dendrites. *Front. Chem.* **2020**, *8*, 109. [CrossRef]
41. Yang, C.P.; Yin, Y.X.; Zhang, S.F.; Li, N.W.; Guo, Y.G. Accommodating lithium into 3D current collectors with a submicron skeleton towards long-life lithium metal anodes. *Nat. Comm.* **2015**, *6*, 8058. [CrossRef] [PubMed]
42. Li, Q.; Zhu, S.; Lu, Y. 3D Porous Cu Current Collector/Li-Metal Composite Anode for Stable Lithium-Metal Batteries. *Adv. Funct. Mater.* **2017**, *27*, 1606422. [CrossRef]
43. Kim, P.J.H.; Seo, J.; Fu, K.; Choi, J.; Liu, Z.; Kwon, J.; Hu, L.; Paik, U. Synergistic protective effect of a BN-carbon separator for highly stable lithium sulfur batteries. *NPG Asia Mater.* **2017**, *9*, e375. [CrossRef]
44. Favors, Z.; Wang, W.; Bay, H.H.; George, A.; Ozkan, M.; Ozkan, C. Stable cycling of SiO_2 nanotubes as high-performance anodes for lithium-ion batteries. *Sci. Rep.* **2014**, *4*, 4605. [CrossRef] [PubMed]
45. Yan, N.; Wang, F.; Zhong, H.; Li, Y.; Wang, Y.; Hu, L.; Chen, Q. Hollow porous SiO_2 nanocubes towards high-performance anodes for lithium-ion batteries. *Sci. Rep.* **2013**, *3*, 1568. [CrossRef] [PubMed]
46. Suriyakumar, S.; Stephan, A.M.; Angulakshmi, N.; Hassan, M.H.; Alkordi, M.H. Metal–organic framework@SiO_2 as permselective separator for lithium–sulfur batteries. *J. Mater. Chem. A* **2018**, *6*, 14623–14632. [CrossRef]
47. Fu, D.; Luan, B.; Argue, S.; Bereau, M.N.; Davidson, I.J. Nano SiO2 particle formation and deposition on polypropylene separators for lithium-ion batteries. *J. Power Sources* **2012**, *206*, 325–333. [CrossRef]
48. Song, R.; Fang, R.; Wen, L.; Shi, Y.; Wang, S.; Li, F. A trilayer separator with dual function for high performance lithium–sulfur batteries. *J. Power Sources* **2016**, *301*, 179–186. [CrossRef]
49. Hou, T.Z.; Chen, X.; Peng, H.G.; Huang, J.Q.; Li, B.Q.; Zhang, Q.; Li, B. Design principles for heteroatom-doped nanocarbon to achieve strong anchoring of polysulfides for lithium–sulfur batteries. *Small* **2016**, *12*, 3283–3291. [CrossRef]
50. He, Y.; Wu, S.; Li, Q.; Zhou, H. Designing a multifunctional separator for high-performance Li–S batteries at elevated temperature. *Small* **2019**, *15*, 1904332. [CrossRef]
51. Huo, H.; Li, X.; Chen, Y.; Liang, J.; Deng, S.; Gao, X.; Doyle-Davis, K.; Li, R.; Guo, X.; Shen, Y.; et al. Bifunctional composite separator with a solid-state-battery strategy for dendrite-free lithium metal batteries. *Energy Storage Mater.* **2020**, *29*, 361–366. [CrossRef]
52. Li, Y.; Wang, W.; Liu, X.; Mao, E.; Wang, M.; Li, G.; Fu, L.; Li, Z.; Eng, A.Y.S.; Seh, Z.W.; et al. Engineering stable electrode-separator interfaces with ultrathin conductive polymer layer for high-energy-density Li–S batteries. *Energy Storage Mater.* **2019**, *23*, 261–268. [CrossRef]
53. Wang, L.; Yang, H.; Liu, X.; Zeng, R.; Li, M.; Huang, Y.; Hu, X. Constructing hierarchical tectorum-like α-Fe_2O_3/PPy nanoarrays on carbon cloth for solid-state asymmetric supercapacitors. *Angew. Chem. Int. Ed.* **2017**, *129*, 1125–1130. [CrossRef]
54. Feng, J.X.; Xu, H.; Ye, S.H.; Ouyang, G.; Tong, Y.X.; Li, G.R. Silica.polypyrrole hybrids as high-performance metal-free electrocatalysts for the hydrogen evolution reaction in neutral media. *Angew. Chem. Int. Ed.* **2017**, *56*, 8120–8124. [CrossRef] [PubMed]
55. Li, W.; Zhang, Q.; Zheng, G.; Seh, Z.W.; Yao, H.; Cui, Y. Understanding the role of different conductive polymers in improving the nanostructured sulfur cathode performance. *Nano Lett.* **2013**, *13*, 5534–5540. [CrossRef]
56. Kong, L.; Li, B.; Peng, H.; Zhang, R.; Xie, J.; Huang, J.; Zhang, Q. Porphyrin-derived graphene-based nanosheets enabling strong polysulfide chemisorption and rapid kinetics in lithium–sulfur batteries. *Adv. Energy Mater.* **2018**, *8*, 1800849. [CrossRef]
57. Li, C.; Zhang, P.; Dai, J.; Shen, X.; Peng, Y.; Zhang, Y.; Zhao, J. Rational method for improving the performance of lithium–sulfur batteries: Coating the separator with lithium fluoride. *ChemElectroChem* **2017**, *4*, 1535–1543. [CrossRef]
58. Wang, Y.; Ai, R.; Wang, F.; Hu, X.; Zeng, Y.; Hou, J.; Zhao, J.; Zhang, Y.; Zhang, Y.; Li, X. Research Progress on Multifunctional Modified Separator for Lithium–Sulfur Batteries. *Polymers* **2023**, *15*, 993. [CrossRef]
59. Zhou, G.; Li, L.; Wang, D.W.; Shan, X.Y.; Pei, S.; Li, F.; Cheng, H.M. A flexible sulfur-graphene-polypropylene separator integrated electrode for advanced Li–S batteries. *Adv. Mater.* **2015**, *27*, 641–647. [CrossRef]

60. Zhao, D.; Qian, X.; Jin, L.; Yang, X.X.; Wang, S.; Shen, X.; Yao, S.; Rao, D.; Zhou, Y.; Xi, X. Separator modified by Ketjen black for enhanced electrochemical performance of lithium–sulfur batteries. *RSC Adv.* **2016**, *17*, 13680–13685. [CrossRef]
61. Chung, S.H.; Manthiram, A. A polyethylene glycol-supported microporous carbon coating as a polysulfide trap for utilizing pure sulfur cathodes in lithium–sulfur batteries. *Adv. Mater.* **2014**, *26*, 7352–7357. [CrossRef] [PubMed]
62. Huang, Y.C.; Yen, Y.J.; Tseng, Y.H.; Chung, S.H. Module-designed carbon-coated separators for high-loading, high-sulfur-utilization cathodes in lithium–sulfur batteries. *Molecules* **2022**, *77*, 228. [CrossRef] [PubMed]
63. Ponraj, R.; Kannan, A.G.; Ahn, J.H.; Lee, J.H.; Kang, J.; Han, B.; Kim, D.W. Effective trapping of lithium polysulfides using a functionalized carbon nanotube-coated separator for lithium–sulfur cells with enhanced cycling stability. *ACS Appl. Mater. Interfaces* **2017**, *9*, 38445–38454. [CrossRef] [PubMed]
64. Wang, L.; Dong, Z.H.; Wang, D.; Zhang, F.X.; Jin, J. Covalent bond glued sulfur nanosheet-based cathode integration for long-cycle-life Li–S batteries. *Nano Lett.* **2013**, *13*, 6244–6250. [CrossRef] [PubMed]
65. Wang, L.; Wang, D.; Zhang, F.; Jin, J. Interface chemistry guided long-cycle life Li–S battery. *Nano Lett.* **2013**, *13*, 4206–4211. [CrossRef] [PubMed]
66. Chen, R.; Zhao, T.; Lu, J.; Wu, F.; Li, L.; Chen, J.; Tan, G.; Ye, Y.; Amine, K. Graphene-based three-dimensional hierarchical sandwich-type architecture for high-performance Li/S batteries. *Nano Lett.* **2013**, *13*, 4642–4649. [CrossRef]
67. Yu, X.; Joseph, J.; Manthiram, A. Suppression of the polysulfide-shuttle behavior in Li–S batteries through the development of a facile functional group on the polypropylene separator. *Mater. Horiz.* **2016**, *3*, 314–319. [CrossRef]
68. Vizintin, A.; Lozinšek, M.; Chellappan, R.K.; Foix, D.; Krajnc, A.; Mali, G.; Drazic, G.; Genorio, B.; Dedryvère, R.; Dominko, R. Fluorinated reduced graphene oxide as an interlayer in Li–S batteries. *Chem. Mater.* **2015**, *27*, 7070–7081. [CrossRef]
69. Lei, T.; Lei, T.; Chen, W.; Lv, W.; Huang, J.; Zhu, J.; Chu, J.; Yan, C.; Wu, C.; Yan, Y.; et al. Inhibiting polysulfide shuttling with a graphene composite separator for highly robust lithium–sulfur batteries. *Joule* **2018**, *2*, 2091–2104. [CrossRef]
70. Zhao, Y.; Gu, Z.; Weng, W.; Zhou, D.; Liu, Z.; Fan, W.; Deng, S.; He, H.; Yao, X. Nitrogen doped hollow carbon nanospheres as efficient polysulfide restricted layer on commercial separators for high-performance lithium-sulfur batteries. *Chinese Chem. Lett.* **2023**, *34*, 107232. [CrossRef]
71. Wu, P.; Hu, H.-Y.; Xie, N.; Wang, C.; Wu, F.; Pan, M.; Li, H.-F.; Wang, X.-D.; Zeng, Z.; Deng, S.; et al. A N-doped graphene-cobalt nickel sulfide aerogel as a sulfur host for lithium–sulfur batteries. *RSC Adv.* **2019**, *9*, 32247–32257. [CrossRef]
72. Li, Y.; Fan, J.; Zhang, J.; Yang, J.; Yuan, R.; Chang, J.; Zheng, M.; Dong, Q. A honeycomb-like Co@N–C composite for ultrahigh sulfur loading Li–S batteries. *ACS Nano* **2017**, *11*, 11417–11424. [CrossRef] [PubMed]
73. Li, Z.; Jiang, Y.; Yuan, L.; Yi, Z.; Wu, C.; Liu, Y.; Strasser, P.; Huang, Y. A highly ordered meso@microporous carbon-supported sulfur@smaller sulfur core–shell structured cathode for Li–S batteries. *ACS Nano* **2014**, *8*, 9295–9303. [CrossRef] [PubMed]
74. Wu, P.; Tan, L.; Wang, X.D.; Liao, P.; Liu, Z.; Hou, P.P.; Zhou, Q.Y.; Jin, X.J.; Chao, M.; Shao, X.R.; et al. Porous 3D nitrogen-doped rGO/Co-Ni-S composite modified separator for high-capacity and stable lithium–sulfur batteries. *Mater. Res. Bull.* **2022**, *145*, 111550. [CrossRef]
75. Fang, R.; Zhao, S.; Pei, S.; Qian, X.; Hou, P.-X.; Cheng, H.-M.; Liu, C.; Li, F. Toward more reliable lithium–sulfur batteries: An all-graphene cathode structure. *ACS Nano* **2016**, *10*, 8676–8682. [CrossRef] [PubMed]
76. Díez, N.; Sevilla, M.; Fuertes, A.B. N/S-Co-doped porous carbon nanoparticles serving the dual function of sulfur host and separator coating in lithium–sulfur batteries. *ACS Appl. Energy Mater.* **2020**, *3*, 3397–3407. [CrossRef]
77. Song, X.; Wang, S.; Chen, G.; Gao, T.; Bao, Y.; Ding, L.X.; Wang, H. Fe-N-doped carbon nanofiber and graphene modified separator for lithium–sulfur batteries. *Chem. Eng. J.* **2018**, *333*, 564–571. [CrossRef]
78. Zhang, Z.; Lai, Y.; Zhang, Z.; Zhang, K.; Li, J. Al_2O_3-coated porous separator for enhanced electrochemical performance of lithium sulfur batteries. *Electrochim. Acta* **2014**, *129*, 55–61. [CrossRef]
79. Han, H.; Niu, S.; Zhao, Y.; Tan, T.; Zhang, Y. TiO_2/porous carbon composite-decorated separators for lithium/sulfur battery. *Nanoscale Res. Lett.* **2019**, *14*, 176. [CrossRef]
80. Gao, Z.; Xue, Z.; Miao, Y.; Chen, B.; Xu, J.; Shi, H.; Tang, T.; Zhao, X. TiO_2@Porous carbon nanotubes modified separator as polysulfide barrier for lithium–sulfur batteries. *J. Alloy. Compd.* **2022**, *906*, 164249. [CrossRef]
81. Li, J.; Huang, Y.; Zhang, S.; Jia, W.; Wang, X.; Guo, Y.; Jia, D.; Wang, L. Decoration of silica nanoparticles on polypropylene separator for lithium–sulfur batteries. *ACS Appl. Mater. Interfaces* **2017**, *9*, 7499–7504. [CrossRef] [PubMed]
82. Liu, D.; Zhang, C.; Zhou, G.; Lv, W.; Ling, G.; Zhi, L.; Yang, Q.H. Catalytic effects in lithium–sulfur batteries: Promoted sulfur transformation and reduced shuttle effect. *Adv. Sci.* **2018**, *5*, 1700270. [CrossRef] [PubMed]
83. Li, S.; Cen, Y.; Xiang, Q.; Aslam, M.K.; Hu, B.; Li, W.; Tang, Y.; Yu, Q.; Liu, Y.; Chen, C. Vanadium dioxide-reduced graphene oxide binary host as an efficient polysulfide plague for high-performance lithium–sulfur batteries. *J. Mater. Chem.* **2019**, *7*, 1658–1668. [CrossRef]
84. Yuan, H.; Peng, H.J.; Li, B.Q.; Xie, J.; Kong, L.; Zhao, M.; Chen, X.; Huang, J.Q.; Zhang, Q. Conductive and catalytic triple-phase interfaces enabling uniform nucleation in high-rate lithium–sulfur batteries. *Adv. Energy Mater.* **2019**, *9*, 1802768. [CrossRef]
85. Qian, X.; Zhao, D.; Jin, L.; Shen, X.; Yao, S.; Rao, D.; Zhou, Y.; Xi, X.M. Hollow spherical Lanthanum oxide coated separator for high electrochemical performance lithium–sulfur batteries. *Mater. Res. Bull.* **2017**, *94*, 104–112. [CrossRef]
86. Tan, L.; Li, X.; Wang, Z.; Guo, H.; Wang, J. Lightweight reduced graphene Oxide@MoS_2 interlayer as polysulfide barrier for high-performance lithium–sulfur batteries. *ACS Appl. Mater. Interfaces* **2018**, *10*, 3707–3713. [CrossRef]

87. Fan, H.J.; Knez, M.; Scholz, R.; Nielsch, K.; Pippel, E.; Hesse, D.; Zacharias, M.; Gösele, U. Monocrystalline spinel nano-tube fabrication based on the kirkendall efect. *Nat. Mater.* **2006**, *5*, 627–631. [CrossRef]
88. Zheng, N.; Jiang, G.; Chen, X.; Mao, J.; Jiang, N.; Li, Y. Battery separators functionalized with edge-rich MoS_2/C hollow microspheres for the uniform deposition of Li_2S in high-performance lithium–sulfur batteries. *Nano-Micro Lett.* **2019**, *11*, 43. [CrossRef]
89. Mao, L.; Mao, J. Ultralow-decay lithium–sulfur batteries: Modified separator with graphene/ZnS(en)$_{0.5}$ exfoliation nanosheets. *J. Solid State Chem.* **2020**, *290*, 121555. [CrossRef]
90. Yao, W.; Zheng, W.; Xu, J.; Tian, C.; Han, K.; Sun, W.; Xiao, S. ZnS-SnS@NC Heterostructure as robust lithiophilicity and sulfiphilicity mediator toward high-rate and long-life lithium–sulfur batteries. *ACS Nano* **2021**, *15*, 7114–7130. [CrossRef]
91. Wang, G.; Jiao, Q.; Zhang, Z.; Zhao, Y.; Lin, C.; Zhang, X.; Ma, H.; Dai, S.; Xu, T. Improved electrochemical behavior of Li–S battery with functional WS_2@PB–PPy–modified separator. *Chem. Eng. J. Adv.* **2021**, *8*, 100145. [CrossRef]
92. Xiao, Y.; Hwang, J.Y.; Sun, Y.K. Transition metal carbide-based materials: Synthesis and applications in electrochemical energy storage. *J. Mater. Chem. A* **2016**, *4*, 10379–10393. [CrossRef]
93. Cui, Z.; Yao, J.; Mei, T.; Zhou, S.; Hou, B.; Li, J.; Li, J.; Wang, J.; Qian, J.; Wang, X. Strong lithium polysulfides chemical trapping of TiC-TiO_2/S composite for long-cycle lithium–sulfur batteries. *Electrochim. Acta* **2019**, *298*, 43–51. [CrossRef]
94. Zhang, Y.; Zhang, P.; Li, B.; Zhang, S.; Liu, K.; Hou, R.; Zhang, X.; Silva, S.R.; Shao, G. Vertically aligned graphene nanosheets on multi-yolk/shell structured TiC@C nanofibers for stable Li–S batteries. *Energy Storage Mater.* **2020**, *27*, 159–168. [CrossRef]
95. Liu, S.; Luo, J.; Xiong, Y.; Chen, Z.; Zhang, K.; Rui, G.; Wang, L.; Hu, G.; Jiang, J.; Mei, T. Taming polysulfides in an Li–S battery with low-temperature one-step chemical synthesis of titanium carbide nanoparticles from waste PTFE. *Front. Chem.* **2021**, *9*, 638557. [CrossRef]
96. Zhao, J.; Yan, Z. Flexible TiC modified separators trapping polysulfide for high performance lithium–sulfur batteries. *J. Alloys Compd.* **2021**, *856*, 156609. [CrossRef]
97. Moon, S.-H.; Kim, J.-H.; Shin, J.-H.; Jang, J.-S.; Kim, S.-B.; Lee, S.-N.; Kwon, S.-H.; Park, K.-W. High absorption and fast polysulfides conversion of duel functional separator based on mesoporous-WC/rGO composite for lithium–sulfur batteries. *J. Alloy. Compd.* **2022**, *904*, 164120. [CrossRef]
98. Zhao, P.; Zhang, Z.; He, H.; Yu, Y.; Li, X.; Xie, W.; Yang, Z.; Cai, J. Cobalt-tungsten bimetallic carbide nanoparticles as efficient catalytic material for high-performance lithium–sulfur batteries. *ChemSusChem* **2019**, *12*, 4866–4873. [CrossRef]
99. Cai, W.; Li, G.; Zhang, K.; Xiao, G.; Wang, C.; Ye, K.; Chen, Z.; Zhu, Y.; Qian, Y. Conductive nanocrystalline niobium carbide as high efficiency polysulfides tamer for lithium–sulfur batteries. *Adv. Funct. Mater.* **2017**, *28*, 1704865. [CrossRef]
100. Abghoui, Y.; Skúlasson, E. Transition Metal Nitride Catalysts for Electrochemical Reduction of Nitrogen to Ammonia at Ambient Conditions. *Procedia Computer Sci.* **2015**, *51*, 1897–1906. [CrossRef]
101. Lan, L.; Chen, D.; Yao, Y.; Peng, X.; Wu, J.; Li, Y.; Ping, J.; Ying, Y. Phase-dependent fluorescence quenching efficiency of MoS_2 nanosheets and their applications in multiplex target biosensing. *ACS Appl. Mater. Interfaces* **2018**, *10*, 42009–42017. [CrossRef] [PubMed]
102. Chen, G.; Song, X.; Wang, S.; Chen, X.; Wang, H. Two dimensional molybdenum nitride nanosheets modified Celgard separator with multifunction for Li–S batteries. *J. Power Sources* **2018**, *408*, 58–64. [CrossRef]
103. Kim, H.S.; Kang, H.J.; Lim, H.; Hwang, H.J.; Park, J.W.; Lee, T.G.; Cho, S.Y.; Jang, S.G.; Jun, Y.S. Boron nitride nanotube-based separator for high-performance lithium–sulfur batteries. *Nanomaterials* **2022**, *12*, 11. [CrossRef] [PubMed]
104. Chen, Y.; Kang, Q.; Jiang, P.; Huang, X. Rapid, high-efficient and scalable exfoliation of high-quality boron nitride nanosheets and their application in lithium–sulfur batteries. *Nano Res.* **2021**, *14*, 2424–2431. [CrossRef]
105. Luo, M.; Bai, Y.; Sun, R.; Wang, Z.; Sun, W.; Lin, P.; Dai, X.; Sun, K. Enhanced performance of lithium–sulfur batteries with Co-doped g-C_3N_4 nanosheet-based separator. *Ind. Eng. Chem. Res.* **2021**, *60*, 1231–1240. [CrossRef]
106. Huang, S.; Huixiang, E.; Yang, Y.; Zhang, Y.; Ye, M.; Li, C.C. Transition metal phosphides: New generation cathode host/separator modifier for Li–S batteries. *J. Mater. Chem. A* **2021**, *9*, 7458–7480. [CrossRef]
107. Chen, X.; Ding, X.; Wang, C.; Feng, Z.; Xu, L.; Gao, X.; Zhai, Y.; Wang, D. A multi-shelled CoP nanosphere modified separator for highly efficient Li–S batteries. *Nanoscale* **2018**, *10*, 13694–13701. [CrossRef]
108. Lin, J.; Zhang, K.; Zhu, Z.; Zhang, R.; Li, N.; Zhao, C. CoP/C nanocubes-modified separator suppressing polysulfide dissolution for high-rate and stable lithium–sulfur batteries. *ACS Appl. Mater. Interfaces* **2019**, *12*, 2497–2504. [CrossRef]
109. Zhao, Z.; Li, H.; Cheng, X.; Ren, R.; Meng, Z.; Wang, X. Multifunctional FeP/spongy carbon modified separator with enhanced polysulfide immobilization and conversion for flame-retardant lithium–sulfur batteries. *Chem. Select* **2021**, *6*, 7098–7102. [CrossRef]
110. Mao, J.; Niu, D.; Huang, G.; Jin, X.; Wei, C.; Cai, J.; Li, Y.; Shi, J. A Ni/Ni_2P heterostructure in modified porous carbon separator for boosting polysulfide catalytic conversion. *Sci. China Mater.* **2022**, *65*, 2453–2462. [CrossRef]
111. Zhou, G.; Tian, H.; Jin, Y.; Tao, X.; Liu, B.; Zhang, R.; Seh, Z.W.; Zhuo, D.; Liu, Y.; Sun, J.; et al. Catalytic oxidation of Li2S on the surface of metal sulfides for Li−S batteries. *PNAS* **2017**, *114*, 840–845. [CrossRef]
112. Peng, H.J.; Zhang, G.; Chen, X.; Zhang, Z.W.; Xu, W.T.; Huang, J.Q.; Zhang, Q. Enhanced electrochemical kinetics on conductive polar mediators for lithium–sulfur batteries. *Angew. Chem. Int. Ed.* **2016**, *55*, 12990–12995. [CrossRef] [PubMed]
113. Liu, Y.; Chen, C.; Valdez, J.; Motta Meira, D.; He, W.; Wang, Y.; Harnagea, C.; Lu, Q.; Guner, T.; Wang, H.; et al. Phase-enabled metal-organic framework homojunction for highly selective CO_2. *Photoreduction. Nat. Comm.* **2021**, *12*, 1231. [CrossRef] [PubMed]

114. Peng, Y.; Li, Y.; Ban, Y.; Jin, H.; Jiao, W.; Liu, X.; Yang, W. Metal-organic framework nanosheets as building blocks for molecular sieving membranes. *Science* **2014**, *346*, 1356–1359. [CrossRef]
115. Wang, B.; Cote, A.P.; Furukawa, H.; O'Keeffe, M.; Yaghi, O.M. Colossal cages in zeolitic imidazolate frameworks as selective carbon dioxide reservoirs. *Nature* **2008**, *453*, 207–211. [CrossRef]
116. Bai, S.; Sheng, T.; Tan, C.; Zhu, Q.; Huang, Y.; Jiang, H.; Hu, S.; Fu, R.; Wu, X. Distinct anion sensing by a 2D self-assembled Cu(I)-based metal–organic polymer with versatile visual colorimetric responses and efficient selective separations via anion exchange. *J. Mater. Chem. A* **2013**, *1*, 2970. [CrossRef]
117. Mori, R. Cathode materials for lithium–sulfur battery: A review. *J. Solid State Electrochem.* **2023**, *27*, 813–839. [CrossRef]
118. Bai, S.; Liu, X.; Zhu, K.; Wu, S.; Zhou, H. Metal-organic framework-based separator for lithium–sulfur batteries. *Nat. Energy* **2016**, *1*, 16094. [CrossRef]
119. Dang, B.; Li, Q.; Luo, Y.; Zhao, R.; Li, J.; Wu, F. Metal-organic framework-based glass fiber separator as an efficacious polysulfide barrier and dendrite suppressor for lithium–sulfur batteries. *J. Alloys Compd.* **2022**, *915*, 165375. [CrossRef]
120. Su, Y.; Wang, W.; Wang, W.; Wang, A.; Huang, Y.; Guan, Y. Cerium-based MOF as a separator coating for high-performance lithium–sulfur batteries. *J. Electrochem. Soc.* **2022**, *169*, 030528. [CrossRef]
121. Ma, B.; Zhang, X.; Deng, X.; Huang, S.; Xiao, M.; Wang, S.; Han, D.; Meng, Y. Construction of KB@ZIF-8/PP composite separator for lithium–sulfur batteries with enhanced electrochemical performance. *Polymers* **2021**, *13*, 4210. [CrossRef]
122. Baker, S.N.; Baker, G.A. Luminescent carbon nanodots: Emergent nanolights. *Angew. Chem. Int. Ed.* **2010**, *49*, 6726–6744. [CrossRef]
123. Fowley, C.; Nomikou, N.; McHale, A.P.; McCaughan, B.; Callan, J.F. Extending the tissue penetration capability of conventional photosensitisers: A carbon quantum dot–protoporphyrin IX conjugate for use in two-photon excited photodynamic therapy. *Chem Commun (Camb)* **2013**, *49*, 8934–8936. [CrossRef]
124. Xu, Q.; Niu, Y.; Li, J.; Yang, Z.; Gao, J.; Ding, L.; Ni, H.; Zhu, P.; Liu, Y.; Tang, Y.; et al. Recent progress of quantum dots for energy storage applications. *Carbon. Neutrality* **2022**, *1*, 13. [CrossRef]
125. Ding, H.; Zhang, Q.; Liu, Z.; Wang, J.; Ma, R.; Fan, L.; Wang, T.; Zhao, J.; Ge, J.; Lu, X.; et al. TiO_2 quantum dots decorated multi-walled carbon nanotubes as the multifunctional separator for highly stable lithium sulfur batteries. *Electrochim. Acta* **2018**, *284*, 314–320. [CrossRef]
126. Pang, Y.; Wei, J.; Wang, Y.; Xia, Y. Synergetic protective effect of the ultralight MWCNTs/NCQDs modified separator for highly stable lithium–sulfur batteries. *Adv. Energy Mater.* **2018**, *8*, 1702288. [CrossRef]
127. Chung, S.H.; Manthiram, A. High-performance Li–S batteries with an ultra-lightweight MWCNT-coated separator. *J. Phys. Chem. Lett.* **2014**, *5*, 197. [CrossRef] [PubMed]
128. Yu, B.; Chen, D.; Wang, Z.; Qi, F.; Zhang, X.; Wang, X.; Hu, Y.; Wang, B.; Zhang, W.; Chen, Y.; et al. Mo_2C quantum dots@graphene functionalized separator toward high-current-density lithium metal anodes for ultrastable Li–S batteries. *Chem. Eng. J.* **2020**, *399*, 125837. [CrossRef]
129. Liu, Z.; Hu, Z.; Jiang, X.; Zhang, Y.; Wang, X.; Zhang, S. Multi-functional ZnS quantum Dots/Graphene aerogel modified separator for high performance lithium–sulfur batteries. *Electrochim. Acta* **2022**, *422*, 140496. [CrossRef]
130. Zhang, J.; Cheng, Y.; Chen, H.; Wang, Y.; Chen, Q.; Hou, G.; Wen, M.; Tang, Y. MoP quantum dot-modified N,P-carbon nanotubes as a multifunctional separator coating for high-performance lithium–sulfur batteries. *ACS Appl. Mater. Interfaces* **2022**, *14*, 16289–16299. [CrossRef]
131. Liu, Q.; Han, X.; Park, H.; Kim, J.; Xiong, P.; Yuan, H.; Yeon, J.S.; Kang, Y.; Park, J.M.; Dou, Q.; et al. Layered double hydroxide quantum dots for use in a bifunctional separator of lithium–sulfur batteries. *ACS Appl. Mater. Interfaces* **2021**, *13*, 179780. [CrossRef] [PubMed]
132. Ran, J.R.; Gao, G.; Li, F.T.; Ma, T.Y.; Du, A.; Qiao, S.Z. Ti_3C_2 MXene co-catalyst on metal sulfide photo-absorbers for enhanced visible-light photocatalytic hydrogen production. *Nat. Commun.* **2017**, *8*, 13907. [CrossRef] [PubMed]
133. Zhao, S.S.; Meng, X.; Zhu, K.; Du, F.; Chen, G.; Wei, Y.J.; Gogotsi, Y.; Gao, Y. Li-ion uptake and increase in interlayer spacing of Nb_4C_3 MXene. *Energy Storage Mater.* **2017**, *8*, 42–48. [CrossRef]
134. Naguib, M.; Kurtoglu, M.; Presser, V.; Lu, J.; Niu, J.; Heon, M.; Hultman, L.; Gogotsi, Y.; Barsoum, M.W. Two-dimensional nanocrystals produced by exfoliation of Ti_3AlC_2. *Adv. Mater.* **2011**, *23*, 4248–4253. [CrossRef] [PubMed]
135. Lukatskaya, M.R.; Mashtalir, O.; Ren, C.E.; Dall'agnese, Y.; Rozier, P.; Taberna, P.L.; Naguib, M.; Simon, P.; Barsoum, M.W.; Gogotsi, Y. Cation intercalation and high volumetric capacitance of two-dimensional titanium carbide. *Science* **2013**, *341*, 1502–1505. [CrossRef]
136. Er, D.; Li, J.; Naguib, M.; Gogotsi, Y.; Shenoy, V.B. Ti_3C_2 MXene as a high capacity electrode material for metal (Li, Na, K, Ca) ion batteries. *ACS Appl. Mater. Interfaces* **2014**, *6*, 11173–11179. [CrossRef]
137. Naguib, M.; Mochalin, V.N.; Barsoum, M.W.; Gogotsi, Y. 25th anniversary article: MXenes: A new family of two-dimensional materials. *Adv. Mater.* **2014**, *26*, 992–1005. [CrossRef]
138. Saeed, M.A.; Shahzad, A.; Rasool, K.; Mateen, F.; Oh, J.; Shim, J.W. 2D MXene: A potential candidate for photovoltaic cells? a critical review. *Adv. Sci.* **2020**, *9*, 2104743. [CrossRef]
139. Chen, X.; Shi, Z.; Tian, Y.; Lin, P.; Wu, D.; Li, X.; Dong, B.; Xu, W.; Fang, X. Two-dimensional Ti_3C_2 MXene-based nanostructures for emerging optoelectronic applications. *Mater. Horiz.* **2021**, *8*, 2929–2963. [CrossRef]

140. Wojciechowski, T.; Rozmysłowska-Wojciechowska, A.; Matyszczak, G.; Wrzecionek, M.; Olszyna, A.; Peter, A.; Mihaly-Cozmuta, A.; Nicula, C.; Mihaly-Cozmuta, L.; Podsiadło, S.; et al. Ti_2C MXene modified with ceramic oxide and noble metal nanoparticles: Synthesis, morphostructural properties, and high photocatalytic activity. *Inorg. Chem.* **2019**, *58*, 7602–7614. [CrossRef]
141. Gao, L.; Chen, H.; Kuklin, A.V.; Wageh, S.; Al-Ghamdi, A.A.; Ågren, H.; Zhang, H. Optical properties of few-layer Ti_3CN MXene: From experimental observations to theoretical calculations. *ACS Nano* **2022**, *16*, 3059–3069. [CrossRef] [PubMed]
142. Naguib, M.; Mashtali, O.; Carle, J.; Presser, V.; Lu, J.; Hultman, L.; Gogotsi, Y.; Barsoum, M.W. Two-dimensional transition metal carbides. *ACS Nano* **2012**, *6*, 1322–1331. [CrossRef] [PubMed]
143. Lin, C.; Zhang, W.; Wang, L.; Wang, Z.; Zhao, W.; Duan, W.; Zhao, Z.; Liu, B.; Jin, J. A few-layered Ti_3C_2 nanosheet/glass fiber composite separator as a lithium polysulphide reservoir for high-performance lithium–sulfur batteries. *J. Mater. Chem.* **2016**, *4*, 5993–5998. [CrossRef]
144. Jiao, L.; Zhang, C.; Geng, C.; Wu, S.; Li, H.; Lv, W.; Tao, Y.; Chen, Z.; Zhou, G.; Li, J.; et al. Capture and catalytic conversion of polysulfides by in situ built TiO_2-MXene heterostructures for lithium–sulfur batteries. *Adv. Energy Mater.* **2019**, *9*, 190021. [CrossRef]
145. Yao, Y.; Wang, S.; Jia, X.; Yang, J.; Li, Y.; Liao, J.; Song, H. Freestanding sandwich-like hierarchically TiS_2–TiO_2/Mxene bi-functional interlayer for stable Li–S batteries. *Carbon* **2022**, *188*, 533–542. [CrossRef]
146. Li, P.; Lv, H.; Li, Z.; Meng, X.; Lin, Z.; Wang, R.; Li, X. The Electrostatic Attraction and Catalytic Effect Enabled by Ionic–Covalent Organic Nanosheets on MXene for Separator Modification of Lithium–Sulfur Batteries. *Adv. Mater.* **2021**, *33*, 2007803. [CrossRef]
147. Wang, J.; Zhai, P.; Zhao, T.; Li, M.; Yang, Z.; Zhang, H.; Huang, J. Laminar MXene-Nafion-modified separator with highly inhibited shuttle effect for long-life lithium–sulfur batteries. *Electrochimica Acta* **2019**, *320*, 134558. [CrossRef]
148. He, Y.; Qiao, Y.; Chang, Z.; Cao, X.; Jia, M.; He, P.; Zhou, H. Developing A "polysulfide-phobic" strategy to restrain shuttle effect in lithium–sulfur batteries. *Angew. Chem. Int. Ed.* **2019**, *58*, 11774–11778. [CrossRef]
149. Huang, J.Q.; Zhang, Q.; Peng, H.J.; Liu, X.Y.; Qian, W.Z.; Wei, F. Ionic shield for polysulfides towards highly-stable lithium–sulfur batteries. *Energy Environ. Sci.* **2013**, *7*, 347–353. [CrossRef]
150. Zhang, Z.; Yi, S.; Wei, Y.; Bian, H.; Wang, R.; Min, Y. Lignin nanoparticle-coated celgard separator for high-performance lithium–sulfur batteries. *Polymers* **2019**, *11*, 1946. [CrossRef]
151. Henriksson, M.; Berglund, L.A.; Isaksson, P.; Lindstrom, T.; Nishino, T. Cellulose nanopaper structures of high toughness. *Biomacromolecules* **2008**, *9*, 1579–1585. [CrossRef] [PubMed]
152. Pavlin, N.; Hribernik, S.; Kapun, G.; Talian, S.D.; Njel, C.; Dedryvère, R.; Dominko, R. The role of cellulose based separator in lithium sulfur batteries. *J. Electrochem. Soc.* **2019**, *166*, A5237–A5243. [CrossRef]
153. Tian, Y.W.; Zhang, Y.J.; Wu, L.; Dong, W.D.; Huang, R.; Dong, P.Y.; Yan, M.; Liu, J.; Mohamed, H.S.H.; Chen, L.H.; et al. Bifunctional Separator with Ultra-Lightweight MnO_2 Coating for Highly Stable Lithium–Sulfur Batteries. *ACS Appl. Mater. Interfaces* **2023**, *15*, 6877–6887. [CrossRef] [PubMed]
154. Chen, K.; Li, Y.; Zhan, H. Advanced separators for lithium–ion batteries. *IOP Conf. Ser. Earth Environ. Sci.* **2022**, *1011*, 012009. [CrossRef]
155. Huang, X. Separator technologies for lithium-ion batteries. *J. Solid. State Electrochem.* **2011**, *15*, 649–662. [CrossRef]
156. Deimede, V.; Elmasides, C. Separators for lithium-ion batteries: A review on the production processes and recent developments. *Energy Technol.* **2015**, *3*, 453–468. [CrossRef]
157. Zhang, L.; Li, X.; Yang, M.; Chen, W. High-safety separators for lithium-ion batteries and sodium-ion batteries: Advances and perspective, energy storage materials. *ScienceDirect* **2021**, *41*, 522–545.
158. Lee, H.; Yanilmaz, M.; Toprakci, O.; Fu, K.; Zhang, X. A review of recent developments in membrane separators for rechargeable lithium-ion batteries. *Energy Environ. Sci.* **2014**, *7*, 3857–3886. [CrossRef]
159. Francis, C.F.J.; Kyratzis, I.L.; Best, A.S. Lithium-ion battery separators for ionic-liquid. electrolytes: A review. *Adv. Mater.* **2020**, *32*, 1904205. [CrossRef]
160. Xie, Y.; Zou, H.; Xiang, H.; Xia, R.; Liang, D.; Shi, P.; Dai, S.; Wang, H. Enhancement on the wettability of lithium battery separator toward nonaqueous electrolytes. *J. Membr. Sci.* **2016**, *503*, 25–30. [CrossRef]
161. Mori, R.; Yokokawa, K. Silicon Coupling Agent Treatment of Cathode Material for Lithium Sulfur Battery. JP Patent No. 2022-155874, 14 October 2022.

materials

MDPI

Article

Manufacturing Shape-Controllable Flexible PEDOT/rGO Composite Electrodes for Planar Micro-Supercapacitors

Haiwei Hu, Yanyan Guo * and Jiang Zhao *

College of Integrated Circuit Science and Engineering, Nanjing University of Posts and Telecommunications, Nanjing 210023, China; 17314416686@163.com
* Correspondence: guoyy@njupt.edu.cn (Y.G.); jzhao@njupt.edu.cn (J.Z.)

Abstract: Flexible electronic products, with their characteristics of flexibility and wearability, have attracted significant attention and have become an important direction in the research and development of the electronics industry. Planar micro-supercapacitors (MSCs) with flexible composite electrodes can provide reliable energy support for these products, propelling their further development. The research employed a quick, effective, and environmentally friendly method of laser scribing to create shape-controllable flexible composite electrodes on composite films of Poly(3,4-ethylenedioxythiophene) and graphene oxide (PEDOT/GO), which were subsequently assembled into MSCs. An analysis of the composite electrode morphology, structure, and elemental distribution was conducted through the utilization of SEM, TEM, and XPS techniques. Following this, a comprehensive evaluation of the electrochemical performance of the flexible MSCs was carried out, which included cyclic voltammetry (CV), galvanostatic charge/discharge (GCD), and assessment of cyclic stability. The analysis of the CV results indicated that the MSCs achieved the areal capacitance of 5.78 mF/cm^2 at 5 mV/s. After 5000 cycles at a current density of 0.05 mA/cm^2, the capacitance retention rate was 85.4%. The high areal capacitance and strong cycle stability of MSCs highlight the potential of PEDOT/reduced graphene oxide (PEDOT/rGO) electrodes in electrode applications.

Keywords: flexibility; micro-supercapacitors; laser scribing; Poly(3,4-ethylenedioxythiophene); graphene oxide; composite electrode; areal capacitance

Citation: Hu, H.; Guo, Y.; Zhao, J. Manufacturing Shape-Controllable Flexible PEDOT/rGO Composite Electrodes for Planar Micro-Supercapacitors. *Materials* **2024**, *17*, 2144. https://doi.org/10.3390/ma17092144

Academic Editor: Jung Bin In

Received: 30 March 2024
Revised: 27 April 2024
Accepted: 30 April 2024
Published: 3 May 2024

1. Introduction

Due to the depletion of fossil fuels and the exacerbation of environmental pollution [1], attention has shifted towards the development of renewable energy sources. However, the production capacity of renewable energies, such as wind and solar energy, is unstable and cannot be directly stored, existing only in the form of electricity [2]. This has accelerated the development of batteries and supercapacitors [3]. Supercapacitors, which are also referred to as electrochemical capacitors, are a modern form of energy storage that falls between conventional capacitors and batteries [4,5]. Supercapacitors are well known for their impressive power density [6,7], making them candidates for rapid power supply and charging [8]. Based on their charge storage mechanism, supercapacitors are categorized into electric double-layer capacitors (EDLCs) and pseudocapacitors [9,10]. Carbon materials, with their excellent conductivity, stable chemical properties, high specific surface area, low cost, and mature processing technology, are considered high-quality electrode materials for EDLCs [11,12]. The electrode materials for pseudocapacitors mainly include metal oxides and conductive polymers [13,14]. Pseudocapacitors store energy through reversible reactions on the surface of the electrode, demonstrating a mechanism that is more intricate than that of EDLCs [15–18].

In recent years, with the rise of new wearable electronic products, there has been an increasing interest in research on flexible planar micro-supercapacitors (MSCs) [19–21]. Graphene, due to its unique properties, has shown great potential as an electrode material for MSCs [22]. The mainstream fabrication techniques for graphene electrode materials include inkjet printing [23], laser scribing [24], and laser-induced approaches [25–27], with laser scribing being significantly more efficient than inkjet printing and comparable to laser-induced methods. Firstly, the laser scribing technique is lauded for its exceptional precision and high resolution, capable of achieving micrometer-level intricate patterns. This precision is crucial for applications requiring high-resolution patterns, such as the fabrication of graphene electrodes in supercapacitors, where exact electrode patterning can significantly impact device performance. In contrast, the preparation of graphene electrodes via inkjet printing requires the pre-processing of graphene powder to formulate the ink. Due to the hydrophobic nature of graphene, it is common to add sodium carboxymethyl cellulose (CMC) to enhance the stability of the dispersion. While CMC helps stabilize the ink, its addition can introduce impurities that may weaken the electrochemical performance of supercapacitors by influencing both the conductivity and the electrochemical reactivity of the electrode material. MSCs based on graphene electrode materials possess advantages such as short charging times, long cycle stability, and greater power density, holding tremendous application prospects in flexible micro-electronic devices [28–31]. Poly(3,4-ethylenedioxythiophene) (PEDOT), as a conductive polymer material, exhibits good conductivity, electrochemical stability, and reversible redox properties [32], making it suitable for use as an electrode material in supercapacitors [33–36]. It is noteworthy that most high-performance MSCs are currently based on composite electrode materials. The use of graphene alone as an electrode material for MSCs is limited by its lower electrochemical activity, affecting the electrode's specific capacitance and reliability. Similarly, using PEDOT alone as an electrode material for MSCs is not ideal, as PEDOT nanoparticles have a lower specific surface area than graphene, meaning there are fewer active areas available for reactions between the electrode and the electrolyte, thus reducing the performance of the capacitor. Therefore, combining graphene and PEDOT into a composite electrode material is expected to overcome their individual limitations and improve the performance of MSCs, pushing the development and application of supercapacitor technology. Fuwei Liu [37] developed a supercapacitor utilizing the rGO/PEDOT/polyaniline composite electrode, which achieved a specific capacitance of 535 F/g and exhibited good cyclic stability. However, the fabrication process for the composite electrode is complex, making it unsuitable for direct use in flexible micro-supercapacitors (MSCs). In another study, Shiyuan Li [38] successfully fabricated an rGO/PEDOT composite electrode using a hydrothermal polymerization technique, which achieved a conductivity of 88.5 S/cm and a specific capacitance of 202.7 F/g. Nevertheless, the fabrication process for this electrode is challenging to control.

In this work, novel flexible PEDOT/rGO composite electrodes with controllable shapes were fabricated through an extremely simple and highly effective laser scribing technique. PEDOT not only provides additional electrochemical activity to enhance the performance of MSCs, but also improves the mechanical strength and flexibility of the electrodes, thereby increasing the reliability of MSCs. The high electrical conductivity of rGO facilitates rapid electron transport on the electrode surface, and its high specific surface area means that there are more active sites available for electrochemical reactions, further enhancing the electrochemical performance of MSCs. The electrochemical testing results of the flexible MSCs indicate that the maximum areal capacitance achieved by the PEDOT/rGO electrode-based MSCs is 5.78 mF/cm^2 at 5 mV/s. Repeated charge/discharge tests conducted on the MSCs with optimal performance show that, under a current density of 0.05 mA/cm^2, the retention of capacitance stands at 85.4% following 5000 cycles. All of these demonstrate that the fabricated flexible PEDOT/rGO composite electrodes hold application potential in electrodes for planar MSCs.

2. Experiment

2.1. Materials

The uniform, precipitate-free, single-layer graphene oxide (GO) dispersion with a concentration of 2 mg/mL, synthesized using an improved Hummers method, was obtained from Tanfeng Technology Co., Ltd., located in Suzhou, China. The Orgacon conductive polymer PEDOT:PSS dispersion, containing an active chemical content of PEDOT:PSS at 1.2% wt, was provided by Agfa-Gevaert N.V., Belgium. Polyvinyl alcohol (PVA) was obtained from Sinopharm Chemical Reagent Co., Ltd., Shanghai, China, and crystalline lithium chloride was sourced from Aladdin Biochemical Technology Co., Ltd., Shanghai, China.

2.2. Fabrication of PEDOT/rGO Electrodes

First, 10 mL of GO dispersion was mixed with no more than 1 mL of PEDOT:PSS dispersion in a beaker, and the mixture was thoroughly stirred on a magnetic stirrer to ensure uniformity. Subsequently, a polyethylene terephthalate (PET) substrate with a diameter of 8 cm was prepared. The PEDOT/GO mixture was carefully drawn into a syringe and evenly dispensed onto the surface of the PET substrate. A coating rod was used to gently spread the mixture, ensuring the formation of a uniform and smooth layer. The mixture was then left to cure naturally at room temperature, forming a PEDOT/GO film. Finally, to pattern the PEDOT/GO film, a laser engraving machine with a wavelength of 405 nm and the maximum output power of 1.5 W was used. The laser power was set to 27.1%. Under these settings, the interdigital PEDOT/rGO electrode was fabricated within two minutes. Figure 1 presents a photograph of the fabricated PEDOT/rGO electrode sample.

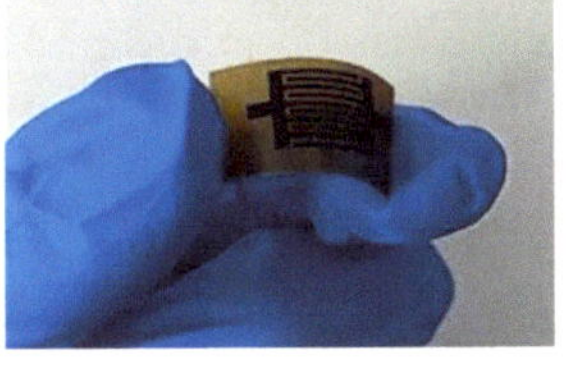

Figure 1. The physical sample of the PEDOT/rGO electrode.

2.3. Electrolyte

Here, 0.6 g of polyvinyl alcohol (PVA) powder was mixed with 5 mL of deionized water. Using a constant temperature magnetic stirrer, the mixture was steadily heated to 85 °C under vigorous stirring by a water bath method, and the stirring was continued until the mixture completely turned into a transparent gel. After cooling to room temperature, the transparent gel was further processed by adding 1 mL of a 0.6 g/mL LiCl solution. Upon thorough stirring, the PVA/LiCl gel electrolyte was obtained.

2.4. Construction of Planar MSCs

The interdigital PEDOT/rGO electrode with a length of 10 mm, finger width of 0.7 mm, and inter-finger spacing of 0.4 mm was uniformly coated with the PVA/LiCl gel electrolyte. To enhance electrical conductivity, copper strips were added to both sides of the electrode. Additionally, a low-temperature conductive silver paste, curing at 80–150 °C, was applied at the interfaces between the copper strips and the electrode to ensure excellent electrical contact performance. Subsequently, the entire assembly was placed in an electrically heated, constant-temperature, air-circulating oven set at 80 °C to accelerate the curing process. Approximately twenty minutes later, the conductive silver paste had solidified. The solidification of the conductive silver paste secured the copper strips, ensuring the reliability and stability of the electrical connections. Finally, to complete the encapsulation process and protect the electrode, a layer of PI tape was carefully applied over the electrode that was coated with the electrolyte. After the encapsulation process, the dimensions of the micro-supercapacitor (MSC) were measured to be 13 mm in length and 8 mm in width. Figure 2 shows the sequential steps involved in the encapsulation process of MSC.

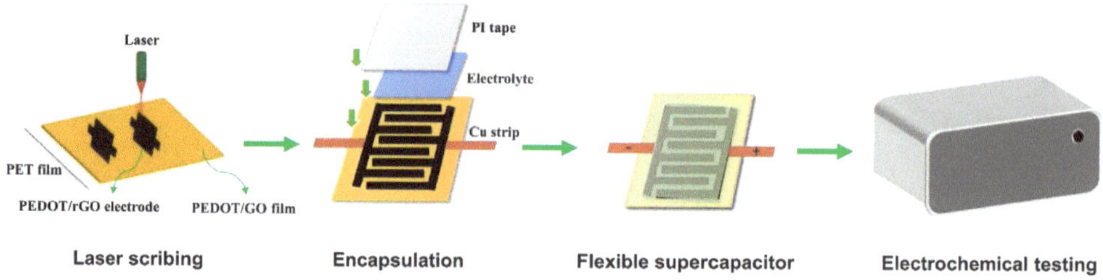

Figure 2. The encapsulation and testing processes of MSCs based on PEDOT/rGO electrodes.

2.5. Characterization

The morphology of the electrodes was analyzed using field emission scanning electron microscopy (FESEM, Regulus 8100, Hitachi, Tokyo, Japan). The GO dispersion doped with PEDOT:PSS dispersion was analyzed using transmission electron microscopy (TEM, Talos F200X G2, Thermo, Waltham, MA, USA). Raman spectra were measured using a Raman spectrometer (Scientific DXR, Thermo, Waltham, MA, USA). The elemental composition was determined by X-ray photoelectron spectroscopy (XPS, Scientific K-Alpha, Thermo, Waltham, MA, USA).

2.6. Electrochemical Measurements

The CHI660E (Chenhua, Shanghai, China) system was used to test the electrochemical performance. Cycling performance tests were carried out on a BTS-3000 (Neware, Shenzhen, China). Analyses of electrochemical involved cyclic voltammetry (CV), galvanostatic charge/discharge (GCD), electrochemical impedance spectroscopy (EIS), and cyclic stability assessment. According to the CV curve, the areal capacitance at different scanning rates was calculated according to the equation $C_A = \int IdV/S\Delta Vv$, where C_A (mF/cm^2) denotes the areal capacitance, $\int IdV$ signifies the integral under the CV curve, S denotes the area of supercapacitor, ΔV (V) represents the voltage window, and v (mV/s) indicates the scanning rate.

3. Results and Discussion

3.1. Surface Morphology Analysis

The alterations in surface morphology of the PEDOT/rGO composite electrode due to laser scribing were analyzed through Scanning Electron Microscopy (SEM), illustrated in Figure 3a–d. Figure 3a presents the surface appearance of the PEDOT/GO composite material before laser treatment on the left, and the surface appearance of the PEDOT/rGO composite electrode on the right. This observation suggests that the laser scribing process successfully converts graphene oxide (GO) to reduced graphene oxide (rGO). Figure 3b demonstrates a rough surface of the composite electrode, showcasing a distinct layered and interconnected porous structure. Figure 3c depict the surface appearance of the composite electrode at increased magnifications, and the SEM image in Figure 3d displays a significant quantity of PEDOT nanoparticles adhering to the rGO framework.

Figure 4 presents the TEM images of a mixture of GO dispersion and PEDOT:PSS dispersion. From Figure 4a, it can be observed that GO is highly electron transparent and exhibits a wrinkled shape. In Figure 4b, PEDOT nanoparticles are observed on the surface of GO. The high-resolution TEM (HRTEM) image in Figure 4c reveals that GO exhibits a few-layer structure of 2–3 layers, with an interlayer distance of about 0.34 nm, which is comparable to the interlayer distance of graphite.

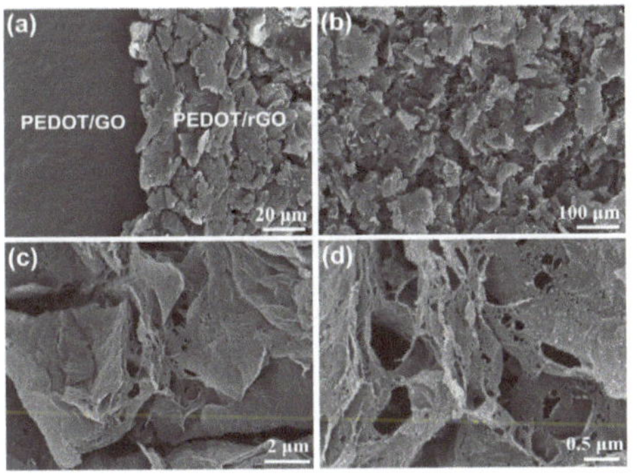

Figure 3. Analysis of material morphology: (**a**) the alteration in surface texture of the PEDOT/GO film due to laser scribing; (**b**) SEM image at low magnification; (**c**,**d**) SEM images at higher magnification.

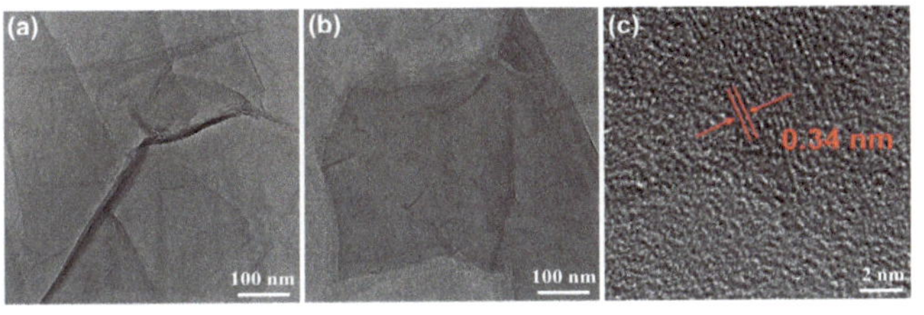

Figure 4. TEM images of the mixture of GO dispersion and PEDOT:PSS dispersion at different magnifications: (**a**,**b**) TEM images with a scale bar of 100 nm; (**c**) TEM image with a scale bar of 2 nm.

3.2. Structure Analysis

Figure 5 presents the Raman spectra of rGO and PEDOT/rGO electrode materials. The D peak, G peak, and 2D peak associated with rGO are located 1342 cm^{-1}, 1578 cm^{-1}, and 2650 cm^{-1}, respectively (where the D and G peaks are the Raman characteristic peaks of carbon atom crystals, and the width of the 2D peak correlates with the uniformity of interlayer spacing and the degree of defects). For the PEDOT/rGO electrode materials, the D and G peaks appear at 1350 cm^{-1} and 1582 cm^{-1}, indicating the presence of carbonaceous components within the composite electrode material [39]. Further analysis reveals that the intensity ratio of the D to G bands (ID/IG) for PEDOT/rGO is 0.97, which is higher than the 0.89 of rGO, suggesting that the presence of PEDOT nanoparticles promotes the formation of disordered multilayer graphene [40]. Additionally, a very broad 2D peak appears at 2700 cm^{-1} in the Raman spectra of PEDOT/rGO, reaffirming that the presence of PEDOT nanoparticles favors the production of disordered multilayer graphene [41].

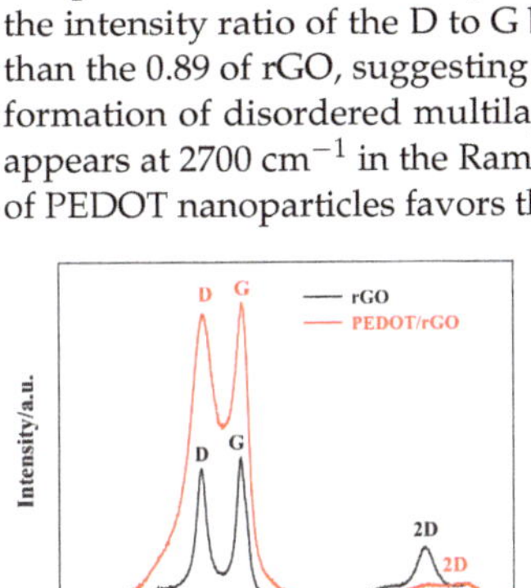

Figure 5. Raman spectra of rGO and PEDOT/rGO electrodes.

Figure 6 displays the XPS analysis of the PEDOT/rGO composite electrode material. The XPS technique reveals information on the chemical composition and chemical bonds of PEDOT/rGO. From the full XPS spectrum shown in Figure 6a, a significantly high peak of

C 1s, along with lower peaks of O 1s, N 1s, and S 2p, can be observed. Figure 6b presents the high-resolution XPS spectrum of C 1s, showcasing three peaks located at 284.8, 286.3, and 289.1 eV, corresponding to the C−C, C−O, and C=O bonds [42]. Figure 6c, the high-resolution XPS spectrum of O 1s, exhibits two typical peaks corresponding to the C=O (530.6 eV) and C−O (533.6 eV) bonds [43]. The high-resolution XPS spectrum of N 1s shown in Figure 6d is fitted with three peaks at 397.9, 399.8, and 401.5 eV, attributed to pyridinic N, pyrrolic N, and graphitic N bonds [44]. Figure 6e shows the high-resolution XPS spectrum of S 2p, with the presence of C−S 2p3/2 (164.0 eV) and C−S 2p1/2 (165.3 eV) bonds, proving the formation of chemical bonds between sulfur and carbon atoms [45].

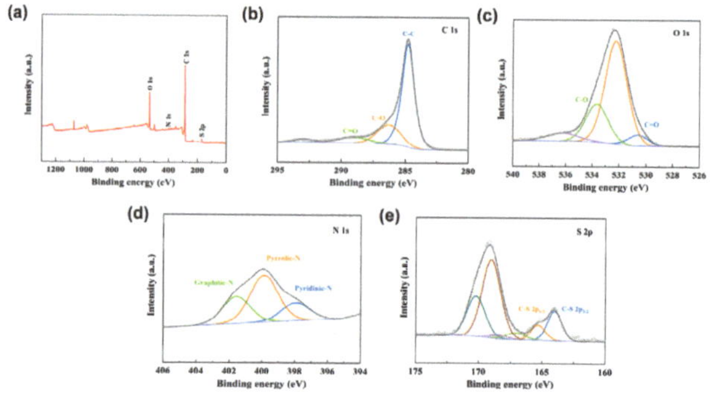

Figure 6. XPS spectra of the composite electrode: (**a**) full spectrum; (**b**) C 1s; (**c**) O 1s; (**d**) N 1s; (**e**) S 2p.

3.3. Electrochemical Performance Analysis

Figure 7 demonstrates the study of the electrochemical performance of MSCs and the optimization of PEDOT/rGO composite electrodes, where the varying content of PEDOT directly affects the performance of individual PEDOT/rGO composite electrodes, with a specific ratio yielding optimal electrode performance. The fabrication of a reduced graphene oxide (rGO) electrode begins with its precursor, a GO film. After patterning the GO film using the laser scribing technique, the surface of the film is converted into an rGO electrode. For convenience, the MSC assembled based on this rGO electrode is designated as pristine rGO MSC. The fabrication of a PEDOT/rGO electrode begins with a composite precursor film. Initially, 10 mL of a GO dispersion with a concentration of 2 mg/mL is mixed with a specific volume of a PEDOT:PSS dispersion containing 1.2% wt active chemical content of PEDOT:PSS. After curing, the mixture forms a PEDOT/GO-V film, where 'V' denotes 100 times the volume of the PEDOT:PSS dispersion in the mixture. Subsequently, the PEDOT/GO-V film is precisely patterned into the desired electrode shape via laser scribing technique, thereby resulting in the formation of a composite. Finally, this composite electrode is assembled into a PEDOT/rGO-V MSC. Figure 7a presents the CV curves of the MSCs. For pristine rGO MSC, the shuttle-like contour of the CV curve indicates that the areal capacitance of pristine rGO MSC is solely provided by the electric double-layer capacitance (EDLC). In contrast, for PEDOT/rGO-V MSC, the fluctuating CV curves reveal that the areal capacitance of PEDOT/rGO-V MSC is provided by both EDLC and pseudocapacitance. Additionally, the integral area of the CV curves increases from pristine rGO MSC to PEDOT/rGO-25 MSC, and then decreases from PEDOT/rGO-25 MSC to PEDOT/rGO-35 MSC. Figure 7b illustrates the GCD curves of the MSCs, which show that the discharge time of PEDOT/rGO-V MSC follows a similar trend of initial increase followed by a decrease, aligning with the changes observed in the integral area of the CV curves. The areal capacitance of the MSCs shown in Figure 7c are calculated based on the CV curves presented in Figure 7a. After uniformly mixing 10 mL of GO dispersion with 0.25 mL of PEDOT:PSS dispersion, a PEDOT/GO-25 film is formed, which contains 20 mg of GO and 3 mg of PEDOT. Based on this specific composition of precursor film, a composite electrode is subsequently fabricated and assembled into a PEDOT/rGO-25 MSC. According to Figure 7c, the PEDOT/rGO-25 MSC exhibits the highest areal capacitance, indicating that the composite electrode made from the PEDOT/GO-25 film exhibits optimal performance. This further suggests that even slight variations in the PEDOT content of the composite precursor film can significantly impact the electrochemical performance of the electrode material. Figure 7d–f display the electrochemical impedance spectroscopy (EIS) graphs for PEDOT/rGO-15, PEDOT/rGO-25, and PEDOT/rGO-35, respectively. Understanding the mass content of PEDOT and GO in the composite precursor film is beneficial for EIS analysis. After mixing 10 mL of GO dispersion with 0.35 mL of PEDOT:PSS dispersion, a PEDOT/GO-35 film is formed, which contains 20 mg of GO and 4.2 mg of PEDOT. Based on this specific composition of precursor film, a composite electrode is subsequently fabricated and assembled into a PEDOT/rGO-35 MSC. In the high-frequency domain, the x-coordinate at which the impedance curve intersects the real axis represents the equivalent series resistance (ESR) of the MSC, incorporating the resistances of both the electrode and the electrolyte. According to the insets in the EIS graphs, which provide a magnified view of this domain, the ESR values are 45.7 Ω for PEDOT/rGO-15 MSC, 52.8 Ω for PEDOT/rGO-25 MSC, and 146.5 Ω for PEDOT/rGO-35 MSC. Analysis of the three EIS graphs reveals that the intercept on the real axis, which represents the ESR of the MSC, progressively increases as the concentration of PEDOT in the precursor film rises. Thus, with the electrolyte held constant, it is evident that the resistance of the electrode increases as the PEDOT concentration rises. Figure 7d–f demonstrate that the resistance of the electrode fabricated from the PEDOT/GO-35 film is significantly higher than those prepared from the other two precursor films. While the variation in ESR between PEDOT/rGO-15 MSC and PEDOT/rGO-25 MSC is slight, a significant increase observed between PEDOT/rGO-25 MSC and PEDOT/rGO-35 MSC clearly demonstrates the impact of higher PEDOT concentrations on the performance of PEDOT/rGO composite materials. The porous structure of rGO in these materials typically enhances ion and electron transport, thereby improving overall conductivity. However, the impact of PEDOT concentration is critical: a moderate concentration of PEDOT slightly increases the resistance, while excessively high concentrations significantly increase it, leading to detrimental effects. Specifically, high concentrations result in the clogging of rGO pores and the accumulation of polymer chains, which impede both ion and electron mobility within the composite material, thereby reducing its conductivity. The reduced conductivity leads to uneven charge distribution across the electrode material, which restricts effective charge storage in specific areas. Consequently, this condition causes wastage of active electrode material and decreases the areal capacitance. If the detrimental effects of reduced conductivity, caused by an excessive concentration of PEDOT nanoparticles in the electrode material, outweigh the benefits of enhanced electrochemical activity, the areal capacitance of the PEDOT/rGO-35 MSC will be lower compared to that of the pristine rGO MSC.

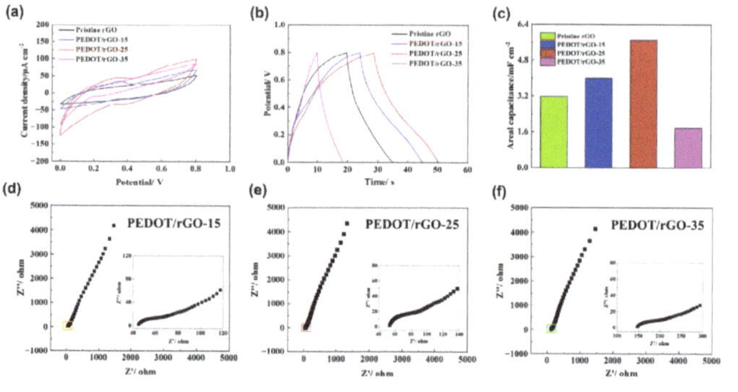

Figure 7. Electrochemical characteristics of the PEDOT/rGO MSC: (**a**) the CV curves at 5 mV/s; (**b**) the GCD curves at 70 μA/cm^2; (**c**) the areal capacitance of MSCs; (**d**) the EIS graphs of the PEDOT/rGO-15 MSC; (**e**) the EIS graphs of the PEDOT/rGO-25 MSC; (**f**) the EIS graphs of the PEDOT/rGO-35 MSC.

Figure 8 presents the results of CV and GCD tests for pristine rGO MSC and PEDOT/rGO-25 MSC. Figure 8a shows the CV curves of pristine rGO MSC, where the curve at 5 mV/s exhibits a shuttle-like contour. This indicates that pristine rGO MSC is capable of accumulating and releasing charge without any chemical reaction, relying solely on electric double-layer capacitance (EDLC) for the storage and release of energy. The EDLC arises from the electrostatic accumulation of electrolyte ions at the graphene electrode surface in pristine rGO MSC. Based on the CV curve at 5 mV/s, the areal capacitance of pristine rGO MSC is calculated to be 3.19 mF/cm^2. Figure 8b shows that the GCD curves of pristine rGO MSC at 30–90 μA/cm^2 form approximately isosceles triangles, indicating that pristine rGO MSC, fabricated from GO film, exhibits good electrode reversibility and capacitive characteristics. Figure 8c displays the CV curves of PEDOT/rGO-25 MSC at 5–100 mV/s, where noticeable redox peaks are observed. This finding indicates that the PEDOT/rGO-25 MSC stores energy through a combination of EDLC and pseudocapacitance. The pseudocapacitance in the PEDOT/rGO-25 MSC originates from rapid and reversible Faradaic processes associated with PEDOT. These processes involve the transfer of electrons and ions at or near the surface of the electrode material, thereby facilitating the efficient storage and release of energy. In the CV testing of the PEDOT/rGO-25 MSC, even as scan rates increased, the displacement of the redox peaks on the CV curves remained very slight. This observation indicates that the electrode material can rapidly facilitate charge transfer reactions even under fast scanning conditions, while the ionic diffusion within the electrolyte is sufficiently swift to promptly adjust to changes in potential. This rapid responsiveness confirms the high reversibility of the electrode reactions across a range of scanning rates, thereby demonstrating excellent electrochemical kinetics at the electrode/electrolyte interface. The areal capacitance of the PEDOT/rGO-25 MSC is calculated to be 5.78 mF/cm^2 based on the CV curve at 5 mV/s, meaning that PEDOT/rGO-25 MSC has achieved an 81% performance improvement compared to pristine rGO MSC. Figure 8d illustrates the GCD curves of the PEDOT/rGO-25 MSC at current densities ranging from 30 to 90 μA/cm^2. Notably, a plateau observed during both the charging and discharging phases corresponds to the oxidation and reduction peaks identified in the CV curves. Furthermore, as the current density increases, the discharge time essentially halves, indicating that even under higher loads, PEDOT/rGO-25 MSC can respond quickly, demonstrating excellent rate performance.

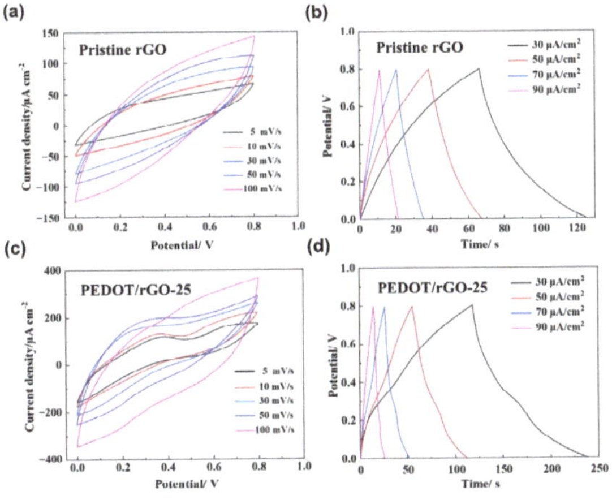

Figure 8. Electrochemical performance tests: (**a**) CV curves of the pristine rGO MSC at 5–100 mV/s; (**b**) GCD curves of the pristine rGO MSC at 30–90 μA/cm^2; (**c**) CV curves of the PEDOT/rGO-25 MSC at 5–100 mV/s; (**d**) GCD curves of the PEDOT/rGO-25 MSC at 30–90 μA/cm^2.

Figure 9 illustrates the flexibility and cyclic stability tests of the PEDOT/rGO-25 MSC. Flexibility testing is performed by bending the MSC to various angles, while CV testing is employed to assess the electrochemical performance of the MSC under these different bending conditions. This method is used to evaluate the stability and reliability of its performance when subjected to changes in physical form. Specifically, CV measurements were performed on the MSC bent at various angles, including 135°, 90° (forming a right angle), 45°, and 0° (flat), with all tests conducted at scan rates of 5 mV/s. The flexibility of the MSC is assessed by comparing the integrated areas of the CV curves obtained at different bending angles. From Figure 9a, it is evident that the four CV curves closely overlap, demonstrating that the PEDOT/rGO-25 MSC exhibits nearly identical performance across all bending angles. This straightforward testing method confirms the robust flexibility of the fabricated PEDOT/rGO-25 MSC. Figure 9b illustrates the cyclic life curve of the PEDOT/rGO-25 MSC following 5000 charge/discharge cycles at a current density of 0.05 mA/cm^2. An analysis of the curve reveals that the capacitance retention is 85.4%; this high level of capacitance maintenance suggests that the PEDOT/rGO-25 MSC exhibits excellent durability and stability.

Table 1 lists three types of previously reported MSCs, including the types of electrodes and electrolytes used, along with their respective areal capacitances. The data clearly demonstrate that MSCs utilizing PEDOT/rGO composite electrodes exhibit a certain leading edge in performance.

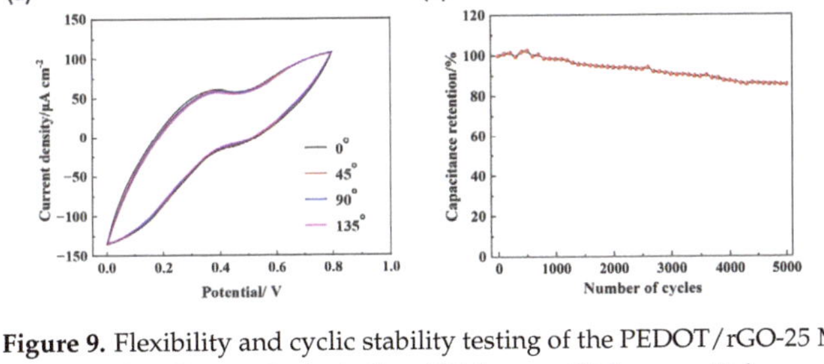

Figure 9. Flexibility and cyclic stability testing of the PEDOT/rGO-25 MSC: (**a**) CV curves obtained at various bending angles including 135 degrees, 90 degrees, 45 degrees, and 0 degrees; (**b**) the cyclic life curve.

Table 1. Performance comparison of the PEDOT/rGO MSCs with other graphene-based MSCs.

Electrode	Electrolyte	Areal Capacitance (mF/cm^2)	Ref.
NiOOH/Ni(OH)$_2$/Graphene	PVA/KOH	0.75	[46]
rGO/CNT	PVA/H_3PO_4	2.6	[47]
Graphene/MWNT	PVA/H_3PO_4	2.54	[48]
PEDOT/rGO	PVA/LiCl	5.78	This work

4. Conclusions

The energy storage mechanism of PEDOT/rGO MSCs is a composite mechanism, relying on both pseudocapacitance and electric double-layer capacitance (EDLC) for efficient energy storage. The pseudocapacitance originates from the rapid and reversible Faradaic processes of PEDOT, involving the transfer of electrons and ions in the surface or near-surface region of the electrode material, thus enabling the storage and release of energy. The conductive polymer framework of PEDOT provides a wealth of active sites and excellent conductivity, allowing it to exhibit high pseudocapacitive performance in electrochemical reactions. On the other hand, EDLC is generated by charge separation in the rGO layers, where charges form electrostatic attractions between the electrode surface and the electrolyte. This mechanism does not involve the chemical transformation of charges, making it a non-Faradaic process. By combining these two mechanisms, PEDOT/rGO MSCs not only achieve high energy storage, but also maintain good cycle stability and rapid charge–discharge capability. The preparation strategy of this composite material, by optimizing the interaction and ratio between PEDOT and rGO, further enhances its electrochemical performance, thereby expanding the application scenarios of flexible MSCs based on PEDOT/rGO composite electrodes.

Author Contributions: Conceptualization, H.H. and Y.G.; methodology, H.H. and Y.G.; software, H.H.; validation, H.H.; formal analysis, Y.G.; investigation, J.Z.; resources, J.Z.; writing—original draft preparation, H.H.; writing—review and editing, Y.G. and J.Z.; visualization, J.Z.; supervision, Y.G. and J.Z.; project administration, Y.G. and J.Z. All authors have read and agreed to the published version of the manuscript.

Funding: This work was supported by the Natural Science Foundation of Jiangsu Province under Grant BK20231229.

Institutional Review Board Statement: Not applicable.

Informed Consent Statement: Not applicable.

Data Availability Statement: Data are contained within the article.

Acknowledgments: The authors acknowledge Shiyanjia Lab (www.shiyanjia.com) for providing XPS testing service.

Conflicts of Interest: The authors declare that they have no known competing financial interests or personal relationships that could have appeared to influence the work reported in this paper.

References

1. Seo, W.; Im, S.; Lee, G. Characteristics of the Received Signal of an Ultrasonic Sensor Installed in a Chamber with Micro-Leakage. *Mech. Sci.* **2021**, *12*, 1051–1060. [CrossRef]
2. Azcárate, C.; Mallor, F.; Mateo, P. Tactical and Operational Management of Wind Energy Systems with Storage Using a Probabilistic Forecast of the Energy Resource. *Renew. Energy* **2017**, *102*, 445–456. [CrossRef]
3. Tasnin, W.; Saikia, L.C. Performance Comparison of Several Energy Storage Devices in Deregulated AGC of a Multi-Area System Incorporating Geothermal Power Plant. *IET Renew. Power Gener.* **2018**, *12*, 761–772. [CrossRef]
4. Zhang, W.; Yin, H.; Yu, Z.; Jia, X.; Liang, J.; Li, G.; Li, Y.; Wang, K. Facile Synthesis of 4, 4′-Biphenyl Dicarboxylic Acid-Based Nickel Metal Organic Frameworks with a Tunable Pore Size towards High-Performance Supercapacitors. *Nanomaterials* **2022**, *12*, 2062. [CrossRef] [PubMed]
5. Zhou, Y.; Liu, C.; Li, X.; Sun, L.; Wu, D.; Li, J.; Huo, P.; Wang, H. Chemical Precipitation Synthesis of Porous Ni2P2O7 Nanowires for Supercapacitor. *J. Alloys Compd.* **2019**, *790*, 36–41. [CrossRef]

6. Reddy, G.R.; Kumar, N.S.; Raju, B.D.P.; Shanmugam, G.; Al-Ghurabi, E.H.; Asif, M. Enhanced Supercapacitive Performance of Higher-Ordered 3d-Hierarchical Structures of Hydrothermally Obtained $ZnCo_2O_4$ for Energy Storage Devices. *Nanomaterials* **2020**, *10*, 1206. [CrossRef]
7. Chen, Y.; Yan, R.; Yu, C.; Zhao, C.; Huang, X.; Wei, L. Investigation on Characteristic Parameters Identification and Evolution of Supercapacitor Energy Storage System From Sparse and Fragmented Monitoring Data. *IEEE Access* **2023**, *11*, 56983–56993. [CrossRef]
8. Zhao, Z.; Wang, X.; Yao, M.; Liu, L.; Niu, Z.; Chen, J. Activated Carbon Felts with Exfoliated Graphene Nanosheets for Flexible All-Solid-State Supercapacitors. *Chin. Chem. Lett.* **2019**, *30*, 915–918. [CrossRef]
9. Wang, X.L.; Jin, E.M.; Chen, J.; Bandyopadhyay, P.; Jin, B.; Jeong, S.M. Facile in Situ Synthesis of $Co(OH)_2$–Ni_3S_2 Nanowires on Ni Foam for Use in High-Energy-Density Supercapacitors. *Nanomaterials* **2022**, *12*, 34. [CrossRef]
10. Wang, Y.; Lu, C.; Cao, X.; Wang, Q.; Yang, G.; Chen, J. Porous Carbon Spheres Derived from Hemicelluloses for Supercapacitor Application. *Int. J. Mol. Sci.* **2022**, *23*, 7101. [CrossRef]
11. Li, M.; Zhu, K.; Zhao, H.; Meng, Z.; Wang, C.; Chu, P.K. Construction of α-MnO_2 on Carbon Fibers Modified with Carbon Nanotubes for Ultrafast Flexible Supercapacitors in Ionic Liquid Electrolytes with Wide Voltage Windows. *Nanomaterials* **2022**, *12*, 2020. [CrossRef] [PubMed]
12. Wu, J.; Xu, L.; Zhou, W.; Jiang, F.; Liu, P.; Zhang, H.; Jiang, Q.; Xu, J. Fishnet-Like, Nitrogen-Doped Carbon Films Directly Anchored on Carbon Cloths as Binder-Free Electrodes for High-Performance Supercapacitor. *Glob. Chall.* **2020**, *4*, 1900086. [CrossRef]
13. Ramkumar, R.; Dhakal, G.; Shim, J.J.; Kim, W.K. NiO/Ni Nanowafer Aerogel Electrodes for High Performance Supercapacitors. *Nanomaterials* **2022**, *12*, 3813. [CrossRef]
14. Shah, S.S.; Alfasane, M.A.; Bakare, I.A.; Aziz, M.A.; Yamani, Z.H. Polyaniline and Heteroatoms–Enriched Carbon Derived from Pithophora Polymorpha Composite for High Performance Supercapacitor. *J. Energy Storage* **2020**, *30*, 101562. [CrossRef]
15. Boyd, S.; Geise, N.R.; Toney, M.F.; Augustyn, V. High Power Energy Storage via Electrochemically Expanded and Hydrated Manganese-Rich Oxides. *Front. Chem.* **2020**, *8*, 715. [CrossRef]
16. Dong, J.Y.; Xu, J.C.; Hui, K.N.; Yang, Y.; Su, S.C.; Li, L.; Zhang, X.T.; Ng, K.W.; Wang, S.P.; Tang, Z.K. Homogeneous Core/Shell $NiMoO_4$@$NiMoO_4$ and Activated Carbon for High Performance Asymmetric Supercapacitor. *Nanomaterials* **2019**, *9*, 1033. [CrossRef]
17. Abbas, Q.; Mateen, A.; Khan, A.J.; Eldesoky, G.E.; Idrees, A.; Ahmad, A.; Eldin, E.T.; Das, H.T.; Sajjad, M.; Javed, M.S. Binder-Free Zinc–Iron Oxide as a High-Performance Negative Electrode Material for Pseudocapacitors. *Nanomaterials* **2022**, *12*, 3154. [CrossRef] [PubMed]
18. Li, T.; Chang, X.; Mei, L.; Shu, X.; Ma, J.; Ouyang, L.; Gu, S. Solvothermal Preparation of Spherical Bi_2O_3nanoparticles Uniformly Distributed on Ti_3C_2Txfor Enhanced Capacitive Performance. *Nanoscale Adv.* **2021**, *3*, 5312–5321. [CrossRef] [PubMed]
19. Mu, C.; Sun, X.; Chang, Y.; Wen, F.; Nie, A.; Wang, B.; Xiang, J.; Zhai, K.; Xue, T.; Liu, Z. High-Performance Flexible All-Solid-State Micro-Supercapacitors Based on Two-Dimensional InSe Nanosheets. *J. Power Sources* **2021**, *482*, 228987. [CrossRef]
20. Raman, V.; Rhee, D.; Selvaraj, A.R.; Kim, J.; Prabakar, K.; Kang, J.; Kim, H.K. High-Performance Flexible Transparent Micro-Supercapacitors from Nanocomposite Electrodes Encapsulated with Solution Processed MoS_2 Nanosheets. *Sci. Technol. Adv. Mater.* **2021**, *22*, 875–884. [CrossRef]
21. Tang, K.; Ma, H.; Tian, Y.; Liu, Z.; Jin, H.; Hou, S.; Zhou, K.; Tian, X. 3D Printed Hybrid-Dimensional Electrodes for Flexible Micro-Supercapacitors with Superior Electrochemical Behaviours. *Virtual Phys. Prototyp.* **2020**, *15*, 511–519. [CrossRef]
22. Wang, M.; Liu, C.; Albolkany, M.K.; Zhao, M.; Zhu, C.; Liu, B. Gram-Scale Synthesis of Porous Graphene via Printing Paper Pyrolysis as Supercapacitor Electrodes. *Energy Technol.* **2021**, *9*, 2001025. [CrossRef]
23. Diao, J.; Yuan, J.; Ding, A.; Zheng, J.; Lu, Z. Flexible Supercapacitor Based on Inkjet-Printed Graphene@Polyaniline Nanocomposites with Ultrahigh Capacitance. *Macromol. Mater. Eng.* **2018**, *303*, 1800092. [CrossRef]
24. Scardaci, V.; Compagnini, G. Raman Spectroscopy Investigation of Graphene Oxide Reduction by Laser Scribing. *C* **2021**, *7*, 48. [CrossRef]
25. Liao, J.; Zhang, X.; Sun, Z.; Chen, H.; Fu, J.; Si, H.; Ge, C.; Lin, S. Laser-Induced Graphene-Based Wearable Epidermal Ion-Selective Sensors for Noninvasive Multiplexed Sweat Analysis. *Biosensors* **2022**, *12*, 397. [CrossRef] [PubMed]
26. Lee, J.U.; Lee, J.H.; Lee, C.W.; Cho, S.C.; Hong, S.M.; Ma, Y.W.; Jeong, S.Y.; Shin, B.S. Green Synthesis of Laser-Induced Graphene with Copper Oxide Nanoparticles for Deicing Based on Photo-Electrothermal Effect. *Nanomaterials* **2022**, *12*, 960. [CrossRef] [PubMed]
27. Paterakis, G.; Vaughan, E.; Gawade, D.R.; Murray, R.; Gorgolis, G.; Matsalis, S.; Anagnostopoulos, G.; Buckley, J.L.; O'Flynn, B.; Quinn, A.J.; et al. Highly Sensitive and Ultra-Responsive Humidity Sensors Based on Graphene Oxide Active Layers and High Surface Area Laser-Induced Graphene Electrodes. *Nanomaterials* **2022**, *12*, 2684. [CrossRef] [PubMed]
28. Zhao, Y.; Du, J.; Li, Y.; Li, X.; Zhang, C.; Zhang, X.; Zhang, Z.; Zhou, J.; Pan, X.; Xie, E. Facile Fabrication of Flexible Graphene-Based Micro-Supercapacitors with Ultra-High Areal Performance. *ACS Appl. Energy Mater.* **2020**, *3*, 8415–8422. [CrossRef]
29. Liu, H.; Moon, K.S.; Li, J.; Xie, Y.; Liu, J.; Sun, Z.; Lu, L.; Tang, Y.; Wong, C.P. Laser-Oxidized Fe_3O_4 Nanoparticles Anchored on 3D Macroporous Graphene Flexible Electrodes for Ultrahigh-Energy in-Plane Hybrid Micro-Supercapacitors. *Nano Energy* **2020**, *77*, 105058. [CrossRef]

30. Liu, F.; Liu, C.; Li, X.; Zhang, L.; Zhao, W.; Zhang, G. Graphene-Based Integrated Planar On-Chip Micro-Supercapacitors with No Internal Connection. *Integr. Ferroelectr.* **2020**, *206*, 96–104. [CrossRef]
31. Zhang, G.; Liu, C.; Liu, L.; Li, X.; Liu, F. Graphene-Based Planar on-Chip Micro-Supercapacitors with Winding Interdigitated Microelectrodes. *Ferroelectrics* **2019**, *547*, 129–136. [CrossRef]
32. Pei, Y.R.; Zhao, M.; Zhou, H.Y.; Yang, C.C.; Jiang, Q. Hollow N-Doped Carbon Nanofibers Provide Superior Potassium-Storage Performance. *Nanoscale Adv.* **2020**, *2*, 4187–4198. [CrossRef] [PubMed]
33. Zhou, H.; Zhai, H.J.; Han, G. Superior Performance of Highly Flexible Solid-State Supercapacitor Based on the Ternary Composites of Graphene Oxide Supported Poly(3,4-Ethylenedioxythiophene)-Carbon Nanotubes. *J. Power Sources* **2016**, *323*, 125–133. [CrossRef]
34. Giuri, A.; Colella, S.; Listorti, A.; Rizzo, A.; Mele, C.; Corcione, C.E. GO/Glucose/PEDOT:PSS Ternary Nanocomposites for Flexible Supercapacitors. *Compos. Part B Eng.* **2018**, *148*, 149–155. [CrossRef]
35. Song, K.; Yang, R.; Chen, X.; Wang, X.; Chen, G.; Zhao, N. The Structures of $CoFe_2O_4$/PEDOT Electrodes Effect on the Stability and Specific Capacity for Electrochemical Energy Storage. *Appl. Surf. Sci.* **2021**, *542*, 148670. [CrossRef]
36. Li, Y.; Xia, Z.; Gong, Q.; Liu, X.; Yang, Y.; Chen, C.; Qian, C. Polyaniline Nanowire Arrays Oriented on the Functionalized RGO/PEDOT/PP Fabric Substrate for High Performance Supercapacitors with Mechanical Flexibility. *Synth. Met.* **2021**, *280*, 116891. [CrossRef]
37. Liu, F.; Xie, L.; Wang, L.; Chen, W.; Wei, W.; Chen, X.; Luo, S.; Dong, L.; Dai, Q.; Huang, Y.; et al. Hierarchical Porous RGO/PEDOT/PANI Hybrid for Planar/Linear Supercapacitor with Outstanding Flexibility and Stability. *Nanomicro Lett.* **2020**, *12*, 17. [CrossRef] [PubMed]
38. Li, S.; Chen, Y.; He, X.; Mao, X.; Zhou, Y.; Xu, J.; Yang, Y. Modifying Reduced Graphene Oxide by Conducting Polymer Through a Hydrothermal Polymerization Method and Its Application as Energy Storage Electrodes. *Nanoscale Res. Lett.* **2019**, *14*, 226. [CrossRef]
39. Yan, J.; Huang, Y.; Zhang, X.; Gong, X.; Chen, C.; Nie, G.; Liu, X.; Liu, P. MoS_2-Decorated/Integrated Carbon Fiber: Phase Engineering Well-Regulated Microwave Absorber. *Nanomicro Lett.* **2021**, *13*, 114. [CrossRef]
40. Choudhary, P.; Biswas, S.; Kandoth, N.; Tayde, D.; Chatterjee, A.; Chattopadhyay, S.; Das, A.; Swarnakar, S.; Pramanik, S.K. Graphene Quantum Dots Alleviate ROS-Mediated Gastric Damage. *iScience* **2022**, *25*, 104062. [CrossRef]
41. Islam, K.; Suhail, A.; Pan, G. A Label-Free and Ultrasensitive Immunosensor for Detection of Human Chorionic Gonadotrophin Based on Graphene FETs. *Biosensors* **2017**, *7*, 27. [CrossRef] [PubMed]
42. Hassanin, H.A.; Taha, A. Sonochemical-Assisted Biogenic Synthesis of Theophrasite β-$Ni(OH)_2$ Nanocluster Using Chia Seeds Extract: Characterization and Anticancer Activity. *Nanomaterials* **2022**, *12*, 1919. [CrossRef]
43. He, G.; Yan, G.; Song, Y.; Wang, L. Biomass Juncus Derived Nitrogen-Doped Porous Carbon Materials for Supercapacitor and Oxygen Reduction Reaction. *Front. Chem.* **2020**, *8*, 226. [CrossRef] [PubMed]
44. Lu, Z.; Wang, J.; Cheng, X.; Xie, W.; Gao, Z.; Zhang, X.; Xu, Y.; Yi, D.; Yang, Y.; Wang, X.; et al. Riemannian Surface on Carbon Anodes Enables Li-Ion Storage at −35 °C. *ACS Cent. Sci.* **2022**, *8*, 905–914. [CrossRef] [PubMed]
45. Zhang, Z.; Yi, Z.; Liu, L.; Yang, J.; Zhang, C.; Pan, X.; Chi, F. 3D Hollow Rgo Microsphere Decorated with ZnO Nanoparticles as Efficient Sulfur Host for High-Performance Li-s Battery. *Nanomaterials* **2020**, *10*, 1633. [CrossRef] [PubMed]
46. Feng, X.; Ning, J.; Wang, D.; Zhang, J.; Dong, J.; Zhang, C.; Shen, X.; Hao, Y. All-Solid-State Planner Micro-Supercapacitor Based on Graphene/NiOOH/$Ni(OH)_2$ via Mask-Free Patterning Strategy. *J. Power Sources* **2019**, *418*, 130–137. [CrossRef]
47. Yang, K.; Cho, K.; Kim, S. Electrochemical Characteristics of Flexible Micro Supercapacitors with Reduced Graphene Oxide-Carbon Nanotubes Composite Electrodes. *Superlattices Microstruct.* **2018**, *118*, 145–151. [CrossRef]
48. Yun, J.; Kim, D.; Lee, G.; Ha, J.S. All-Solid-State Flexible Micro-Supercapacitor Arrays with Patterned Graphene/MWNT Electrodes. *Carbon* **2014**, *79*, 156–164. [CrossRef]

energies

Article

Honeycomb-like Hierarchical Porous Carbon from Lignosulphonate by Enzymatic Hydrolysis and Alkali Activation for High-Performance Supercapacitors

Xin Zhang [1,2], Shi Liu [1], Yuqi Zhao [1], Haicun Yang [1] and Jinchun Li [1,2,3,*]

[1] School of Materials Science and Engineering, Changzhou University, Changzhou 213164, China
[2] National-Local Joint Engineering Research Center of Biomass Refining and High-Quality Utilization, Changzhou 213164, China
[3] Jiangsu Key Laboratory of Environmentally Friendly Polymeric Materials, Changzhou University, Changzhou 213164, China
* Correspondence: ljc999@cczu.edu.cn

Abstract: Porous carbon materials (PCs) were prepared via hydrothermal carbonization from calcium lignosulfonate (CL) based on enzymatic hydrolysis and alkali activation. The effects of enzymatic hydrolysis and different KOH feeding ratios on the structure and electrochemical properties of enzymatic hydrolysis CL (EHCL)-derived PCs were evaluated in detail. The results showed that the EHCL-derived PCs showed a higher SSA than that of CL. When the mass ratio of KOH/EHCL was 3/2, the PCs exhibited a honeycomb-like microscopic morphology with a specific surface area of up to 1771 m^2/g and a 3D hierarchical porous structure composed of abundant micropores, mesopores, and macropores. As an electrode in a supercapacitor, the highest specific capacitance was 147 F/g at a current density of 0.25 A/g, and it maintained 78% of the initial value at a high current density of 10 A/g. The excellent electrochemical cycle and structural stability were confirmed on the condition of a higher capacitance retention of 95.2% after 5000 times of galvanostatic charge/discharge. This work provides a potential application of CL in high-performance supercapacitors.

Keywords: calcium lignosulfonate; enzymatic hydrolysis; alkali activation; hierarchical porosity; supercapacitors

Citation: Zhang, X.; Liu, S.; Zhao, Y.; Yang, H.; Li, J. Honeycomb-like Hierarchical Porous Carbon from Lignosulphonate by Enzymatic Hydrolysis and Alkali Activation for High-Performance Supercapacitors. *Energies* **2023**, *16*, 3824. https://doi.org/10.3390/en16093824

Academic Editors: Huang Zhang and Yuan Ma

Received: 18 March 2023
Revised: 20 April 2023
Accepted: 27 April 2023
Published: 29 April 2023

1. Introduction

As a basic demand for social progress and development, the energy demand in transportation, electronics, smart devices, and other fields of the world has significantly increased in recent years, and new energy technologies have gradually become the focus of researchers [1,2]. Relatively clean energy conversion technologies such as new batteries and supercapacitors are considered to be the key to solving the future energy crisis [3,4]. Among these, supercapacitors with higher power densities, high-efficiency energy storage systems, and excellent durability have attracted relatively strong attention, and have been widely used in electronic equipment in daily life, national defense, and military fields [5,6]. These green and sustainable energy storage devices can effectively avoid the excessive consumption of fossil raw materials—especially high-efficiency, environmentally friendly, and low-cost supercapacitors with excellent performance—because most carbon source materials are currently related to fossil fuels such as ethanol, acetylene, and pyridine [7,8]. Biomass energy has a large storage capacity, low development cost, and low environmental pollution in nature, and it is crucial to switch to bio-renewable carbon sources [9]. Generally, the performance of supercapacitors is mainly determined by their electrode materials, and high-performance electrode materials can endow superelectric devices with a higher energy density [10]. Carbon materials have been the research focus in the field of electrode materials due to their high stability and preferred electrical conductivity. Importantly, their structural adjustability makes it possible to obtain porous carbon materials with a high specific surface area (SSA) and an ideal pore structure [11–13]. Compared with other new carbon-based electrode materials (nano-onions, nanotubes, nanospheres, aerogels, graphene, etc.), porous carbon has the advantages of a wide source of precursors, designability of diversified morphology, relatively simple preparation process, and low cost, resulting in the main direction for the future development of supercapacitor electrode materials [14–16].

Compared with fossil fuels and their derivatives, the use of sustainable, inexpensive, and environmentally friendly bio-based materials to prepare porous carbons is more attractive [17,18]. Biomass materials as new energy are mainly composed of polysaccharides, lignin, proteins, chitin, nanocellulose, and wood, which can be modified by physical or chemical methods to prepare renewable carbon materials in the presence of an activator [19–23]. As an amorphous polymer, lignin is a type of natural biomass fuel with abundant reserves on earth, and its molecular structure endows it with a high carbon content. At present, lignin and its derivatives are the primary by-products of the paper industry, with an amazing output in the world every year. The main treatment method is to directly burn it to recycle new energy, but this also causes serious environmental pollution [24,25]. The molecular unit of lignin is based on hydroxyphenylpropane, which is linked by carbon–carbon bonds or other chemical bonds [26]. This type of high-carbon macromolecular structure makes lignin widely used in the field of flame-retardant polymers [27,28]. Moreover, the chemical composition of lignin includes three propane-type phenol units, and this macromolecule with a high carbon content is an ideal precursor for preparing porous carbon materials. In recent years, low-cost lignin-derived carbon materials have also been used in dye adsorption, titanium dioxide capture, and energy storage technology to explore potential ways to solve environmental and climate problems [29–31]. Compared with pure lignin and lignosulfonates, the enzymatic hydrolysis of lignin has no influence on its main chemical structure; the carbon skeleton can be doped with some nitrogen and oxygen at the same time, which has been demonstrated to be an effective approach to enhance the hydrophilicity and electrochemical performance of lignin-derived carbon materials [32,33].

Unlike the metal oxide click materials used in most pseudocapacitors, carbon-based materials mostly feature double-layer capacitance, in which electrostatic charges are accumulated and released in the electric double-layer formed at the interface between the electrode and the electrolyte [34–36]. This type of electric double-layer capacitor (EDLC) is expected to overcome the poor cycle stability and poor conductivity of pseudocapacitors, and has a high-cost performance [37]. Nevertheless, SSA and hydrophilicity are important indicators of EDLC electrode materials, and hydrophilic porous structures with a high SSA can increase the contact area of electrodes and electrolytes and provide enough reservoirs and smooth channels for electrolyte diffusion and ion transport [38,39]. Although traditional porous carbon materials (PCs) have a high specific surface area, the pore size distribution (0.5–1.1 nm) of their microporous structure is usually narrow, which leads to a poor rate performance due to the nanoporous structure, severely limiting the ion diffusion [40]. Materials such as carbon nanotubes (CNTs), graphene, and carbon fibers (CFs) have been often used as a support for flexible electrode materials, and the surface needs to be chemically modified to reconstruct a suitable porous structure [41–43]. In terms of improving the double-layer capacitance rate performance, novel PCs with hierarchical porosity have more advantages in the field of double-layer capacitors than those with a uniform pore size distribution [44]. Specifically, macropores can provide reservoirs for ion buffering, mesopores can shorten the distance that ions are transported to internal channels, and micropores can provide more ion accumulation sites. The preparation of PCs with hierarchical porosity by the template method is the common method reported in the literature. The template materials involved include zeolite, silica, and metal–organic frameworks [45–47]. However, the complex synthetic steps and expensive price of these template materials make them unable to be mass-produced.

It has been fully proved that chemical activation also provides a facile and effective method to obtain porous structures with a high SSA. For example, under an inert gas environment, carbon precursors can be mixed with KOH, K_2CO_3, etc., and then pyrolyzed to produce a porous structure through multiple chemical and physical reactions at high temperatures [48]. Among them, template-free KOH pore-forming technology is regarded as relatively easy and inexpensive to produce a carbon material with hierarchical porosity, in which the pore structure and electrochemical properties tend to be more dependent on the precursor and KOH feed ratio [49]. In addition to the type of activator and activation process parameters, the intrinsic structure difference of precursors also leads to the diversification of the PCs' microstructure during carbonization [50,51]. Generally speaking, precursors with smaller molecular weights are beneficial to the formation of carbon materials with a high porosity, improving the graphitization degree when under direct carbonization. The decomposing enzyme contained in the bacterial strain can effectively reduce the molecular weight of lignin. It has been reported that the biotransformation of lignin to lipids, adipic acid, and polyhydroxyalkanoate can be easily realized in the presence of some bacteria [52].

Calcium lignosulfonate (CL), as the main by-product of the paper industry, is often used as a multi-component polymer anionic surfactant. CL may also be preferred as a more ideal precursor of porous carbon materials than lignin, which endows it with a potential application in porous carbon electrode materials for supercapacitors [53]. A large number of natural small molecular compounds and calcium ions in CL act as a template and activating synergist, respectively, which promotes the development of the pore structure in PCs [54]. Laccase is a type of commonly used lignin-degrading enzyme, which is widely distributed in plants, insects, fungi, and bacteria. Enzymatic hydrolysis lignin shares a similar chemical composition to that of native lignin, and nitrogen and oxygen can be doped to the carbon skeleton through biological fermentation during the enzymatic hydrolysis process, thus enhancing the hydrophilicity of carbon materials and introducing pseudocapacitance [55]. For CL, in addition to the above advantages, the deep degradation of CL by laccase can also improve the uniformity of calcium ion distribution in CL, obtaining a more ideal pore structure.

Hence, in this work, hierarchical PCs were prepared from low-cost calcium lignosulfonate (CL) by combining enzymatic hydrolysis and KOH activation. Firstly, CL was enzymatically hydrolyzed with laccase to depolymerize the macromolecules and promote carbonization. Secondly, the enzymatic hydrolysis CL (EHCL) obtained was carbonized in the presence of KOH to generate 3D hierarchical PCs composed of abundant cross-linked macropores with incalculable micropores and mesopores based on the template-free method. The prepared PCs were subsequently used as EDLC electrodes in liquid electrolytes, and the effect of enzymatic hydrolysis and the lignin/KOH feed ratio on the microstructure and electrochemical performance was investigated in detail. This work promotes the application of low-cost biomass resources in supercapacitors. The preparation of 3D-structured hierarchical PCs is an ideal approach to broaden the application field of lignin and its derivatives.

2. Materials and Methods

2.1. Materials

The calcium lignosulphonate (CL; industrial grade) was supplied by Gongyi Shengshi Refractories Co., Ltd. (Gongyi, Henan, China). Potassium hydroxide (AR), sulfuric acid (98 wt%), and hydrochloric acid (36–38 wt%) were purchased from Sinopharm Chemical Reagent Co., Ltd. (Shanghai, China). Laccase (99 wt%) was bought from Tianjin Mingshi Biotechnology Co., Ltd. (Tianjin, China). Poly(vinylidene fluoride) (PVDF; 9000 HD) was supplied by Arkema Inc. (Serquigny, France). All raw materials and reagents were used directly without any pretreatment.

2.2. Preparation of Enzymatic Hydrolysis Calcium Lignin (EHCL)

In a 500 mL three-necked flask, 10 g of CL was mixed with 100 mL of HCl (pH = 4), and then 1 mL of laccase was added. After full dissolution, the solution was continuously stirred at 50 °C for 6 h under nitrogen protection. After the reaction, the product—enzymatic hydrolysis calcium lignin (EHCL)—was filtrated and repeatedly washed with ultrapure water, followed by drying to a constant weight at 80 °C.

2.3. Preparation of KOH-Activated PCs

In an alumina ark, 3 g of EHCL and 50 mL of 5 wt% H_2SO_4 were introduced. After thorough mixing, the ark was placed in an oven at 180 °C for 24 h for pre-carbonization and then naturally cooled to room temperature to afford hydrochar. After being washed and purified by ultrapure water, 2.4 g of acid-treated EHCL hydrochar and KOH (KOH/EHCL = 1:2, 2:3, 1:1, and 3:2) were mixed and then evenly ground in a planetary ball mill. During the grinding process, about 50 mL of ultrapure water was added to form a uniform slurry. Subsequently, under nitrogen protection, the slurry was heated up to 300 °C and kept for 2 h, and then further heated up to 800 °C at 5 °C/min. After carbonization at 800 °C for 3 h, the residual samples were cooled, washed with HCl (4 mol/L) several times, and thoroughly washed with ultrapure water to neutrality. The prepared PCs were completely dried at 120 °C in an oven and named EHCL-Ka, EHCL-Kb, EHCL-Kc, and EHCL-Kd. Meanwhile, CL was directly activated with KOH under a mass ratio of 3/2 (KOH/CL) to prepare the comparative sample, which was named CL-K.

2.4. Preparation of Solid Electrolyte

The supercapacitor electrodes were prepared according to the method reported previously [32]. Typically, the EHCL-Ks ('s' indicates a, b, c, and d), poly(vinylidene fluoride) (PVDF), and acetylene black (mass ratio of 8:1:1) were mixed with 10 mL of ethanol to form a homogeneous slurry and then molded together with foamed nickel at 3 mPa. Finally, the volatile product was thoroughly removed in a vacuum oven at 100 °C to obtain the solid electrolyte.

2.5. Characterization

Fourier transform infrared (FTIR; Avatar 460, Thermo Nicolet, Madison, WI, USA) was performed to confirm the chemical composition at a frequency range of 500–4000 cm^{-1} in transmission mode. Raman spectra (Raman, DXR, Thermo Scientific, Waltham, MA, USA) and X-ray diffraction (XRD; XRD-6000, Shimadzu, Kyoto, Japan) were used to confirm the physical structures. The microstructure observation of carbon materials was performed using scanning electron microscopy (SEM; JSM-6360LA, JEOL, Tokyo, Japan). The SSA values of different carbon materials were determined by nitrogen adsorption–desorption at 77 K on a TriStar II 3020 micropore analyzer (Micromeritics, Norcross, GA, USA) after degassing at 150 °C for 12 h. The SSA and PSD values were calculated based on the Brunauer–Emmett–Teller (BET) density functional theory (DFT) method. The average molecular weights of CL and EHCL were determined by gel permeation chromatography (GPC). A sodium nitrate aqueous solution of 0.1 mol/L and sodium polystyrene sulfonate with a narrow distribution were used as the eluent and reference, respectively.

The electrochemical performance measurement was tested using three–electrode systems under ambient conditions on an electrochemical workstation (CHI660E, Shanghai Chenhua Instruments, Shanghai, China), and the cyclic voltammetry (CV), galvanostatic charge/discharge (GCD), and electrochemical impedance spectroscopy (EIS) curves were recorded. The mass loading of the active material in the PC electrode was about 1.0 mg/cm^2. A total of 6 M KOH and platinum foil containing Hg/HgO were used as the reference and counter electrode, respectively, and formed a three–electrode system with the PC electrode. The CV curves were recorded at an operating voltage from −0.2 to −1 V, EIS was conducted in the frequency range of 10 mHz to 100 kHz, and the GCD curves were recorded on a LAND CT2001A tester. The gravimetric specific capacitance was obtained from the GCD tests via the following equation (Equation (1)), where C was the specific capacitance (F/g), I was the discharge current (A) during the GCD test, Δt was the discharge time, ΔV was the potential change during the discharge, and m was the weight of the PC electrode.

$$C = (I \times \Delta t)/(\Delta V \times m) \tag{1}$$

The electrochemical performance of EHCL-Kd was also evaluated by a two−electrode system using an assembled symmetrical supercapacitor. In a coin cell, two identical working electrodes were separated by a filter paper separator (the active material load of a single electrode is 3 mg). The electrochemical performances were also determined by the CV and GCD results. The specific capacitances were calculated using the following equation (Equation (2)), where m was the mass of the loading of EHCL-Kd on a single working electrode.

$$C = 2(I \times \Delta t)/(\Delta V \times m) \tag{2}$$

The energy density E (Wh/kg) and power density P (W/kg) were calculated based on the following equations (Equations (3) and (4)).

$$E = C \times \Delta V^2/(2 \times 4 \times 3.6) \tag{3}$$

$$P = 3600 \times E/\Delta t \tag{4}$$

3. Results

SEM characterization was carried out to investigate the morphology of the different precursors and carbon materials. As seen in Figure 1a,b, the surface of CL presented an uneven, porous (1–4 μm), and rough morphology, and the CL-derived hydrochar presented many broken carbon balls (Figure 1c,d). After enzymatic hydrolysis, EHCL exhibited a relatively smooth surface with an obviously decreased particle size (Figure 1e,f), and the structure of the EHCL-derived hydrochar was significantly looser compared with that of CL (Figure 1g,h). The hydrochar derived from CL and EHCL exhibited dramatically different microstructures, which indicated that enzymatic hydrolysis had a significant effect on the carbonization process. The EHCL-derived hydrochar was almost completely composed of carbon blocks with a regular carbon skeleton, higher surface area, and smooth surface. The microstructures of CL-K and EHCL-Kd were further compared (Figure 1i–l). CL-K exhibited collapsed structures with irregular holes, and the inner surface of these holes was distributed with individual small holes with irregular structures (Figure 1i,j). Compared with the EHCL-derived carbon materials, the particles of EHCL-Kd were further loosened and refined, and the single particle showed a distinct porous structure (Figure 1k). In the high-resolution image (Figure 1l), EHCL-Kd presented a honeycomb-like microstructure with a smooth pore wall, interconnected pores, and hierarchical porosity. Macropores with diameters of 100–450 nm could be observed from the high-resolution SEM images, and were interconnected by a large number of mesopores with diameters of 20–80 nm. The SEM results directly demonstrated that enzymatic hydrolysis combined with an alkali catalysis was the necessary step to convert CL into 3D hierarchical porous carbons through the hydrothermal carbonization process.

The morphology and microstructure of EHCL-Kd were also analyzed by the TEM method, and Figure 2 shows the TEM images of EHCL-Kd with different magnifications. In Figure 2a, EHCL-Kd presents a two-dimensional gauze-like structure similar to graphene, which promoted the formation of mesopores. In Figure 2b, at a higher magnification, parallel graphene lattice stripes can be observed around the mesopores, which were formed from the aromatic ring of CL at a high temperature. This two-dimensional structure rich in nanopores could optimize the transport and storage of ions, and the graphite structure could improve the electron transport ability of PCs.

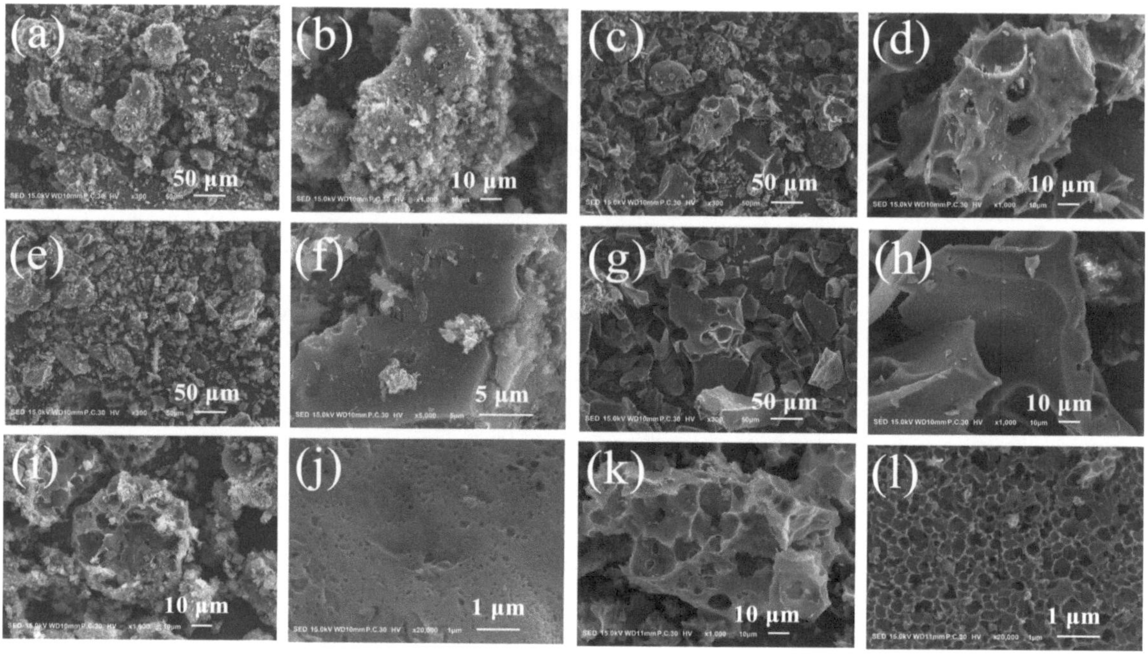

Figure 1. SEM images: (**a**,**b**) CL; (**c**,**d**) hydrochar derived from CL pre-carbonized at 180 °C; (**e**,**f**) EHCL; (**g**,**h**) hydrochar derived from EHCL pre-carbonized at 180 °C; (**i**,**j**) CL-K; and (**k**,**l**) EHCL-Kd.

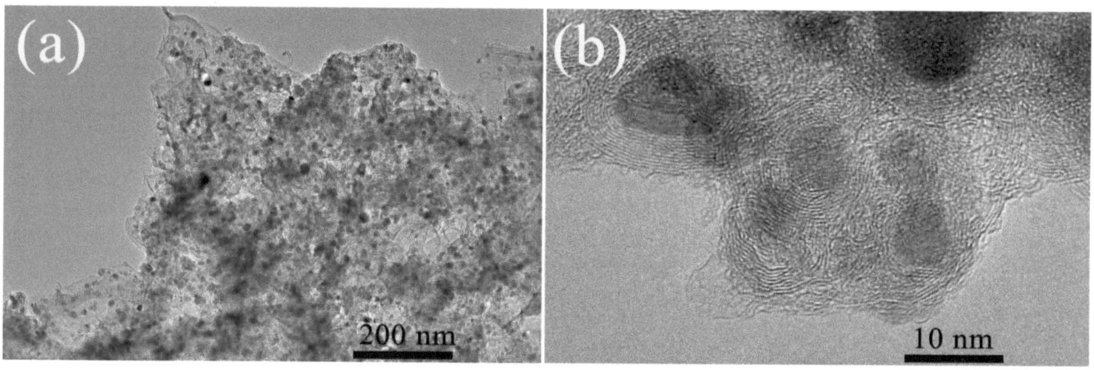

Figure 2. TEM images of EHCL-Kd: (**a**) 50,000 and (**b**) 1,000,000 magnifications.

The above morphology analysis results indicated that the carbon skeleton of EHCL-Kd contained abundant mesopores and micropores as well as interconnected macropores, in which the micropores and mesopores were evenly distributed on the surface of the macropores. This unique 3D structure with a high SSA enlarged the electrode/electrolyte contact interface for EDLCs. Importantly, with the contribution of hierarchical porosity, the high SSA could significantly improve the electrochemical performance. Micropores, mesopores, and macropores play a role in promoting charge accumulation, accelerating ion diffusion, and providing storage buffering space for ions.

The changes in the chemical groups of CL before and after enzymatic hydrolysis were compared by FTIR, and the results are shown in Figure 3a. For CL, 3370 cm^{-1} was the O−H stretching vibration; 2949 and 2837 cm^{-1} belonged to the antisymmetric and symmetric stretching vibration of C-H in the methoxy group, respectively; and the benzene ring skeleton vibration gave two peaks at 1595 and 1515 cm^{-1}. The area near 1421 cm^{-1} was related to the C-H bending vibration. The two sharp peaks at 1163 and 1044 cm^{-1} corresponded with the sulfonic acid group, and the C-S stretching vibration was found at 641 cm^{-1}. In the spectra of EHCL, the characteristic peak corresponding with O-H near 3400 cm^{-1} still existed after enzymatic hydrolysis, and the characteristic peaks at 2937 and 2844 cm^{-1} corresponding with C-H became more obvious. Meanwhile, a new peak of C=O appeared at 1699 cm^{-1}, which was mainly from the laccase molecule. The peak at 1600 cm^{-1} was assigned to the benzene ring skeleton connected with the polar groups; 1508 cm^{-1} was the superposition peak of the benzene ring, C=C, and O-H. The area in the range of 1320–1470 cm^{-1} was the overlapping peaks of a series of groups, mainly including the O-H bending and the in-plane bending of C-H in C=CH. The peak at 1217 cm^{-1} corresponded with the asymmetric stretching of =C-O-C connecting the benzene ring, and 831 cm^{-1} was caused by the out-of-plane vibration of C-H in the multi-substituted benzene ring. Comparing the FTIR spectra of CL and EHCL, it was found that characteristic peaks such as C-H, the benzene ring, and C-O related to organic components in EHCL were more obvious. Therefore, it could be speculated that enzymatic hydrolysis led to the degradation of chlorine and a reduction in the average particle size. It could be seen from the GPC results of CL and EHCL that enzymatic hydrolysis reduced the average molecular weight of CL by about 11%, and the polydispersity index decreased from 1.15 to 1.05 (Figure S1).

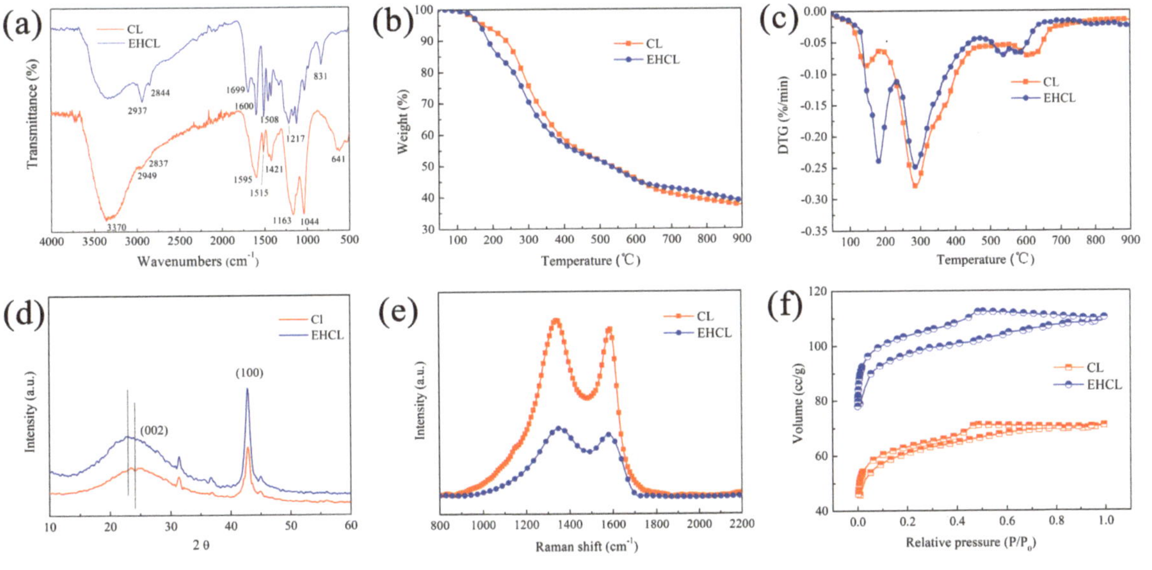

Figure 3. (**a**) FTIR spectra, (**b**) TG curves, and (**c**) DTG curves of CL and EHCL; (**d**) XRD patterns, (**e**) Raman spectra, and (**f**) N_2 adsorption–desorption isotherms of carbon materials derived from CL and EHCL.

The differences in the carbonization process of CL and EHCL were compared by TGA (Figure 3b,c). In general, the activation degree of the lignin chemical bonds in CL was relatively high, so the TGA curve had a wide range of weight loss at 100–900 °C, containing three stages [56]. The first stage, lower than 200 °C, was mainly due to the removal of moisture in the CL, which resulted from the cleavage of some hydroxyl-containing side chains. The violent decomposition stage after 200 °C lasted until about 490 °C, which was mainly due to the decomposition of a large number of heat-labile aliphatic chains. As these side chains contained hydroxyl groups and carbonyl-containing groups, the decomposition generated a large amount of H_2O, CO_2, CO, etc., resulting in a higher weight loss ratio.

Finally, the third stage, in the range of about 550–800 °C, appeared when most of the oxygen-containing groups cracked and deoxygenated. In this main carbonization stage, the residual substances and C-C and C-H in the benzene ring were further degraded to form charcoal residue through the deprivation of hydrogen [57]. After enzymatic hydrolysis, the weight loss ratio in the first stage of EHCL significantly increased, which was possible because a large number of long branches were destroyed into shorter ones during the enzymatic hydrolysis process with lower thermal stability. At the same time, the weight loss ratio of the second stage was relatively lower than that of CL. In addition, the carbonization stage of EHCL advanced; two weight loss peaks corresponding with the pyrolysis appeared on the DTG curve of EHCL and a weak deep carbonization peak was found at about 800 °C. All the changes in the TG and DTG curves demonstrated that the enzymatic modification of CL could promote carbonation.

The crystal structures of the CL- and EHCL-derived hydrochars were compared with an XRD analysis. Figure 3d shows the XRD patterns of the above two hydrochars. Both of them showed broad amorphous carbon and sharp graphitized carbon diffraction peaks at around 23° (002) and 44° (100) [58]. The peak at 23° of EHCL shifted to a lower diffraction angle compared with that of CL, which proved that the interlayer spacing of the carbon layer increased, possibly resulting from the doping of N elements during the enzymatic hydrolysis process. The crystal structures of the two hydrochars were further proved by Raman spectroscopy. As shown in Figure 3e, the Raman spectra of the two hydrochars both showed D and G peaks at about 1345 and 1580 cm^{-1}, corresponding with disordered and graphitized carbon, respectively [59]. It was calculated that the value of ID/IG reduced from 1.1 of CL to 0.96 of EHCL, indicating that enzymatic modification could increase the graphitization degree of carbon materials due to the lower molecular weight and distribution of EHCL [60].

The structural differences between the CL- and EHCL-derived hydrochars were further compared by nitrogen adsorption–desorption isotherm experiments, and the results are shown in Figure 3f. It could be seen that both carbon materials showed typical type II adsorption isotherms with a lower N_2 adsorption quantity, according to the IUPAC classification. In the low-pressure region of $p/p_0 < 0.01$, the sharp increment in the slope of the curves suggested that there were a certain number of micropores in both carbon materials. A hysteresis loop was observed in the region of p/p_0 of 0.4~0.6 without limiting adsorption in the high-pressure region, indicating the formation of mesopores and macropores. Remarkably, the larger hysteresis loop of EHCL indicated that the carbon skeleton structure was looser, and more mesopores and macropores were associated with the EHCL-derived carbon material. The EHCL-derived carbon material exhibited a higher SSA than that of CL (Table 1), and the BET SSA increased from 194 to 307 m^2/g. In addition, it was found that enzymatic hydrolysis could simultaneously increase the micropores and mesopores as well as the average pore size.

Table 1. Porosity characteristics of different carbon materials.

Sample	S_{BET} (m^2/g)	V_{total} (cm^3/g)	V_{micro} (m^3/g)	V_{meso} (m^3/g)	V_{micro}/V_{total} (%)	D (nm)
CL	194	0.11	0.08	0.01	73	2.97
EHCL	307	0.18	0.13	0.03	72	3.94
EHCL-Ka	779	0.51	0.25	0.02	49	3.73
EHCL-Kb	1576	0.62	0.38	0.05	61	2.88
EHCL-Kc	1605	0.81	0.56	0.08	69	3.10
EHCL-Kd	1771	1.17	0.74	0.12	73	3.22
CL-K	1217	0.53	0.33	0.06	62	2.18

Under the same carbonization conditions, the effect of enzymatic hydrolysis on carbonization products was compared (Figure 4). The XRD and Raman spectra showed the amorphous structure of the two samples (Figure 4a,b). It can be seen from Figure 3c that EHCL-Kd exhibited a smaller average particle size than CL-K, which was consistent with the SEM results. The difference between the structure of EHCL-Kd and CL-K was revealed by adsorption–desorption isotherms (Figure 4d). The isotherm curve of CL-K exhibited a type IV hysteresis loop without an apparent saturated adsorption plateau, corresponding with the irregular pore structure. The XRD and Raman results were also in accordance with the SEM images. Moreover, CL-K owned a lower quantity of absorbed product, suggesting that the SSA was lower than that of EHCL-Kd (Table 1). Conversely, EHCL-Kd transformed into a type I isotherm with a higher adsorption capacity, which was due to the more regular honeycomb-like pore structure with hierarchical porosity (Figure 1k,l).

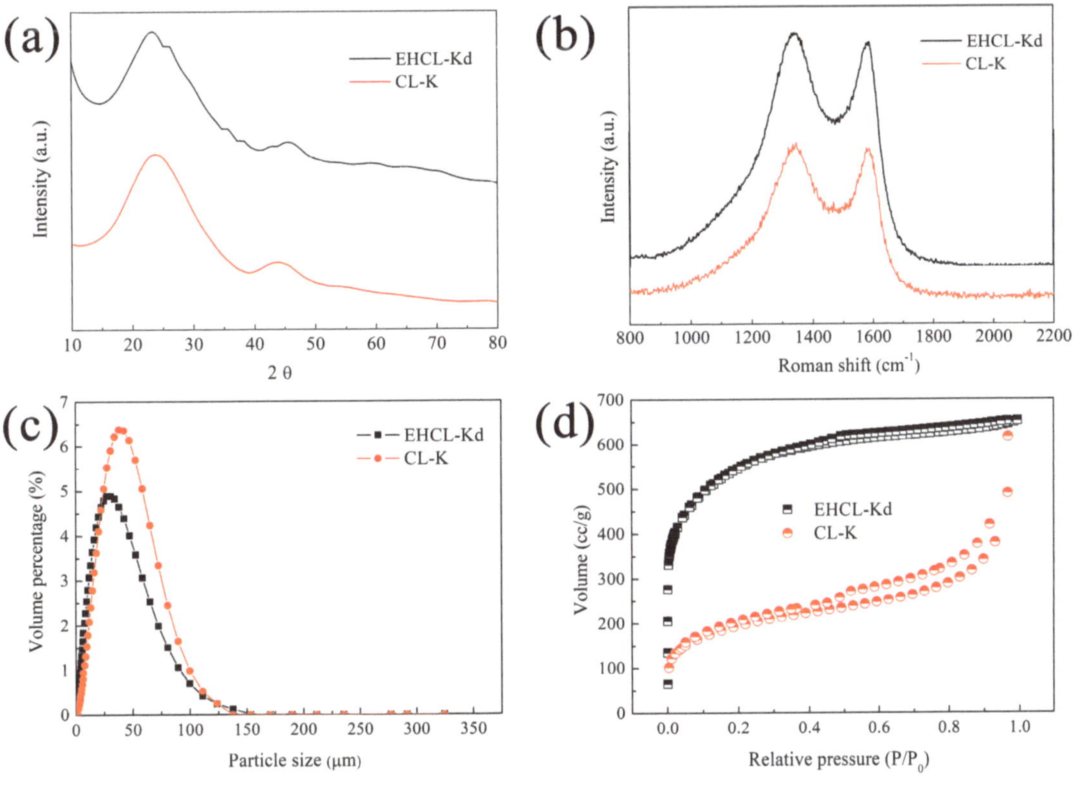

Figure 4. (**a**) XRD patterns; (**b**) Raman spectra; (**c**) particle size distribution; and (**d**) N_2 adsorption–desorption isotherms of CL-K and EHCL-Kd.

The effect of the KOH/EHCL feed ratio on the structure of EHCL-derived PCs was also researched. Figure 5a shows the XRD patterns of the PCs in the wide-angle region under different KOH/EHCL feed ratios, and all spectra showed amorphous and graphitized carbon diffraction peaks at 22° (002) and 44° (100), as expected. However, the diffraction peak intensity at 44° was markedly decreased when compared with that of the EHCL-derived carbon materials, which may have been due to the increase in disorder resulting from the formation of micropores by KOH activation. Increasing the KOH/EHCL feed ratio gradually enhanced the peak intensity at 44°, which increased the graphitization degree. The structure of the EHCL-derived PCs was further confirmed by Raman spectroscopy (Figure 5b). D and G peaks appeared at 1330 and 1560 cm^{-1} in all spectra, corresponding with disordered carbon and SP^2-hybridized graphitized carbon, respectively. Meanwhile, with the increase in KOH dosage, the I_D/I_G decreased from 0.98 to 0.91, confirming that the graphitization degree gradually increased. The collapse and rearrangement of a framework during pyrolysis is beneficial to improve the graphitization degree and the electrical conductivity of PCs.

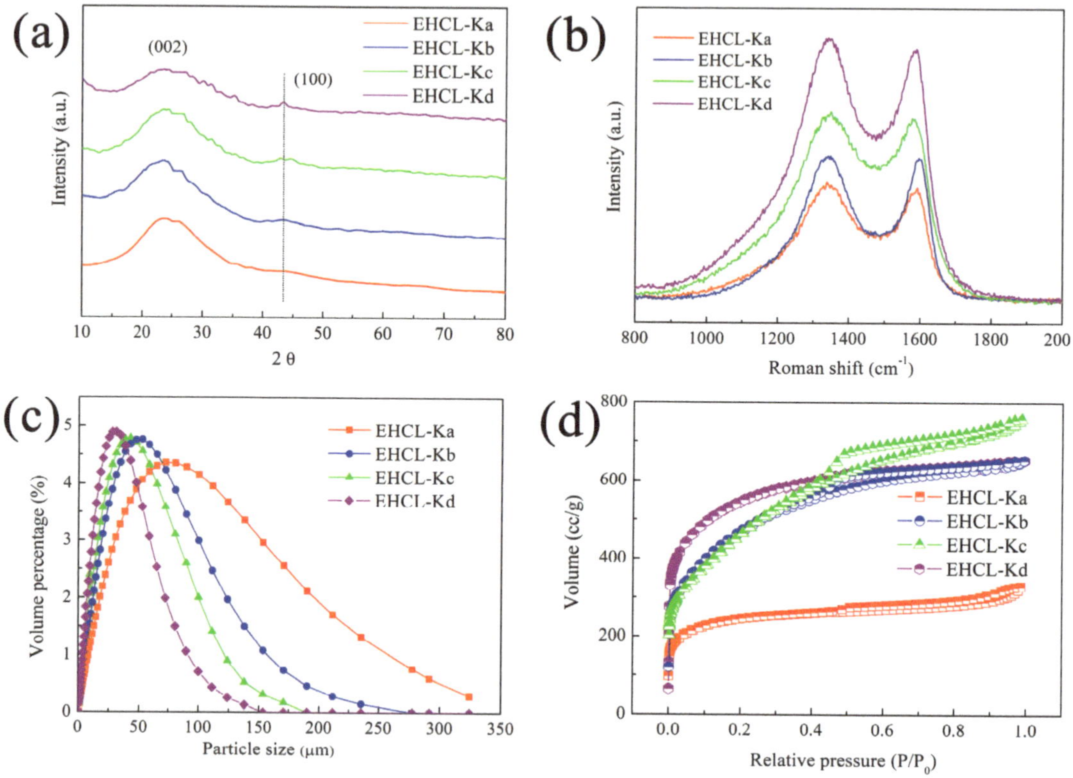

Figure 5. (**a**) XRD patterns; (**b**) Raman spectra; (**c**) particle size distribution; and (**d**) N_2 adsorption–desorption isotherms of EHCL-Ks.

Figure 5c shows the particle size distribution of different PCs. When increasing the KOH dosage, the particle size distribution of carbon materials significantly decreased with a gradual increase in the pore volume, which was beneficial to improve the contact area between the electrode and electrolyte, promoting the diffusion of ions in the internal pores of the electrode. The diversity in the structure of PCs under different KOH dosages was further revealed by nitrogen adsorption technology.

Figure 5d shows the isotherm nitrogen adsorption–desorption curves of different PCs, and the relevant pore data are listed in Table 1. Compared with the EHCL-derived carbon materials, the KOH activation significantly increased the N_2 adsorption capacity, and the curve gradually transformed from type II to type IV, and finally to type I, with evident plateaus with the increase in the KOH/EHCL feed ratio. It suggested that KOH could increase the content of pores, especially the micropore content in carbon materials. As the amount of KOH increased, some micropores were squeezed and deformed, and collapsed by the nearby macropores or mesopores due to the uneven pore size distribution, causing the curve to transform into a type IV characteristic. When the amount of KOH was further increased, the adsorption capacity of the curve in the low-pressure region ($p/p_0 < 0.01$) significantly increased, proving that the proportion of microporous structures in the PCs was further increased. The relatively uniform distribution could stabilize the micropore so that it was not prone to collapse, which corresponded with the characteristics of type I. It can also be seen from Table 1 that the BET SSA and pore volume increased from

779 m^2/g and 0.51 cm^3/g to 1771 m^2/g and 1.17 cm^3/g, respectively, with the increase in the KOH/EHCL ratio. EHCL-Kd showed the highest contents of the total pore, micropore, and mesopore volume than the others, in which the increased mesopore number facilitated the diffusion of ions through the pores [61].

The electrochemical performances of EHCL-based PC electrodes were investigated by a three–electrode configuration, and CV, GCD, and EIS tests were performed in a 6 M KOH aqueous solution. The CV curves of carbon material electrodes at different KOH/EHCL feed ratios are shown in Figure 6a; all the curves showed a similar rectangular shape, with broadening oxidation-reduction peaks appearing in the range of −0.9 to −0.3 v. The above results implied that the EHCL-derived PC electrodes were dominated by EDLC behavior, whilst the corresponding pseudocapacitive behavior was caused by some oxygen-containing functional groups [62]. Among all samples, the CV curve of EHCL-Kd showed the largest area, indicating the best capacitance performance. This was mainly attributed to EHCL-Kd exhibiting the largest SSA, with a unique hierarchical pore size distribution, higher mesopore volume, and an appropriate graphitization degree. Figure 6b shows the GCD curves of different PC electrodes; all curves were approximately isosceles triangles, which further proved the dominating EDLC behavior. Unlike the charging process, the discharge lines of all samples were not perfectly linear and the curvature corresponded with a broadening peak in the CV curve. As expected, EHCL-Kd exhibited the longest charge–discharge time at a current density of 1 A/g, which was consistent with the CV test results. Based on Equation (1), the specific capacitance changes of different PC electrodes under different current densities were calculated, according to the corresponding GCD results (Figure 6c). EHCL-Kd exhibited the largest specific capacitance of 147 F/g at 0.25 A/g and maintained 78% of the initial value at 10 A/g, which indicated its excellent rate performance. Electrolyte ions were first stored in the macropores in EHCL-Kd, and then quickly entered the electrode through the mesopores and smooth pore walls [63].

The capacitive behavior of different PC electrodes during charge storage was further tested by EIS (Figure 6d). In the low-frequency region (magnified figure), the EHCL-Kd electrode distinctly exhibited an almost perpendicular line to the real axis, indicating a relatively low mass transfer hindrance. The higher mesoporous ratio became a low-resistance pathway for electrolyte ion diffusion, and the excellent EDLC properties of the electric layer enabled the fast transport of ions in the electrolyte to the inner pores. A smaller semicircle radius of EHCL-Kd in the high-frequency region indicated its lower interfacial charge transfer resistance. The factors affecting ohmic resistance are not only the resistance of the electrolyte, current collector, active material, coin cell, and separator, but also the contact resistance between the current collector and active material, which could be evaluated by the first intercept value on the real axis. It was known from the EIS curves that the ohmic values of the EHCL-based supercapacitors were 0.83, 0.72, 0.62, and 0.51 Ω, respectively. In addition, the equivalent series resistance of different carbon electrodes could be calculated as 3.5, 2.8, 2.5, and 1.4 Ω, respectively, according to the shorter Warburg-type line (45° slope) in the high-frequency region, indicating a faster ion diffusion rate in the EHCL-Kd electrode due to the larger SSA and hierarchical porous structure [64].

The effect of the 3D structure of EHCL-Kd on the electrochemical performance was further confirmed (Figure 7). The CV curves of EHCL-Kd maintained a rectangular-like shape at different scan rates from 10 to 200 mV/s, which indicated the main EDLC characteristics and preferred rate capability. The GCD curves of EHCL-Kd exhibited in Figure 7b showed high linearity and symmetry at different current densities, demonstrating its excellent electrochemical reversibility and coulombic efficiency. As shown in Figure 7c, the cyclic stability of EHCL-Kd was also evaluated by consecutive GCD for 15,000 cycles at 10 A/g. After 15,000 cycles, the specific capacitance maintained 95.3% of the initial value, indicating that the charge–discharge process was highly reversible. The GCD curve remained close to its initial shape after 15,000 cycles, which also indicated high cycle durability. In Figure 7d, when using EHCL-Kd as the electrode in an aqueous electrolyte for the symmetric supercapacitor, the highest energy density was 3.3 Wh/kg at a power density of 25 W/kg, and maintained 2.0 Wh/kg at 2 kW/kg.

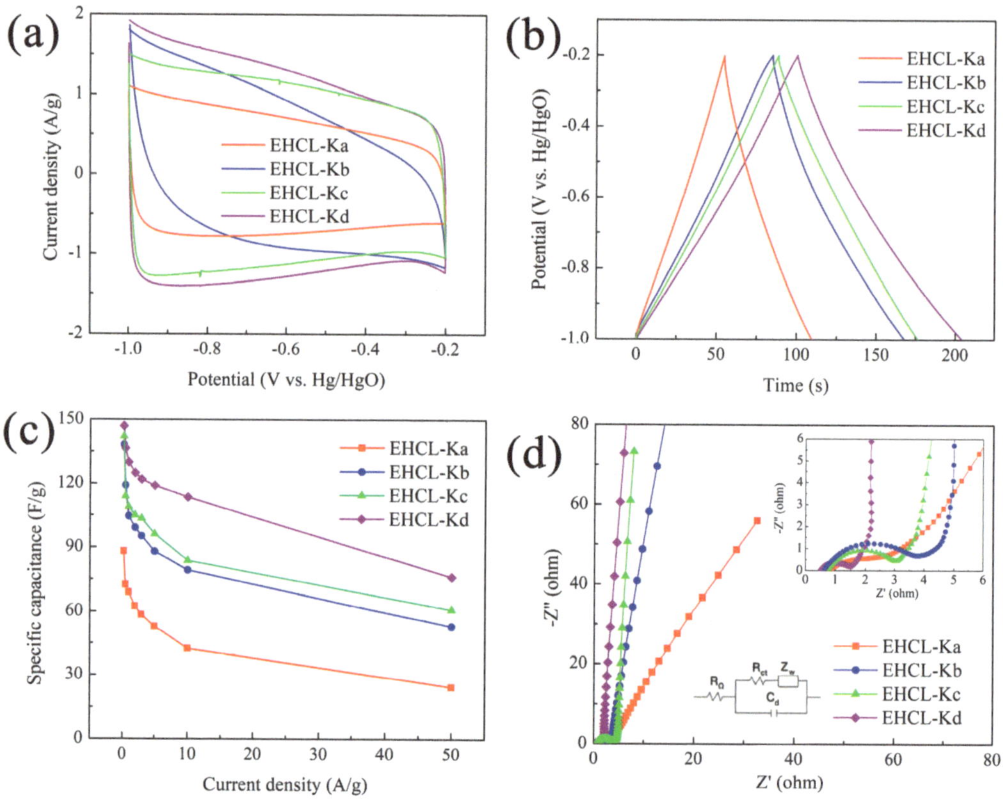

Figure 6. (**a**) CV curves at 100 mV/s; (**b**) GCD curves at 1 A/g; (**c**) rate performance in a three–electrode system in 6 M KOH; and (**d**) Nyquist plots of EHCL-Ks.

Based on symmetrical two–electrode supercapacitors, the electrochemical properties of the EHCL-Kd electrode were also evaluated (Figure 8). In the scanning rate range of 10–200 mVs, the CV curve of the device based on EHCL-Kd had a symmetrical rectangular shape. At the same time, due to its good capacitance behavior and rate performance, the current response rate of the device was faster when the voltage was reversed (Figure 8a). As shown in Figure 8b, when the current density gradually increased from 0.5 A/g to 10 A/g, the GCD curve of the device presented approximate isosceles triangles, which further reflected its excellent capacitance performance. Figure 8c shows the cyclic stability of the device. As expected, after 10,000 cycles, the GC curve had no obvious change and its specific capacitance retention rate remained above 98%, which further proved the cyclic stability of the device.

The corresponding electrochemical performances of CL-K were carried out to compare them with those of EHCL-Kd (Figure S2). The irregular hole structure and lower SSA of CL-K led to a poor electrochemical performance. These results strongly suggested that enzymatic hydrolysis combined with the alkaline activation method was an essential step to convert CL into 3D hierarchical PCs for high-performance supercapacitor electrodes. Furthermore, the above electrochemical properties were compared with those of the reported biochar electrodes in a three–electrode system, and are summarized in Table 2 [65–69]. It was found that, in terms of specific capacitance, rate performance, and cycle stability, our work was superior to that of other reported bio-based carbon material electrodes.

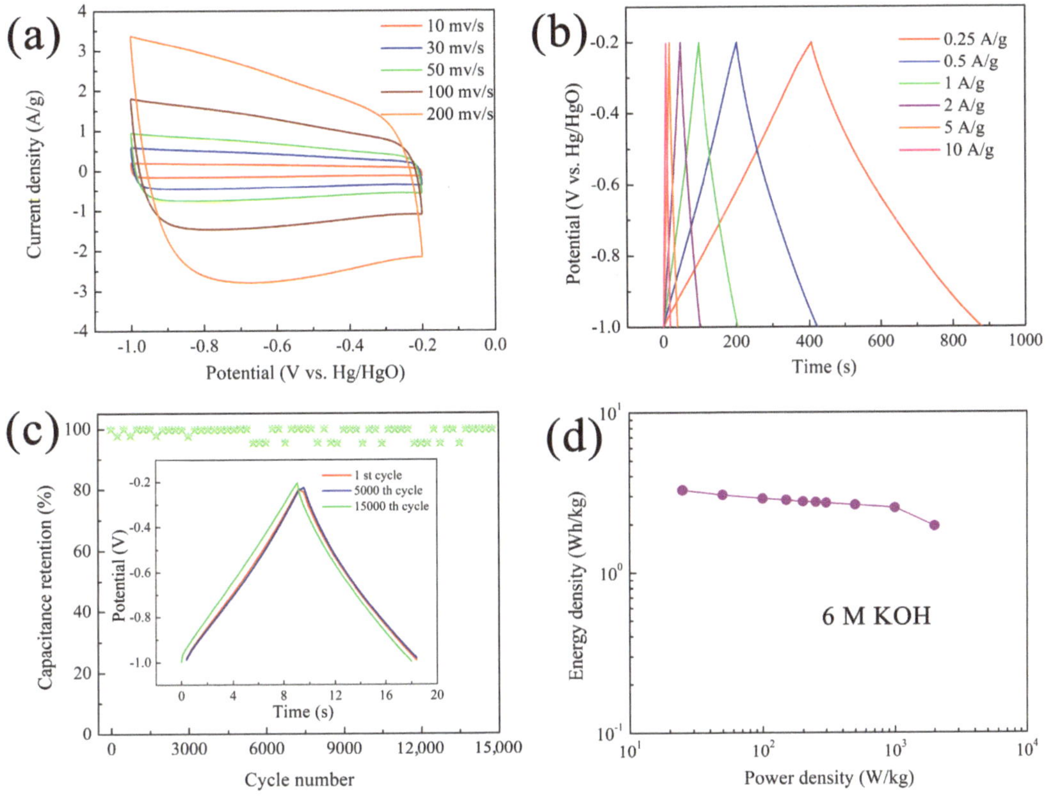

Figure 7. (**a**) CV curves, (**b**) GCD curves, and (**c**) the long-term durability of the EHCL-Kd electrode in a three–electrode system in 6 M KOH (the inset shows the GCD lines of the 1st, 5000th, and 15,000th cycle); (**d**) Ragone plots for the EHCL-Kd–based supercapacitors.

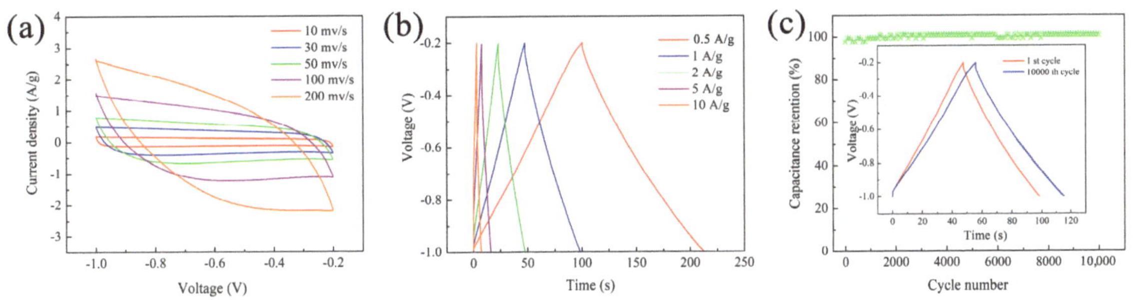

Figure 8. (**a**) CV curves, (**b**) GCD curves, and (**c**) the long-term durability of the EHCL-Kd electrode in a two–electrode system in 6 M KOH (the inset shows the GCD lines of the 1st and 10,000th cycle).

Table 2. Comparison of capacitive performance of EHCL in a three–electrode system.

Biomass Precursor	Current Density (A/g)	Capacitance (F/g)	Rate Capability	Capacitance Retention	Ref.
Sugarcane bagasse	0.5	142	63.4% (10 A/g)	93.9 (5000 cycles at 10 A/g)	[65]
Lignin	0.5	129	84% (5 A/g)	99% (10,000 cycles at 5 A/g)	[66]
Redcedar wood	0.5	115	96% (2.5 A/g)	>90% (5000 cycles at 10 A/g)	[67]
Millet straw	0.2	144	68% (2 A/g)	>95% (10,000 cycles at 10 A/g	[68]
Wood	0.05	101	42% (10 A/g)	85% (10,000 cycles at 10 A/g)	[69]
EHCL	0.25	147	78% (10 A/g)	95.3% (15,000 cycles at 10 A/g)	This work

4. Conclusions

Several 3D hierarchical PCs were prepared by a low-cost and template-free method using CL as the precursor. Enzymatic hydrolysis depolymerized the macromolecular CL, promoting carbon formation and increasing the SSA and graphitization degree of the carbon materials. Alkali activation enabled the carbonized products to exhibit a unique honeycomb-like 3D hierarchical nanoporous structure whilst significantly increasing the SSA, micropore, and mesopore content. The obtained EHCL-Kd (mass ratio of KOH/EHCL = 3/2) possessed the highest SSA (1771 m^2/g), hierarchical porous structure, and good electrochemical performance. Based on the above structural advantages, the EHCL-Kd electrode exhibited a high capacitance (147 F/g at 0.25 A/g), significant rate capability (capacitance retention of 78% at 10 A/g), and good long-term cycling stability, with a capacitance retention of 95.3% after 15,000 cycles in a 6 M KOH aqueous electrolyte. This study demonstrated that the combination of enzymatic hydrolysis and alkali activation may be a promising route to apply renewable lignin derivatives in high-performance supercapacitors.

Supplementary Materials: The following supporting information can be downloaded at: https://www.mdpi.com/article/10.3390/en16093824/s1, Figure S1: GPC curves of CL and EHCL; Figure S2: (a) CV curves at 100 mV/s, (b) GCD curves at 1A/g, (c) rate performance in a three–electrode system in 6 M KOH and (d) Nyquist plots of CL-K and EHCL-Kd.

Author Contributions: Conceptualization, X.Z. and S.L.; Formal Analysis, X.Z. and Y.Z.; Funding Acquisition, J.L.; Investigation, S.L. and Y.Z.; Methodology, H.Y.; Project Administration, J.L.; Resources, J.L.; Validation, X.Z.; Writing—Original Draft, X.Z. and Y.Z.; Writing—Review and Editing, J.L., H.Y. and S.L. All authors have read and agreed to the published version of the manuscript.

Funding: This research was funded by the Jiangsu Key Laboratory of Environmentally Friendly Polymeric Materials (PML2003), The Top-Notch Academic Programs Project of Jiangsu Higher Education Institutions (TAPP), and The Priority Academic Program Development of Jiangsu Higher Education Institutions (PAPD).

Data Availability Statement: Not applicable.

Conflicts of Interest: The authors declare no conflict of interest.

References

1. Liu, C.-F.; Liu, Y.-C.; Yi, T.-Y.; Hu, C.-C. Carbon materials for high-voltage supercapacitors. *Carbon* **2019**, *145*, 529–548. [CrossRef]
2. Zhou, Y.; Qi, H.; Yang, J.; Bo, Z.; Huang, F.; Islam, M.S.; Lu, X.; Dai, L.; Amal, R.; Wang, C.H.; et al. Two-birds-one-stone: Multifunctional supercapacitors beyond traditional energy storage. *Energy Environ. Sci.* **2021**, *14*, 1854–1896. [CrossRef]
3. Hussain, A.; Arif, S.M.; Aslam, M. Emerging renewable and sustainable energy technologies: State of the art. *Renew. Sust. Energy Rev.* **2017**, *71*, 12–28. [CrossRef]
4. Wang, L.; Morabito, M.; Payne, C.T.; Robinson, G. Identifying institutional barriers and policy implications for sustainable energy technology adoption among large organizations in California. *Energy Policy* **2020**, *146*, 111768. [CrossRef]
5. Moustakas, K.; Loizidou, M.; Rehan, M.; Nizami, A.S. A review of recent developments in renewable and sustainable energy systems: Key challenges and future perspective. *Renew. Sust. Energy Rev.* **2020**, *119*, 109418. [CrossRef]
6. Poonam; Sharma, K.; Arora, A.; Tripathi, S.K. Review of supercapacitors: Materials and devices. *J. Energy Storage* **2019**, *21*, 801–825. [CrossRef]
7. Özarslan, S.; Atelge, M.R.; Kaya, M.; Ünalan, S. Production of dual functional carbon material from biomass treated with NaOH for supercapacitor and catalyst. *Energy Storage* **2021**, *3*, e257. [CrossRef]

8. Shao, Y.; El-Kady, M.F.; Sun, J.; Li, Y.; Zhang, Q.; Zhu, M.; Wang, H.; Dunn, B.; Kaner, R.B. Design and Mechanisms of Asymmetric Supercapacitors. *Chem. Rev.* **2018**, *118*, 9233–9280. [CrossRef] [PubMed]
9. Gou, G.; Huang, F.; Jiang, M.; Li, J.; Zhou, Z. Hierarchical porous carbon electrode materials for supercapacitor developed from wheat straw cellulosic foam. *Renew. Energy* **2020**, *149*, 208–216. [CrossRef]
10. Shah, S.A.R.; Naqvi, S.A.A.; Riaz, S.; Anwar, S.; Abbas, N. Nexus of biomass energy, key determinants of economic development and environment: A fresh evidence from Asia. *Renew. Sustain. Energy Rev.* **2020**, *133*, 110244. [CrossRef]
11. Wang, F.; Wu, X.; Yuan, X.; Liu, Z.; Zhang, Y.; Fu, L.; Zhu, Y.; Zhou, Q.; Wu, Y.; Huang, W. Latest advances in supercapacitors: From new electrode materials to novel device designs. *Chem. Soc. Rev.* **2017**, *46*, 6816–6854. [CrossRef] [PubMed]
12. Miao, L.; Song, Z.; Zhu, D.; Li, L.; Gan, L.; Liu, M. Recent advances in carbon-based supercapacitors. *Mater. Adv.* **2020**, *1*, 945–966. [CrossRef]
13. Luo, X.-Y.; Chen, Y.; Mo, Y. A review of charge storage in porous carbon-based supercapacitors. *New Carbon Mater.* **2021**, *36*, 49–68. [CrossRef]
14. Yin, J.; Zhang, W.; Alhebshi, N.A.; Salah, N.; Alshareef, H.N. Synthesis Strategies of Porous Carbon for Supercapacitor Applications. *Small Methods* **2020**, *4*, 1900853. [CrossRef]
15. Gong, Y.; Li, D.; Fu, Q.; Zhang, Y.; Pan, C. Nitrogen Self-Doped Porous Carbon for High-Performance Supercapacitors. *ACS Appl. Energy Mater.* **2020**, *3*, 1585–1592. [CrossRef]
16. Zuliani, J.E.; Tong, S.; Jia, C.Q.; Kirk, D.W. Contribution of surface oxygen groups to the measured capacitance of porous carbon supercapacitors. *J. Power Sources* **2018**, *395*, 271–279. [CrossRef]
17. Li, Z.; Guo, D.; Liu, Y.; Wang, H.; Wang, L. Recent advances and challenges in biomass-derived porous carbon nanomaterials for supercapacitors. *Chem. Eng. J.* **2020**, *397*, 125418. [CrossRef]
18. Bi, Z.; Kong, Q.; Cao, Y.; Sun, G.; Su, F.; Wei, X.; Li, X.; Ahmad, A.; Xie, L.; Chen, C.-M. Biomass-derived porous carbon materials with different dimensions for supercapacitor electrodes: A review. *J. Mater. Chem. A* **2019**, *7*, 16028–16045. [CrossRef]
19. Niu, W.; Xiao, Z.; Wang, S.; Zhai, S.; Qin, L.; Zhao, Z.; An, Q. Synthesis of nickel sulfide-supported on porous carbon from a natural seaweed-derived polysaccharide for high-performance supercapacitors. *J. Alloys Compd.* **2021**, *853*, 157123. [CrossRef]
20. Hu, Y.; Xie, K.; Wang, H.; Yuan, C.; Cao, B.; Qian, L.; Wang, S.; Fazeli Zafar, F.; Ding, K.; Wang, Q. Preparation and property of N-doped porous carbon material by one-step pyrolysis of protein-rich algal biomass. *J. Anal. Appl. Pyrolysis* **2021**, *157*, 105221. [CrossRef]
21. Zheng, S.; Zhang, J.; Deng, H.; Du, Y.; Shi, X. Chitin derived nitrogen-doped porous carbons with ultrahigh specific surface area and tailored hierarchical porosity for high performance supercapacitors. *J. Bioresour. Bioprod.* **2021**, *6*, 142–151. [CrossRef]
22. Xiao, J.; Li, H.; Zhang, H.; He, S.; Zhang, Q.; Liu, K.; Jiang, S.; Duan, G.; Zhang, K. Nanocellulose and its derived composite electrodes toward supercapacitors: Fabrication, properties, and challenges. *J. Bioresour. Bioprod.* **2022**, *7*, 245–269. [CrossRef]
23. Yan, B.; Feng, L.; Zheng, J.; Zhang, Q.; Jiang, S.; Zhang, C.; Ding, Y.; Han, J.; Chen, W.; He, S. High performance supercapacitors based on wood-derived thick carbon electrodes synthesized via green activation process. *Inorg. Chem. Front.* **2022**, *9*, 6108–6123. [CrossRef]
24. Liao, J.J.; Latif, N.H.A.; Trache, D.; Brosse, N.; Hussin, M.H. Current advancement on the isolation, characterization and application of lignin. *Int. J. Biol. Macromol.* **2020**, *162*, 985–1024. [CrossRef] [PubMed]
25. Meng, Y.; Lu, J.; Cheng, Y.; Li, Q.; Wang, H. Lignin-based hydrogels: A review of preparation, properties, and application. *Int. J. Biol. Macromol.* **2019**, *135*, 1006–1019. [CrossRef] [PubMed]
26. Neiva, D.M.; Rencoret, J.; Marques, G.; Gutiérrez, A.; Gominho, J.; Pereira, H.; del Río, J.C. Lignin from Tree Barks: Chemical Structure and Valorization. *ChemSusChem* **2020**, *13*, 4537–4547. [CrossRef]
27. Yang, H.; Yu, B.; Xu, X.; Bourbigot, S.; Wang, H.; Song, P. Lignin-derived bio-based flame retardants toward high-performance sustainable polymeric materials. *Green Chem.* **2020**, *22*, 2129–2161. [CrossRef]
28. Liang, D.; Zhu, X.; Dai, P.; Lu, X.; Guo, H.; Que, H.; Wang, D.; He, T.; Xu, C.; Robin, H.M.; et al. Preparation of a novel lignin-based flame retardant for epoxy resin. *Mater. Chem. Phys.* **2021**, *259*, 124101. [CrossRef]
29. Supanchaiyamat, N.; Jetsrisuparb, K.; Knijnenburg, J.T.N.; Tsang, D.C.W.; Hunt, A.J. Lignin materials for adsorption: Current trend, perspectives and opportunities. *Bioresour. Technol.* **2019**, *272*, 570–581. [CrossRef]
30. Shao, L.; Liu, N.; Wang, L.; Sang, Y.; Wan, H.A.; Zhan, P.; Zhang, L.; Huang, J.; Chen, J. Facile preparation of oxygen-rich porous polymer microspheres from lignin-derived phenols for selective CO_2 adsorption and iodine vapor capture. *Chemosphere* **2022**, *288*, 132499. [CrossRef]
31. Wang, D.; Lee, S.H.; Kim, J.; Park, C.B. "Waste to Wealth": Lignin as a Renewable Building Block for Energy Harvesting/Storage and Environmental Remediation. *ChemSusChem* **2020**, *13*, 2807–2827. [CrossRef] [PubMed]
32. Guo, N.; Li, M.; Sun, X.; Wang, F.; Yang, R. Enzymatic hydrolysis lignin derived hierarchical porous carbon for supercapacitors in ionic liquids with high power and energy densities. *Green Chem.* **2017**, *19*, 2595–2602. [CrossRef]
33. Jian, W.; Zhang, W.; Wu, B.; Wei, X.; Liang, W.; Zhang, X.; Wen, F.; Zhao, L.; Yin, J.; Lu, K.; et al. Enzymatic Hydrolysis Lignin-Derived Porous Carbons through Ammonia Activation: Activation Mechanism and Charge Storage Mechanism. *ACS Appl. Mater. Interfaces* **2022**, *14*, 5425–5438. [CrossRef] [PubMed]
34. Yedluri, A.K.; Kim, H.-J. Wearable super-high specific performance supercapacitors using a honeycomb with folded silk-like composite of $NiCo_2O_4$ nanoplates decorated with $NiMoO_4$ honeycombs on nickel foam. *Dalton Trans.* **2018**, *47*, 15545–15554. [CrossRef] [PubMed]

35. Kumar, Y.A.; Kim, H.-J. Effect of Time on a Hierarchical Corn Skeleton-Like Composite of CoO@ZnO as Capacitive Electrode Material for High Specific Performance Supercapacitors. *Energies* **2018**, *11*, 3285. [CrossRef]
36. Moniruzzaman, M.; Anil Kumar, Y.; Pallavolu, M.R.; Arbi, H.M.; Alzahmi, S.; Obaidat, I.M. Two-Dimensional Core-Shell Structure of Cobalt-Doped@MnO_2 Nanosheets Grown on Nickel Foam as a Binder-Free Battery-Type Electrode for Supercapacitor Application. *Nanomaterials* **2022**, *12*, 3187. [CrossRef]
37. Schütter, C.; Pohlmann, S.; Balducci, A. Industrial Requirements of Materials for Electrical Double Layer Capacitors: Impact on Current and Future Applications. *Adv. Energy Mater.* **2019**, *9*, 1900334. [CrossRef]
38. Lee, D.-Y.; An, G.-H.; Ahn, H.-J. High-surface-area tofu based activated porous carbon for electrical double-layer capacitors. *J. Ind. Eng. Chem.* **2017**, *52*, 121–127. [CrossRef]
39. Wei, J.; Li, Y.; Dai, D.; Zhang, F.; Zou, H.; Yang, X.; Ji, Y.; Li, B.; Wei, X. Surface Roughness: A Crucial Factor To Robust Electric Double Layer Capacitors. *ACS Appl. Mater. Interfaces* **2020**, *12*, 5786–5792. [CrossRef]
40. Karaman, C.; Karaman, O.; Atar, N.; Yola, M.L. Sustainable electrode material for high-energy supercapacitor: Biomass-derived graphene-like porous carbon with three-dimensional hierarchically ordered ion highways. *Phys. Chem. Chem. Phys.* **2021**, *23*, 12807–12821. [CrossRef]
41. Luo, H.; Lu, H.; Qiu, J. Carbon fibers surface-grown with helical carbon nanotubes and polyaniline for high-performance electrode materials and flexible supercapacitors. *J. Electroanal. Chem.* **2018**, *828*, 24–32. [CrossRef]
42. Shao, F.; Bian, S.-W.; Zhu, Q.; Guo, M.-X.; Liu, S.; Peng, Y.-H. Fabrication of Polyaniline/Graphene/Polyester Textile Electrode Materials for Flexible Supercapacitors with High Capacitance and Cycling Stability. *Chem.-Asian J.* **2016**, *11*, 1906–1912. [CrossRef] [PubMed]
43. Li, J.; Qiu, S.; Liu, B.; Chen, H.; Xiao, D.; Li, H. Strong interaction between polyaniline and carbon fibers for flexible supercapacitor electrode materials. *J. Power Sources* **2021**, *483*, 229219. [CrossRef]
44. Cai, Y.; Luo, Y.; Dong, H.; Zhao, X.; Xiao, Y.; Liang, Y.; Hu, H.; Liu, Y.; Zheng, M. Hierarchically porous carbon nanosheets derived from Moringa oleifera stems as electrode material for high-performance electric double-layer capacitors. *J. Power Sources* **2017**, *353*, 260–269. [CrossRef]
45. Schwieger, W.; Machoke, A.G.; Weissenberger, T.; Inayat, A.; Selvam, T.; Klumpp, M.; Inayat, A. Hierarchy concepts: Classification and preparation strategies for zeolite containing materials with hierarchical porosity. *Chem. Soc. Rev.* **2016**, *45*, 3353–3376. [CrossRef]
46. Liu, X.; Liu, X.; Sun, B.; Zhou, H.; Fu, A.; Wang, Y.; Guo, Y.-G.; Guo, P.; Li, H. Carbon materials with hierarchical porosity: Effect of template removal strategy and study on their electrochemical properties. *Carbon* **2018**, *130*, 680–691. [CrossRef]
47. Cai, G.; Yan, P.; Zhang, L.; Zhou, H.-C.; Jiang, H.-L. Metal–Organic Framework-Based Hierarchically Porous Materials: Synthesis and Applications. *Chem. Rev.* **2021**, *121*, 12278–12326. [CrossRef]
48. Gao, F.; Zang, Y.-H.; Wang, Y.; Guan, C.-Q.; Qu, J.-Y.; Wu, M.-B. A review of the synthesis of carbon materials for energy storage from biomass and coal/heavy oil waste. *New Carbon Mater.* **2021**, *36*, 34–48. [CrossRef]
49. Guan, T.; Zhao, J.; Zhang, G.; Wang, J.; Zhang, D.; Li, K. Template-Free Synthesis of Honeycomblike Porous Carbon Rich in Specific 2–5 nm Mesopores from a Pitch-Based Polymer for a High-Performance Supercapacitor. *ACS Sustain. Chem. Eng.* **2019**, *7*, 2116–2126. [CrossRef]
50. Jeon, J.-W.; Zhang, L.; Lutkenhaus, J.L.; Laskar, D.D.; Lemmon, J.P.; Choi, D.; Nandasiri, M.I.; Hashmi, A.; Xu, J.; Motkuri, R.K.; et al. Controlling Porosity in Lignin-Derived Nanoporous Carbon for Supercapacitor Applications. *ChemSusChem* **2015**, *8*, 428–432. [CrossRef]
51. Li, H.; Yuan, D.; Tang, C.; Wang, S.; Sun, J.; Li, Z.; Tang, T.; Wang, F.; Gong, H.; He, C. Lignin-derived interconnected hierarchical porous carbon monolith with large areal/volumetric capacitances for supercapacitor. *Carbon* **2016**, *100*, 151–157. [CrossRef]
52. Si, M.; Yan, X.; Liu, M.; Shi, M.; Wang, Z.; Wang, S.; Zhang, J.; Gao, C.; Chai, L.; Shi, Y. In Situ Lignin Bioconversion Promotes Complete Carbohydrate Conversion of Rice Straw by Cupriavidus basilensis B-8. *ACS Sustain. Chem. Eng.* **2018**, *6*, 7969–7978. [CrossRef]
53. Liu, Y.; Huang, G.; Li, Y.; Yao, Y.; Zhang, F.; Xing, B.; Zhang, C. N–O–S Co-doped Hierarchical Porous Carbons Derived from Calcium Lignosulfonate for High-Performance Supercapacitors. *Energy Fuels* **2020**, *34*, 3909–3922. [CrossRef]
54. Huang, G.; Liu, Y.; Chen, H.; Xing, B.; Li, Y.; Yao, Y.; Jia, J.; Liu, Q.; Zhang, C. A novel grafting-template method to prepare three-dimensional hierarchical porous carbon with high surface area and electrical conductivity for superior-performance supercapacitors. *J. Power Sources* **2021**, *482*, 228922. [CrossRef]
55. Shi, F.; Tong, Y.; Li, H.; Li, J.; Cong, Z.; Zhai, S.; An, Q.; Wang, K. Synthesis of oxygen/nitrogen/sulfur codoped hierarchical porous carbon from enzymatically hydrolyzed lignin for high-performance supercapacitors. *J. Energy Storage* **2022**, *52*, 104992. [CrossRef]
56. Yang, H.; Yan, R.; Chen, H.; Lee, D.H.; Zheng, C. Characteristics of hemicellulose, cellulose and lignin pyrolysis. *Fuel* **2007**, *86*, 1781–1788. [CrossRef]
57. Zhou, H.; Long, Y.; Meng, A.; Li, Q.; Zhang, Y. The pyrolysis simulation of five biomass species by hemi-cellulose, cellulose and lignin based on thermogravimetric curves. *Thermochim. Acta* **2013**, *566*, 36–43. [CrossRef]
58. Woo, S.-W.; Dokko, K.; Nakano, H.; Kanamura, K. Preparation of three dimensionally ordered macroporous carbon with mesoporous walls for electric double-layer capacitors. *J. Mater. Chem.* **2008**, *18*, 1674–1680. [CrossRef]

59. Yoo, E.; Nakamura, J.; Zhou, H. N-Doped graphene nanosheets for Li–air fuel cells under acidic conditions. *Energy Environ. Sci.* **2012**, *5*, 6928–6932. [CrossRef]
60. Zhang, K.; Liu, M.; Zhang, T.; Min, X.; Wang, Z.; Chai, L.; Shi, Y. High-performance supercapacitor energy storage using a carbon material derived from lignin by bacterial activation before carbonization. *J. Mater. Chem. A* **2019**, *7*, 26838–26848. [CrossRef]
61. Li, Y.; Li, Z.; Shen, P.K. Simultaneous Formation of Ultrahigh Surface Area and Three-Dimensional Hierarchical Porous Graphene-Like Networks for Fast and Highly Stable Supercapacitors. *Adv. Mater.* **2013**, *25*, 2474–2480. [CrossRef] [PubMed]
62. Wang, K.; Yan, R.; Zhao, N.; Tian, X.; Li, X.; Lei, S.; Song, Y.; Guo, Q.; Liu, L. Bio-inspired hollow activated carbon microtubes derived from willow catkins for supercapacitors with high volumetric performance. *Mater. Lett.* **2016**, *174*, 249–252. [CrossRef]
63. Wang, K.; Meng, Q.; Zhang, Y.; Wei, Z.; Miao, M. High-Performance Two-Ply Yarn Supercapacitors Based on Carbon Nanotubes and Polyaniline Nanowire Arrays. *Adv. Mater.* **2013**, *25*, 1494–1498. [CrossRef] [PubMed]
64. Wang, H.; Xu, Z.; Kohandehghan, A.; Li, Z.; Cui, K.; Tan, X.; Stephenson, T.J.; King'ondu, C.K.; Holt, C.M.B.; Olsen, B.C.; et al. Interconnected Carbon Nanosheets Derived from Hemp for Ultrafast Supercapacitors with High Energy. *ACS Nano* **2013**, *7*, 5131–5141. [CrossRef]
65. Hao, P.; Zhao, Z.; Tian, J.; Li, H.; Sang, Y.; Yu, G.; Cai, H.; Liu, H.; Wong, C.P.; Umar, A. Hierarchical porous carbon aerogel derived from bagasse for high performance supercapacitor electrode. *Nanoscale* **2014**, *6*, 12120–12129. [CrossRef] [PubMed]
66. Park, J.H.; Rana, H.H.; Lee, J.Y.; Park, H.S. Renewable flexible supercapacitors based on all-lignin-based hydrogel electrolytes and nanofiber electrodes. *J. Mater. Chem. A* **2019**, *7*, 16962–16968. [CrossRef]
67. Jiang, J.; Zhang, L.; Wang, X.; Holm, N.; Rajagopalan, K.; Chen, F.; Ma, S. Highly ordered macroporous woody biochar with ultra-high carbon content as supercapacitor electrodes. *Electrochim. Acta* **2013**, *113*, 481–489. [CrossRef]
68. Ding, Y.; Wang, T.; Dong, D.; Zhang, Y. Using Biochar and Coal as the Electrode Material for Supercapacitor Applications. *Front. Energy Res.* **2020**, *7*, 159. [CrossRef]
69. Wan, C.; Jiao, Y.; Li, J. Core–shell composite of wood-derived biochar supported MnO_2 nanosheets for supercapacitor applications. *RSC Adv.* **2016**, *6*, 64811–64817. [CrossRef]

nanomaterials

MDPI

Article

Electrochemical Performance of Chemically Activated Carbons from Sawdust as Supercapacitor Electrodes

Meruyert Nazhipkyzy [1,2,3,*], Mukhtar Yeleuov [1,3], Shynggyskhan T. Sultakhan [1,2], Anar B. Maltay [1,2], Aizhan A. Zhaparova [1], Dana D. Assylkhanova [1] and Renata R. Nemkayeva [4]

1 Institute of Combustion Problems, Almaty 050012, Kazakhstan
2 Department of Chemical Physics and Material Science, Al-Farabi Kazakh National University, Almaty 050040, Kazakhstan
3 Department of Materials Science, Nanotechnology and Engineering Physics, Satbayev University, Almaty 050013, Kazakhstan
4 National Nanotechnology Laboratory of Open Type, Al-Farabi Kazakh National University, Almaty 050040, Kazakhstan
* Correspondence: meruert82@mail.ru or meruyert.nazhipkyzy@kaznu.edu.kz

Abstract: Activated carbons (ACs) have been the most widespread carbon materials used in supercapacitors (SCs) due to their easy processing methods, good electrical conductivity, and abundant porosity. For the manufacture of electrodes, the obtained activated carbon based on sawdust (karagash and pine) was mixed with conductive carbon and polyvinylidene fluoride as a binder, in ratios of 75% activated carbon, 10% conductive carbon black, and 15% polyvinylidene fluoride (PVDF) in an N-methyl pyrrolidinone solution, to form a slurry and applied to a titanium foil. The total mass of each electrode was limited to vary from 2.0 to 4.0 mg. After that, the electrodes fitted with the separator and electrolyte solution were symmetrically assembled into sandwich-type cell construction. The carbon's electrochemical properties were evaluated using cyclic voltammetry (CV) and galvanostatic charge–discharge (CGD) studies in a two-electrode cell in 6M KOH. The CV and CGD measurements were realized at different scan rates (5–160 mV s^{-1}) and current densities (0.1–2.0 A g^{-1}) in the potential window of 1 V. ACs from KOH activation showed a high specific capacitance of 202 F g^{-1} for karagash sawdust and 161 F g^{-1} for pine sawdust at low mass loading of 1.15 mg cm^{-2} and scan rate of 5 mV s^{-1} in cyclic voltammetry test and 193 and 159 F g^{-1} at a gravimetric current density of 0.1 A g^{-1} in the galvanostatic charge–discharge test. The specific discharge capacitance is 177 and 131 F g^{-1} at a current density of 2 A g^{-1}. Even at a relatively high scan rate of 160 mV s^{-1}, a decent specific capacitance of 147 F g^{-1} and 114 F g^{-1} was obtained, leading to high energy densities of 26.0 and 22.1 W h kg^{-1} based on averaged electrode mass. Surface properties and the porous structure of the ACs were studied by scanning electron microscopy, energy-dispersive X-ray analysis, Raman spectroscopy, and the Brunauer–Emmett–Teller method.

Keywords: sawdust; karagash; pine; activated carbon; supercapacitor

Citation: Nazhipkyzy, M.; Yeleuov, M.; Sultakhan, S.T.; Maltay, A.B.; Zhaparova, A.A.; Assylkhanova, D.D.; Nemkayeva, R.R. Electrochemical Performance of Chemically Activated Carbons from Sawdust as Supercapacitor Electrodes. *Nanomaterials* **2022**, *12*, 3391. https://doi.org/10.3390/nano12193391

Academic Editors: Huang Zhang, Yuan Ma and Carlos M. Costa

Received: 26 August 2022
Accepted: 21 September 2022
Published: 28 September 2022

1. Introduction

Renewable energy sources have been urged for continual improvement because of the increased demand for energy and depletion of fossil fuels. Therefore, the development of feasible energy storage devices is crucial to creating a more prosperous society. Due to their attractive advantages, including higher energy density and capacitance than conventional capacitors together with their high power density, and better life cycle than batteries, supercapacitors (SCs) are thought to be one of the most promising electrochemical energy storage solutions of the future [1–10]. To put it other words, SCs bridge the technology gap between batteries and dielectric capacitors due to their high power density along with high stability and reversibility [11,12]. It is known that the substantial capacitance produced by these systems results from a combination of the double-layer capacitance and pseudocapacitance associated with the participation of adsorbed intermediates in the surface redox-type processes [13]. The highly tunable properties of SCs and possibilities of coupling SCs with additional energy storage devices also serve as advantages for some applications such as hybrid and/or electric vehicles, wearable electronics, and stretchable devices [14–16].

It must be underlined that electrodes, current collectors, electrolyte, and spacers are the main components that should be brought up for maintaining efficient energy storage on SCs. Especially, electrodes that operate at the electrode/electrolyte interface are critically required to maximize the power density and cyclability of SCs [17]. Conducting polymers, metal oxides/nitrides, graphene, and several carbon materials are frequently employed as supercapacitor electrode materials. Carbon-based capacitors have drawn a lot of interest since they can have diversified morphologies and modification possibilities together with high stability and conductivity [18]. In the literature, graphene, carbon nanotube, and carbon nanofibers are reported as SC electrodes. It is known that graphene and carbon nanotube-based SCs offer great specific capacitance, excellent energy and power density, and can also increase the voltage window range, all of which enhance the material's electrochemical performance; however, their low-cost mass production is still a great challenge [19–21]. Although the specific capacitance of the carbon fibers is very high, their unique structure makes it possible for them to be well-combined with conductive polymers such as polypyrrole, resulting in higher specific capacitance and energy density [22–24]. At this point, biomass-based porous carbons such as activated carbons seem attractive in terms of cost and sustainability [25]. Moreover, the amount of the charge stored, and its propagation during the charging/discharging, can differ significantly by altering the porous structure, size, and shape of pores and their functionalities [14].

With this perspective, the requirements for a cost-effective and environmentally benign approach from nonedible, abundant biomass sources have been anticipated to produce electrodes for SCs.

In [26], the electrode material was made from activated banana peels at a carbonization temperature of 900 °C using potassium hydroxide as an activating agent. It showed a specific capacity of 165 F g^{-1} at a current density of 0.5 A g^{-1} in a two-electrode cell.

The authors of work [27] carried out hydrothermal precarbonization and then pyrolysis of cucumber, transforming it into graphene-like carbon nanosheets. The obtained carbon nanosheets showed a specific capacity of 143 F g^{-1} at a current density of 0.2 A g^{-1} in an aqueous electrolyte 6 M KOH in the two-electrode cell setup.

In [28], biobased carbon materials from potato waste obtained by hydrothermal and thermochemical conversion technologies and used as electrode materials for supercapacitors showed an electrochemical performance of up to 134.15 F g^{-1}.

Yakaboylu et al. [29] used the miscanthus grass biomass as the precursor for the sheet-like activated carbon synthesis and a specific capacitance up to 188 F g^{-1} at 0.1 A g^{-1} was achieved.

The waste wood-dust of Dalbergia sisoo (Sisau) was tested as a supercapacitor electrode base and showed a lesser specific capacitance of 104.4 F g^{-1} [30].

Activated potassium citrate obtained from 750 °C showed the best capacitance performance, exhibiting the highest specific capacitance of 242.5 F g^{-1} at 0.2 A g^{-1} [31].

Considering the performance of biomass-based carbonaceous electrodes has not reached an excellent level yet, determination of the ideal characteristics of each biomass material is essential to explore the appropriate methods for the activation technique as physical activation and chemical activation. Moreover, the carbonization pathway and parameters have a marked effect on the texture of the resultant activated carbon. For instance, hydrothermal carbonization (HTC) is a preferred technique to alter the morphology and porosity of biomass-based precursors under mild processing temperatures and self-generated pressure carbonization using pure water as a solvent [32–34]. Additionally, the combination of HTC with chemical activation may develop the desired specific surface area, porosity for specific applications, and enhance the yield of the carbons [35–37]. In this study, activated carbons were prepared by two different sawdust samples, karagash (type of elm which expanded in Kazakhstan) and pine, to compare the structural features and to determine the influence of the precursor and carbonization temperature on the capacitive performance of the porous carbon electrodes. Hydrothermal pretreatment and subsequent high-temperature activation were used to produce activated carbons and KOH served as a chemical activator in the process. To the best of our knowledge, no studies on the production of AC using combined HTC and KOH using these two regional biomass samples have been reported, specifically for electrode material in supercapacitor devices. Through the structural characterization as well as the chemical properties via electrochemical tests, the energy storage performance of the produced activated carbons as an electrode material was explored. The products were tested as electrodes in a two-electrode cell and the results from measurement of the impedance responses of the equivalent circuit model were obtained to explain physically significant parameters that showed the capacitive characteristics of the carbon electrodes.

This study showed that the use of electrode composites consisting of activated carbon derived from an available biomass precursor leads to improved performance of energy storage systems, in particular, electrochemical capacitors.

2. Experimental Procedure

2.1. Materials and Chemicals

Karagash sawdust was collected from Almaty (Kazakhstan). Then, it was crushed. The chemicals used in this study were of analytical grade: hydrochloric acid (HCl, 36.6%), potassium hydroxide (KOH, ≥85%, Sigma Aldrich, St. Louis, MO, USA), polyvinylidene fluoride (PVDF, EQ-Lib-PVDF, MTI Corporation), conductive carbon black (EQ-Lib-SuperC45, MTI Corporation), 1-methyl-2-pyrrolidone (NMP, ≥99.0%, Sigma Aldrich, St. Louis, MO, USA), argon (99.993% Ikhsan Technogas Ltd.), and titanium foil (MF-Ti-Foil-700L-105, MTI Corporation). Distilled water was used to prepare solutions and wash samples

2.2. Preparation of Activated Carbon

Activated carbons obtained from sawdust of karagash and pine were prepared by the following method. Sawdust (karagash, pine) was dried at 105 °C overnight. For the hydrothermal process, 5 g of sawdust was suspended in 100 mL of an aqueous solution of KOH (5 wt. %) in an autoclave and treated at 120 °C for 2 h. Then, the resulting suspension solution was separated by filtration through glass cloth (100 mesh). The resulting product was dried without washing at 105 °C to constant weight. This mass was crushed and carbonized at 700, 800, and 900 °C for 1 h in an inert argon gas at a heating rate of 5 °C min^{-1}. The carbonized samples were washed with 2 M HCl solution and washed with distilled water until neutral acidity, and then dried at 100 °C to constant weight. The result was porous carbon from sawdust. Furthermore, the resulting material was ground by a mill in a pulsed mode. Next, the resulting material was separated by a 260 mesh sieve.

These obtained samples were designated as activated sawdust samples: AKS-700, AKS-800, AKS-900 and APS-700, APS-800, APS-900.

2.3. Characterization

Scanning electron microscopy (SEM, JEOL, model JSM-6490LA, FEI, USA) was used to determine the morphology of the samples. Energy-dispersive X-ray analysis (EDAX, JSM-6490LA, FEI, USA) was used for elemental analysis of the samples.

The specific surface area was characterized by the method of physical nitrogen adsorption at −196 °C with a surface area and adsorption analyzer (SORBTOMETR-M, Catakon, Novosibirsk, Russia). The surface area was calculated using the Brunauer–Emmett–Teller (BET) model.

2.4. Electrochemical Measurements

2.4.1. Preparation of Electrodes

Activated carbon electrodes were prepared in the following way: activated carbon (75 wt.%) was mixed with a binder based on polyvinylidene fluoride (PVDF) (15 wt.%), carbon black (CB) (10 wt.%). N-methyl-2- pyrrolidone (NMP) was used as a solvent for PVDF and was added into the mixture (AC:PVDF:CB) in an amount of 5.3 mL per gram of the mixture. The mixture was stirred for 20 min. The prepared suspension was applied to a titanium foil current collector on an area of 1×2 cm^2. Then, the prepared electrodes were dried at 120 °C for 12 h.

2.4.2. Electrochemical Measurements

To measure the electrochemical characteristics of the fabricated electrodes based on karagash and pine sawdust, a two-electrode cell with 6 M KOH was used as an electrolyte. The masses of the active material of the two electrodes were the same, and a symmetrical device was assembled from them to separate the two electrodes electrically. Cyclic voltammetry and galvanostatic charge–discharge measurements were carried out using an electrochemical workstation (galvanostat-potentiostat P-40X with an FRA-24M electrochemical impedance measurement module). The specific capacity was calculated from the curves of cyclic voltammetry and galvanostatic charge–discharge.

The CV data were obtained in a potential window of −0.1–0.9 V at scan rates of 5–160 mV s^{-1} in a 6 M KOH electrolyte. Equation (1) was applied to calculate the specific capacitance (C_s) using cyclic voltammetry data:

$$C = \frac{A}{2\, m\, \nu\, \Delta V} \quad (1)$$

where A is the hysteresis area (C); m is the mass of the active substance of one electrode (g), that is, the mass of activated carbon or composite, which is 75% of the mixture; ν is the scan rate (V s^{-1}); and ΔV is the potential window (V).

The specific capacity was also calculated from the GCD curves of the two-electrode cell using the equation:

$$C_s = \frac{2I \times t}{m \times (V_2 - V_1)} \quad (2)$$

where I is the discharge current (A); t is the discharge time (s); m is the mass of the active substance of one electrode (g); and V_2–V_1 is the potential window (V).

For the impedance test, an alternating voltage with a constant amplitude of 10 mV was applied to the electrode with the frequency changing from 300 kHz to 10 MHz.

3. Results and Discussion

3.1. Activated Carbon Surface Characterization

Surface properties are important for materials for double-layer capacitors. The morphology of the APS-700, APS-800, APS-900 and AKS-700, AKS-800, AKS-900 was studied using SEM (Figure 1).

The irregular particle is clearly visible under SEM, as observed in Figure 2, and the images further reveal the well-developed porous structure of the samples. From SEM images, it is clearly seen that the pore sizes of AC from karagash sawdust were in the range of ~300 nm up to ~19 μm, while pore sizes of AC from pine sawdust were in the range of ~500 nm up to ~15 μm.

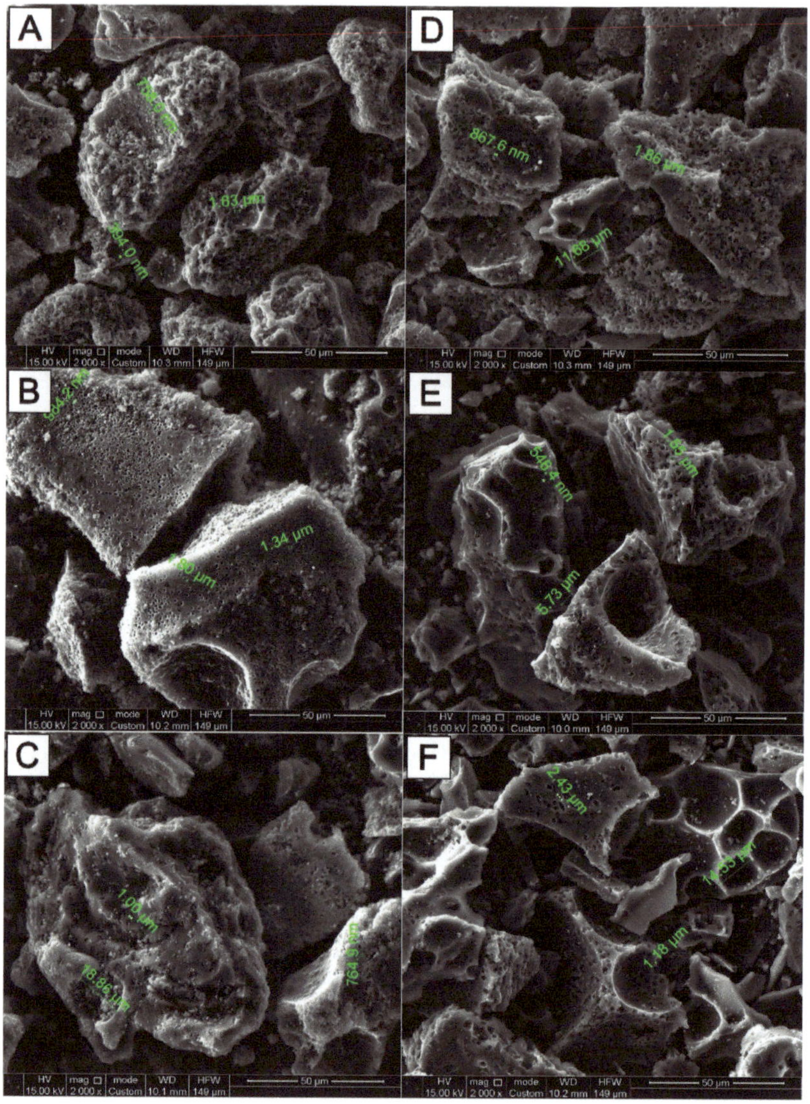

Figure 1. SEM images of carbon materials activated at different temperatures. (**A**) AKS-700 °C; (**B**) AKS-800 °C; (**C**) AKS-900 °C; (**D**) APS-700 °C; (**E**) APS-800 °C; (**F**) APS-900 °C.

The pore development on the surfaces of the AC might be due to the course of dehydration of an activating agent, i.e., KOH. In addition to this, during the washing process, remaining alkali was removed, creating an empty space or pores. These structures could contribute to the ion transport by acting as charge storage interfaces.

Table 1 summarizes the specific surface area of the carbon samples which was calculated by BET method [38].

Table 1. The specific surface area of the activated carbons based on karagash and pine sawdust.

Sample	AKS-700	AKS-800	AKS-900	APS-700	APS-800	APS-900
S_{BET}, (m^2 g^{-1})	505	557	532	494	715	666

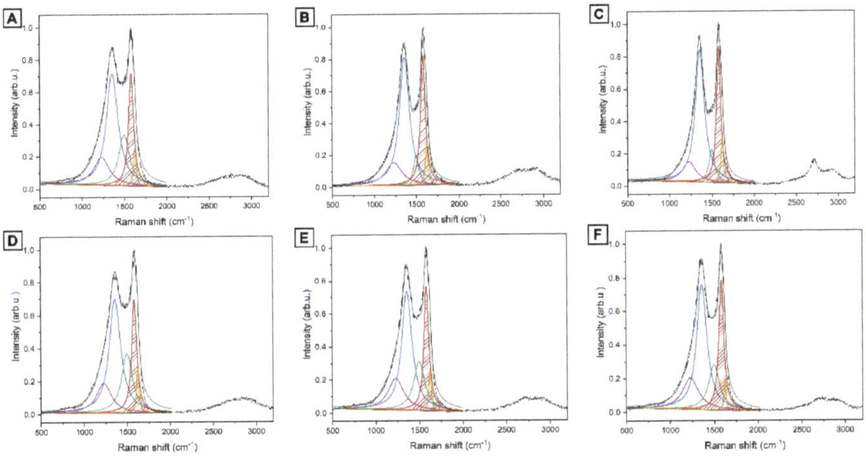

Figure 2. Raman spectra of a carbon material. (**A**) AKS-700 °C; (**B**) AKS-800 °C; (**C**) AKS-900 °C; (**D**) APS-700 °C; (**E**) APS-800 °C; (**F**) APS-900 °C.

AC from pine sawdust has the highest specific surface area of 715 m^2 g^{-1} compared to 557 m^2 g^{-1} for the AC from karagash sawdust.

The elemental analysis of the samples was carried out by the EDAX method.

The elemental contents of the sawdust (karagash and pine) were evaluated by elemental analysis, as summarized in Table 2, indicating that the AC materials from the karagash sawdust are composed of C, O, K, Ca, Fe, and Ni, but the AC materials from the pine sawdust are composed of C, O, K.

Table 2. Elemental composition of samples AKS and APS (atomic concentration, at. %).

	C	O	K	Ca	Fe	Ni
AKS-700	90.29	7.30	1.80	0.57	–	–
AKS-800	85.37	14.63	–	–	1.80	0.57
AKS-900	84.69	10.90	1.17	–	1.41	1.83
APS-700	87.14	12.86	–	–	–	–
APS-800	93.87	6.13	–	–	–	–
APS-900	87.065	12.95	–	–	–	–

The Raman spectra of sawdust based on karagash and pine are represented in Figure 2. Degree of graphitization was calculated by the Formula (3) [39]

$$G_f = \frac{A(G)}{\sum_{500}^{2000} A} \cdot 100\% \quad (3)$$

where G_f—degree of graphitization; $A(G)$—area of G peak; and $\sum A$—full area of the spectrum.

The dependence of the graphitization degree from temperature is shown in Figure 3.

The degree of graphitization calculated from the spectra increases with increasing carbonization temperature, and in the case of karagash, it reached a maximum value of ~25%. These carbonaceous materials with partial graphitization are highly suitable for application as electrodes due to their high electronic conductivities.

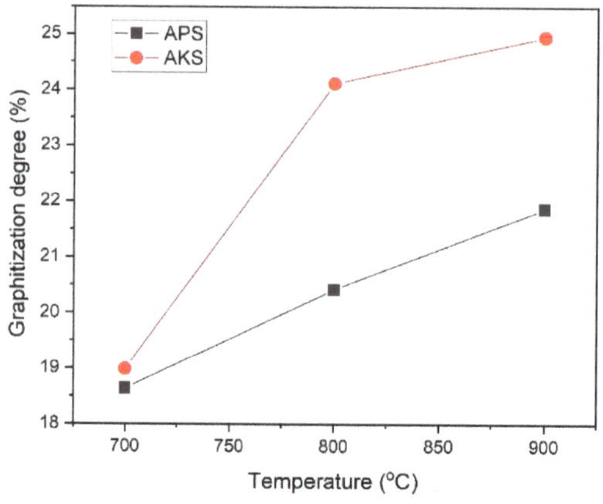

Figure 3. Dependence of graphitization degree from temperature.

3.2. Electrochemical Characterization

The results of a comparative study of electrodes based on karagash and pine sawdust obtained at different activation temperatures (700, 800, 900 °C) are given in Tables 3 and 4. It can be seen from the obtained data that the activation temperature is an important affecting parameter on the electrochemical characteristics of both AKS and APS-based electrodes. It has been stated that AKS-700 and APS-700 electrodes show the highest specific capacitance at various CV sweep rates (Table 3) and at various GCD current densities (Table 4) compared to activated carbons obtained at an activation temperature of 800 and 900 °C. This behavior can be mainly associated with a decrease in the capacitance of the electrical double layer due to a decrease in the available surface of the pore space for electrolyte ions with an increase in the activation temperature.

Table 3. The specific capacitance of AKS and APS obtained from CV data at different scan rates.

Samples	Scan Rate, mV s^{-1}					
	5	10	20	40	80	160
	Specific Capacitance, F g^{-1}					
AKS-700	202	194	185	175	162	147
AKS-800	128	111	91	65	42	28
AKS-900	121	113	105	102	99	92
APS-700	161	155	148	139	127	114
APS-800	146	142	135	126	115	104
APS-900	115	108	99	87	73	59

Table 4. The specific capacitance of AKS and APS calculated by the GCD data at different current densities.

Samples	Current Density, mA g^{-1}				
	2000	1000	500	250	100
	Specific Capacitance, F g^{-1}				
AKS-700	178	183	187	189	193
AKS-800	106	111	117	131	148
AKS-900	102	104	105	106	104
APS-700	131	140	147	153	159
APS-800	109	115	120	123	126
APS-900	91	104	112	118	118

A general view of the experimental curves of cyclic voltammetry at different sweep rates (5–160 mV s^{-1}) for electrode materials based on AKS-700 and APS-700 is shown in Figure 4A,B. The voltammetric characteristics of the AKS and APS electrodes were tested in the potential range of 0.0–1.0 V in a two-electrode cell configuration. The CV curves for both pairs of electrodes made on the basis of AKS-700 and APS-700 are almost rectangular in shape. This indicates that the materials act as electrical double-layer supercapacitors. It can be seen that the CV curve of the AKS-700 sample shows a larger surface area of the rectangle compared to the APS-700, which indicates the larger capacity of the AKS-700 sample (Figure 4).

Figure 5 shows the galvanostatic charge–discharge characteristics for electrode material based on AKS-700 and APS-700 in the potential range between 0 and 1 V in a 6 M KOH solution at different charge/discharge currents. The curves show approximately isosceles triangles, which are typical of electrical double-layer capacitance, not pseudocapacitance. This is also consistent with the information obtained from the CV curves (without a peak indicating Faraday processes) (Figure 4). It is clearly seen that the charge–discharge time for the AKS-700 electrodes is longer than for the APS-700 electrodes, indicating a higher specific capacity (Figure 5).

In Figure 5A, at a gravimetric current density of 100 mA g^{-1}, the irreversible phenomenon, a long charge time with short discharge time, appeared. The irreversible phenomenon that the discharge time is shorter than the charging time is common in many supercapacitors at low current densities. It is very likely that the internal resistance of the device is relatively large, resulting in a significant voltage drop. In addition, the supercapacitor still has a severe self-discharge phenomenon, and even some other micro short circuits will lead to capacity loss and shorten the discharge time. This is one of the main drawbacks of the supercapacitor. The difference between charge and discharge times is negligible at high current densities. For this reason, supercapacitors are used in applications where high power density is required.

Figure 6A shows the specific capacitance of electrodes based on AKS-700 and APS-700 calculated from GCD curves at different discharge current densities. The specific capacitance of the obtained electrode materials is significantly higher at high current densities and scan rates even without additional modification of the materials [40–42]. The figure clearly shows that the specific capacitance of the AKS-700 electrode decreases slightly with increasing current density. This trend is characteristic of supercapacitors and reflects the fact that as the current density increases, some active surface areas become unsuitable for charge storage. However, in the studied electrodes, a slight decrease in capacitance with an increase in the charge–discharge current characterizes the best charge-exchange processes compared to traditional materials. Typically, pseudocapacitive materials show a higher capacitance loss as current density increases.

To study the fundamental behavior of the carbon in EDLCs, electrochemical impedance spectroscopy was performed by using an open-circuit potential. The semicircle at high frequency indicated the resistance from the diffusion/mass transfer of the ions through the porous carbon (Figure 7).

Solution resistance (Rs), which corresponds to the beginning of the semicircle, of the sample AKS-700 is 0.17 Ohm, whereas the Rs of the sample APS-700 is 0.23 Ohm. The magnitude of charge transfers resistance (Rct) can be obtained from the diameter of the semicircle. Rct of APS-700 is approximately 0.6 Ohm.

The electrochemical performance of activated carbon as the electrode active material of a supercapacitor depends on the porous structure, electrical conductivity, particle size distribution, etc. All these properties are mainly affected by the synthesis process and the nature of the precursors. Despite the fact that the synthesis of both precursors (AKS-700, APS-700) is the same, their initial properties are different. This is the reason why AKS and APS show different electrochemical characteristics. Further research should be aimed at understanding the influence of the initial properties of the starting materials on the characteristics of the final electrode materials.

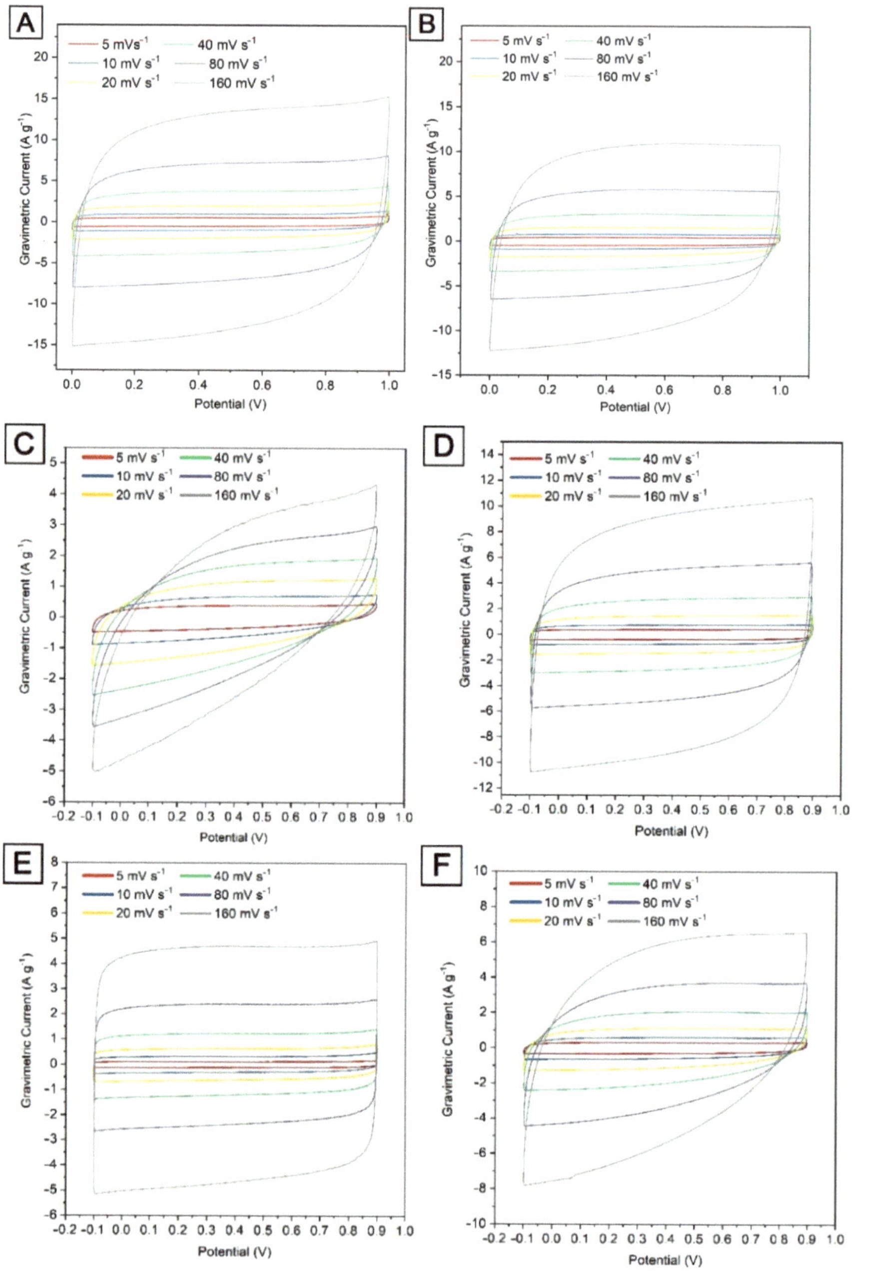

Figure 4. Cyclic voltammetry curves for electrodes based on (**A**) AKS-700, (**B**) APS-700, (**C**) AKS-800, (**D**) APS-800, (**E**) AKS-900 and (**F**) APS-900 at different scanning rates: 5, 10, 20, 40, 80, 160 mV s^{-1}.

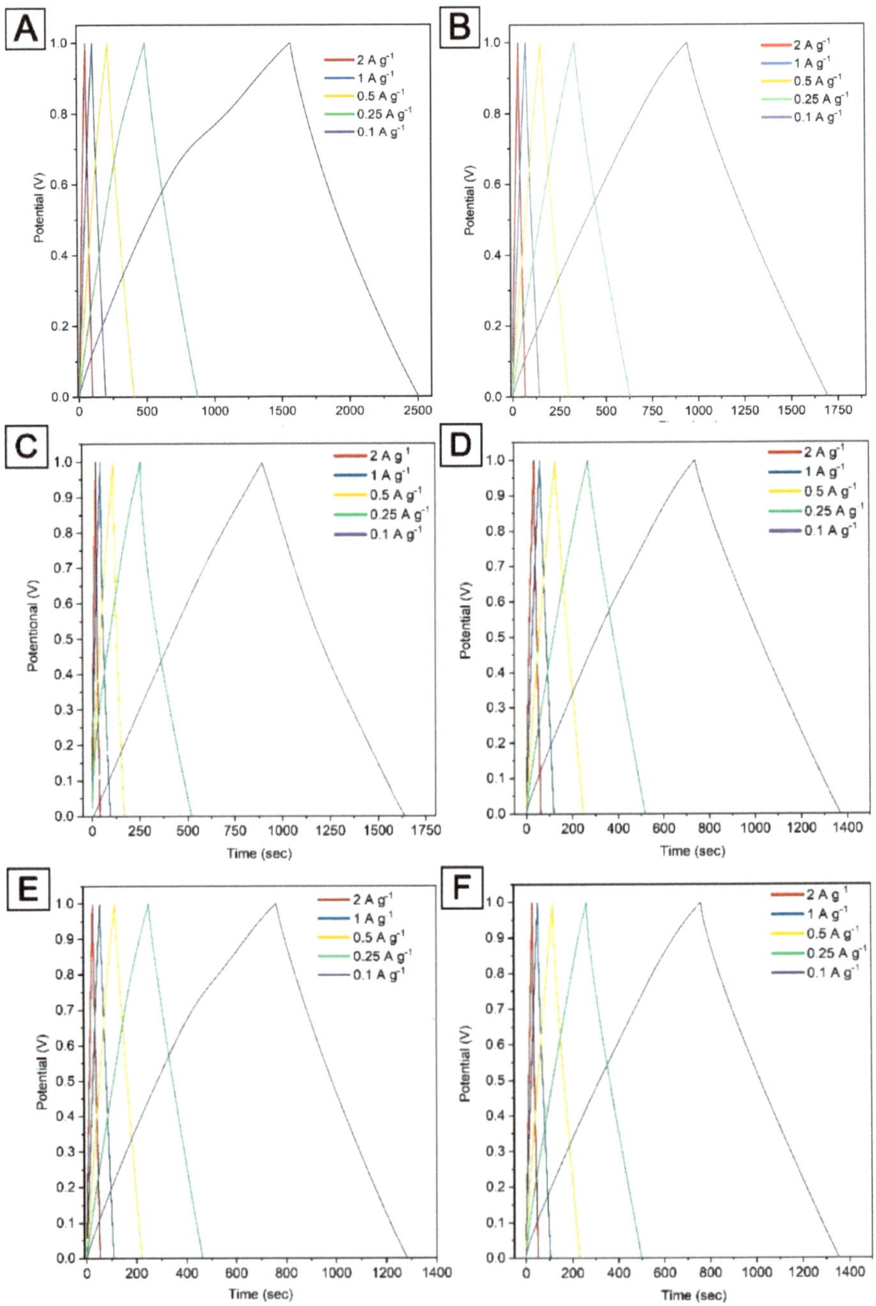

Figure 5. Curves of galvanostatic charge–discharge at different current densities (0.1, 0.25, 0.5, 1.0, 2.0 A g^{-1}) for electrodes based on (**A**) AKS-700, (**B**) APS-700, (**C**) AKS-800, (**D**) APS-800, (**E**) AKS-900 and (**F**) APS-900 with 6 M KOH electrolyte.

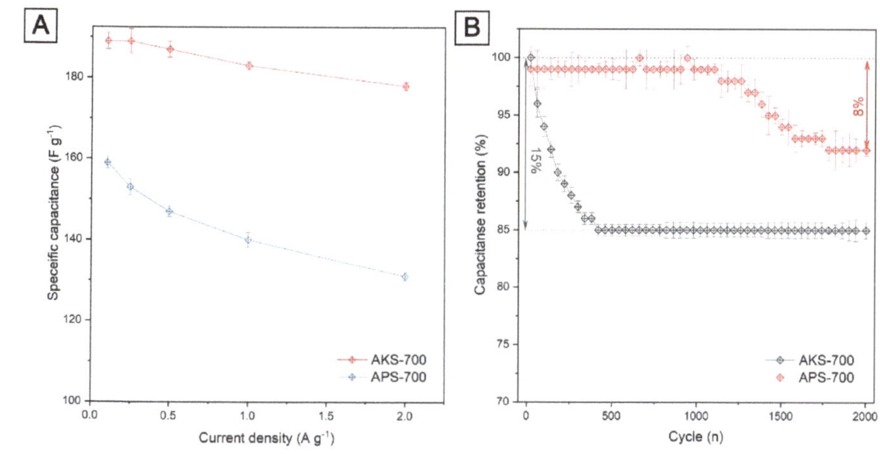

Figure 6. (**A**) The values of the specific capacitance of the electrodes measured at different current densities; (**B**) capacity retention of electrodes based on AKS-700 and APS-700 for 2000 cycles at constant current charge–discharge cycling, current density 1 A g^{-1}.

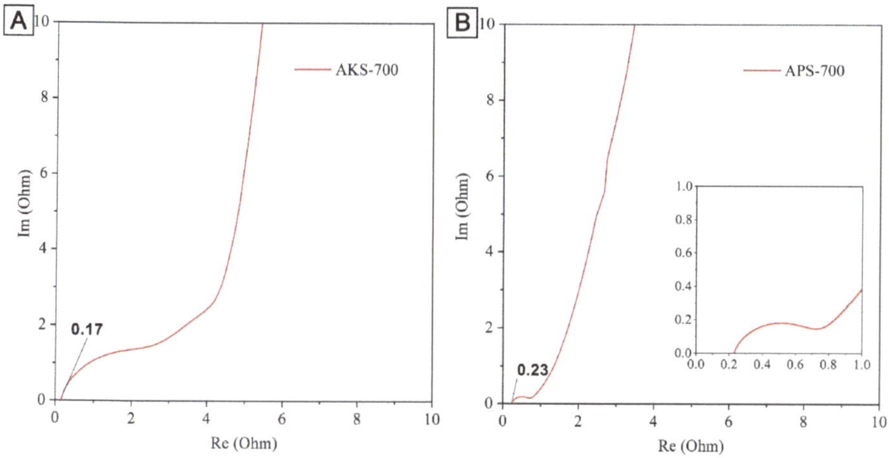

Figure 7. Nyquist plot for activated sawdust electrodes (**A**) AKS-700 and (**B**) APS-700.

4. Conclusions

Activated carbon from karagash (type of elm), which are expanding in Kazakhstan region, and pine sawdust was used as electrode for supercapacitors. Activated carbon from karagash as an electrode material for supercapacitors showed high rate capability and stable cyclability compared to the activated carbon from pine. Our experiment showed that karagash is a promising electrode material for supercapacitors.

This study provided a new approach to using the sawdust of karagash and pine trees to pro[illegible]luce carbon material for EDLC application due to the low cost of biowaste ma[illegible] [illegible]d simplicity of the activation process. The electrochemical property studies ind[illegible] [illegible]promising use of activated carbon from the sawdust of karagash and pine tre[illegible] [illegible]effective energy storage device, which curtails global energy demand and en[illegible]ntal concerns. At a relatively high scan rate of 160 mV s^{-1}, a decent specific [illegible]e of 147 F g^{-1} and 114 F g^{-1} was obtained, leading to high energy densities of [illegible].1 W h kg^{-1} based on averaged electrode mass.

Author Contributions: Conceptualization, M.N. and M.Y.; validation and formal analysis: A.B.M. and A.A.Z.; investigation: S.T.S. and D.D.A.; writing—original draft preparation: M.N. and M.Y.; writing—review and editing, M.N. and R.R.N.; visualization: S.T.S. and A.A.Z.; supervision, M.Y.; project administration: M.N. All authors have read and agreed to the published version of the manuscript.

Funding: The work was supported by a grant from the Science Committee of the Ministry of Education and Science of the Republic of Kazakhstan: AP08856321 "Obtaining fiber composite materials by electrospinning and creating electrodes based on them for supercapacitors".

Acknowledgments: The authors also acknowledge Malika Tulegenova and Zanzamgul Sattygulova for their technical assistance in the laboratory.

Conflicts of Interest: The authors declare no conflict of interest.

References

1. Huang, J.; Liang, Y.; Hu, H.; Liu, S.; Cai, Y.; Dong, H.; Zheng, M.; Xiao, Y.; Liu, Y. Ultrahigh-surface-area hierarchical porous carbon from chitosan: Acetic acid mediated efficient synthesis and its application in superior supercapacitors. *J. Mater. Chem. A* **2017**, *5*, 24775–24781. [CrossRef]
2. Xie, K.; Qin, X.; Wang, X.; Wang, Y.; Tao, H.; Wu, Q.; Yang, L.; Hu, Z. Carbon nanocages as supercapacitor electrode materials. *Adv. Mater.* **2012**, *24*, 347–352. [CrossRef] [PubMed]
3. Chen, Z.; Peng, X.; Zhang, X.; Jing, S.; Zhong, L.; Sun, R. Facile synthesis of cellulose-based carbon with tunable N content for potential supercapacitor application. *Carbohydr. Polym.* **2017**, *170*, 107–116. [CrossRef] [PubMed]
4. Zhang, M.; He, L.; Shi, T.; Zha, R. Nanocasting and direct synthesis strategies for mesoporous carbons as supercapacitor electrodes. *Chem. Mater.* **2018**, *30*, 7391–7412. [CrossRef]
5. Jiang, S.; Wang, F.; Cheong, J.Y.; Lee, J.; Ahn, J.; Duan, G.; Chen, H.; Zhang, Q.; Kim, I.-D. Pyrolysis of enzymolysis-treated wood: Hierarchically assembled porous carbon electrode for advanced energy storage devices. *Adv. Funct. Mater.* **2021**, *31*, 2101077–21010786.
6. Borchardt, L.; Zhu, Q.; Casco, M.E.; Berger, R.; Zhuang, X.; Kaskel, S.; Feng, X.; Xu, Q. Toward a molecular design of porous carbon materials. *Mater. Today* **2017**, *20*, 592–610. [CrossRef]
7. Xiao, J.; Li, H.; Zhang, H.; He, S.; Zhang, Q.; Liu, K.; Jiang, S.; Duan, G.; Zhang, K. Nanocellulose and its derived composite electrodes toward supercapacitors: Fabrication, properties, and challenges. *J. Bioresour. Bioprod.* **2022**, 25. [CrossRef]
8. Wang, Y.M.; Lin, X.; Liu, T.; Chen, H.; Chen, S.; Jiang, Z.; Liu, J.; Huang, J.; Liu, M. Wood-derived hierarchically porous electrodes for high-performance all-solid-state supercapacitors. *Adv. Funct. Mater.* **2018**, *28*, 1806207. [CrossRef]
9. Long, Y.; An, X.; Zhang, H.; Yang, J.; Liu, L.; Tian, Z.; Yang, G.; Cheng, Z.; Cao, H.; Liu, H.; et al. Highly graphitized lignin-derived porous carbon with hierarchical N/O co-doping "core-shell" superstructure supported by metal-organic frameworks for advanced supercapacitor performance. *Chem. Eng. J.* **2022**, *451*, 138877. [CrossRef]
10. Zheng, S.; Zhang, J.; Deng, H.; Du, Y.; Shi, X. Chitin derived nitrogen-doped porous carbons with ultrahigh specific surface area and tailored hierarchical porosity for high performance supercapacitors. *J. Bioresour. Bioprod.* **2021**, *6*, 142–151. [CrossRef]
11. Yan, B.; Zheng, J.; Feng, L.; Du, C.; Jian, S.; Yang, W.; Wu, Y.A.; Jiang, S.; He, S.; Chen, W. Wood-derived biochar as thick electrodes for high-rate performance supercapacitors. *Biochar* **2022**, *4*, 50. [CrossRef]
12. Mapleback, B.; Dao, V.; Webb, L.; Rider, A. Composite Structural Supercapacitors: High-Performance Carbon Nanotube Supercapacitors through Ionic Liquid Localisation. *Nanomaterials* **2022**, *12*, 2558. [CrossRef] [PubMed]
13. Sarangapani, S.; Tilak, B.; Chen, C.P. Materials for electrochemical capacitors: Theoretical and experimental constraints. *J. Electrochem. Soc.* **1996**, *143*, 3791. [CrossRef]
14. Fic, K.; Platek, A.; Piwek, J.; Frackowiak, E. Sustainable materials for electrochemical capacitors. *Mater. Today* **2018**, *21*, 437–454. [CrossRef]
15. Elsaidy, A.; Majcherkiewicz, J.N.; Puértolas, B.; Salgueiriño, V.; Nóvoa, X.R.; Correa-Duarte, M.A. Synergistic Interaction of Clusters of Iron Oxide Nanoparticles and Reduced Graphene Oxide for High Supercapacitor Performance. *Nanomaterials* **2022**, *12*, 2695. [CrossRef]
16. Wang, D.-G.; Liang, Z.; Gao, S.; Qu, C.; Zou, R. Metal-organic framework-based materials for hybrid supercapacitor application. *Coord. Chem. Rev.* **2020**, *404*, 213093. [CrossRef]
17. Farzana, R.; Rajarao, R.; Ramachandra, B.; Sahajwalla, V. Performance of an activated carbon supercapacitor electrode synthesised from waste Compact Discs (CDs). *J. Ind. Eng. Chem.* **2018**, *65*, 387–396. [CrossRef]
18. Timperman, L.; Vigeant, A.; Anouti, M. Eutectic mixture of protic ionic liquids as an electrolyte for activated carbon-based supercapacitors. *Electrochim. Acta* **2015**, *155*, 164–173. [CrossRef]
19. Sun, Y.; Tang, J.; Qin, F.; Yuan, J.; Zhang, K.; Li, J.; Zhud, D.-M.; Qine, L.-C. Hybrid lithium-ion capacitors with asymmetric graphene electrodes. *J. Mater. Chem. A* **2017**, *5*, 13601–13609. [CrossRef]
20. Kshetri, T.; Thanh, T.; Singh, S.; Kim, N.; Lee, J. Hierarchical material of carbon nanotubes grown on carbon nanofibers for high performance electrochemical capacitor. *Chem. Eng. J.* **2018**, *345*, 39–47. [CrossRef]

21. Li, X.; Chen, M.; Wang, L.; Xu, H.; Zhong, J.; Zhang, M.; Wang, Y.; Zhang, Q.; Mei, L.; Wang, T.; et al. Nitrogen-doped carbon nanotubes as an anode for a highly robust potassium-ion hybrid capacitor. *Nanoscale Horiz.* **2020**, *5*, 1586–1595. [CrossRef] [PubMed]
22. Lee, H.-M.; Kwac, L.; An, K.; Park, C.; Kim, B. Electrochemical behavior of pitch-based activated carbon fibers for electrochemical capacitors. *Energy Convers. Manag.* **2016**, *125*, 347–352. [CrossRef]
23. Yuan, W.; Han, G.; Xiao, Y.; Chang, Y.; Liu, C.; Li, M.; Li, Y.; Zhang, Y. Flexible electrochemical capacitors based on polypyrrole/carbon fibers via chemical polymerization of pyrrole vapor. *Appl. Surf. Sci.* **2016**, *377*, 274–282. [CrossRef]
24. Zhou, X.; Tong, W.; Huang, F.; Chen, L.; Wu, H. Methylene blue enhanced bamboo activated carbon as high performance supercapacitor electrode materials. *Ind. Crops Prod.* **2022**, *180*, 114786.
25. Özsin, G.; Kılıç, M.; Apaydın-Varol, E.; Pütün, A. Chemically activated carbon production from agricultural waste of chickpea and its application for heavy metal adsorption: Equilibrium, kinetic, and thermodynamic studies. *Appl. Water Sci.* **2019**, *9*, 56. [CrossRef]
26. Fasakin, O.; Dangbegnon, J.K.; Momodu, D.Y.; Madito, M.J.; Oyedotun, K.O.; Eleruja, M.A.; Manyala, N. Synthesis and characterization of porous carbon derived from activated banana peels with hierarchical porosity for improved electrochemical performance. *Electrochim. Acta* **2018**, *262*, 187–196. [CrossRef]
27. Gopalakrishnan, A.; Badhulika, S. Ultrathin Graphene-like 2D porous carbon nanosheets and its excellent capacitance retention for supercapacitor. *J. Ind. Eng. Chem.* **2018**, *68*, 257–266. [CrossRef]
28. Hoffman, V.; Jung, D.; Alhnidi, J.A.; Mackle, L.; Kruse, A. Bio-Based Carbon Materials from Potato Waste as Electrode Materials in Supercapacitors. *Energies* **2020**, *13*, 2406. [CrossRef]
29. Gunes, A.Y.; Changle, J.; Yumak, T.; Zondlo, J.; Wang, J.; Sabolsky, E. Engineered hierarchical porous carbons for supercapacitor applications through chemical pretreatment and activation of biomass precursors. *Renew. Energy* **2021**, *163*, 276–287.
30. Shrestha, D. Activated carbon and its hybrid composites with manganese (IV) oxide as effectual electrode materials for high performance supercapacitor. *Arab. J. Chem.* **2022**, *15*, 103946. [CrossRef]
31. Zheng, J.; Yan, B.; Feng, L.; Zhang, Q. Potassium citrate assisted synthesis of hierarchical porous carbon materials for high performance supercapacitors. *Diam. Relat. Mater.* **2022**, *128*, 109247. [CrossRef]
32. White, R.J.; Budarin, V.; Luque, R.; Clark, J.; Macquarrie, D. Tuneable porous carbonaceous materials from renewable resources. *Chem. Soc. Rev.* **2009**, *38*, 3401–3418. [CrossRef] [PubMed]
33. Fang, Y.; Gu, D.; Zou, Y.; Wu, Z. A low-concentration hydrothermal synthesis of biocompatible ordered mesoporous carbon nanospheres with tunable and uniform size. *Angew. Chem. Int. Ed.* **2010**, *49*, 7987–7991. [CrossRef] [PubMed]
34. Tang, D.; Lu, L.; Luo, Z.; Yang, B.; Ke, J.; Lei, W.; Zhen, H.; Zhuang, Y.; Sun, J.; Chen, K.; et al. Heteroatom-Doped Hierarchically Porous Biochar for Supercapacitor Application and Phenol Pollutant Remediation. *Nanomaterials* **2022**, *12*, 2586. [CrossRef] [PubMed]
35. Alshareef, S.A.; Alqadami, A.A.; Khan, M.A.; Alanazi, H.S.; Siddiqui, M.R.; Jeon, B.-H. Simultaneous co-hydrothermal carbonization and chemical activation of food wastes to develop hydrochar for aquatic environmental remediation. *Bioresour. Technol.* **2022**, *347*, 126363. [CrossRef]
36. Diaz, E.; Sanchis, I.; Coronella, C.; Mohedano, A.F. Activated Carbons from Hydrothermal Carbonization and Chemical Activation of Olive Stones: Application in Sulfamethoxazole Adsorption. *Resources* **2022**, *11*, 43. [CrossRef]
37. Suhdi, S.; Wang, S.-C. The Production of Carbon Nanofiber on Rubber Fruit Shell-Derived Activated Carbon by Chemical Activation and Hydrothermal Process with Low Temperature. *Nanomaterials* **2021**, *11*, 2038. [CrossRef]
38. Brunauer, S.; Emmett, P.H.; Teller, E. Adsorption of gases in multimolecular layers. *J. Am. Chem. Soc.* **1938**, *60*, 309–319. [CrossRef]
39. Yerlanuly, Y.; Zhumadilov, R.; Nemkayeva, R.; Uzakbaiuly, B.; Beisenbayev, A.R.; Bakenov, Z.; Ramazanov, T.; Gabdullin, M.; Ng, A.; Brus, V.; et al. Physical properties of carbon nanowalls synthesized by the ICP-PECVD method vs. the growth time. *Sci. Rep.* **2021**, *11*, 19287. [CrossRef]
40. Chengkai, X.; Subramani, S.; Seulgi, J.; Kim, D. A sulfur self-doped multifunctional biochar catalyst for overall water splitting and a supercapacitor from Camellia japonica flowers. *Carbon Energy* **2022**, *4*, 491–505.
41. Ahmed, S.; Ahmed, A.; Rafat, M. Investigation on activated carbon derived from biomass Butnea monosperma and its application as a high performance supercapacitor electrode. *J. Energy Storage* **2019**, *26*, 100988. [CrossRef]
42. Cao, L.; Li, H.; Xu, Z.; Gao, R.; Wang, S.; Zhang, G.; Jiang, S.; Xu, W.; Hou, H. Camellia pollen-derived carbon with controllable N content for high-performance supercapacitors by ammonium chloride activation and dual N-doping. *Chem. Nano. Mat.* **2021**, *7*, 34–43. [CrossRef]

energies

MDPI

Article

Waterborne $LiNi_{0.5}Mn_{1.5}O_4$ Cathode Formulation Optimization through Design of Experiments and Upscaling to 1 Ah Li-Ion Pouch Cells

Lander Lizaso [1], Idoia Urdampilleta [1,2], Miguel Bengoechea [1], Iker Boyano [1], Hans-Jürgen Grande [1,3], Imanol Landa-Medrano [1,*], Aitor Eguia-Barrio [1] and Iratxe de Meatza [1,4]

1 CIDETEC, Basque Research and Technology Alliance (BRTA), 20014 Donostia-San Sebastian, Spain
2 Department of Applied Chemistry, University of Basque Country (UPV/EHU), 20018 Donostia-San Sebastian, Spain
3 Advanced Polymers and Materials: Physics, Chemistry and Technology Department, University of the Basque Country (UPV/EHU), Avda. Tolosa 72, 20018 Donostia-San Sebastian, Spain
4 Department of Organic and Inorganic Chemistry, Universidad del País Vasco (UPV/EHU), 48080 Bilbao, Spain
* Correspondence: ilanda@cidetec.es

Abstract: High-voltage spinel $LiNi_{0.5}Mn_{1.5}O_4$ (LNMO) is a promising candidate as a lithium-ion battery cathode material to fulfill the high-energy density demands of the electric vehicle industry. In this work, the design of the experiment's methodology has been used to analyze the influence of the ratio of the different components in the electrode preparation feasibility of laboratory-scale coatings and their electrochemical response. Different outputs were defined to evaluate the formulations studied, and Derringer–Suich's methodology was applied to obtain an equation that is usable to predict the desirability of the electrodes depending on the selected formulation. Afterward, Solver's method was used to figure out the formulation that provides the highest desirability. This formulation was validated at a laboratory scale and upscaled to a semi-industrial coating line. High-voltage 1 Ah lithium-ion pouch cells were assembled with LNMO cathodes and graphite-based anodes and subjected to rate-capability tests and galvanostatic cycling. 1 C was determined as the highest C-rate usable with these cells, and 321 and 181 cycles above 80% SOH were obtained in galvanostatic cycling tests performed at 0.5 C and 1 C, respectively. Furthermore, it was observed that the LNMO cathode required an activation period to become fully electrochemically active, which was shorter when cycled at a lower C-rate.

Keywords: lithium-ion batteries; LNMO; high voltage cells; design of experiments; pouch cells

Citation: Lizaso, L.; Urdampilleta, I.; Bengoechea, M.; Boyano, I.; Grande, H.-J.; Landa-Medrano, I.; Eguia-Barrio, A.; de Meatza, I. Waterborne $LiNi_{0.5}Mn_{1.5}O_4$ Cathode Formulation Optimization through Design of Experiments and Upscaling to 1 Ah Li-Ion Pouch Cells. *Energies* **2023**, *16*, 7327. https://doi.org/10.3390/en16217327

Academic Editors: Huang Zhang and Yuan Ma

Received: 15 September 2023
Revised: 18 October 2023
Accepted: 25 October 2023
Published: 29 October 2023

1. Introduction

Lithium-ion batteries (LIBs) are, nowadays, the dominating battery technology [1]. They were initially implemented in small electronic devices [2]. Still, in past years, they have enabled the substitution of classic combustion engines and promoted the electrification of new models of vehicles [2,3]. The automotive industry is pushing to produce even better batteries, which leads to a continuous evolution of the chemistries implemented in electric vehicles [4,5]. A few years ago, nickel-cobalt-manganese/aluminum oxides ($LiNi_xMn_yCo_zO_2$/$LiNi_{0.8}Co_{0.15}Al_{0.05}O_2$, NMC/NCA) were the cathode materials designed for the transition from an oil-consuming to an electric engine-based fleet due to their high energy density [6,7]. Nevertheless, in recent years, lithium iron phosphate ($LiFePO_4$, LFP) has gained interest, as its robustness, safety, and cycling stability are higher and the absence of Co and Ni makes it cheaper [8,9]. The benefits of LFP are significant, but the energy density of this material is limited by its lower working voltage (3.2 V) [10,11].

Other materials have attracted the attention of the research community or have even been implemented in commercial cells. The first commercial LIBs consisted of lithium cobalt oxide ($LiCoO_2$, LCO) as the cathode material [12]. It can provide high-rate and cycling performances, but its high content of Co makes it more expensive than other chemistries. Furthermore, its low thermal stability is another disadvantage for its implementation in electric vehicles [3]. $LiNiO_2$ (LNO) is an interesting candidate for high-energy applications as it could be analogous to NMC and NCA (consisting of Ni as the sole transition metal). Thus, it has a high specific capacity (the highest among those discussed) due to its high Ni content at a higher working voltage than that of LFP. However, the negative aspects of NMC and NCA are more noticeable in this material. Its thermal and electrochemical stability are low [13], and the costs of Ni, together with some geopolitical disagreements with one of its leading exporters (Indonesia) [14,15], make it recommendable to limit the content of Ni in the cathode material. On the other hand, spinel $LiMn_2O_4$ (LMO) is a much cheaper material, based on an abundance of manganese, and it works at high voltage, which is positive for high-energy applications [16]. However, the main disadvantage of LMO is its low structural stability, which has its origin in the Jahn–Teller distortion [17] and manganese dissolution due to the $2Mn^{3+} \rightarrow Mn^{4+} + Mn^{2+}$ disproportionation reaction [16]. A usual strategy to avoid the former is doping the material with other metals [17].

In this context, $LiNi_{0.5}Mn_{1.5}O_4$ (LNMO) can be considered an interesting alternative to LMOI Its working potential is ca. 4.7 V vs. Li, higher than that of LMO, and the presence of nickel provides a higher capacity (theoretically 147 $mAh \cdot g^{-1}$ [18], even if the practical capacity is usually lower). In addition, the mitigation of the Jahn–Teller effect makes this material a good alternative for high-energy applications [19]. Furthermore, the absence of cobalt in its composition is another significant advantage from a commercial perspective. Nevertheless, one of the main advantages of this material, i.e., its high working potential, becomes its main disadvantage when facing this cathode with graphite anode to obtain a high working voltage. This voltage is excessive for the state-of-the-art electrolytes and undergoes oxidative decomposition [20]. In addition, the surface of LNMO is transformed into the highly insulating rock salt, NiO [21]. A recent work by Jusys et al. studied gas evolution by means of DEMS of LIBs based on a LNMO cathode, graphite anode, and LP30 electrolyte, evidencing that Ni^{4+} catalyzed the dehydrogenation of the organic carbonates even after the decay of current to zero, indicating that the degradation processes can occur in the absence of Faradaic currents [22]. The main strategies to avoid this reaction consist of modifying the surface of LNMO [23–27], doping this material with other elements [28–34], using electrolyte additives [35–39] or alternative electrolytes [40] and separators [41,42], and following electrochemical methods [43] to avoid degradation reactions. Furthermore, it has been evidenced that it is possible to boost the energy density of the cells based on a LNMO cathode by combining it with high-capacity Si-based anodes if the electrolyte is carefully chosen [44].

Another disadvantage of LNMO is its limited electronic conductivity, which hinders its use in commercially attractive high-loading electrodes [45]. Conductive additives are therefore needed to work with this material. In addition, high-loading electrodes can also suffer from lower adhesion strength to the current collector compared with low-loading electrodes, which is mitigated with the increase in binder content. Both conductive and binder additives are electrochemically inactive; the higher their concentration in the formulation, the lower the energy density of the electrodes. Thus, it is necessary to minimize their presence without affecting the electrochemical performance and manufacturability of the electrodes. In fact, as Frith et al. [46] mentioned in a recent work, "When carrying out research focusing on industrial product development, researchers should develop products that solve a problem rather than develop a solution that needs to find a problem to solve". This means not limiting the research to the evaluation of the materials at the laboratory scale but proposing formulations and/or techniques that can be realistically upscaled to industrial-scale coating lines. Ue et al. also suggested a similar idea [47]: "the knowledge generated at academia must be brought to our society assessed by the technological criteria used in industry". This was the driving force motivating this work, where a cathode formulation based on LNMO active material was optimized and upscaled to a semi-industrial coating line. The second motivation was to propose a formulation using water as solvent instead of the current state-of-the-art organic solvent-based systems to decrease the carbon footprint of LIBs. There are interesting works that report alternative LNMO electrode manufacturing methods; dry-coating [48] and chemical vapor deposition [25] are examples of them. Nevertheless, we believe that it might be too early to produce electrodes via these techniques and expect that the current electrode manufacturing industry can be more easily adapted to water-based processing. Di Giorgio et al. compared the LNMO electrodes produced in aqueous (with carboxymethyl cellulose, CMC, as binder) and organic media (with polyvinylidene fluoride, PVDF, as binder) [43]. Waterborne electrodes outperformed those subjected to organic processing in terms of cycling performance. This improvement was attributed to the more stable electrode surface for the CMC-based electrodes.

However, there are even more ambitious strategies to boost the energy density of the lithium-based cells. On the one hand, lithium-sulfur batteries (LSBs) are a cheap, sustainable, and promising alternative for some applications to the state-of-the-art LIBs [49]. Their gravimetric energy density is significantly high (theoretically 2600 $Wh\cdot kg^{-1}$ [50]) and may find a business niche in applications where this feature is a determining factor. The cell chemistry of LSBs is based on the reduction of S8 to Li_2S upon discharge and its subsequent oxidation to S_8 upon charge, while lithium metal is usually used in the anode [51]. However, the intermediate products of these reactions, i.e., lithium polysulfides, are soluble in electrolytes and undergo shuttling, decreasing the coulombic efficiency of the cells [52]. Thus, the most significant efforts in past years have been focused on mitigating this inconvenience [53–55]. Even if the defunction of Oxis Energy Ltd. (Abingdon, UK) was a hard blow for the adepts of this technology, there are optimistic results published at prototype level [56]. Apart from LSBs, there are other types of lithium metal batteries (LMBs). In fact, lithium metal was the first anode material used in lithium batteries, which was later replaced by carbonaceous materials [57]. Among these carbonaceous materials, it is possible to find state-of-the-art graphite, which, in some cases, can be found blended with silicon-based materials in commercial cells [6]. Nevertheless, the high gravimetric and volumetric energy densities, achievable with metallic lithium, keep that old goal alive [58]. Even if the use of metallic lithium can lead to safety problems due to its reactivity and dendrite formation, they are being commercialized combined with solid electrolytes and special casings [59].

In any case, LSBs and LMBs are examples of mid-future technologies, and this work aims to analyze the short-term alternatives to current LIBs. As already mentioned, LNMO is an interesting solution to maximize the energy density of this type of batteries, which can become more sustainable by processing the electrodes using water instead of organic solvents. Thus, we propose a formulation optimization to maximize the electrochemical performance and the feasibility of LNMO electrodes. This electrode optimization was performed following the design of experiments (DoE) approach. This decision aimed to delimit the area of analysis and explore this area in an orderly manner, maximizing the information that could be obtained with the minimum number of samples by identifying the influential parameters affecting different factors and pointing out an optimum combination for the system studied [60,61]. A work by Rynne et al. [62] provides an excellent flowchart of the steps that should be followed to conduct an efficient DoE. The objectives of the study must initially be defined, followed by the definition of the responses, the parameters, and the plan. Afterward, the experimental part of the study is conducted to obtain the experimental results. Finally, it is time to refine the data to obtain a model that accurately describes the system and draw conclusions. This mathematical model can also be used to predict combinations that might even improve the response of those studied [62].

In this work, we report the use of a DoE matrix to obtain a mathematical model that predicts the best formulation of cathodes with LNMO as the active material. After obtaining and validating this formulation, its upscaling to a semi-industrial coating line is described. The electrodes obtained from this coating were used to assemble 1 Ah pouch cells as demonstrators which were tested under different rate capabilities.

2. Materials and Methods

2.1. Electrode Manufacturing

The positive electrodes consisted of $LiNi_{0.5}Mn_{1.5}O_4$ (LNMO, NANOMYTE SP-10 from NEI corporation, Somerset, NJ, USA) as the active material, carbon black (C-NERGY Super C45 from IMERYS Carbon & Graphite, Paris, France) as the conductive additive, carboxymethyl cellulose (CMC, Walocel 2000 from DOW Chemical Company, Midland, MI, USA) as the disperser and the thickener, and non-fluorinated acrylate polymer (from ZEON Chemicals, Louisville, KY, USA) as the binder. The percentage for each component in the formulation varied to prepare the samples subjected to the current study. They were mixed using water as the solvent by means of a mechanical stirrer. The solid-to-liquid (S/L) ratio in these formulations was between 1/1 and 1/1.15. Using a doctor blade table-top coater, the resulting slurries were coated onto a 20 µm-thick aluminum current collector (Hydro). The loading of such coatings was 1.6 mAh·cm^{-2} using 130 mAh·g^{-1} as the reference specific capacity of LNMO. The water of the as-coated electrodes was evaporated in a convection oven at 60 °C and the coatings were later calendered at a 2 g·cm^{-3} density using a laboratory-scale calendering machine (DPM solutions). The thickness of the coatings was measured using a Mitutoyo 389–271 C micrometer.

The electrodes used to assemble the pouch cells were manufactured at CIDETEC, using a semi-industrial coating line. On one hand, the cathodes consisted of the formulation which provided the highest desirability among those studied at the laboratory scale (see Section 3 for further detail). Three 250 g solid slurries were prepared and coated on the same aluminum current collector used in the laboratory coatings. The loading of the coating was controlled using a knife system with a variable gap, and the coating was passed through three convection ovens at 60, 70, and 80 °C to dry it. In addition, 9 m of double-side coating with a load of 1.5 mAh·cm^{-2} (per side) were obtained and calendered at a 2.0 g·cm^{-3} density. A picture of the cathode coating is displayed in Figure S1a.

The anodes used to assemble the pouch cells were also manufactured at CIDETEC, using an electrode coating line. They consisted of 94% graphite (MEG-2C, SGL Carbon, Wiesbaden, Germany) as the active material, 2% carbon black (C-NERGY Super C45 from IMERYS Carbon & Graphite) as the conductive additive, CMC (Walocel 2000 from DOW) as the disperser and the thickener, and SBR (TRD102A from JSR Micro NV, Leuven, Belgium) as the binder. The slurry was coated onto a 15 mm-thick copper current collector (Showa Denko Carbon, Tokio, Japan) and dried at 60, 70, and 80 °C. The loading and density of the doble-side anode were 1.7 mAh·cm^{-2} (per side, based on a specific capacity of 355 mAh/g) and 1.4 g·cm^{-3}, respectively.

2.2. Physicochemical Characterization

The LNMO powder and the as-prepared and calendered electrodes were analyzed by X-ray diffraction (XRD) to confirm the absence of degradation reactions for LNMO upon electrode processing, using a Bruker D8 Discover diffractometer (Cu Kα radiation, λ = 0.154 nm) equipped with a LynxEye PSD detector. The diffractograms were recorded between 2θ = 10° and 85° at 0.003° s^{-1}. The obtained data were fitted using the FULLPROF program [63]. Furthermore, the correct dispersion of the different components in the electrodes was confirmed means of field emission scanning electron microscopy (FE-SEM, ULTRA plus ZEISS). The rheological behavior of the slurries was characterized using a DHR2 rheometer from TA instruments. Finally, the adhesion strength of the coated electrodes on the current collector was measured using a 90° peel test (LS1 model, Lloyd Instrument, Bognor Regis, UK) on three stripes of 2 cm width.

2.3. Cell Assembly

The analysis of the optimum cathode LNMO-based formulation was performed in the half coin cell (HCC) configuration. The coatings prepared with the different formulations were cut with a diameter of 16.6 mm and dried for 16 h at 120 °C under vacuum conditions to eliminate traces of water. The cell cases were cleaned with ethanol in an ultrasonic bath for 15 min and dried at 60 °C for 1 h. A disc of lithium (50 µm-thick, from Albemarle Corporation, Charlotte, NC, USA) with a diameter of 18.2 mm was used both as a counter and reference electrode in each cell. CR2025-type Hohsen HCCs were assembled in a dry room with a dew point of −50 °C. The separator (Celgard ECT-2015), dried at 60 °C under vacuum conditions, was impregnated with 50 µL of 1 mol·L^{-1} lithium hexafluorophosphate in (3:7 vol%) ethylene carbonate:ethyl methyl carbonate + 1 wt% vinylene carbonate (1 M $LiPF_6$ in EC:EMC (3:7 wt.) + 1 wt.% VC) electrolyte, provided by Solvionic.

Electrodes for the soft-packaged pouch cells were directly cut from the obtained coatings and calendered on the semi-industrial coating line in a semiautomatic die-cutting unit (MTI Corp., Richmond, CA, USA). The dimensions of the cathodes and the anodes were 9.8 cm × 5.9 cm and 10.0 cm × 6.1 cm, respectively. A picture of a cathode is shown in Figure S1b. The drying protocols and separator used were the same as those in HCCs, and the cells were assembled in the dry room under the same conditions. The cells consisted of six cathodes and seven anodes per cell, which were stacked manually using a guiding tool to guarantee stack alignment. The separator was laminated to the cathodes before stacking with the anodes.

After stacking, the flanges (tabs) of the electrodes were ultrasonically welded to terminal tabs (100 µm-thick Al (+) and Ni-plated Cu (−)), and then placed between two half-shells of an aluminum laminated foil (ALF) pouch material and heat sealed on three sides. Afterward, the cells were filled with 7 mL of 1 M $LiPF_6$ in EC:EMC (3:7 wt) + 1 wt.% of succinic anhydride (SA) electrolyte, which was provided by Solvionic. Then, the remaining side was thermally sealed under −850 mbar using a vacuum chamber sealer. The cells were subjected to a routine analysis of voltage and AC resistance at 1 kHz using a Hioki BT3554 multimeter. The average resistance and voltage of the cells were 7.6 mOhm ± 0.2 mOhm and 0.10 V ± 0.03 V, respectively. Then, the cells underwent the formation protocol (described in Section 2.4) and were brought back to the dry room for the degassing step. In this step, one side of the cells was opened and thermally sealed again under vacuum conditions. Finally, the routine voltage (4.23 V ± 0.11 V) and resistance (9.4 mOhm ± 0.64 mOhm) check-up was repeated. A picture of the assembled pouch cells is displayed in Figure S1c.

2.4. Electrochemical Testing

All the experiments were performed in a Basytec Cell Test System potentiostat at a room temperature of 25 °C ± 1 °C controlled by air conditioning. The HCCs were left for 16 h at open circuit potential to guarantee the adequate wetting of the electrodes and the separator. Afterward, they were subjected to three galvanostatic cycles at 0.2 C, one at 0.5 C, three at 1 C, and one at 0.5 C, all of them between 5 V and 3.5 V. The protocol continued with a fast discharge C-rate analysis, three at 2 C, three at 5 C, three at 8 C, and three at 10 C. The charges of all these cycles were performed at 1 C, and a check-up cycle at 0.5 C was intercalated between each discharge C-rate three-cycle block. Finally, the recovery capacity of the cells was analyzed by performing three cycles at 1 C. All the cycles in the protocol described included a constant voltage (CV) step to 0.05 C at 5 V. The testing protocol is summarized in Table S1.

The pouch cells were subjected to a formation cycle at 0.05 C, and two at 0.1 C, all of them between 4.8 and 3.6 V. A CV step to 0.025 C (for the 0.05 C cycle) and to 0.05 C (in the 0.1 C cycles) was set at 4.7 V to finalize the charge step. After completing the three cycles in a full-discharged state, the cells were disconnected from the testing channels, moved to the dry room, and degassed. After sealing under vacuum, the cells were connected back to the potentiostat and divided into two groups of three cells each. The first group was subjected to galvanostatic cycling at 0.5 C charge and discharge C-rate to analyze the capacity retention of the cells. On the other hand, the other three cells were subjected to more exigent rate-capability and cycling experiments: the first five cycles consisted of discharges at rates of 1 C, 2 C, 3 C, 4 C, and 5 C with charges at 0.5 C, followed by four cycles with charges at 0.1 C., 0.2 C, 0.5 C, and 1 C and discharges at a constant C-rate (1 C).

Afterward, the cells were cycled at parallel 1 C discharge and charge C-rates. All these tests were conducted between 4.8 and 3.6 V.

2.5. Definition of the Formulations and the Parameters for Their Evaluation

The positive electrode formulation to be optimized was composed of four different components: LNMO, carbon black, CMC, and binder. The mathematical calculations to obtain these formulations were conducted using Equation (1).

$$y = \sum_{i=1}^{q} \beta_i x_i + \sum_{i<j}^{q} \beta_{ij} x_i x_j \quad (1)$$

The experimental approach used was that of a mixture design of extreme vertices plus centroid, with each point in the experimental space being carried out in triplicate. To determine the composition of the test points, a parallelepiped was defined (within a tetrahedron), composed of 8 vertices plus a centroid. The proposed mathematical model was a mixed model (linear plus quadratic) using Equation (1). Afterward, the upper and lower limits in the formulation were defined for each of the components. This information has been compiled in Table S2. These upper and lower limits were selected based on the literature [36,37,48], previous experience, and recommendations by the component providers. The selected upper and lower limits defined the experimental area in the current study. To determine the composition of the samples to be prepared and analyzed, a matrix (Table 1) consisting of eight corners and a centroid was defined.

Table 1. Design matrix for positive electrode formulations.

Point Type	Formulation Number	LNMO wt.%	Carbon Black wt.%	CMC wt.%	Binder wt.%
Corners	1	91	4	2	3
	2	91	4	3	2
	3	90	4	3	3
	4	91	5	2	2
	5	89	5	3	3
	6	90	6	2	2
	7	89	6	2	3
	8	89	6	3	2
Centroid	9	90	5	2.5	2.5

After determining the formulations (inputs) to be used in the experimental part of this study, it was necessary to define the parameters (outputs) to evaluate these formulations. As these formulations were designed to be implemented in the cathodes of LIBs, their electrochemical performance must be an essential output to decide which is more suitable for this application. Furthermore, as electrodes for high energy applications with high working voltages, the specific capacities achievable with the different formulations at different C-rates (0.2, 1, and 2 C) were thought to be adequate outputs for evaluating electrochemical performance. The higher the capacity, the higher the energy, and, therefore, the more desirable the response is. On the other hand, electrode preparation feasibility should not be left out of the evaluation, as a formulation with excellent electrochemical performance but that is hardly processable is not attractive for electrode manufacturers. For this aim, the outputs selected for this feasibility group were the slope of the viscosity vs. the shear rate curve in the rheological analysis, the viscosity at 10 s^{-1} shear rate, and the adhesion strength after coating and calendering. Higher viscosities, slopes, and adhesion strengths are desired for electrode manufacturing [64–67]. The list of parameters used to evaluate the formulations is summarized in Table 2. In addition, the schematic representation of the analysis flow followed in the current work is represented in Figure 1.

Table 2. List of experimental outputs.

Parameter Group	Parameter	Unit	Sample	Optimization Criteria
Feasibility	Viscosity at shear rate = 10 s^{-1}	Pa·s	Slurry	Nominal-better
	Slope of the viscosity with the shear rate	Pa·s^2	Slurry	Higher-better
	Peel resistance	N·m^{-1}	Electrode	Higher-better
Electrochemical performance	Discharge capacity at 0.2 C	mAh·g^{-1}	Half coin cell	Higher-better
	Discharge capacity at 2 C	mAh·g^{-1}	Half coin cell	Higher-better
	Discharge capacity after fast cycles (at 1 C)	mAh·g^{-1}	Half coin cell	Higher-better

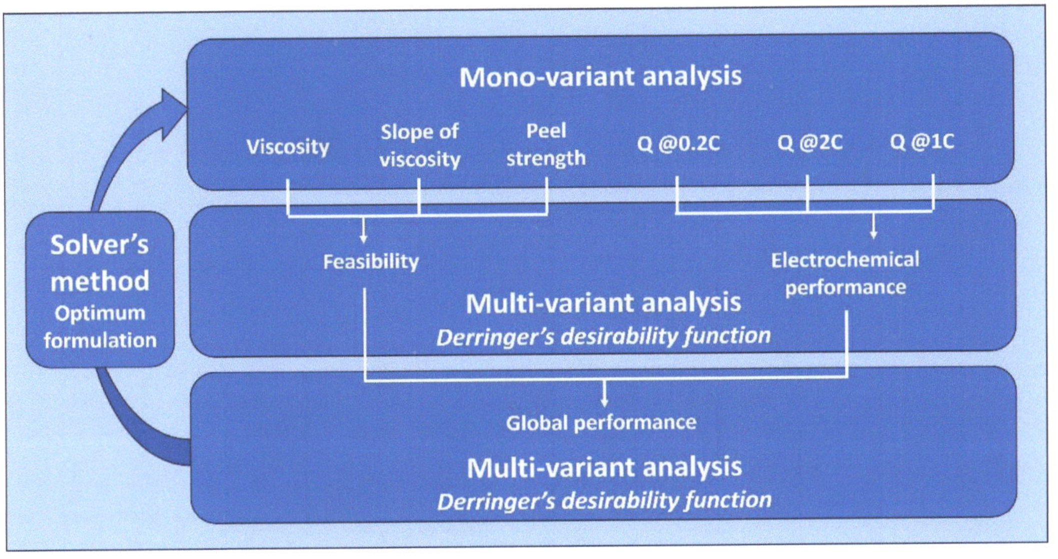

Figure 1. Schematic representation of the analysis flow followed in the current work.

3. Results and Discussion

3.1. Mono-Variant Analysis

After defining the parameters for evaluating the formulations, the experimental part of the study was initiated. The slurries were prepared and subjected to a rheological analysis. The viscosity vs. shear rate curves obtained in this analysis can be observed in Figure S2. All the curves showed a shear thinning behavior, i.e., the viscosity was decreased with the shear rate [68], making them suitable to be coated on a coating line with a knife system [69]. The viscosity values at 10 s^{-1} and the slope for each viscosity vs. shear rate curve can be observed in Table 3. The highest viscosities were obtained with formulations #2 and #9, while the highest slopes were calculated from the rheological analysis of formulations #6 and #7. A quick revision of these results suggests that the increase in the carbon black concentration entails an increase in the slope of the curve, resulting in more stable slurries.

Afterward, the slurries were coated onto the current collector and calendered to the target density. The electrodes were subjected to XRD and FE-SEM characterization to discard degradation reactions or morphology modifications in the LNMO upon processing (see Figures S3 and S4 for further detail). Then, the peel strength of the coatings with the different formulations was measured; the results obtained are compiled in Table 3. The highest peel strength was obtained with formulation #7, consisting of 89% LNMO, 6% carbon black, 2% CMC, and 3% binder. On the other hand, the lowest value was provided by formulation #3, with 90% LNMO, 4% carbon black, 3% CMC, and 3% binder. The latter result is surprising as this formulation consists of the highest CMC and binder concentrations among those studied (together with formulation #5). This evidences that, to boost a specific parameter, it is not enough to increase the concentration of the components intuitively promoting this parameter; the interaction of these components with the rest is also determining.

Table 3. Results obtained for each parameter selected for the evaluation with different formulations. The reported results are an average of three replicates.

Formulation	Feasibility			Electrochemical Performance		
	Viscosity (Pa·s)	Slope Viscosity (Pa·s^2)	Peel (N·m^{-1})	0.2 C (mAh·g^{-1})	2 C (mAh·g^{-1})	1 C (mAh·g^{-1})
1	9.60	0.36	36.8	128.28	116.47	114.64
2	27.14	0.36	20.9	127.22	108.36	108.39
3	20.94	0.33	17.4	128.02	110.32	107.72
4	13.01	0.45	24.9	128.99	120.93	117.04
5	20.09	0.40	32.4	126.58	112.07	97.36
6	15.93	0.57	29.5	125.93	116.76	117.89
7	13.54	0.54	44.6	128.25	117.10	101.81
8	19.61	0.44	31.9	128.62	118.96	110.05
9	24.10	0.49	30.9	129.29	120.10	114.31

Finally, the electrodes obtained from these coatings were subjected to electrochemical testing in HCCs. The capacities at 0.2 C, 2 C, and 1 C for each formulation are included in Table 3. The capacity at 0.2 C is very similar for all the samples (slightly lower for formulation #6), but more differences arise when subjecting the cells to higher C-rates. The highest capacity at 2 C was obtained with formulation #4, followed by formulation #9. On the other hand, formulations #6 and #4 (in this order) provided the highest capacity at 1 C. Thus, it is likely that the multivariant analysis will reveal a high electrochemical desirability for formulation #4.

The mono-variant analysis aims to determine the influence of the electrode formulations on the individual response to each of the parameters analyzed. Thus, Minitab software (version 16.2.3) was used to obtain a formal functional relation between the experimental outputs and the corresponding formulations (Equation (2)).

$$Output\ value = k_{LNMO} \cdot C_{LNMO} + k_{CB} \cdot C_{CB} + k_{CMC} \cdot C_{CMC} + k_{Binder} \cdot C_{Binder} + k_{LNMO-CB} \cdot C_{LNMO} \cdot C_{CB} + k_{LNMO-CB} \cdot C_{LNMO} \cdot C_{CMC} + k_{LNMO-Binder} \cdot C_{LNMO} \cdot C_{Binder} + k_{CB-Binder} \cdot C_{CB} \cdot C_{Binder} \quad (2)$$

Here, k_i is the correlation coefficient and C_i the concentration of each component. The results obtained with this method are depicted in Table S3. The statistical correlation of the functions and the output can be evaluated based on the linear regression of these functions. The electrode components most significantly influencing the output of each parameter and the regression coefficient are shown in Table 4.

The results in Table 4 evidence that the hypothesis formulated when analyzing the output of the slope of the rheological curve, i.e., the increase in the slope with the concentration of carbon black, was correct. Interestingly, the component most significantly affecting the peel strength of the coatings was the carbon black, which was penalized with the increase in its concentration. Furthermore, the increase in the CMC benefited the rheological response of the slurry. On the other hand, the electrochemical response was clearly improved with the increase in the LNMO and carbon black concentrations. Nevertheless, the regression coefficient was 39.11 for the discharge capacity at 0.2 C as an output. This was attributed to the absence of remarkable differences in the specific capacity when the cells were discharged at this low C-rate, as previously discussed.

Table 4. Mono-variant analysis summary: the component most significantly affecting each parameter based on the functions obtained and the corresponding regression coefficient (R^2). ↑ and ↓ are used to express increase and decrease, respectively.

Parameter	Input	Output	R^2
Viscosity at shear rate = 10 s^{-1}	CMC ↑	↑	99.79
Slope of the viscosity with the shear rate	CB ↑	↑	99.57
Peel resistance	CB ↓	↑	95.64
Discharge capacity at 0.2 C	LNMO + CB ↑	↑	39.11
Discharge capacity at 2 C	LNMO + CB ↑	↑	90.57
Discharge capacity after fast cycles (at 1 C)	LNMO + CB ↑	↑	98.10

3.2. Multi-Variant Analysis

The mono-variant analysis allows the identification of the components more significantly affecting the defined outputs. Still, it is necessary to conduct a multi-variant analysis to evaluate the inputs and outputs jointly and adjust the formulation to obtain electrodes with optimized feasibility and electrochemical performance. In that regard, desirability, a concept introduced by E. C. Harrington [70] and later developed by Derringer and Suich [71], allows the compiling of individual outputs in a single quantitative value. The first step in this analysis consists of the conversion of the individual evaluation criteria of the parameters in elemental desirability functions (d_i) with values between 0 (non-desirable) and 1 (most desirable). The functions for each of the parameters are shown in Figure S5.

The shape of the functions is similar for the slope of the rheology curve, peel strength, and specific capacities at 0.2 C, 2 C, and 1 C. There is a minimum value at which d_i is 0 below which the response is not desirable, a value above which the response is evaluated as optimum (desirable, d_i = 1), and a linear regression between these two values. On the other hand, the viscosity of the slurry cannot be evaluated following the same rule; too high (>30 Pa·s) or too low (<5 Pa·s) viscosities make it difficult to successfully coat the slurry in a coating line and, thus, were defined as d_i = 0. Nevertheless, viscosity values between 10 and 25 Pa·s are excellent for coating and were defined as d_i = 1. Finally, values between d_i = 0 and d_i = 1 were fitted following linear regression.

The next step consisted of weighting each parameter group (i.e., feasibility and electrochemical performance) and each of the parameters included in these groups. Electrochemical performance was preferred over feasibility, weighting these factors with 75 and 25% of the overall desirability, respectively. Among the three parameters considered in the feasibility group, it was decided that the peel strength is the most important one (50% of the feasibility), as it is crucial for the electrode cutting and manipulation during cell assembly, as well as contributing to the electrochemical performance (avoiding the delamination of the electrodes). The viscosity slope and the viscosity at 10 s^{-1} shear rate were weighted with 30% and 15% of the feasibility, respectively. On the other hand, the weighting of the electrochemical performance group decreased with the increasing C-rate. These were 50%, 30%, and 20% for the specific capacity at 0.2 C, 1 C, and 2 C, respectively. The desirability of each parameter, group of parameters, and global desirability are shown in Table S4. The highest global desirability value was obtained with formulation #9. Nevertheless, this only indicates that this sample is the most desirable among those studied, not necessarily the most desirable in the defined experimentation area. To determine this most desirable formulation, Solver's method will be used in the sext section.

The obtained values were used to obtain a mathematical model to predict global desirability (D), which is referred to here as Equation (3). In addition, the desirability heat map that can be observed in Figure 2 was obtained.

$$D = -0.0017\,[\mathrm{LNMO}] - 1.08\,[\mathrm{C45}] - 3.13\,[\mathrm{CMC}] - 1.13\,[\mathrm{Binder}] + 0.01\,[\mathrm{LNMO}]\,[\mathrm{C45}] + 0.03\,[\mathrm{LNMO}]\,[\mathrm{CMC}] + 0.01\,[\mathrm{LNMO}]\,[\mathrm{Binder}] + 0.05\,[\mathrm{C45}]\,[\mathrm{Binder}] \quad (3)$$

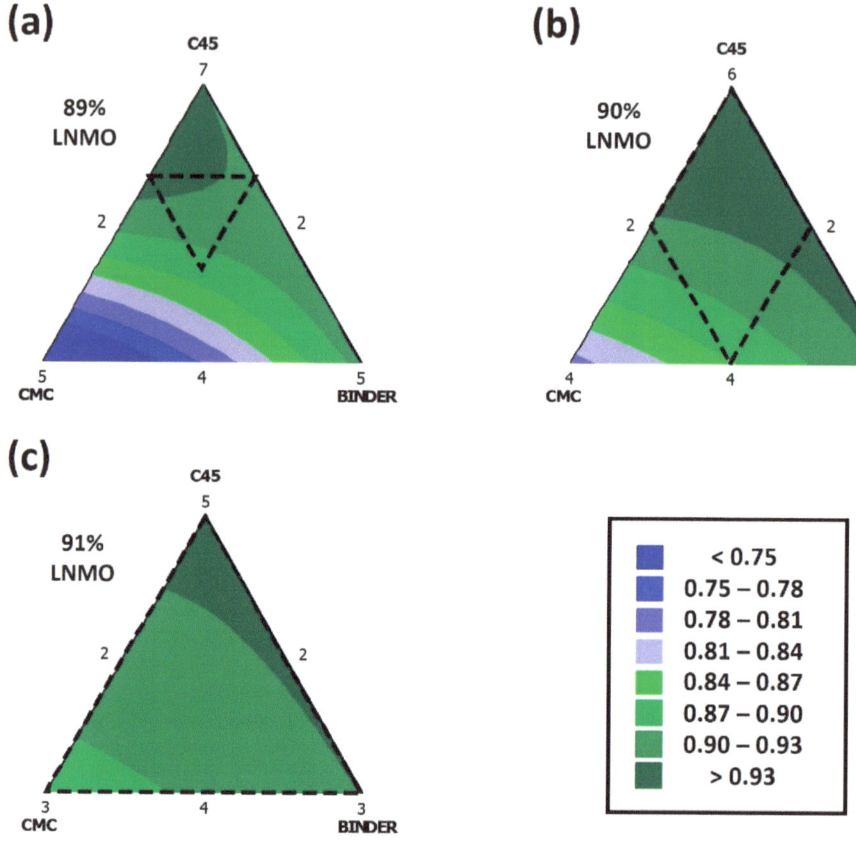

Figure 2. Heat map of the global desirability for the experimental area defined in the current study for (**a**) 89%, (**b**) 90%, and (**c**) 91% LNMO. A legend is displayed for the interpretation of the heat maps.

The regression coefficient for Equation (3) is 97.83%. The differences between the experimental and the mathematical values are shown and discussed in Table S5.

3.3. SOLVER's Method to Obtain the Cathode with the Best Formulation

Based on Equation (3), a 10th formulation was designed with the maximum global desirability value in the working region consisting of 89.5% LNMO, 6% C45, 2.5% CMC, and 2% binder. Furthermore, another formulation (#11) was predicted to provide even higher global desirability but was placed out of the working region. It consisted of 89.5% LNMO, 6% C45, 3% CMC, and 1.5% binder. The mathematical desirability for these samples was 0.952 and 0.967, respectively. Both coatings were prepared and subjected to the same physicochemical characterization as those used to obtain the desirability prediction function. The results have been included in Table 5.

The feasibility, electrochemical, and global desirability values for formulation #10 were 0.861, 0.914, and 0.949, respectively. The latter value is close to the desirability calculated from the mathematical model (0.952), evidencing the benefits of this method. As expected, the experimental global desirability value for formulation #10 is higher than those obtained for the samples used to build the prediction model. On the other hand, the viscosity value obtained with formulation #11 (34.53 Pa·s) was above the highest acceptable value defined in the multi-variant analysis (30 Pa·s). Thus, its feasibility desirability is 0 and was discarded. This shows the importance of conducting corroboration experiments. It might occur that the outputs for certain parameters of the formulations calculated by Solver's method are close to the lower desirability limits and can be affected by the difference between the calculated and experimental values.

Table 5. Experimental outputs and desirability values of the formulations obtained by Solver's method, based on the previous multi-variant analysis. The reported results are an average of three replicates.

Formulation	#10: 89.5% LNMO, 6% C45, 2.5% CMC and 2% Binder		#11: 89.5% LNMO, 6% C45, 3% CMC, and 1.5% Binder	
	Output	**Desirability**	**Output**	**Desirability**
Viscosity (Pa·s)	16.82	1	34.53	0.000
Viscosity slope (Pa·s^2)	0.530	0.767	0.582	0.940
Peel strength (N·m^{-1})	29.6	0.490	28.9	0.473
Specific capacity at 0.2 C (mAh·g^{-1})	128.31	0.610	124.73	0.490
Specific capacity at 2 C (mAh·g^{-1})	120.84	1.000	115.58	0.780
Specific capacity at 1 C (mAh·g^{-1})	117.83	0.930	118.06	0.940

3.4. Upscaling and Pouch Cell Testing

Formulation #10 was upscaled to the coating line, as described in Section 2.1. The slurry consisted of 250 g of solids, which is much higher than the 30 g solids slurries prepared at laboratory scale. Therefore, the mixing conditions, such as the stirring force to which the slurry components were subjected, were slightly different. This required an adjustment of the solid-to-liquid (S/L) ratio (1/0.73 at laboratory scale, 1/1.09 at pilot plant scale). As expected, the rheological analysis (Figure S2) evidenced that the viscosity was lower for the pilot plant slurry, which had a higher fraction of liquid phase per solvent amount than the laboratory scale slurry. However, the difference in the viscosity was minimal for the shear rate of the region analyzed, which could be attributed to a better dispersion of the components. Therefore, the better the dispersion, the more water the solid particles take, and the higher the viscosity of the slurry. After calendaring, the coating was subjected to the adhesion strength measurement. The 90° peel strength test result was slightly lower with the pilot plant coating at 26.3 N·m^{-1} versus the 29.6 N·m^{-1} obtained at laboratory scale. In any case, the difference between the two values was minimal. Furthermore, to conduct the peel test, it is necessary to manually clean one of the sides of the double-side coated electrode with water and a wipe, which can affect the adhesion strength measured on the other side.

The coatings on the electrodes were cut to later assemble the pouch cells used for the electrochemical validation of the upscaling process. The cells comprised six 9.8 cm × 5.9 cm double-side cathodes and seven 10 cm × 6.1 cm double-side anodes. After being filled with electrolyte, the cells were subjected to a formation protocol consisting of one 0.05 C cycle, followed by two 0.1 C cycles. The galvanostatic profiles of these three cycles are depicted in Figure 3a. In the first charge, ~0.1 Ah was consumed to form the solid-electrolyte interphase (SEI) below 4.4 Ah. Afterward, a pseudo-plateau at 4.4 V and two defined plateaus at 4.6 and 4.7 V were registered. In the subsequent discharge, complementary plateaus were obtained at 4.35, 4.55, and 4.6 V, respectively. References in the literature [25,32,72,73] report the presence of two main redox peaks between 4.6 and 4.8 V; two features of charge and two (associated with each of the former) of discharge. These signals are attributed to the redox activity of nickel (Ni^{2+}/Ni^{4+}). On the other hand, they also evidence that the contribution of manganese (Mn^{3+}/Mn^{4+}) to the cell capacity can be observed to plateau at ~4 V. The absence of a feature with this potential suggests that the redox activity observed at 4.35 (discharge) and 4.4 V (charge) can be attributed to the electrochemical activity of Mn. The capacities obtained in the first charge and discharge of the cells were 1.03 and 0.85 Ah, respectively, for a Coulombic efficiency of 82.6%. The charge–discharge capacities in the second and third cycles were 0.83–0.82 Ah and 0.83–0.81 Ah, respectively. The increasing, but relatively low, coulombic efficiencies in these two cycles (97–98%) suggest that part of the lithium is still being consumed in the formation of the SEI. These two cycles presented the same plateaus, confirming that the same electrochemical reactions are occurring.

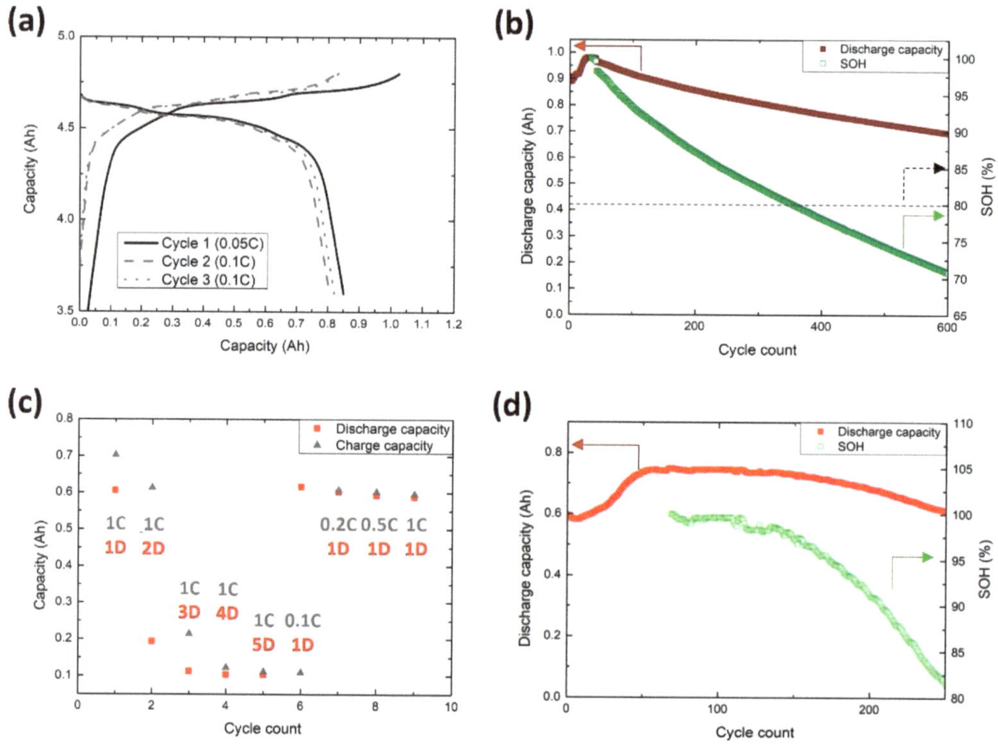

Figure 3. Electrochemical results obtained with the LNMO/graphite pouch cells. (**a**) Galvanostatic curves obtained in the formation step, (**b**) capacity evolution and capacity retention curves at 0.5 C cycling, (**c**) results of the rate-capability test, and (**d**) capacity evolution and capacity retention curves at 1 C. Discharge (D) and charge (C) C-rates are specified for each cycle in (**c**).

After finishing the formation, three of the cells were degassed and subjected to 0.5 C cycling; the discharge capacity values and capacity retention with the cycle count are displayed in Figure 3b. Interestingly, the capacity was increased until the maximum value was reached in the 33rd cycle. This phenomenon was not observed in the half cells tested in the formulation optimization study, in which the upper voltage limit was higher (5 V in coin cells, 4.8 V in pouch cells). Nevertheless, it is possible to find the same behavior in other works [24,74,75]. In pouch configuration, the upper voltage limit was decreased to avoid excessive gassing due to electrolyte degradation. In Figure S7a, the galvanostatic profiles of the 2nd and the 38th cycles are compared; it can be observed that the upper voltage limit was reached earlier in the 2nd cycle, and that the CV step was longer than in the 38th cycle. In addition, at first sight, the discharges almost overlap up to 0.55 Ah, with an earlier voltage decay in the 2nd cycle. Our first thought was to attribute this unexpected behavior to a partial consumption of the charge in SEI formation; but, if this were the case, we would obtain low Coulombic efficiencies in these first cycles, which was not the case (Figure S7b). In fact, Coulombic efficiency was above 100% in many of the cycles before the previously mentioned 33rd cycle. A Coulombic efficiency above 100% means that more Li^+ is being transferred from the graphite electrode to the LNMO electrode than in the opposite way. Nevertheless, the $dQ{\cdot}dV^{-1}$ vs. voltage representation of the curves in Figure S7a, displayed in Figure S7c, provides a different point of view. In this type of graph, each plateau in the galvanostatic profile is represented as a peak. The flatter and longer the plateau, the sharper and larger the peak in the dQ·dV^{-1} vs. voltage representation. Negative dQ·dV^{-1} values are used to represent the discharge, while charge curves are plotted in the positive part of the *y*-axis. The first 0.1 C cycle in the formation protocol is included as a reference. One significant difference that can be observed when comparing the 2nd and the 38th cycles is the appearance of an addition peak at 4.67 V during charge. This does not necessarily involve the occurrence of a new reaction and could be attributed to the decoupling of two signals that occurred together in the 2nd cycle. In fact, this signal is also observed in the formation cycle, which is conducted at a lower C-rate. In these conditions, it is easier to decouple different electrochemical processes into different plateaus. On the other hand, based on the galvanostatic curve in Figure S7a, it was assumed that there was not any overpotential in the discharge reaction. Nevertheless, this overpotential is evidenced in Figure S7c, as all the signals in the negative *y*-axis occur at lower voltage in the 2nd cycle. The overpotential is even lower in the formation step and conducted at a lower C-rate. However, the most significant difference is registered at 4.4–4.5 V in this region; the activity has been attributed to the redox reactions of manganese. The previously discussed pseudo-plateaus result in a wide range of features, both upon discharge and charge, for the formation and the 2nd 0.5 C cycles. Contrarily, two evident peaks are observed in the 38th cycle. Thus, it can be concluded that the increase in the capacity that can be observed during the first 33 cycles is attributed to the activation of the redox activity of Mn. The cells cycled at 0.5 C completed 321 cycles (plus the first 32 cycles before reaching the maximum capacity) before reaching the referential 80% SOH commonly used as a reference to determine the end of life in Li-ion technology. In addition, they completed 592 cycles (+32) before reaching 70% SOH.

On the other hand, the other three cells were subjected to rate-capability experiments (Figure 3c). The first cycle was performed at 0.5 C charge and 1 C discharge C-rates. The capacities obtained were 0.70 and 0.60 Ah, respectively. Compared with the last 0.1 C cycle in the formation cycle (0.81 Ah), the capacity was 26% lower at 1 C. The limitations of the discharge rate capabilities were further evidenced by the increasing the discharge C-rate in the following cycles: 0.19, 0.11, 0.10, and 0.10 Ah, which were obtained at 2 C, 3 C, 4 C, and 5 C, respectively. Thus, it was determined that 1 C was the most usable discharge current with an acceptable performance, as the drop in the capacity was excessive when the cells were subjected to higher C-rates. The charge rate capability analysis was performed by analyzing lower C-rates (0.1 C, 0.2 C, 0.5 C, and 1 C) with discharges at 1 C. In these discharges, the cells provided 0.62, 0.60, 0.59, and 0.59 Ah. Therefore, the discharge current obtained was almost the same after charging the cells at the different C-rates. Afterward, the cells were cycled using a parallel 1 C rate: the evolution of the capacity and the capacity retention with the number of cycles is shown in Figure 3d. Again, there was an increase in the discharge capacity with the increasing cycle count, as observed in the 0.5 C cycling. Nevertheless, it seems that the incomplete charge was even more significant at higher charge C-rates. Thus, the capacity was continuously increased until the 50th cycle, and the maximum capacity was obtained in the 69th cycle. Thus, the activation of Mn is dependent on the C-rate at which the cells are tested, with extended delay with increasing the C-rate. The cells completed another 181 cycles at 1 C without reaching the reference 80% SOH capacity (+68 cycles at the beginning of life before reaching the maximum capacity). This lower cycling performance compared with the cells cycled at 0.5 C is expected, as a higher degradation of the materials usually occurs at higher C-rates. Lower capacity retention with an increasing C-rate has also been reported in other works [29,38]. The capacity fade at both C-rates can be more accelerated than in LIBs with other cell chemistries, such as NMC and, principally, LFP [76]. However, as mentioned by Yao et al. [48], this is something common in the literature related with LNMO, as the technology is not mature enough yet. The higher cycling stability is not a matter of just the cathode material itself, as the stability of the other components in the cell can also condition the capacity retention. This becomes particularly important when working with high-voltage materials; it is important that the rest of the cell components are stable at such a working voltage [35,40]. Furthermore, the formation of a stable cathode-electrolyte interphase (CEI) is mandatory for enhancing the cycle life of the LNMO cells [48,77]. The main objective of this work was to implement DoE as a methodology for defining a robust formulation and upscaling it to a semi-industrial coating line, with the pouch cells being assembled-and-tested demonstrators of the latter process. Future works could be directed toward analyzing the main failure mechanism in these cells and focusing on mitigating it.

4. Conclusions

The water-based formulation of the emerging Li-ion cathode chemistry, LNMO, was optimized by means of DoE methodology. It was evidenced that the fraction of the different components directly impacts the performance of the feasibility and electrochemical outputs defined. In addition, the desirability of the different formulations studied was compared following Derringer–Suich's methodology, which showed that the formulation with the highest desirability in the study consisted of 90% LNMO, 5% C45, 2.5% CMC, and 2.5% binder. The results were used to build up a regression line to predict desirability in the region of study, and, interestingly, the formulation with the highest mathematical desirability was not the same as that with the highest experimental desirability. This was attributed to small experimental deviations. Furthermore, Solver's method was used to predict a formulation with the maximum desirability in the region of study and outside this region. The higher desirability of the formulation in the region of study (89.5% LNMO, 6% C45, 2.5% CMC, and 2% binder) was confirmed experimentally and the recipe was upscaled to a semi-industrial coating line. The double-side electrodes obtained were used to assemble 1 Ah pouch cells and then tested. It was revealed that the activation of manganese redox activity needs an activation period, after which the maximum capacity of the cells is obtained. Furthermore, more activation cycles are necessary for this activation with increasing C-rate, based on our results with cells cycled at 0.5 C and 1 C. The cells cycled at the 0.5 C rate completed over 300 cycles before reaching 80% SOH, while the cycling performance was lower with those subjected to 1 C. Indeed, rate capability results evidenced a significant decay in the electrochemical performance when subjecting the cells to C-rates higher than 1 C. Overall, the optimization and upscaling of the LNMO electrode recipe resulted in the obtaining of high-energy pouch cells of 1 Ah that could be used to gain knowledge in the chemistry of this lithium-ion cathode material.

Supplementary Materials: The following are available online at https://www.mdpi.com/article/10.3390/en16217327/s1, Figure S1: Pictures of (**a**) the cathode coating after passing through the knife, (**b**) a cathode, and (**c**) assembled pouch cells; Figure S2: Rheological studies with the different formulations: viscosity vs. shear rate curves; Figure S3: X-ray diffraction patterns of the (**a**) LNMO powder and (**b**) a LNMO electrode. Patterns were fitted to LNMO phase using FULLPROF software and its theoretical reflections and the difference between the experimental and the fitting curves are included; Figure S4: FE-SEM images of the LNMO powder and a LNMO electrode: LNMO powder with magnification (**a**) X1000 and (**b**) X10000 and LNMO electrode with magnification (**c**) X1000 and (**d**) X10000; Figure S5: Derringer–Suich desirability functions for the different experimental outputs under evaluation: (**a**) viscosity, (**b**) viscosity slope, (**c**) peel strength, (**d**) 0.2 C capacity, (**e**) 2 C capacity, and (**f**) 1 C capacity; Figure S6: Rheological analyses (viscosity vs. shear rate curves) of the laboratory-scale slurry with formulation #10 and slurry prepared for upscaling to the coating line with the same formulation; Figure S7: (**a**) Voltage vs. capacity representation of the 2nd and the 38th cycles with the pouch cells cycled at 0.5 C. (**b**) Coulombic efficiency of the cells subjected to 0.5 C C-rate. (**c**) dQ/dV vs. voltage representation of the 2nd and the 38th cycles with the pouch cells cycled at 0.5 C; Table S1: Testing protocol applied to the LNMO HCCs; Table S2: Components of the positive electrodes and upper and lower limits defined for the optimization of the slurry; Table S3: Functional relations between the concentration of the electrode components and the experimental outputs; Table S4: Desirability of the different samples for each parameter, group of parameters and the global desirability; Table S5: Experimental mathematical desirability values obtained for the 9 formulations in this study.

Author Contributions: Conceptualization: L.L, I.U. and I.d.M.; methodology, L.L, I.U., H.-J.G. and I.d.M.; software, L.L.; validation, L.L., I.U. and A.E.-B.; formal analysis, L.L. and M.B.; investigation, L.L., I.U., I.B., A.E.-B. and I.d.M.; resources, I.d.M.; data curation, L.L, I.U. and I.L.-M.; writing—original draft preparation, I.U. and I.L.-M.; writing—review and editing, L.L., I.U., I.L.-M. and I.d.M.; supervision, M.B., A.E.-B. and I.d.M.; project administration, I.B. and I.d.M.; funding acquisition, I.d.M. All authors have read and agreed to the published version of the manuscript.

Funding: This research was funded by the European Union Seventh Framework Program (FP7/2007–2013), grant number 609201 (MARS-EV project).

Data Availability Statement: The data presented in this study are available on request from the corresponding author. The raw data are not publicly available due to the related patent.

Acknowledgments: The authors acknowledge SGL Carbon (Christian Schreiner) for the graphite supply, Lithops srl (Michaela Aneta Dumitrescu) for the anode electrode and Solvionic (Rongying Lin, Sebastien Fantini) for the electrolytes. We would like to thank the electrode manufacturing and pouch cell assembly teams at the CIDETEC pilot plant.

Conflicts of Interest: The authors declare no conflict of interest.

References

1. Winter, M.; Barnett, B.; Xu, K. Before Li Ion Batteries. *Chem. Rev.* **2018**, *118*, 11433–11456. [CrossRef]
2. International Energy Agency. Global EV Outlook 2021 Accelerating Ambitions despite the Pandemic. 2021. Available online: https://www.iea.org/reports/global-ev-outlook-2021 (accessed on 24 October 2023).
3. Li, M.; Lu, J.; Chen, Z.; Amine, K. 30 Years of Lithium-Ion Batteries. *Adv. Mater.* **2018**, *30*, 1800561. [CrossRef]
4. Volta Foundation The Battery Report. 2022. Available online: https://intercalationstation.substack.com/p/the-battery-report-2022 (accessed on 24 October 2023).
5. Marinaro, M.; Bresser, D.; Beyer, E.; Faguy, P.; Hosoi, K.; Li, H.; Sakovica, J.; Amine, K.; Wohlfahrt-Mehrens, M.; Passerini, S. Bringing Forward the Development of Battery Cells for Automotive Applications: Perspective of R&D Activities in China, Japan, the EU and the USA. *J. Power Sources* **2020**, *459*, 228073. [CrossRef]
6. Tidblad, A.A.; Edström, K.; Hernández, G.; de Meatza, I.; Landa-Medrano, I.; Jacas Biendicho, J.; Trilla, L.; Buysse, M.; Ierides, M.; Horno, B.P.; et al. Future Material Developments for Electric Vehicle Battery Cells Answering Growing Demands from an End-User Perspective. *Energies* **2021**, *14*, 4223. [CrossRef]
7. Qin, Z.; Zhang, Y.; Luo, W.; Zhang, T.; Wang, T.; Ni, L.; Wang, H.; Zhang, N.; Liu, X.; Zhou, J.; et al. A Universal Molten Salt Method for Direct Upcycling of Spent Ni-Rich Cathode towards Single-Crystalline Li-Rich Cathode. *Angew. Chem. Int. Ed.* **2023**, *62*, e202218672. [CrossRef]
8. Fichtner, M. Recent Research and Progress in Batteries for Electric Vehicles. *Batter. Supercaps* **2022**, *5*, e202100224. [CrossRef]
9. Ruffo, G.H. Tesla Slide Reveals $25,000 Electric Car Will Have LFP Batteries. Available online: https://insideevs.com/news/446245/tesla-slide-25000-car-lfp-batteries/ (accessed on 15 September 2023).
10. Moretti, A.; Sharova, V.; Carvalho, D.V.; Boulineau, A.; Porcher, W.; de Meatza, I.; Passerini, S. A Comparison of Formation Methods for Graphite//$LiFePO_4$ Cells. *Batter. Supercaps* **2019**, *2*, 240–247. [CrossRef]
11. Aiken, C.P.; Logan, E.R.; Eldesoky, A.; Hebecker, H.; Oxner, J.M.; Harlow, J.E.; Metzger, M.; Dahn, J.R. $Li[Ni_{0.5}Mn_{0.3}Co_{0.2}]O_2$ as a Superior Alternative to $LiFePO_4$ for Long-Lived Low Voltage Li-Ion Cells. *J. Electrochem. Soc.* **2022**, *169*, 050512. [CrossRef]
12. Editorial: The Journey of an Electrifying (r)Evolution. *Nat. Commun.* **2021**, *12*, 4153. [CrossRef]
13. Su, L.; Jarvis, K.; Charalambous, H.; Dolocan, A.; Manthiram, A. Stabilizing High-Nickel Cathodes with High-Voltage Electrolytes. *Adv. Funct. Mater.* **2023**, *33*, 2213675. [CrossRef]
14. WTO. Panel Rules against Indonesia's Export Limitations on Raw Material. Available online: https://ec.europa.eu/commission/presscorner/detail/en/ip_22_7314 (accessed on 15 September 2023).
15. Indonesia's Uncertain Climb up the Nickel Value Chain. Available online: https://www.lowyinstitute.org/the-interpreter/indonesia-s-uncertain-climb-nickel-value-chain (accessed on 15 September 2023).
16. Xu, W.; Li, Q.; Sui, F.; Guo, S.; Qi, R.; Yan, C.; Chen, L.; Xia, S.; Guo, J.; Li, Z.; et al. Unveiling the Role of Ni Doping in the Electrochemical Performance Improvement of the $LiMn_2O_4$ Cathodes. *Appl. Surf. Sci.* **2023**, *624*, 157142. [CrossRef]
17. Capsoni, D.; Bini, M.; Chiodelli, G.; Massarotti, V.; Mustarelli, P.; Linati, L.; Mozzati, M.C.; Azzoni, C.B. Jahn-Teller Transition in Al^{3+} Doped $LiMn_2O_4$ Spinel. *Solid. State Commun.* **2003**, *126*, 169–174. [CrossRef]
18. Fey, G.T.K.; Lu, C.Z.; Prem Kumar, T. Preparation and Electrochemical Properties of High-Voltage Cathode Materials, $LiMyNi_{0.5-}YMn_{1.5}O_4$ (M = Fe, Cu, Al, Mg; y = 0.0–0.4). *J. Power Sources* **2003**, *115*, 332–345. [CrossRef]
19. Liu, G.; Zhang, J.; Zhang, X.; Du, Y.; Zhang, K.; Li, G.; Yu, H.; Li, C.; Li, Z.; Sun, Q.; et al. Study on Oxygen Deficiency in Spinel LiNi0.5Mn1.5O4 and Its Fe and Cr-Doped Compounds. *J. Alloys Compd.* **2017**, *725*, 580–586. [CrossRef]
20. Hu, L.; Amine, K.; Zhang, Z. Fluorinated Electrolytes for 5-V Li-Ion Chemistry: Dramatic Enhancement of LiNi0.5Mn1.5O4/Graphite Cell Performance by a Lithium Reservoir. *Electrochem. Commun.* **2014**, *44*, 34–37. [CrossRef]

21. Lim, G.; Shin, D.; Chae, K.H.; Cho, M.K.; Kim, C.; Sohn, S.S.; Lee, M.; Hong, J. Regulating Dynamic Electrochemical Interface of $LiNi_{0.5}Mn_{1.5}O_4$ Spinel Cathode for Realizing Simultaneous Mn and Ni Redox in Rechargeable Lithium Batteries. *Adv. Energy Mater.* **2022**, *12*, 2202049. [CrossRef]
22. Jusys, Z.; Binder, M.; Schnaidt, J.; Behm, R.J. A Novel DEMS Approach for Studying Gas Evolution at Battery-Type Electrode | electrolyte Interfaces: High-Voltage $LiNi_{0.5}Mn_{1.5}O_4$ Cathode in Ethylene and Dimethyl Carbonate Electrolytes. *Electrochim. Acta* **2019**, *314*, 188–201. [CrossRef]
23. Kim, J.H.; Pieczonka, N.P.W.; Lu, P.; Liu, Z.; Qiao, R.; Yang, W.; Tessema, M.M.; Sun, Y.K.; Powell, B.R. In Situ Formation of a Cathode-Electrolyte Interface with Enhanced Stability by Titanium Substitution for High Voltage Spinel Lithium-Ion Batteries. *Adv. Mater. Interfaces* **2015**, *2*, 1500109. [CrossRef]
24. Huang, X.; Chen, K.; Liu, Y. Enhancement of $LiNi_{0.5}Mn_{1.5}O_4$ Cathode Materials through Interfacial Modification of Amorphous Al2O3 in Lithium Ion Batteries. *J. Electrochem. Soc.* **2019**, *166*, A5081–A5089. [CrossRef]
25. Sun, P.; Ma, Y.; Zhai, T.; Li, H. High Performance LiNi0.5Mn1.5O4 Cathode by Al-Coating and Al3+-Doping through a Physical Vapor Deposition Method. *Electrochim. Acta* **2016**, *191*, 237–246. [CrossRef]
26. Deng, H.; Nie, P.; Luo, H.; Zhang, Y.; Wang, J.; Zhang, X. Highly Enhanced Lithium Storage Capability of $LiNi_{0.5}Mn_{1.5}O_4$ by Coating with Li_2TiO_3 for Li-Ion Batteries. *J. Mater. Chem. A Mater.* **2014**, *2*, 18256–18262. [CrossRef]
27. Yang, X.; Yang, T.; Liang, S.; Wu, X.; Zhang, H. Modification of $LiNi_{0.5}Mn_{1.5}O_4$ High Potential Cathode from the Inner Lattice to the Outer Surface with Cr^{3+}-Doping and Li^+-Conductor Coating. *J. Mater. Chem. A Mater.* **2014**, *2*, 10359–10364. [CrossRef]
28. Hagh, N.M.; Amatucci, G.G. Effect of Cation and Anion Doping on Microstructure and Electrochemical Properties of the $LiMn_{1.5}Ni_{0.5}O_{4-\delta}$ Spinel. *J. Power Sources* **2014**, *256*, 457–469. [CrossRef]
29. Höweling, A.; Glatthaar, S.; Nötzel, D.; Binder, J.R. Evidence of Loss of Active Lithium in Titanium-Doped $LiNi_{0.5}Mn_{1.5}O_4$/Graphite Cells. *J. Power Sources* **2015**, *274*, 1267–1275. [CrossRef]
30. Lim, J.M.; Oh, R.G.; Kim, D.; Cho, W.; Cho, K.; Cho, M.; Park, M.S. Design of Surface Doping for Mitigating Transition Metal Dissolution in $LiNi_{0.5}Mn_{1.5}O_4$ Nanoparticles. *ChemSusChem* **2016**, *9*, 2967–2973. [CrossRef]
31. Liu, M.H.; Huang, H.T.; Lin, C.M.; Chen, J.M.; Liao, S.C. Mg Gradient-Doped $LiNi_{0.5}Mn_{1.5}O_4$ as the Cathode Material for Li-Ion Batteries. *Electrochim. Acta* **2014**, *120*, 133–139. [CrossRef]
32. Kim, J.-H.; Myung, S.-T.; Yoon, C.S.; Oh, I.-H.; Sun, Y.-K. Effect of Ti Substitution for Mn on the Structure of $LiNi_{0.5}Mn_{1.5-x}Ti_xO_4$ and Their Electrochemical Properties as Lithium Insertion Material. *J. Electrochem. Soc.* **2004**, *151*, A1911. [CrossRef]
33. Höweling, A.; Stenzel, D.; Gesswein, H.; Kaus, M.; Indris, S.; Bergfeldt, T.; Binder, J.R. Variations in Structure and Electrochemistry of Iron- and Titanium-Doped Lithium Nickel Manganese Oxyfluoride Spinels. *J. Power Sources* **2016**, *315*, 269–276. [CrossRef]
34. Sun, H.; Kong, X.; Wang, B.; Luo, T.; Liu, G. $LiNi_{0.5}Mn_{1.45}Zn_{0.05}O_4$ with Excellent Electrochemical Performance for Lithium Ion Batteries. *Int. J. Electrochem. Sci.* **2017**, *12*, 8609–8621. [CrossRef]
35. Hofmann, A.; Höweling, A.; Bohn, N.; Müller, M.; Binder, J.R.; Hanemann, T. Additives for Cycle Life Improvement of High-Voltage LNMO-Based Li-Ion Cells. *ChemElectroChem* **2019**, *6*, 5255–5263. [CrossRef]
36. Milien, M.S.; Beyer, H.; Beichel, W.; Klose, P.; Gasteiger, H.A.; Lucht, B.L.; Krossing, I. Lithium Bis(2,2,2-Trifluoroethyl)Phosphate $Li[O_2P(OCH_2CF_3)_2]$: A High Voltage Additive for LNMO/Graphite Cells. *J. Electrochem. Soc.* **2018**, *165*, A2569–A2576. [CrossRef]
37. Solchenbach, S.; Wetjen, M.; Pritzl, D.; Schwenke, K.U.; Gasteiger, H.A. Lithium Oxalate as Capacity and Cycle-Life Enhancer in LNMO/Graphite and LNMO/SiG Full Cells. *J. Electrochem. Soc.* **2018**, *165*, A512–A524. [CrossRef]
38. Li, Y.; Veith, G.M.; Browning, K.L.; Chen, J.; Hensley, D.K.; Paranthaman, M.P.; Dai, S.; Sun, X.G. Lithium Malonatoborate Additives Enabled Stable Cycling of 5 V Lithium Metal and Lithium Ion Batteries. *Nano Energy* **2017**, *40*, 9–19. [CrossRef]
39. Lee, T.J.; Lee, J.B.; Yoon, T.; Kim, D.; Chae, O.B.; Jung, J.; Soon, J.; Ryu, J.H.; Kim, J.J.; Oh, S.M. Tris(Pentafluorophenyl)Silane as an Electrolyte Additive for 5 V $LiNi_{0.5}Mn_{1.5}O_4$ Positive Electrode. *J. Electrochem. Soc.* **2016**, *163*, A898–A903. [CrossRef]
40. Feng, T.; Yang, G.; Zhang, S.; Xu, Z.; Zhou, H.; Wu, M. Low-Temperature and High-Voltage Lithium-Ion Battery Enabled by Localized High-Concentration Carboxylate Electrolytes. *Chem. Eng. J.* **2022**, *433*, 134138. [CrossRef]
41. Arbizzani, C.; De Giorgio, F.; Mastragostino, M. Characterization Tests for Plug-in Hybrid Electric Vehicle Application of Graphite/LiNi0.4Mn1.6O4 Cells with Two Different Separators and Electrolytes. *J. Power Sources* **2014**, *266*, 170–174. [CrossRef]
42. Banerjee, A.; Ziv, B.; Shilina, Y.; Ziegelbauer, J.M.; Liu, H.; Harris, K.J.; Botton, G.; Goward, G.R.; Luski, S.; Aurbach, D.; et al. Review—Multifunctional Separators: A Promising Approach for Improving the Durability and Performance of Li-Ion Batteries. *J. Electrochem. Soc.* **2019**, *166*, A5369–A5377. [CrossRef]
43. De Giorgio, F.; Laszczynski, N.; von Zamory, J.; Mastragostino, M.; Arbizzani, C.; Passerini, S. Graphite//LiNi0.5Mn1.5O4 Cells Based on Environmentally Friendly Made-in-Water Electrodes. *ChemSusChem* **2017**, *10*, 379–386. [CrossRef]
44. Zhang, K.; Tian, Y.; Wei, C.; An, Y.; Feng, J. Building Stable Solid Electrolyte Interphases (SEI) for Microsized Silicon Anode and 5V-Class Cathode with Salt Engineered Nonflammable Phosphate-Based Lithium-Ion Battery Electrolyte. *Appl. Surf. Sci.* **2021**, *553*, 149566. [CrossRef]
45. Li, W.; Cho, Y.G.; Yao, W.; Li, Y.; Cronk, A.; Shimizu, R.; Schroeder, M.A.; Fu, Y.; Zou, F.; Battaglia, V.; et al. Enabling High Areal Capacity for Co-Free High Voltage Spinel Materials in next-Generation Li-Ion Batteries. *J. Power Sources* **2020**, *473*, 228579. [CrossRef]
46. Frith, J.T.; Lacey, M.J.; Ulissi, U. A Non-Academic Perspective on the Future of Lithium-Based Batteries. *Nat. Commun.* **2023**, *14*, 420. [CrossRef]

47. Ue, M.; Sakaushi, K.; Uosaki, K. Basic Knowledge in Battery Research Bridging the Gap between Academia and Industry. *Mater. Horiz.* **2020**, *7*, 1937–1954. [CrossRef]
48. Yao, W.; Chouchane, M.; Li, W.; Bai, S.; Liu, Z.; Li, L.; Chen, A.X.; Sayahpour, B.; Shimizu, R.; Raghavendran, G.; et al. A 5 V-Class Cobalt-Free Battery Cathode with High Loading Enabled by Dry Coating. *Energy Environ. Sci.* **2023**, *16*, 1620–1630. [CrossRef]
49. Colombo, R.; Garino, N.; Versaci, D.; Amici, J.; Para, M.L.; Quartarone, E.; Francia, C.; Bella, F.; Bodoardo, S. Designing a Double-Coated Cathode with High Entropy Oxides by Microwave-Assisted Hydrothermal Synthesis for Highly Stable Li–S Batteries. *J. Mater. Sci.* **2022**, *57*, 15690–15704. [CrossRef]
50. Wang, T.; Zhang, Q.; Zhong, J.; Chen, M.; Deng, H.; Cao, J.; Wang, L.; Peng, L.; Zhu, J.; Lu, B. 3D Holey Graphene/Polyacrylonitrile Sulfur Composite Architecture for High Loading Lithium Sulfur Batteries. *Adv. Energy Mater.* **2021**, *11*, 2100448. [CrossRef]
51. Medenbach, L.; Adelhelm, P. Cell Concepts of Metal–Sulfur Batteries (Metal = Li, Na, K, Mg): Strategies for Using Sulfur in Energy Storage Applications. *Top. Curr. Chem.* **2017**, *375*, 81. [CrossRef] [PubMed]
52. Deng, Y.; Li, J.; Li, T.; Gao, X.; Yuan, C. Life Cycle Assessment of Lithium Sulfur Battery for Electric Vehicles. *J. Power Sources* **2017**, *343*, 284–295. [CrossRef]
53. Wang, T.; He, J.; Zhu, Z.; Cheng, X.; Zhu, J.; Lu, B.; Wu, Y. Heterostructures Regulating Lithium Polysulfides for Advanced Lithium-Sulfur Batteries. *Adv. Mater.* **2023**, 2303520. [CrossRef]
54. Wang, T.; He, J.; Cheng, X.B.; Zhu, J.; Lu, B.; Wu, Y. Strategies toward High-Loading Lithium-Sulfur Batteries. *ACS Energy Lett.* **2023**, *8*, 116–150. [CrossRef]
55. Colombo, R.; Versaci, D.; Amici, J.; Bella, F.; Para, M.L.; Garino, N.; Laurenti, M.; Bodoardo, S.; Francia, C. Reduced Graphene Oxide Embedded with ZnS Nanoparticles as Catalytic Cathodic Material for Li-S Batteries. *Nanomaterials* **2023**, *13*, 2149. [CrossRef]
56. Leonet, O.; Landa-Medrano, I.; Shafique, A.; Yari, S.; Rangasamy, V.; Vanhulsel, A.; Safari, M.; Van Bael, M.K.; Hardy, A.; Urdampilleta, I.; et al. Electrochemical Investigation of the Ageing of a 400 Wh/kg Li-S Pouch-Cell. *Electrochim. Acta* **2023**, *464*, 142889. [CrossRef]
57. Muchuweni, E.; Mombeshora, E.T.; Muiva, C.M.; Sathiaraj, T.S. Lithium-Ion Batteries: Recent Progress in Improving the Cycling and Rate Performances of Transition Metal Oxide Anodes by Incorporating Graphene-Based Materials. *J. Energy Storage* **2023**, *73*, 109013. [CrossRef]
58. Chen, S.; Dai, F.; Cai, M. Opportunities and Challenges of High-Energy Lithium Metal Batteries for Electric Vehicle Applications. *ACS Energy Lett.* **2020**, *5*, 3140–3151. [CrossRef]
59. Jetybayeva, A.; Aaron, D.S.; Belharouak, I.; Mench, M.M. Critical Review on Recently Developed Lithium and Non-Lithium Anode-Based Solid-State Lithium-Ion Batteries. *J. Power Sources* **2023**, *566*, 232914. [CrossRef]
60. Niri, M.F.; Apachitei, G.; Lain, M.; Copley, M.; Marco, J. Machine Learning for Investigating the Relative Importance of Electrodes' N:P Areal Capacity Ratio in the Manufacturing of Lithium-Ion Battery Cells. *J. Power Sources* **2022**, *549*, 232124. [CrossRef]
61. Teo, J.H.; Strauss, F.; Tripković, Đ.; Schweidler, S.; Ma, Y.; Bianchini, M.; Janek, J.; Brezesinski, T. Design-of-Experiments-Guided Optimization of Slurry-Cast Cathodes for Solid-State Batteries. *Cell Rep. Phys. Sci.* **2021**, *2*, 100465. [CrossRef]
62. Rynne, O.; Dubarry, M.; Molson, C.; Lepage, D.; Prébé, A.; Aymé-Perrot, D.; Rochefort, D.; Dollé, M. Designs of Experiments for Beginners—A Quick Start Guide for Application to Electrode Formulation. *Batteries* **2019**, *5*, 72. [CrossRef]
63. Rodríguez-Carvajal, J. Recent Advances in Magnetic Structure Determination by Neutron Powder Diffraction. *Phys. B Phys. Condens. Matter* **1993**, *192*, 55–69. [CrossRef]
64. Kusano, T.; Ishii, M.; Tani, M.; Hiruta, O.; Matsunaga, T.; Nakamura, H. Rheological Behavior of Concentrated Slurry and Wet Granules for Lithium Ion Battery Electrodes. *Adv. Powder Technol.* **2020**, *31*, 4491–4499. [CrossRef]
65. Wood, D.L.; Quass, J.D.; Li, J.; Ahmed, S.; Ventola, D.; Daniel, C. Technical and Economic Analysis of Solvent-Based Lithium-Ion Electrode Drying with Water and NMP. *Dry. Technol.* **2018**, *36*, 234–244. [CrossRef]
66. Sahore, R.; Wood, D.L.; Kukay, A.; Grady, K.M.; Li, J.; Belharouak, I. Towards Understanding of Cracking during Drying of Thick Aqueous-Processed $LiNi_{0.8}Mn_{0.1}Co_{0.1}O_2$ Cathodes. *ACS Sustain. Chem. Eng.* **2020**, *8*, 3162–3169. [CrossRef]
67. Guo, S.; Li, H.; Li, Y.; Han, Y.; Chen, K.; Xu, G.; Zhu, Y.; Hu, X. SiO_2-Enhanced Structural Stability and Strong Adhesion with a New Binder of Konjac Glucomannan Enables Stable Cycling of Silicon Anodes for Lithium-Ion Batteries. *Adv. Energy Mater.* **2018**, *8*, 1800434. [CrossRef]
68. Escudier, M.P.; Gouldson, I.W.; Pereira, A.S.; Pinho, F.T.; Poole, R.J. On the Reproducibility of the Rheology of Shear-Thinning Liquids. *J. Non-Newton. Fluid. Mech.* **2001**, *97*, 99–124. [CrossRef]
69. De Meatza, I.; Urdampilleta, I.; Boyano, I.; Castrillo, I.; Landa-Medrano, I.; Sananes-Israel, S.; Eguia-Barrio, A.; Palomares, V. From Lab to Manufacturing Line: Guidelines for the Development and Upscaling of Aqueous Processed NMC622 Electrodes. *J. Electrochem. Soc.* **2023**, *170*, 010527. [CrossRef]
70. Harrington, D.A. *Electrochemical Impedance Spectroscopy*; 2004; ISBN 9781118684030. Available online: https://www.ntnu.edu/documents/140124/0/EIS+Literature/cd116109-3fbf-4db5-9f99-ff5ab68cf950 (accessed on 24 October 2023).
71. Derringer, G.; Suich, R. Simultaneous Optimization of Several Response Variables. *J. Qual. Technol.* **1980**, *12*, 214–219. [CrossRef]
72. Zheng, J.; Xiao, J.; Nie, Z.; Zhang, J.-G. Lattice Mn 3+ Behaviors in $Li_4Ti_5O_{12}/LiNi_{0.5}Mn_{1.5}O_4$ Full Cells. *J. Electrochem. Soc.* **2013**, *160*, A1264–A1268. [CrossRef]
73. Jehnichen, P.; Wedlich, K.; Korte, C. Degradation of High-Voltage Cathodes for Advanced Lithium-Ion Batteries–Differential Capacity Study on Differently Balanced Cells. *Sci. Technol. Adv. Mater.* **2019**, *20*, 1–9. [CrossRef]

74. Mereacre, V.; Stüble, P.; Ghamlouche, A.; Binder, J.R. Enhancing the Stability of $LiNi_{0.5}Mn_{1.5}O_4$ by Coating with $LiNbO_3$ Solid-State Electrolyte: Novel Chemically Activated Coating Process versus Sol-Gel Method. *Nanomaterials* **2021**, *11*, 548. [CrossRef]
75. Stüble, P.; Mereacre, V.; Geßwein, H.; Binder, J.R. On the Composition of $LiNi_{0.5}Mn_{1.5}O_4$ Cathode Active Materials. *Adv. Energy Mater.* **2023**, *13*, 2203778. [CrossRef]
76. Houache, M.S.E.; Yim, C.H.; Karkar, Z.; Abu-Lebdeh, Y. On the Current and Future Outlook of Battery Chemistries for Electric Vehicles—Mini Review. *Batteries* **2022**, *8*, 70. [CrossRef]
77. Li, L.; Zhao, R.; Xu, T.; Wang, D.; Pan, D.; Zhang, K.; Yu, C.; Lu, X.; He, G.; Bai, Y. Stabilizing a High-Voltage $LiNi_{0.5}Mn_{1.5}O_4$ Cathode towards All Solid State Batteries: A Li-Al-Ti-P-O Solid Electrolyte Nano-Shell with a Host Material. *Nanoscale* **2019**, *11*, 8967–8977. [CrossRef] [PubMed]

nanomaterials

MDPI

Article

Micron-Sized SiO_x-Graphite Compound as Anode Materials for Commercializable Lithium-Ion Batteries

Minki Jo [1,†], Soojin Sim [2,†], Juhyeong Kim [1], Pilgun Oh [3,4] and Yoonkook Son [1,*]

1 Department of Electric Engineering, Chosun University, 309, Pilmun-daero, Dong-gu, Gwangju 61452, Korea; mkjo@chosun.ac.kr (M.J.); juhyeong@chosun.kr (J.K.)
2 Department of Energy Engineering, School of Energy and Chemical Engineering, Ulsan National Institute of Science and Technology (UNIST), Ulsan 44919, Korea; thegreatest25@gmail.com
3 Department of Smart Green Technology Engineering, Pukyong National University, Busan 485471, Korea; poh@pknu.ac.kr
4 Department of Nanotechnology Engineering, Pukyong National University, Busan 485471, Korea
* Correspondence: y_son@chosun.ac.kr
† These authors contributed equally to this work.

Abstract: The electrode concept of graphite and silicon blending has recently been utilized as the anode in the current lithium-ion batteries (LIBs) industry, accompanying trials of improvement of cycling life in the commercial levels of electrode conditions, such as the areal capacity of approximately 3.3 mAh/cm^2 and volumetric capacity of approximately 570 mAh/cm^3. However, the blending concept has not been widely explored in the academic reports, which focused mainly on how much volume expansion of electrodes could be mitigated. Moreover, the limitations of the blending electrodes have not been studied in detail. Therefore, herein we investigate the graphite blending electrode with micron-sized SiO_x anode material which is one of the most broadly used Si anode materials in the industry, to approach the commercial and practical view. Compared to the silicon micron particle blending electrode, the SiO_x blending electrode showed superior cycling performance in the full cell test. To elucidate the cause of the relatively less degradation of the SiO_x blending electrode as the cycling progressed in full-cell, the electrode level expansion and the solid electrolyte interphase (SEI) thickening were analyzed with various techniques, such as SEM, TEM, XPS, and STEM-EDS. We believe that this work will reveal the electrochemical insight of practical SiO_x-graphite electrodes and offer the key factors to reducing the gap between industry and academic demands for the next anode materials.

Keywords: SiO_x; SiO_x-graphite compound; anode materials; lithium-ion batteries

Citation: Jo, M.; Sim, S.; Kim, J.; Oh, P.; Son, Y. Micron-Sized SiO_x-Graphite Compound as Anode Materials for Commercializable Lithium-Ion Batteries. *Nanomaterials* **2022**, *12*, 1956. https://doi.org/10.3390/nano12121956

Academic Editors: Huang Zhang and Yuan Ma

Received: 20 May 2022
Accepted: 5 June 2022
Published: 7 June 2022

1. Introduction

The growing demand for electric vehicles has sparked great interest in lithium-ion batteries (LIBs) with high energy density for a longer driving range. To meet the higher energy density of LIBs, various electrode materials and formats have been introduced. Especially for the anode materials, Si has many attractive advantages such as its approximately 10 times larger theoretical capacity compared to graphite (372 mAh/g), comparatively low reduction potential, and abundance on earth [1–5]. However, unlike graphite, which exhibits only a capacity decrease of less than 30% during 2000 cycles despite full cell evaluation under typical electrolyte conditions [6], the volume expansion of Si and the solid electrolyte interphase (SEI) formation derived by decomposition of non-aqueous liquid electrolyte during lithiation lead to rapid and gradual capacity degradation of Si anode in cycling life [7,8]. Moreover, under commercial electrode conditions based on the graphite anode, such as the areal capacity of 3.3 mAh/cm^2 (loading level: 10 mg/cm^2) and volumetric capacity of 570 mAh/cm^3 (electrode density: 1.6 g/cc), the cell degradation of the Si anode induced by volume expansion in the electrode arises more severely [9,10]. To avoid the fading mechanisms in Si anode, diverse strategies have been innumerably reported [11]. For example, morphological design providing void space is one of the most well-known strategies, and representative designs include hollow nanoparticles (2725 mAh/g initial capacity and 52% capacity retention after 700 cycles), nanotubes (1780 mAh/g initial capacity and 88% capacity retention after 6000 cycles) and yolk-shells (2833 mAh/g initial capacity and 74% capacity retention after 1000 cycles) [7,12–15]. Another familiar strategy is forming a robust buffer matrix with various materials, such as carbon (1950 mAh/g initial capacity and ~100% capacity retention after 100 cycles) and metal oxide materials (~1000 mAh/g initial capacity and ~65% capacity retention after 1000 cycles) [16–19]. Beyond the physical/chemical material design of Si, Si-graphite-blending-based materials were recently introduced to compensate for the limitation of Si via diluting its content in the anode. This strategy has the advantage to overcome the difficulties of Si anode via an easy and simple blending approach. Therefore, currently, up to 5 wt% Si mixed with graphite anodes is utilized for commercial anodes, satisfying the harsh commercial electrode conditions [9]. The roles of graphite in Si-graphite anode are alleviating the volume expansion, enhancing electrical conductivity, and increasing the packing density of the electrode [10]. In this regard, practical investigation of Si-graphite anode with commercial loading level and electrode density is a significant field of academic and industrial battery society to increase the content of Si in anode for higher energy density.

Among the diverse approaches for Si materials design, SiO_x is one of the most widely used designs because of its great cycle stability and scalable synthetic routes [20–22]. The SiO_x anode shows stable cycling life, although it has low initial coulombic efficiency (ICE) with the intrinsic problem. The low ICE of the SiO_x anode is caused via the formation of lithium dioxide (Li_2O) or lithium silicate (usually, Li_4SiO_4 or $Li_2Si_2O_5$) [23]. Lowering the oxygen content in SiO_x reduces the formation of Li_2O, Li_4SiO_4, and $Li_2Si_2O_5$. However, Li_2O and lithium silicate have also been supposed as robust buffer matrices to improve the cycling performance of SiO_x anode [24,25]. Hereafter, oxygen content in SiO_x anode would be a trade-off relationship between ICE and cycle performance. To study the potential beyond this limitation of SiO_x anode as a practical approach for the LIBs industry, the graphite blending concept is proposed in this study. In the graphite blending concept, graphite not only reduces the expansion of the electrode but also improves the electronic conductivity of the electrode. Here, we prepared the blending electrode of SiO_x and graphite to gravimetric and areal capacities of 440 mAh/g and 3.5 mAh/cm^2, respectively. The electrode density of the blending electrode was approx.1.6 g/cc, and a minimum amount of 3 wt% binders was utilized. We investigated the detailed degradation mechanisms of SiO_x-graphite blending anode in full cells with reliable ex-situ analyses, such as SEM, TEM, and XPS. We believe that our findings will be helpful to analyze the degradation mechanism of Si-graphite anode for LIBs and enlighten the considerations of future practical use of Si-graphite anode for containing more higher ratio of Si.

2. Materials and Methods

All electrode materials were purchased from commercial battery materials incorporation. To achieve the gravimetric capacity of 440 mAh/g, the blending ratios of SiO_x: natural graphite = 6.9:93.1 (*w/w*) for SiO_x blending and silicon micron particles (SiMPs): natural graphite = 4.0:96.0 (*w/w*) for SiMPs were utilized. The SiO_x and SiMPs blending electrodes consisted of active material: carboxymethyl cellulose (CMC): styrene-butadiene rubber (SBR) = 97:1.5:1.5 (*w/w*) were prepared. For the full cell evaluations, the cathode electrodes using NCM622 cathode material were prepared with electrode ratio of active material, super-P, and PVDF = 94:3:3. For the direct comparison of all electrodes, the loading levels of all anodes were adjusted to achieve approx. 3.5 mAh/cm^2. For the used electrolyte, 1.3M $LiPF_6$ was dissolved in ethylene carbonate (EC), ethyl-methyl carbonate (EMC), and dimethyl carbonate (DEC) which had a volume ratio of 3:5:2 with the addition of 0.2 wt% $LiBF_4$, 10.0 wt% fluoroethylene carbonate (FEC), and 0.5 wt% vinylene carbonate (VC) (Panax Etec). A microporous polyethylene (PE) (Celgard) was used as a separator.

For the half-cell test, punched anodes of a diameter of 15 mm were used. All the cells were assembled with 2032 coin-type cells in an argon-filled glove box. For the formation of cells, 2 cycles of charging and discharging were performed at 0.2 C. Cycling life evaluation was performed with charging and discharging at 0.5 C. The cut-off voltage of 0.01–1.5 V for constant current mode (CC mode) and cut-off C-rate of 0.02 C for constant voltage (CV mode) were utilized. For the full-cell test, punched anodes of a diameter of 15 mm and punched cathodes of a diameter of 14 mm were used. The N/P ratio of full cells was 1.04 ± 0.1. As for the formation condition of full-cell, 3 cycles were performed with charging and discharging at 0.2 C. The cycle condition of the full-cell was 0.5 C charging and discharging for 500 cycles. The cut-off conditions of CC and CV modes were 4.4–2.75 V and 0.02 C-rate, respectively, in the full-cell test.

The morphology of the samples was observed using scanning electron microscopy (SEM) (Verios 460, FEI). The transmission electron microscopy (TEM) specimens were prepared using a focused ion beam (FIB, Quanta 3D FEG, FEI). The TEM images and high-resolution TEM images were captured using a JEOL JEM-2100 that was operated at 200 kV. Scanning transmission electron microscope (STEM) bright-field (BF) images and STEM- energy-dispersive X-ray spectroscopy (EDS) scan were operated with Oxford Aztec TEM. X-ray diffraction (XRD) patterns were obtained using a Rigaku D/Max-2200/PC and investigated with Cu-Kα radiation at 40 kV and 40 mA in a step of 0.02 degrees. The X-ray photoelectron spectroscopy (XPS) was operated using a Thermo Fisher K-alpha radiation of energy beam (1486.6 eV). Binding energies of all elements were calibrated concerning the C 1s peak at 284.4 eV. To analyze the core part of active materials, depth profiling was conducted under conditions of 1 keV Ar^+ for 10 min in XPS.

3. Results and Discussion

The micron-sized SiO_x particles have an average diameter of 6.46 μm shown in Figure 1. Note that the size and distribution of SiMPs are similar to that of SiO_x particles. The SiO_x was a totally amorphous phase in Figure S1, which is dissimilar to preceding analysis reports on SiO_x structure which has a separated phase of crystalline Si and SiO_2 [26]. Amorphous Si structure could be more promising than crystalline Si structure since Si-Si bond breaking in amorphous Si structure shows more stable fracture rearrangement behavior during lithiation regarding the formation cycle [27]. The chemical composition of SiO_x was investigated with the X-ray photoelectron spectroscopy (XPS) spectrum to confirm the different oxidation states of Si atoms in SiO_x depending on the depth in Figure S2. The Si 2p peaks corresponded to Si^{4+} (103.6 eV), Si^{3+} (102.5 eV), Si^{2+} (101.4 eV), and Si^0 (99.4 eV), respectively [28]. At both surface and core of SiO_x, Si^{4+} represents the highest major peak intensity. The other major peaks were Si^0 and Si^{2+}. The O 1s peak (532.9 eV) corresponded to the oxidation state in SiO_2, and the other peaks (531.15 and 531.84 eV) represent the SiO_x [29]. From the surface to the core, in the O 1s spectrum, it was confirmed that the peak shift to lower binding energy is due to the characteristics of SiO_x rather than that of SiO_2. In the Si 2p spectrum, from the surface to the core, it was observed that the peak shift representing Si^{4+} toward low binding energy and the characteristics of low Si oxidation peaks were remarkable due to the decrease in the oxidation state of Si. In addition, Si and SiO_2 were found to be well mixed in the SiO_x phase despite native SiO_2 on the surface [30].

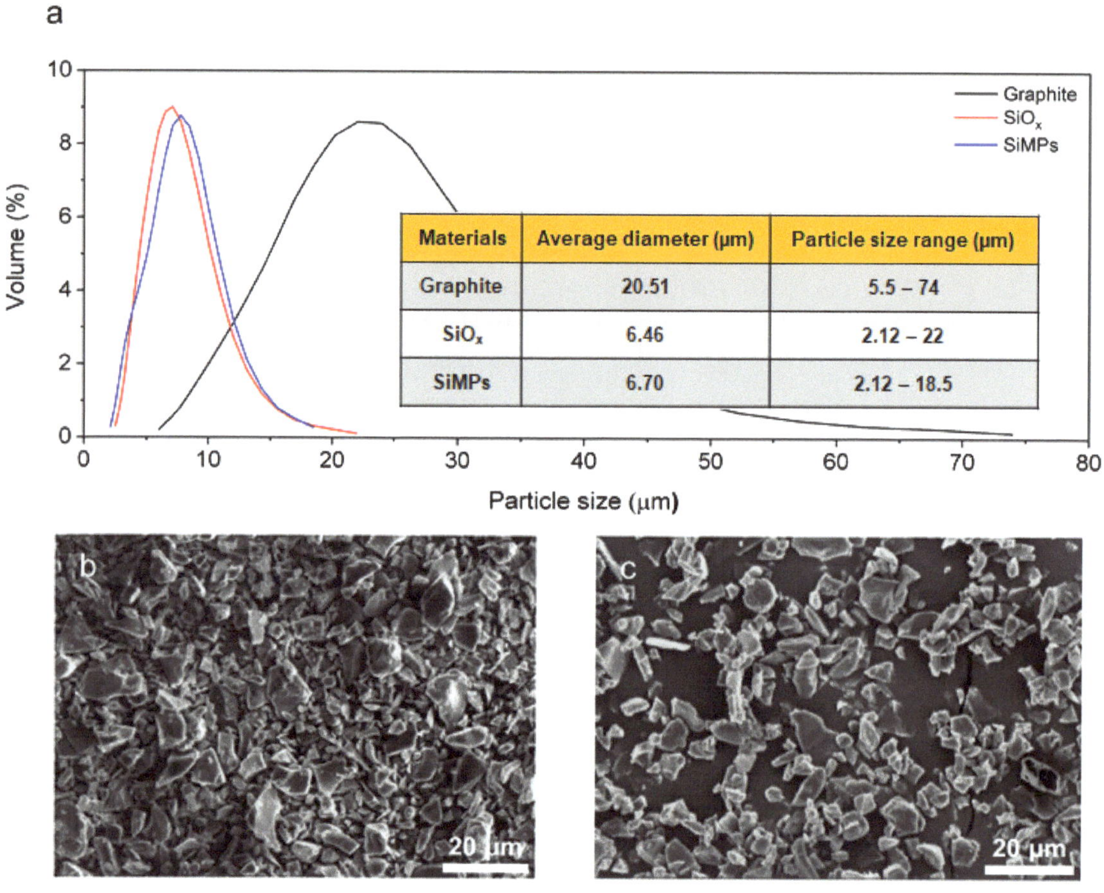

Figure 1. (**a**) Particle size distribution of graphite, SiO_x, and SiMPs. SEM images of (**b**) SiMPs and (**c**) SiO_x.

Figure 2a,b show half-cell formation voltage profiles of SiMPs and SiO_x blending electrodes, respectively. The first delithiation capacity showed approximately 440 mAh/g for both blending electrodes and the initial coulombic efficiency (ICE) of SiMPs and SiO_x blending electrodes were 92% and 84%, respectively. The voltage profile of SiMPs blending shows the voltage plateau of Li-Si redox reaction at 0.42 V and that of SiO_x blending shows no plateau which is typical in the previous SiO_x anode report [31,32]. Figure 2c,d show the cycling performance of SiMPs and SiO_x blending electrodes in half and full cells. In the half-cell test, both electrodes showed similar stable cycling stability up to the 50th cycle and the capacities rapidly dropped after the 50th cycle because of fading mechanisms on Li metal due to exhaustion of electrolyte and solid electrolyte interphase (SEI) thickening [33–35]. Since the half cells tests were limited to only 50 cycles, we conducted a full-cell test. As shown in Figure S3, the first discharge capacities and ICE of SiMPs and SiO_x blendings in full-cell were 3.54 mAh/cm^2 and 89% for SiMPs blending and 3.45 mAh/cm^2 and 80% for SiO_x blending. Note that areal loading and full cell design were adjusted to obtain a similar level of areal capacities. The cycling retentions of SiMPs and SiO_x blending electrodes at the 200th cycle in full-cell were 51% and 70%, respectively, and at the 500th cycle 29% and 56%, respectively. This indicates that SiO_x is more suitable than SiMPs in practical graphite blending electrodes.

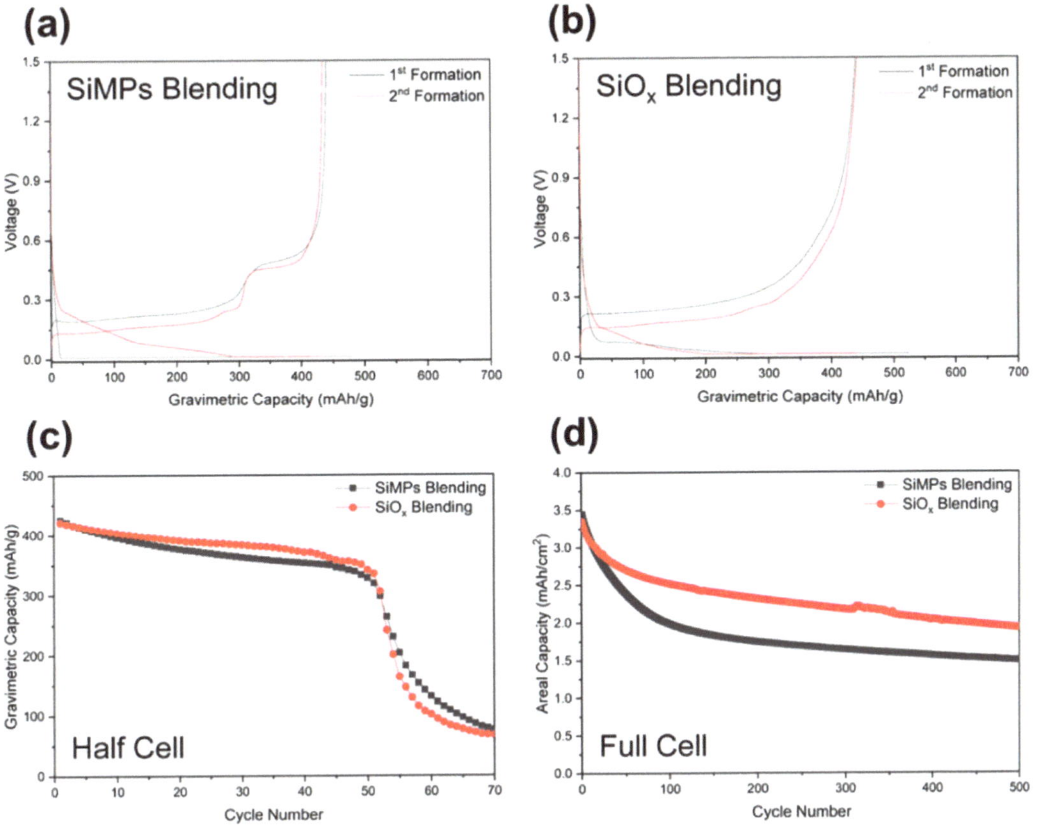

Figure 2. Half-cell formation voltage profiles of (**a**) SiMPs blending and (**b**) SiO_x blending. Comparison of cycling performance of SiMPs blending and SiO_x blending in (**c**) half-cell and (**d**) full-cell.

We also measured the extent of swelling of the blending electrodes depending on the full cell cycling in Figure S4. All the samples were measured at a fully discharged state by a micrometer. The thickness of the SiO_x blending electrode expanded from 75 μm to 77 μm (after 250 cycles) and 90 μm (after 500 cycles), representing 103% and 120% expansion rates relative to the initial thickness, respectively. In contrast, the thickness of SiMP blending electrodes swelled from 63 μm to 122 μm (after 250 cycles) and 125 μm (after 500 cycles), representing 194% and 198% expansion rates relative to the initial thickness, respectively. As shown in Figure 3, there were no serious cracks in the electrodes due to volume expansion in both blending samples during the cycling, which are believed to be caused by the surrounding graphite buffer effect. Regarding the macroscopic surface state of the electrode, the SiO_x blending electrode showed a relatively clean surface state during 500 cycles, while the SiMP blending electrodes showed that large by-products ranging from several micrometers to tens of micrometers were formed on the electrode surface after 250 cycles. Based on these results, the SiO_x blending electrode effectively blocks the electrode swelling with the intraparticle and interparticle buffer matrices, which are the SiO_2 and lithium silicate in SiO_x structure and surrounding graphite buffer, respectively.

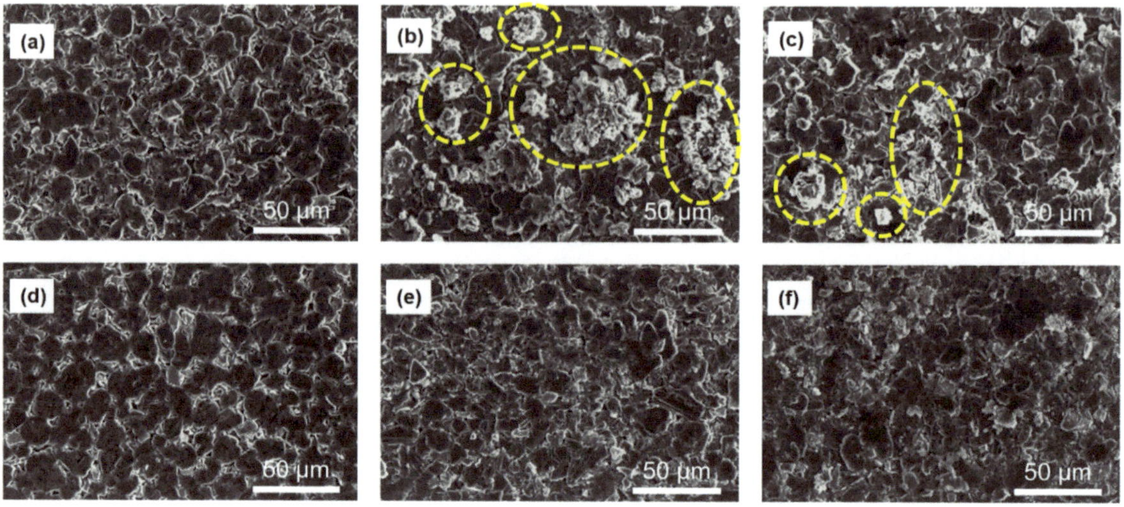

Figure 3. Top-view SEM images of (**a**–**c**) SiMP blending and (**d**–**f**) SiO_x blending electrodes in full-cell: (**a**) pristine, (**b**) after 250 cycles, (**c**) after 500 cycles, respectively.

To elucidate the cause of the relatively less degradation of the SiO_x blending electrode as the cycling progressed in full-cell, we measured cross-sectional high-resolution transmission electron microscopy (HR-TEM) images of SiO_x blending anodes after 50, 100, and 250 cycles in Figure 4. The thickness of the SEI layer after 50, 100, and 250 cycles were gradually increased to 80, 90, and 230 nm, respectively. Likewise in the TEM result, electrochemical impedance spectroscopy (EIS) indirectly indicates an increase in SEI by an increase in internal resistance (Figure S5). The right shift of the semi-circle in the Nyquist plot shows increased overpotential in electrolyte due to its depletion, and the increased size of the semi-circle is indicative of increased overpotential due to SEI information. However, although it exhibited gradual thickened SEI, the SEI showed dense and robust characteristics in HR-TEM images, which is contrary to the conventional feature of coarse and loose structure in ordinary silicon anodes [36,37]. This result corresponds with electrode swelling data and the surface state of the electrode in Figures 3 and S4. Figure S6 shows the STEM-BF image and STEM-EDS mapping of the SiO_x blending electrode after 250 cycles. The distinct boundary between the SEI layer and SiO_x could be observed in the EDS mapping. The SEI mainly contained elements of carbon, phosphorus, and fluorine, since the main components of electrolytes are organic solvents, lithium salts, and additives. We also confirmed the chemical compositions of SEI layers with ex-situ XPS analysis in Figure 5. In the C 1s peaks, we could observe carbon bonding of -O-C-O_2-, -C-O-C-, and CH_x. There are no large variations in C 1s peaks [38,39]. In the F 1s peaks, we could observe relative amounts of Li_xPF_y and LiF. In the P 1s peaks, we could observe a relative amount of Li_xPF_y and $Li_xPO_yF_z$ [40,41]. The LiF peaks are gradually increased after several tens of cycles. The decomposition of LiPF6 or fluoroethylene carbonate (FEC) forms LiF. From these analysis results, we can confirm that the gradually thickened SEI layer in Figure 4 was proven to be fluorine-rich (F-rich). Generally, in the case of Si-based materials, with increasing cycling, the SEI layer becomes thicker or forms by-products, including Si, due to the continuous side reaction of the electrolyte, which reduces the capacity and increases internal resistance, resulting in capacity fading. However, such F-rich layers have been generally ascertained to be robust, strong, and provide stable lithium-ion transport channels, resulting in good electrochemical performances [42]. Therefore, we believe that the superior cyclability of the SiO_x blending electrode is strongly related to its uniform formation of the F-rich SEI layer.

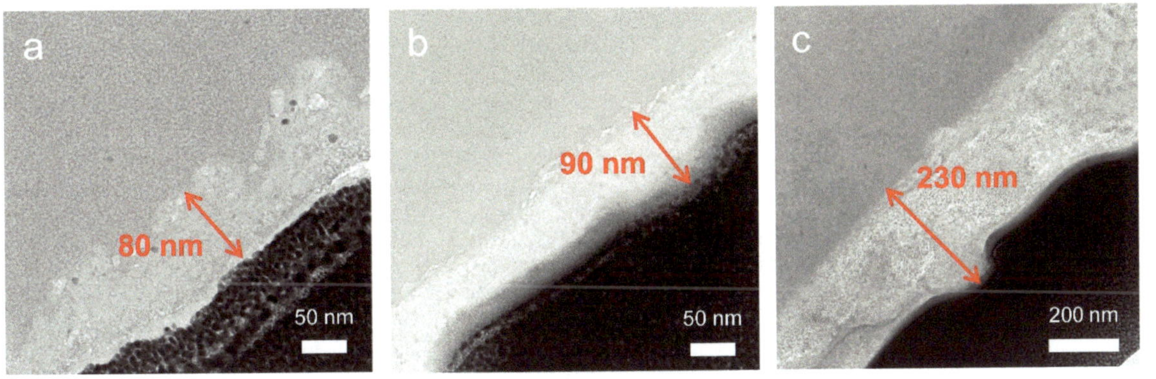

Figure 4. Cross-sectional TEM images of SiO_x blending electrode after (**a**) 50 cycles, (**b**) 100 cycles, and (**c**) 250 cycles.

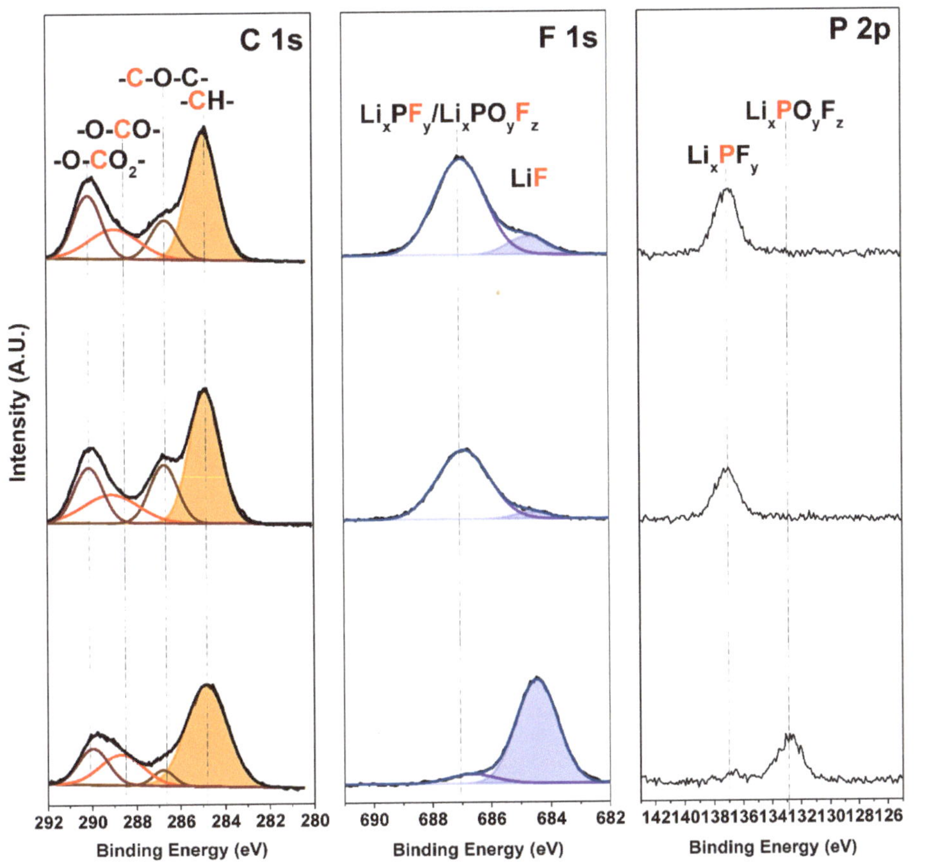

Figure 5. XPS analysis of SiO_x blending electrode after 5, 50, and 350 cycles from top to bottom.

4. Conclusions

In this study, the superiority of the SiO_x and graphite blending electrode compared to the SiMPs and graphite blending electrode was investigated through full-cell-based evaluation. Continuous SEI thickening due to volume change and electrolyte decomposition during cycling may cause capacity to decrease as internal resistance increases. However, in the case of SiO_x blending, the F-rich SEI layer formation, which is relatively robust and can serve as a stable transport channel for lithium ions, resulted in superior electrochemical performance compared to the SiMP blending electrode. The strategy of SiO_x and graphite blending electrodes shows much more effective physicochemical and electrochemical characteristics in commercial-grade electrodes than SiMPs and graphite blending electrodes, and may clearly suggest a commercial breakthrough direction for Si-based anode materials.

Supplementary Materials: The following supporting information can be downloaded at: https://www.mdpi.com/article/10.3390/nano12121956/s1, Figure S1: (a) Cross-section HR-TEM image and diffraction patterns of SiO_x. (b) XRD patterns of SiO_x, commercialized SiO, and SiO_2; Figure S2: XPS spectra of SiO_x surface and core in O 1s and Si 2p; Figure S3: Formation voltage profiles of (a) SiMPs blending and (b) SiO_x blending in full cells; Figure S4: Cross-section SEM images of (a–c) SiMPs blending and (d–f) SiO_x blending electrodes during 200 cycles in full-cell: (a,d) pristine, (b,e) after 250 cycles, (c,f) after 500 cycles, respectively; Figure S5: EIS analysis of SiO_x blending electrode in the full cell. Data were collected after formation, 10, and 50 cycles; Figure S6: Scanning transmission electron microscope (STEM)-bright field (BF) image and STEM-the energy dispersive X-ray spectroscopy (EDS) mapping of SiO_x blending electrode after full-cell 250 cycles.

Author Contributions: Conceptualization, M.J. and S.S.; methodology, M.J., J.K. and S.S.; formal analysis, P.O.; investigation, M.J., J.K. and S.S.; data curation, P.O.; writing—original draft preparation, M.J. and S.S.; writing—review and editing, Y.S.; supervision, Y.S.; project administration, Y.S.; All authors have read and agreed to the published version of the manuscript.

Funding: This research was supported by the Basic Science Research Program through the National Research Foundation of Korea (NRF) funded by the Ministry of Education (No. 2021R1I1A3049294); This work was supported by the National Research Foundation of Korea (NRF) grant funded by the Korea government (MSIT; Ministry of Science and ICT) (No. 2020R1A4A1019463).

Institutional Review Board Statement: Not applicable.

Informed Consent Statement: Not applicable.

Data Availability Statement: Not applicable.

Conflicts of Interest: The authors declare no conflict of interest.

References

1. Park, C.-M.; Kim, J.-H.; Kim, H.; Sohn, H.-J. Li-alloy based anode materials for Li secondary batteries. *Chem. Soc. Rev.* **2010**, *39*, 3115–3141. [CrossRef] [PubMed]
2. Szczech, J.R.; Jin, S. Nanostructured silicon for high capacity lithium battery anodes. *Energy Environ. Sci.* **2011**, *4*, 56–72. [CrossRef]
3. Zhang, W.-J. A review of the electrochemical performance of alloy anodes for lithium-ion batteries. *J. Power Sources* **2011**, *196*, 13–24. [CrossRef]
4. Obrovac, M.; Chevrier, V. Alloy negative electrodes for Li-ion batteries. *Chem. Rev.* **2014**, *114*, 11444–11502. [CrossRef]
5. Su, X.; Wu, Q.; Li, J.; Xiao, X.; Lott, A.; Lu, W.; Sheldon, B.W.; Wu, J. Silicon-based nanomaterials for lithium-ion batteries: A review. *Adv. Energy Mater.* **2014**, *4*, 1300882. [CrossRef]
6. Yang, G.; Shi, J.; Shen, C.; Wang, S.; Xia, L.; Hu, H.; Luo, H.; Xia, Y.; Liu, Z. Improving the cyclability performance of lithium-ion batteries by introducing lithium difluorophosphate (LiPO2F2) additive. *RSC Adv.* **2017**, *7*, 26052–26059. [CrossRef]
7. Wu, H.; Chan, G.; Choi, J.W.; Ryu, I.; Yao, Y.; McDowell, M.T.; Lee, S.W.; Jackson, A.; Yang, Y.; Hu, L. Stable cycling of double-walled silicon nanotube battery anodes through solid–electrolyte interphase control. *Nat. Nanotechnol.* **2012**, *7*, 310–315. [CrossRef]
8. Wu, H.; Cui, Y. Designing nanostructured Si anodes for high energy lithium ion batteries. *Nano Today* **2012**, *7*, 414–429. [CrossRef]
9. Chae, S.; Choi, S.H.; Kim, N.; Sung, J.; Cho, J. Integration of graphite and silicon anodes for the commercialization of high-energy lithium-ion batteries. *Angew. Chem. Int. Ed.* **2020**, *59*, 110–135. [CrossRef]
10. Son, Y.; Kim, N.; Lee, T.; Lee, Y.; Ma, J.; Chae, S.; Sung, J.; Cha, H.; Yoo, Y.; Cho, J. Calendering-compatible macroporous architecture for silicon–graphite composite toward high-energy lithium-ion batteries. *Adv. Mater.* **2020**, *32*, 2003286. [CrossRef]
11. Feng, K.; Li, M.; Liu, W.; Kashkooli, A.G.; Xiao, X.; Cai, M.; Chen, Z. Silicon-based anodes for lithium-ion batteries: From fundamentals to practical applications. *Small* **2018**, *14*, 1702737. [CrossRef]
12. Chan, C.K.; Peng, H.; Liu, G.; McIlwrath, K.; Zhang, X.F.; Huggins, R.A.; Cui, Y. High-performance lithium battery anodes using silicon nanowires. *Nat. Nanotechnol.* **2008**, *3*, 31–35. [CrossRef] [PubMed]

13. Yao, Y.; McDowell, M.T.; Ryu, I.; Wu, H.; Liu, N.; Hu, L.; Nix, W.D.; Cui, Y. Interconnected silicon hollow nanospheres for lithium-ion battery anodes with long cycle life. *Nano Lett.* **2011**, *11*, 2949–2954. [CrossRef]
14. Liu, N.; Wu, H.; McDowell, M.T.; Yao, Y.; Wang, C.; Cui, Y. A yolk-shell design for stabilized and scalable Li-ion battery alloy anodes. *Nano Lett.* **2012**, *12*, 3315–3321. [CrossRef] [PubMed]
15. Son, Y.; Son, Y.; Choi, M.; Ko, M.; Chae, S.; Park, N.; Cho, J. Hollow silicon nanostructures via the Kirkendall effect. *Nano Lett.* **2015**, *15*, 6914–6918. [CrossRef]
16. Lee, H.-Y.; Lee, S.-M. Graphite–FeSi alloy composites as anode materials for rechargeable lithium batteries. *J. Power Sources* **2002**, *112*, 649–654. [CrossRef]
17. Magasinski, A.; Dixon, P.; Hertzberg, B.; Kvit, A.; Ayala, J.; Yushin, G. High-performance lithium-ion anodes using a hierarchical bottom-up approach. *Nat. Mater.* **2010**, *9*, 353–358. [CrossRef]
18. Lee, J.-I.; Ko, Y.; Shin, M.; Song, H.-K.; Choi, N.-S.; Kim, M.G.; Park, S. High-performance silicon-based multicomponent battery anodes produced via synergistic coupling of multifunctional coating layers. *Energy Environ. Sci.* **2015**, *8*, 2075–2084. [CrossRef]
19. Floraki, C.; Sapountzis, A.; Vernardou, D. APCVD graphene-based composite electrodes for Li-ion batteries. *Energies* **2022**, *15*, 926. [CrossRef]
20. Dai, F.; Yi, R.; Gordin, M.L.; Chen, S.; Wang, D. Amorphous $Si/SiO_x/SiO_2$ nanocomposites via facile scalable synthesis as anode materials for Li-ion batteries with long cycling life. *RSC Adv.* **2012**, *2*, 12710–12713. [CrossRef]
21. Park, E.; Park, M.S.; Lee, J.; Kim, K.J.; Jeong, G.; Kim, J.H.; Kim, Y.J.; Kim, H. A highly resilient mesoporous SiO_x lithium storage material engineered by oil–water templating. *ChemSusChem* **2015**, *8*, 688–694. [CrossRef]
22. Liu, Z.; Yu, Q.; Zhao, Y.; He, R.; Xu, M.; Feng, S.; Li, S.; Zhou, L.; Mai, L. Silicon oxides: A promising family of anode materials for lithium-ion batteries. *Chem. Soc. Rev.* **2019**, *48*, 285–309. [CrossRef]
23. Chang, W.-S.; Park, C.-M.; Kim, J.-H.; Kim, Y.-U.; Jeong, G.; Sohn, H.-J. Quartz (SiO_2): A new energy storage anode material for Li-ion batteries. *Energy Environ. Sci.* **2012**, *5*, 6895–6899. [CrossRef]
24. Guerfi, A.; Charest, P.; Dontigny, M.; Trottier, J.; Lagacé, M.; Hovington, P.; Vijh, A.; Zaghib, K. SiO_x–graphite as negative for high energy Li-ion batteries. *J. Power Sources* **2011**, *196*, 5667–5673. [CrossRef]
25. Yan, N.; Wang, F.; Zhong, H.; Li, Y.; Wang, Y.; Hu, L.; Chen, Q. Hollow porous SiO_2 nanocubes towards high-performance anodes for lithium-ion batteries. *Sci. Rep.* **2013**, *3*, 1568. [CrossRef]
26. Yi, R.; Dai, F.; Gordin, M.L.; Chen, S.; Wang, D. Micro-sized Si-C composite with interconnected nanoscale building blocks as high-performance anodes for practical application in lithium-ion batteries. *Adv. Energy Mater.* **2013**, *3*, 295–300. [CrossRef]
27. McDowell, M.T.; Lee, S.W.; Harris, J.T.; Korgel, B.A.; Wang, C.; Nix, W.D.; Cui, Y. In situ TEM of two-phase lithiation of amorphous silicon nanospheres. *Nano Lett.* **2013**, *13*, 758–764. [CrossRef]
28. Thøgersen, A.; Selj, J.H.; Marstein, E.S. Oxidation effects on graded porous silicon anti-reflection coatings. *J. Electrochem. Soc.* **2012**, *159*, D276. [CrossRef]
29. Paparazzo, E. XPS and auger spectroscopy studies on mixtures of the oxides SiO_2, Al_2O_3, Fe_2O_3 and Cr_2O_3. *J. Electron. Spectrosc. Relat. Phenom.* **1987**, *43*, 97–112. [CrossRef]
30. Sim, S.; Oh, P.; Park, S.; Cho, J. Critical thickness of SiO_2 coating layer on core@ shell bulk@ nanowire Si anode materials for Li-ion batteries. *Adv. Mater.* **2013**, *25*, 4498–4503. [CrossRef]
31. Fukui, H.; Ohsuka, H.; Hino, T.; Kanamura, K. A Si−O−C composite anode: High capability and proposed mechanism of lithium storage associated with microstructural characteristics. *ACS Appl. Mater. Interfaces* **2010**, *2*, 998–1008. [CrossRef]
32. Kitada, K.; Pecher, O.; Magusin, P.C.; Groh, M.F.; Weatherup, R.S.; Grey, C.P. Unraveling the reaction mechanisms of SiO anodes for Li-ion batteries by combining in situ 7Li and ex situ 7Li/29Si solid-state NMR spectroscopy. *J. Am. Chem. Soc.* **2019**, *141*, 7014–7027. [CrossRef] [PubMed]
33. Xu, W.; Wang, J.; Ding, F.; Chen, X.; Nasybulin, E.; Zhang, Y.; Zhang, J.-G. Lithium metal anodes for rechargeable batteries. *Energy Environ. Sci.* **2014**, *7*, 513–537. [CrossRef]
34. Lu, D.; Shao, Y.; Lozano, T.; Bennett, W.D.; Graff, G.L.; Polzin, B.; Zhang, J.; Engelhard, M.H.; Saenz, N.T.; Henderson, W.A. Failure mechanism for fast-charged lithium metal batteries with liquid electrolytes. *Adv. Energy Mater.* **2015**, *5*, 1400993. [CrossRef]
35. Lin, D.; Liu, Y.; Cui, Y. Reviving the lithium metal anode for high-energy batteries. *Nat. Nanotechnol.* **2017**, *12*, 194–206. [CrossRef]
36. Haruta, M.; Kijima, Y.; Hioki, R.; Doi, T.; Inaba, M. Artificial lithium fluoride surface coating on silicon negative electrodes for the inhibition of electrolyte decomposition in lithium-ion batteries: Visualization of a solid electrolyte interphase using in situ AFM. *Nanoscale* **2018**, *10*, 17257–17264. [CrossRef] [PubMed]
37. Prasanna, K.; Subburaj, T.; Jo, Y.N.; Santhoshkumar, P.; Karthikeyan, S.; Vediappan, K.; Gnanamuthu, R.; Lee, C.W. Chitosan complements entrapment of silicon inside nitrogen doped carbon to improve and stabilize the capacity of Li-ion batteries. *Sci. Rep.* **2019**, *9*, 3318. [CrossRef] [PubMed]
38. Chan, C.K.; Ruffo, R.; Hong, S.S.; Cui, Y. Surface chemistry and morphology of the solid electrolyte interphase on silicon nanowire lithium-ion battery anodes. *J. Power Sources* **2009**, *189*, 1132–1140. [CrossRef]
39. Ruffo, R.; Hong, S.S.; Chan, C.K.; Huggins, R.A.; Cui, Y. Impedance analysis of silicon nanowire lithium ion battery anodes. *J. Phys. Chem. C* **2009**, *113*, 11390–11398. [CrossRef]
40. Choi, N.-S.; Yew, K.H.; Lee, K.Y.; Sung, M.; Kim, H.; Kim, S.-S. Effect of fluoroethylene carbonate additive on interfacial properties of silicon thin-film electrode. *J. Power Sources* **2006**, *161*, 1254–1259. [CrossRef]

41. Markevich, E.; Salitra, G.; Aurbach, D. Fluoroethylene carbonate as an important component for the formation of an effective solid electrolyte interphase on anodes and cathodes for advanced Li-ion batteries. *ACS Energy Lett.* **2017**, *2*, 1337–1345. [CrossRef]
42. Park, S.; Jeong, S.Y.; Lee, T.K.; Park, M.W.; Lim, H.Y.; Sung, J.; Cho, J.; Kwak, S.K.; Hong, S.Y.; Choi, N.-S. Replacing conventional battery electrolyte additives with dioxolone derivatives for high-energy-density lithium-ion batteries. *Nat. Commun.* **2021**, *12*, 838. [CrossRef] [PubMed]

batteries

MDPI

Article

Modeling Silicon-Dominant Anodes: Parametrization, Discussion, and Validation of a Newman-Type Model

Axel Durdel *, Sven Friedrich, Lukas Hüsken and Andreas Jossen

Chair of Electrical Energy Storage Technology, Department of Energy and Process Engineering, School of Engineering and Design, Technical University of Munich, Arcisstraße 21, 80333 Munich, Germany; lukas.huesken@tum.de (L.H.); andreas.jossen@tum.de (A.J.)

* Correspondence: axel.durdel@tum.de

Abstract: Silicon is a promising anode material and can already be found in commercially available lithium-ion cells. Reliable modeling and simulations of new active materials for lithium-ion batteries are becoming more and more important, especially regarding cost-efficient cell design. Because literature lacks an electrochemical model for silicon-dominant electrodes, this work aims to close the gap. To this end, a Newman p2D model for a lithium-ion cell with a silicon-dominant anode and a nickel-cobalt-aluminum-oxide cathode is parametrized. The micrometer silicon particles are partially lithiated to 1200 mAh g_{Si}^{-1}. The parametrization is based on values from the electrode manufacturing process, measured values using lab cells, and literature data. Charge and discharge tests at six different C-rates up to 2C serve as validation data, showing a root-mean-squared error of about 21 mV and a deviation in discharge capacity of about 1.3%, both during a 1 C constant current discharge. Overall, a validated parametrization for a silicon-dominant anode is presented, which, to the best of our knowledge, is not yet available in literature. For future work, more in-depth studies should investigate the material parameters for silicon to expand the data available in the literature and facilitate further simulation work.

Keywords: silicon; p2D model; parametrization; validation; half-cell balancing

Citation: Durdel, A.; Friedrich, S.; Hüsken, L.; Jossen, A. Modeling Silicon-Dominant Anodes: Parametrization, Discussion, and Validation of a Newman-Type Model. *Batteries* **2023**, *9*, 558. https://doi.org/10.3390/batteries9110558

Academic Editors: Huang Zhang and Yuan Ma

Received: 18 October 2023
Revised: 10 November 2023
Accepted: 13 November 2023
Published: 15 November 2023

1. Introduction

Since their market launch in 1991 [1], lithium-ion batteries (LIBs) have developed into an everyday energy storage device, powering almost all mobile electronic devices [2]. Graphite has been the anode material of choice, despite a comparably low gravimetric capacity of 372 mAh g_{Gr}^{-1}. Even as early as in the 1960s, about 30 years before the market launch of commercial LIBs, silicon (Si) has been known to be able to store 3579 mAh g_{Si}^{-1} or even up to 4212 mAh g_{Si}^{-1} via electrochemical alloying to $Li_{15}Si_4$ or $Li_{22}Si_5$, respectively. This is more than ten times the capacity of graphite [3]. At the same time, silicon has a comparable or even lower mass density than graphite [3], rendering it a potentially interesting anode material. However, the volume of silicon changes drastically during (de)lithiation, introducing several problems, e.g., mechanical cracking and disintegration, loss of electric contact, or breakdown of the solid-electrolyte interphase (SEI) [3].

Silicon's high specific capacity is utilized in composite anodes consisting of both graphite and silicon. By adding small amounts of silicon to conventional graphite anodes, their capacity can be increased. Such composite anodes have been investigated and commercialized already [4–9]. Pure-silicon and Si-dominant anodes, on the other hand, usually are mechanically unstable due to the large volume changes of up to 300% upon (de)lithiation [10]. Ko et al. [10] summarized several approaches to manage this and to enable pure-silicon and silicon-dominant anodes as a successor to graphite anodes. The approaches include size-controlled materials, patterned thin films, shape-preserving shell designs, porous structures, and graphene composites. The silicon anode used in this work belongs to the category of porous structures. In the work of Ko et al. [10], most of the porous structures were synthesized via electrochemical etching of a silicon wafer resulting in nanometer-sized pores. The silicon in this work, however, consists of micrometer-sized silicon particles prepared via jet milling of photovoltaic grade silicon lumps [11]. Obrovac and Krause [12] proposed partial lithiation as another approach to increase cycle life and maintain electrode integrity while still providing higher gravimetric capacities than graphite at full lithiation. This concept of partial lithiation also holds true for the electrode design used in this work, as shown by Jantke et al. [11]. Haufe et al. [13] investigated the failure mechanism of this electrode design with partially lithiated Si. They found that the main failure mechanism is the continuous formation of SEI. Another major contribution is found in particle decoupling when operating at low discharge voltages.

With an appropriate model, an estimation about (im)practical cell designs without the need for time- and cost-intensive experiments is possible. Doyle et al. [14] proposed the dual insertion model, also known as the pseudo-2D (p2D) model or Newman model, which is capable of describing the electrochemical behavior of a rocking-chair cell. Since its publication in 1993, the Newman model and its variations have become a famous and widely used approach for physicochemical modeling of LIBs. For example, it has been used to describe relaxation phenomena [15], short-circuit events [16], or to aid cell design [17]. Additionally, the original publication by Doyle et al. [14] has gained increased research interest over the years, receiving 20% of all citations in the last two years. However, the model first and foremost considers insertion reactions. Hence, care should be taken when using the p2D model for alloying materials, because surface processes (electrode kinetics) as well as transport processes inside the particles (solid phase diffusion) might differ from the implemented processes in the standard p2D model.

Table 1 lists a selection of different simulation models all related to silicon. The models found in the literature cover a wide range of different application scenarios and a variety of different model classes, such as finite element method model, density functional theory models, or impedance-based models, can be found.

Table 1. Overview of selected silicon models from literature.

Ref.	Focal Area	Silicon	Dimensionality
[18]	Voltage hysteresis via asymmetric reactions	pure	0D model
[19]	Current distribution & inhomogeneous lithation	composite	p2D
[20]	Kinetic limitations & rate capability	pure	0D impedance model
[21]	Phase boundaries & diffusivity	pure	2D SPM
[22]	Volume change & (de)insertion process	dominant	1D SPM
[23]	Volume change & electrolyte displacement	dominant	p2D
[24]	Mechanical stress & stress-induced diffusion	pure	1D SPM
[25]	Mechanical stress & geometry dependency	dominant	1D thin-film & SPM
[26]	Coupling of electrochemistry and mechanics	pure	2D MSM [a]
[27,28]	Capacity fade via SEI interaction	composite	1D model
[29]	Voltage hysteresis via mechanical stress	pure	0D model
[30]	First principles model of silicon lithiation	pure	0D DFT calculation
This work	Parametrization & validation of an electrode in a full-cell setup	dominant	p2D

p2D = pseudo two-dimendional model; SPM = single-particle model; MSM = microstructure model; SEI = solid electrolyte interphase; DFT = density functional theory; [a] microstructure of a single thin-film and a single nanowire.

When it comes to electrochemical modeling of silicon electrodes, the literature provides different approaches, most of which rely on the Butler–Volmer equation as a reaction kinetic and Fickian solid phase diffusion. Jiang et al. [18], for example, presented a model depicting the different lithiation stages of silicon. At the same time, different activation energies allow for the representation of crystallization effects and different particle sizes can be accounted for by varying equilibrium potentials. Lory et al. [19] presented an adapted p2D model for a Si-graphite composite anode. The model considers three different active material phases; graphite, a carbon matrix, and nano-silicon embedded in the carbon matrix. Thus, the influence of silicon on the cell behavior can be modeled and investigated. Swamy and Chiang [20] published an impedance-based equivalent circuit model to compare the charge transfer overpotentials of pure silicon particles with those of pure graphite particles with respect to the C-rate and the individual particle radius. According to their results, overpotentials may become rate limiting in dependence on the exchange current density.

Regarding the solid phase diffusion, Obrovac and Krause [12] proposed that a partially lithiated silicon particle would show a core–shell structure with the lithiated amorphous phase as the outer layer and a crystalline core. Such a core–shell structure was also detected by McDowell et al. [31] during a lithation process of a single particle using in-situ transmission electron microscopy measurements. To properly capture the impacts of this core–shell structure in a model, Wang et al. [21] used a two-dimensional single particle diffusion model which also considers a reaction front at the phase boundary. Doing so, they could replicate the experimental results. Such a phase boundary is not readily available for the standard p2D model. However, Bordes et al. [32] investigated polycrystalline silicon particles via transmission electron microscopy and found that lithiation preferentially occurs along defects. The particle then consists of several smaller crystalline cores separated by amorphous "veins". Such a structure is similar to the porous structure of the electrode where the liquid diffusion is described by an effective diffusion coefficient [14]. In this analogy, the pore space filled with electrolyte is equivalent to the amorphous Si phase, and the solid-phase structure of the electrode is equivalent to the crystalline cores which hinder the overall transport. Thus, from a modeling perspective, such a structure can be described by an effective diffusion coefficient without the need for a moving phase boundary, which can be accomplished with the standard p2D model.

Other models in the literature focus not on electrochemistry but on mechanics. This work does not focus on mechanics and the interested reader is referred to the literature given in Table 1, especially references [22,24–26,29].

The aim of this work is to provide a parametrized and validated p2D model for a silicon-dominant electrode, since to the best knowledge of the authors, the literature lacks such a model. The anode coating in this work contains graphite, but it is largely electrochemically inactive and mainly serves as a conductive additive [11]. Thus, the electrode is considered silicon-dominant. The loading of $3.61\,\mathrm{mAh\,cm^{-2}}$ is application-oriented. Furthermore, the silicon particle diameter is in the micrometer range [11], which has hardly been studied in literature, especially not simulatively. Finally, it is important to mention that the silicon is partially lithiated to only approximately 30% (i.e., approximately $1200\,\mathrm{mAh\,g_{Si}^{-1}}$). This benefits the mechanical integrity and thus the cycling stability of the electrode [11].

For a systematic presentation of its contents, this work is structured as follows. First, the properties of the electrodes used are presented. Second, a literature review for material parameters that could not be determined otherwise is discussed. Third, measurements conducted at coin cells and T-cells are presented and discussed. The equilibrium potential measurements of the silicon electrode are of particular interest here. Fourth, the reconstruction of the full-cell quasi open-circuit potential (qOCP) from measured half-cell qOCP curves is discussed, as the curves are a required model input. In a fifth step, a parameter fitting is performed and validated against the full-cell potential at C-rates up to 2C for charging and discharging. Finally, the model accuracy is evaluated and recommendations for further work are drawn.

2. Experimental Procedure

The experimental methods in this work are split into three parts: Electrode manufacturing, cell assembly, and operational procedures. This work includes all aspects of these three steps necessary to perform parametrization and simulation of a silicon-dominant anode. A more detailed description to each of these steps can be found in the Supplementary Materials.

Electrode manufacturing was performed by the *iwb* at the Technical University of Munich. Silicon material was provided by the manufacturer. The preparation of the coating slurry as well as coating and drying of electrode sheets was performed according to the recipe provided by Jantke et al. [11]. The anode coating comprises of 69.70 wt% micrometer Si particles, 19.90 wt% graphite, and 10.4 wt% binders and additives. Note that the graphite is electrochemically inactive and only serves as a conductive agent [11], hence the naming of a silicon-dominant anode. The cathode active material is a commercially available nickel-rich nickel-cobalt-aluminum-oxide (NCA) with a composition of $LiNi_{0.8}Co_{0.15}Al_{0.05}O_{1.985}(BO_3)_{0.01}$. As already mentioned, a more detailed description of further electrode properties related to the manufacturing step can be found in Section S1 along with Tables S1 and S2 of the Supplementary Materials.

The assembled cells are circular coin cells (CR2032) for characterization as well as Swagelok® (Solon, OH, USA) T-cells for validation. Coin cells were assembled as half cells vs. lithium (Li) metal as well as full cells. The data retrieved from the coin cells was used for characterization, especially to retrieve the half-cell equilibrium potentials using an electrode balancing process, which is discussed below. Additionally, Swagelok® T-cells using a three-electrode configuration were assembled as full cells with a Li metal reference electrode. The T-cells were used for validation at higher C-rates. Detailed information on the cell assembly process as well as the separator and electrolyte used are given in Section S2 of the Supplementary Materials.

The cells then underwent one formation and nine stabilization cycles. Subsequently, a qOCP measurement at C/50 and C/10 was conducted for both coin and T-cells. The formation and stabilization cycles can be evaluated to retrieve an estimate of the capacity losses during formation, while the qOCP measurement yields the usable capacity after formation. This information will later be used to perform an x-axis transformation. The qOCP test is a constant current (CC) charge/discharge cycle at low C-rates of C/50 and C/10 which typically is used to measure the open-circuit voltage of a battery. In addition to the qOCP measurement using CC phases only, a pulsed open-circuit potential (pOCP) measurement was conducted. The pOCP test also makes use of a CC charge/discharge cycle at low C-rates of C/50 and C/10. However, current-carrying phases are interrupted at every 5% state of charge (SoC) and the cell is allowed to relax for 2 h. The cell voltage after this relaxation period at different SoC is interpolated to measure the open-circuit voltage of a battery. The tests are compared to one another in order to investigate the qOCP as well as the polarization behavior at low currents. All tests were carried out at 25 °C and are assumed to be isothermal because of the large thermal capacity of the cell housing.

Subsequent to the qOCP measurement, a rate capability test was conducted using the T-cells in order to later validate the model at higher C-rates. The procedure includes CC charging and discharging at different C-rates of C/10, C/5, C/3, C/2, 1C, 2C, 3C, 5C, and 10C. Constant voltage (CV) phases were used for preconditioning so that every (dis)charge step starts with defined and reproducible conditions. More detailed descriptions of the exact procedures can be found in Section S3 along with Table S3 in the Supplementary Materials.

3. Modeling

In this work, a p2D model is used to describe the electrochemical behavior of a full cell incorporating a silicon-dominant anode. For modeling and simulations, the entire system is assumed to be isothermal at 25 °C at all times. The model equations are depicted in Table A1 and are solved using COMSOL (Stockholm, Sweden) Multiphysics® 5.4. The solver settings are unchanged from the default solver settings recommended by the software for the given set of equations. The final parameters as used for the simulations are listed in Table A2. Some parameters are given by the cell design, such as the porosity or the active material volume fractions. Since these parameters are given, they are discussed in Section S1 of Supplementary Materials. The equilibrium potentials of the anode and cathode were measured in half-cell setups vs. Li metal. The remaining parameters are discussed based on a literature review, namely the electrolyte properties, solid-phase diffusivity, electrical conductivity, exchange current density, and charge transfer coefficients. A summary of these unknown parameters of the silicon electrode is given in Table 2.

Table 2. Overview of silicon material parameters as can be found in the literature. The selected parameters are those model parameters that could not be retrieved otherwise.

Parameter	Unit	Value Range			Value Selected for This Work	References
Electrical conductivity	$S\,m^{-1}$	0.2×10^{-9}	to	5×10^{3}	33	[23,33–37]
Solid-phase diffusivity	$m^2\,s^{-1}$	1×10^{-20}	to	1×10^{-11} [a]	2×10^{-15}	[21,38,39]
Exchange current density	$A\,m^{-2}$	1×10^{-3}	to	1×10^{5}	2.2 [b]	[19,20,22,38,40,41]
Charge transfer coefficient	–	0.2	to	3.42	0.5	[22,42,43]

[a] value might be even higher along defects [32]; [b] after fitting procedure, more information can be found below.

3.1. Equilibrium Potential

Sethuraman et al. [42] discussed the existence of an equilibrium potential and different methods for measuring it. They state that the literature is not in agreement on what the equilibrium potential might look like. Some references discuss a path dependence and represent the equilibrium potential via two potential curves, one for the lithiation process and one for the delithiation process [22,44]. However, a single potential curve could also represent the equilibrium state for both the lithation and delithiation process [30,42]. Sethuraman et al. [42] presented such an approach to simulate the cell potential based on a single equilibrium potential curve. This approach is based on an asymmetric Butler–Volmer equation and a uni-directional side reaction. Additionally, they state that the equilibrium potential might yet be influenced by stress effects.

Because no uni-directional side reaction and no additional stress component are considered in the present work, the equilibrium potential in this work is implemented using two potential curves, similar to what is reported in the literature [22,44]. The equilibrium potential curves used in this work are determined experimentally and shown in Figure 1.

3.2. Voltage Hysteresis

Silicon shows a pronounced hysteresis effect in the measured qOCP curves, as can be seen in Figure 1. By using two potential curves as described in the previous section, the voltage hysteresis effect is already accounted for. This is true because this work's focus is on full cycles and not on partial cycling and also not on the mechanical behavior. The model uses lithiation and delithiation qOCP curves of both electrodes and switches between the curves whenever the cell current switches sign.

If hysteresis was to be explicitly considered using a single curve only, literature provides several sophisticated approaches. Tjandra et al. [45] discussed different approaches to model the hysteresis based on the Preisach model and eventually used the discrete Preisach model to capture the hysteresis effect during full and partial cycling, i.e., $\Delta SoC = 100\%$ and $\Delta SoC < 100\%$. Lu et al. [29] explained the voltage hysteresis in silicon based on mechanical stress. They proposed a model that accounts for an additional stress-induced overpotential via an adjusted Butler–Volmer equation. They assumed homogeneous lithium distribution due to very thin films (0.1 μm) and their model results match experimental data. Another possibility to implement the strong relaxation behavior into the model could be an empirically altered shape of the Butler–Volmer kinetics. The linear region usually attributed to small overpotentials could be broad for Si, followed by an almost abrupt and steep slope of the nonlinear regions. If the overall kinetics would follow such a description, a very small current density would be expected to already require quite large overpotentials for the charge transfer to occur, while the additionally required overpotentials for higher current densities would be almost negligible. As mentioned before, hysteresis is accounted for by two (de)lithiation curves in this work.

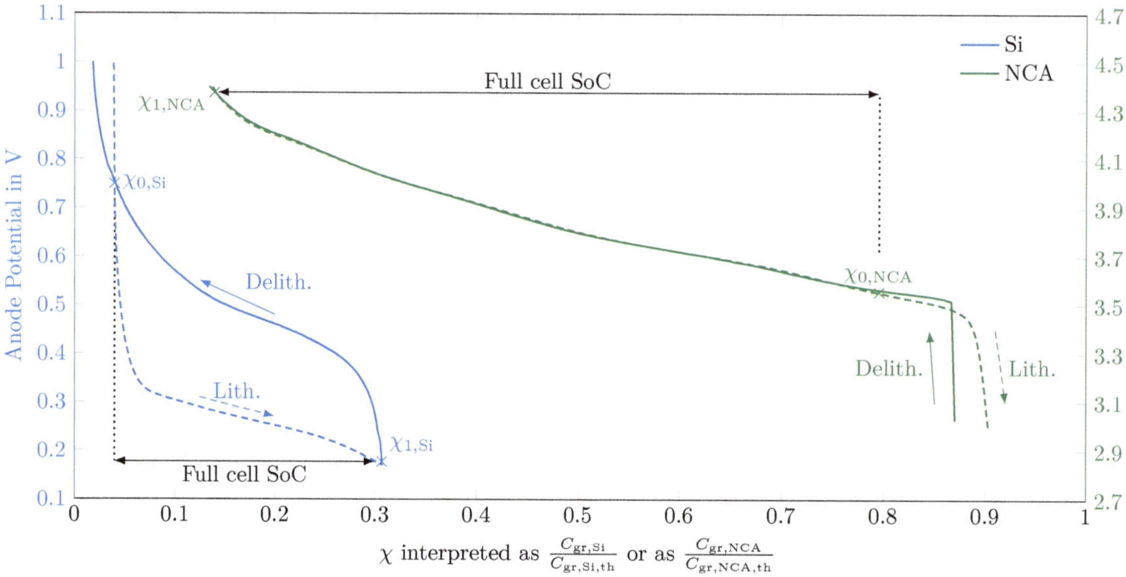

Figure 1. Quasi open-circuit potential of the measured Si and NCA half cell for the (de)lithiation as used in the model. The data are balanced to the full-cell potential (see Figures 2 and A1) and transformed to refer to the calculated degree of lithiation (DoL) χ of the respective electrode. The degree of lithiation (DoL) is the capacity stored in the material relative to the fully lithiated state with $C_{\mathrm{gr,Si,th}} = 3579\,\mathrm{mAh\,g^{-1}}$ and $C_{\mathrm{gr,NCA,th}} = 279\,\mathrm{mAh\,g^{-1}}$. Further information on the axis transformation process can be found in Section S5 along with Table S4 and Figure S1 of the Supplementary Materials.

3.3. Electrolyte Properties

The electrolyte supplier specified the bulk ionic conductivity κ_l to be $6.398\,\mathrm{mS\,cm^{-1}}$. The electrolyte properties were assumed to be similar to th published by Valøen and Reimers [46] and only the ionic conductivity was scaled to match the supplier information.

3.4. Electrical Conductivity

Regarding the constituents of both electrodes, conductive agents such as carbon black are added to the electrode to enhance the otherwise poor electrical conductivity of the active material. Pure graphite particles (in a compacted powder sample) show a comparably high intrinsic conductivity in the order of $10^3\,\mathrm{S\,m^{-1}}$ [47]. This is approximately four orders of magnitude greater than the values reported for the two active materials used in the present work. A composite electrode consisting of active material and graphite as a conductive agent would thus show an increase in its electrical conductivity when compared to pure active material. However, since the properties of such a composite electrode again depend on morphological parameters such as the porosity or the particle size distribution, and especially the volume or weight ratio between the two constituents, finding a definitive value for a given electrode can be challenging.

Bulk values for the electrical conductivity of NCA have been reported to be in the range from $10^{-1.5}\,\mathrm{S\,m^{-1}}$ to $1\,\mathrm{S\,m^{-1}}$, depending on the stoichiometric coefficient χ in $Li_{1-\chi}Ni_{0.8}Co_{0.15}Al_{0.05}O_2$ at 30 °C [48]. For NCA composite electrodes, on the other hand, an effective value of $100\,\mathrm{S\,m^{-1}}$ [49] has been reported. The presence of conductive agent in the porous electrode increases the electrical conductivity by approximately two orders of magnitude. Given the fact that the electrode used in the present work also contains conductive agent, the effective value of $100\,\mathrm{S\,m^{-1}}$ is used here rather than the bulk value.

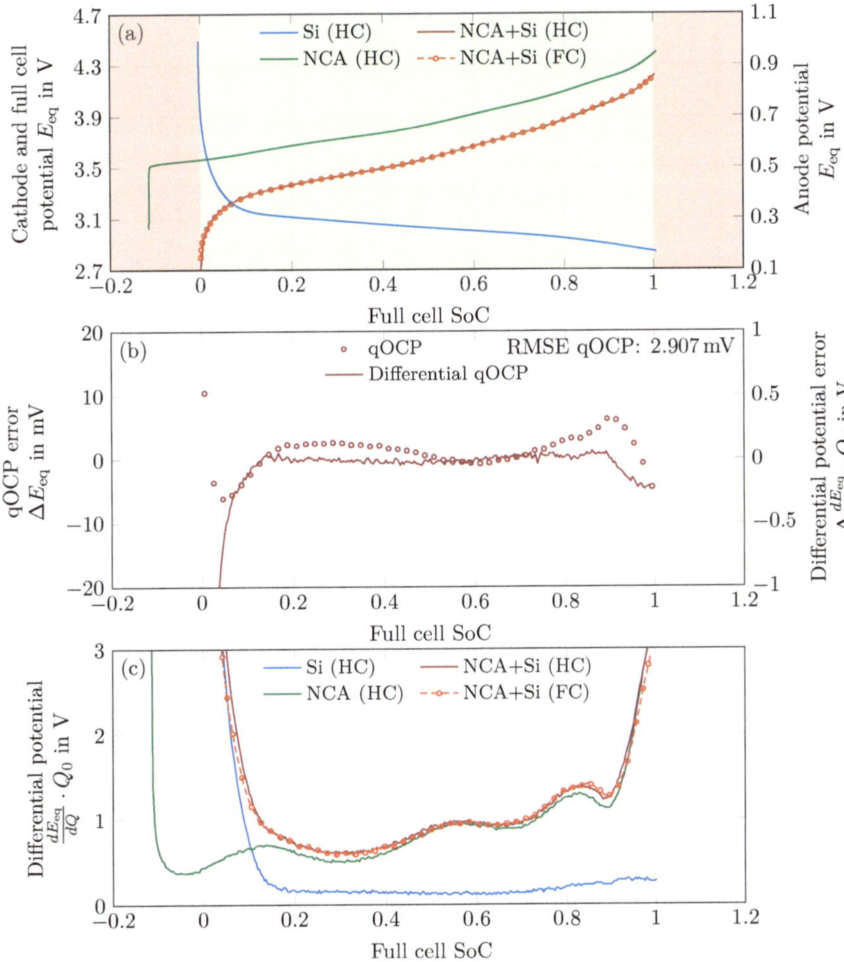

Figure 2. Approximation of the quasi open-circuit potential (qOCP) under constant current charge at C/50 and 25 °C. Differential voltage analysis (**c**) was used to reconstruct the measured qOCP vs. full-cell SoC (**a**) under charge of the full cell (see NCA + Si (FC)) via the (de)lithiation paths of the respective half cells (see NCA (HC) and Si (HC)). The approximation errors vs. SoC as well as the RMSE of the qOCP are shown in subplot (**b**)).

On the anode side, finding a viable value for the electrical conductivity of the Si active material or porous matrix for a given composition is just as difficult. Since Si is a semi-conductor, its intrinsic electrical conductivity strongly depends on (im)purities from doping as well as on temperature. Pure crystalline silicon has a low electrical conductivity of $4.5 \times 10^{-4}\,\mathrm{S\,m^{-1}}$ [50] (p. 258), which can be increased by several orders of magnitude by doping [51] while following Arrhenius characteristics with increasing temperature [52]. Chockla et al. [33] for example manufactured silicon nanowires without any conductive carbon and measured an as-made electrical conductivity of $0.2\,\mathrm{nS\,m^{-1}}$. They could improve these values to $1400\,\mathrm{nS\,m^{-1}}$ via annealing under a reducing atmosphere at elevated temperatures. McDowell and Cui [34] also investigated the electrical conductivity of nanowires without conductive carbon, but found values of around $400\,\mathrm{S\,m^{-1}}$. They found similar values of $600\,\mathrm{S\,m^{-1}}$ for silicon thin films. An important difference is that Chockla et al. [33] could not even lithiate the untreated nanowire because of its low electrical conductivity, while McDowell and Cui [34] conducted their measurement in the lithiated state. The difference in conductivity depending on the DoL was reported by Pollak et al. [35], who found amorphous silicon thin films to be approximately 3.5 orders of magnitude more conductive in the fully lithiated state than in the delithiated state. The highest conductivity value they measured is approximately $5 \times 10^3\,\mathrm{S\,m^{-1}}$. In another study, Kim et al. [36] found the electrical conductivity of bare ⌀10 µm Si particles to be $2\,\mathrm{S\,m^{-1}}$, which could be improved to $1.81 \times 10^3\,\mathrm{S\,m^{-1}}$ via copper deposition on etched Si particles. The conductivity of the untreated particles is two orders of magnitude smaller than for the nanowires and thin films investigated by McDowell and Cui [34]. This might indicate a dependence on the electrode morphology, but might also be due to different degrees of lithiation or different phase compositions of the silicon, e.g., crystalline or amorphous. Among the values already mentioned, Chandrasekaran and Fuller [23] assumed a bulk value of $33\,\mathrm{S\,m^{-1}}$ for their Si electrode including conductive carbon. Wang [37] used this value in simulations and found it to be a suitable value via comparison to experiments. Hence, the electrical conductivity of $33\,\mathrm{S\,m^{-1}}$ is also used in this work. Due to the high fraction of conductive agent, the true value of the electrode used in this work might be even higher.

Based on the literature discussed above, care should be taken when selecting material parameters from the literature, especially for materials as diverse as silicon. Values may vary by several orders of magnitude depending on the morphology and composition of the sample as well as the degree of lithiation. Additionally, not necessarily all experimental information might be given. Experimental studies on the electrical conductivity of Si-based electrodes using different electrode compositions, i.e., pure silicon as well as different silicon to graphite ratios, at different degrees of lithiation would be very valuable for future modeling.

3.5. Solid-Phase Concentration

The maximum concentration of lithium inside the active material particles $c_{\mathrm{s,max}}$ is calculated based on the gravimetric capacity C_{gr} and the density ρ according to

$$c_{\mathrm{s,max}} = \frac{C_{\mathrm{gr,th}} \cdot \rho}{\mathcal{F}}. \tag{1}$$

The theoretic gravimetric capacities for Si and NCA are $3579\,\mathrm{mAh\,g^{-1}}$ and $279\,\mathrm{mAh\,g^{-1}}$, respectively, while the densities of $2336\,\mathrm{kg\,m^{-3}}$ and $4730\,\mathrm{kg\,m^{-3}}$ are supplier information. The density for Si is the density in the purely crystalline state, and the density for NCA was obtained from powder X-ray diffraction data. These density values yield concentration values of $311{,}943\,\mathrm{mol\,m^{-3}}$ and $49{,}239\,\mathrm{mol\,m^{-3}}$, respectively. Note that these values are considered to be a good guess but will be adjusted in order to fit the model to the measurement data.

In Equation (1), the density is not a function of the DoL but has a constant value that corresponds to the pristine particle state, i.e., no Li inside Si and NCA fully lithiated. In reality, especially for Si, the density of a particle will change with the DoL because of the lithiation-induced volume expansion of up to 300% for Si [10]. Chandrasekaran et al. [22] presented an approach to account for this effect which also takes the changing particle radius and thus a variable diffusion path length into account. In this case, the maximum solid-phase concentration in the fully lithiated state would be $77{,}986\,\mathrm{mol\,m^{-3}}$ instead of $311{,}943\,\mathrm{mol\,m^{-3}}$. However, since no volume change of the active material particles is considered in the present work, the change in the maximum solid-phase concentration in dependence on the DoL is of no further interest here.

3.6. Solid-Phase Diffusivity

For NCA particles, Dees et al. [40,53] determined the diffusion coefficient of Li inside NCA using galvanostatic intermittent titration technique (GITT) measurements to be in the range of $10^{-14}\,\mathrm{m^2\,s^{-1}}$ to $10^{-17}\,\mathrm{m^2\,s^{-1}}$ depending on the DoL. A value of $6 \times 10^{-15}\,\mathrm{m^2\,s^{-1}}$ is used for the simulations in this work, i.e., the value at 3.77 V vs. Li metal. This is because it is centered between the minimal and maximal cathode half-cell potentials for the cell setup used in this work (3.0 V and 4.4 V NCA vs. Li metal).

The solid diffusion coefficient of Li in Si varies from $10^{-13}\,m^2\,s^{-1}$ to $10^{-20}\,m^2\,s^{-1}$ [21,38]. A detailed and more recent literature review by Sivonxay et al. [39] suggests even higher values up to $10^{-11}\,m^2\,s^{-1}$. They investigated Li diffusion properties in Li_xSi and Li_xSiO_2 via first-principle calculations which yielded values between $10^{-11}\,m^2\,s^{-1}$ and $10^{-15}\,m^2\,s^{-1}$. Wang et al. [21] compared numerical simulations with in situ lithiation experiments in order to determine the diffusion coefficient of Li in Si. They reported the coefficient in the amorphous as well as in the Li-rich phase after the formation cycle to be in the order of $2 \times 10^{-15}\,m^2\,s^{-1}$. However, Bordes et al. [32] showed that fast Li diffusion pathways exist along defects. Such defects may occur due to stacking faults in a single crystal but can also occur along grain boundaries in polycrystalline Si particles. They propose that due to this phenomenon, Li diffusion may be much faster than in a single crystal without defects, regardless of the particle size. This suggests that, at least in polycrystalline particles as in this work, the diffusion coefficient might be greater than that reported by Wang et al. [21]. Nonetheless, their value of $2 \times 10^{-15}\,m^2\,s^{-1}$ is used for the simulations in the present work, as it has been validated against experimental data.

As mentioned before, the value for solid-phase diffusivity reported in the literature spans several orders of magnitude. Hence, a detailed analysis of the solid-phase diffusivity in silicon considering its various types such as nanowires, thin-films, or particles as well as its different amorphization states would be of great value when it comes to physicochemical modeling. However, obtaining a consistent value for the solid-phase diffusivity is a problem that has not been solved yet, not even for long-used materials such as graphite [54].

3.7. Electrode Kinetics

The Butler–Volmer electrode kinetics described by Equation (A7) requires an exchange current density i_0. Since no experimental determination of the exchange current densities was carried out in the scope of this work, values from the literature are used. For the exchange current density of ⌀9.4 µm NCA particles, a value of $1\,A\,m^{-2}$ has been reported [40]. For NCA particles dipped in PVdF binder, the exchange current density during lithiation can reach higher values of up to $10\,A\,m^{-2}$ [55].

For the exchange current density of Si electrodes, however, a wide range of values spanning several orders of magnitude from $10^{-3}\,A\,m^{-2}$ to $10^{5}\,A\,m^{-2}$ can be found. Bucci et al. [38] reported a value in the order of $10^{5}\,A\,m^{-2}$ for silicon thin film electrodes of about 100 µm thickness. Lory et al. [19] assumed the exchange current density to be non-limiting and chose a constant value of $2\,A\,m^{-2}$ for their model. Note that, in their model, silicon is assumed to have no direct contact with the electrolyte because it is embedded inside carbon particles of 12 µm diameter. The value thus represents the exchange current density between carbon and silicon. Li et al. [41] derived values between $0.7\,A\,m^{-2}$ to $1.3\,A\,m^{-2}$ based on potentiostatic intermittent titration technique (PITT) measurements on silicon thin films of up to 1000 nm. Swamy and Chiang [20] found values of approximately $1\,A\,m^{-2}$ by fitting impedance data of 500 µm thick silicon wafers to an equivalent circuit model. Chandrasekaran et al. [22] report values of about $10^{-2}\,A\,m^{-2}$ to show a good fit between their single particle model with a diameter of 60 nm and experiments conducted using silicon nanowires with a pristine diameter of 89 nm [56]. Finally, Wang [37] fitted a multi-radius model to measurements on nanowires and found a value of $10^{-3}\,A\,m^{-2}$ to be a good fit for the exchange current density. This value is well below the reported values determined using measurements, which might indicate that some cross-dependencies due to model assumptions influence the value found via fitting. Thus, this value should be taken with care. None of the references stated a clear dependency on the DoL. With the exception of Chandrasekaran et al. [22], none of the references discussed charge transfer coefficients and assumed them to be symmetric, i.e., $\alpha_a = \alpha_c = 0.5$. Additionally, different sample geometries that have been investigated in the literature also make it difficult to get a true value for the exchange current density when only determined for a single state. This is due to the uncertainty in the active surface area of different samples, which depends on the sample morphology and undergoes changes due to volume expansion. Furthermore, the different measurement approaches used to determine the exchange current density might impede cross-reference comparisons. For simulations in this work, $1\,\mathrm{A\,m^{-2}}$ is considered to be a reasonable initial guess because it is on neither end of the range found in the literature. For future simulation works, two different experimental results would be valuable: On the one hand, a consistent value for a single sample obtained via different measurement approaches would indicate under which conditions cross-reference comparisons are valid. On the other hand, an extensive investigation of various samples using a single measurement approach could serve as a baseline for a consistent data base.

The exchange current densities in the present model are expressed as a function of the local concentrations (see Equation (A9)). In this case, the exchange current density is dependent on the reaction rate constant k and on the concentrations c_l and $c_\mathrm{s,surf}$ at the silicon–electrolyte interface. This equation is used because it contains limitations for surface concentrations near the minimum and maximum solid-phase concentrations. This representation at least simulates a DoL-dependence in a most basic way. In order to convert the given exchange current densities to such reaction rate constants, Equation (A9) is solved for k. No distinction between anodic and cathodic reaction rates is made here, thus $k_\mathrm{c}^{\alpha_\mathrm{a}} k_\mathrm{a}^{\alpha_\mathrm{c}} = k^{\alpha_\mathrm{a}+\alpha_\mathrm{c}}$. A further assumption for this calculation is $\alpha_\mathrm{a} = \alpha_\mathrm{c} = 0.5$. Hence, the anodic and cathodic reactions are assumed to be symmetrical. A more detailed discussion on this topic can be found in Section S4 of the Supplementary Materials. Equation (A9) is solved for in such manner that the reaction rate k is a function of $c_\mathrm{s,surf}$. Accordingly, values for c_l and $c_\mathrm{s,max}$ are needed. For $c_\mathrm{s,max}$, the values calculated using Equation (1) are used. For c_l, a constant value of $1000\,\mathrm{mol\,m^{-3}}$ is assumed, i.e., the value in the equilibrium state. By assuming a constant value for c_l, the impact of gradients during operation is excluded for the transformation from exchange current densities to reaction rates. Doing so, the transformed reaction rates are not dependent on c_l and thus the dependency of i_0 on c_l in Equation (A9) is not accounted for twice. Since $c_\mathrm{s,surf}$ may vary from the maximum value $c_\mathrm{s,max}$, the reaction rate constant for a constant exchange current density may vary, too. To avoid division by zero when $c_\mathrm{s,surf}$ becomes 0 or $c_\mathrm{s,max}$, $c_\mathrm{s,surf}$ is limited to the range $[0.01c_\mathrm{s,max}; 0.99c_\mathrm{s,max}]$ here. The reaction rate k is solved for using these values and then averaged so that it is no longer a function of $c_\mathrm{s,surf}$. Averaging is necessary so that the dependency of i_0 on $c_\mathrm{s,surf}$ in Equation (A9) is not accounted for twice. The averaged values are $2.95 \times 10^{-12}\,\mathrm{m\,s^{-1}}$ for silicon and $1.90 \times 10^{-11}\,\mathrm{m\,s^{-1}}$ for NCA. However, because of the uncertainty of the exchange current densities found in the literature, these values will be subject to a fitting process. The final values after the fitting can be found in Table A2.

4. Results and Discussion

To maintain readability, only the results of one representative lab cell are shown and discussed. This decision is based on the variations in gravimetric capacity between the tested cells of 3.18% during cell formation and 1.94% during the qOCP tests, indicating that cell manufacturing, cell formation and stabilization are reproducible.

4.1. Open-Circuit Potential Measurements

The qOCP and pOCP measurements are shown in Figure 3. Black curves refer to C/50 and red curves refer to C/10. The cell potential is plotted as a function of the normalized cell capacity for the charge and discharge step that was recorded during the individual tests. The reversible discharge capacity at C/50 is 5.5605 mAh. With respect to the smaller cathode electrode sample with a diameter of 14 mm, this yields a utilized areal capacity of $3.61\,\mathrm{mAh\,cm^{-2}}$.

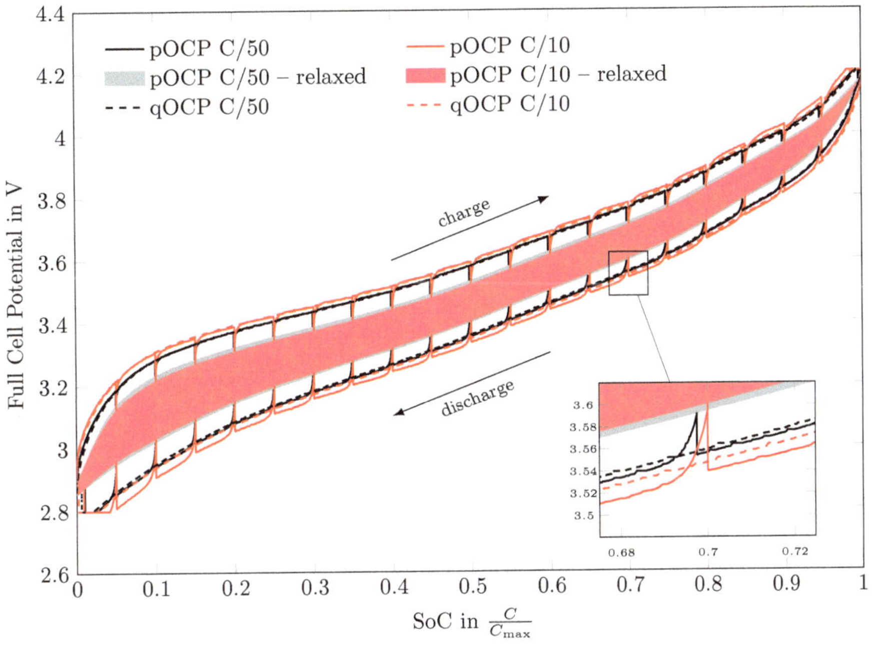

Figure 3. quasi open-circuit potential (qOCP) data (dashed lines) and pulsed open-circuit potential (pOCP) data (solid lines) for a current rate of C/50 (black and gray) as well as C/10 (red). The relaxed potentials of the 2 h open-circuit phases of the pOCP test were interpolated and are displayed as a gray/red area, indicating the voltage hysteresis effect between charging and discharging. The inset shows one exemplary open-circuit phase in more detail. The data are plotted as a function of the normalized charge capacity of the respective test. The tests were conducted at 25 °C.

Overall, the potential during the CC phases of the pOCP measurement plotted with solid lines are in good agreement with the potential of the pure CC load, i.e., the qOCP measurement, plotted with dashed lines. However, even at C-rates as low as C/50, the cell potential still relaxes by up to 90 mV during 2 h of open-circuit conditions. The value at the end of this relaxation phase is called the "relaxed potential" in the following. The relaxed potentials at different SoC were interpolated and are plotted as areas, indicating the voltage hysteresis effect between charging and discharging. Both observations hold true for C-rates of C/50 and C/10. The pronounced relaxation has been described in the literature [42,44,57], but to the best knowledge of the authors, its origin has not been identified yet. The fact that the cell potentials at C/10 and C/50 are quite similar, but still differ significantly from the relaxed potential, indicates that the effect observed here is at least not related to ohmic losses but rather to electrode kinetics. Possibilities to implement such behavior into the model have been discussed above.

As previously mentioned, the qOCP data at C/50 are used as a model input. Thus, the model is not expected to be able to capture the strong relaxation behavior but the behavior under load, as shown below. Because individual lithiation and delithiation curves are implemented, no dedicated hysteresis model is used here.

4.2. Electrode Balancing

One of the model outputs is the full-cell potential, which is a superposition of the half-cell potentials including overpotentials. Thus, an accurate reconstruction of the full-cell potential from the measured half-cell potentials is needed to provide a correct model input. During this reconstruction process, the half-cell measurements are stretched and shifted along the capacity axis in such way that their superposition results in the full-cell measurement.

By doing so, the electrodes are being matched to each other so that they give the reference full-cell measurement. Complementary to the measured voltage curve, its derivative with respect to the capacity—the differential voltage analysis (DVA)—can be used to retrieve a satisfying reconstruction. Certain features in the DVA like the transition between between two two-phase regions as is the case for graphite or the steep limiting region at either end of the utilizable stoichiometry can be used to compare the superposition of anode and cathode curves with that of the full-cell curve, and eventually minimize the difference. The procedure was described by Smith et al. [58] and was later used by Dahn et al. [59] to facilitate the analysis of degradation mechanisms in lithium-ion batteries.

In this work, especially the DVA was used to fit the half cells to the full cell. Additionally, the process was carried out for the full-cell charge and discharge step individually, analogous to [7].

The result of the balancing process is shown in Figure 2a for the coin full-cell charging. The results for the discharge are given in Figure A1 in the Appendix B.

As a result of the balancing process, the measured half-cell as well as the superposed full-cell potential is a function of the full-cell SoC instead of the electrode capacity. The balanced half-cell potential curves show an overhang beyond the limits of the full-cell SoC, which is due to the different cut-off potentials for the tests on full cells and half cells.

Figures 2b and A1b show a maximum absolute qOCP error of less than 11 mV and 15 mV for the charging and discharging process, respectively. The corresponding root-mean-squared error (RMSE) is 2.9 mV and 5.4 mV. The small RMSE indicates that the superposition of the half cells using the identified full-cell SoC is satisfactory to represent the full-cell potential at low C-rates. This can also be seen in the overlay of the measured and reconstructed full-cell curves (light and dark red) in Figures 2c and A1c. Here, the differential potentials are shown and normalized by multiplying by the total capacity of the respective (dis)charge step Q_0.

4.3. *Transformation to the Degree of Lithiation*

At this point, the balancing process is completed and the results are dependent on the full-cell SoC. However, the model solves for the solid-phase concentrations c_s of both electrodes. These concentrations can be normalized using the maximum solid-phase concentration $c_{s,max}$, yielding the DoL χ, as described in Equation (2). Hence, a representation of the equilibrium potential as a function of the DoL instead of the full-cell SoC would be advantageous.

$$\chi = \frac{c_s}{c_{s,max}} = \frac{C_{gr}}{C_{gr,th}} \tag{2}$$

The DoL can not only be calculated based on concentrations taken from the model but also based on capacities taken from measurements. Here, the gravimetric capacity C_{gr} from the tests and the theoretical maximum gravimetric capacity $C_{gr,th}$ of the material are used. For the maximum gravimetric capacity, the maximum theoretical value of 3579 mAh g^{-1} [5] is used. For the current gravimetric capacity, the current capacity from the battery cycler is divided by the active material mass of the sample. Additionally, unlike the concentration, the measured capacity is no absolute value but the integration of the electric current. An integration constant or rather an initial value is needed to retrieve a proper value for the DoL. This initial value is assumed to be the irreversible loss during formation and will be discussed in the following.

To obtain such a representation as a function of the DoL, an estimate based on the losses during the first formation and stabilization cycles $\Delta\chi_{loss}$ as well as the utilization during the qOCP test $\Delta\chi_{qOCP}$ is made. These values are then used for shifting and scaling the curves to perform the transformation from SoC to DoL. A detailed description of this transformation including a table with all necessary interim and final values is given in Section S5 along with Table S4 of the Supplementary Materials.

Figure 1 shows the final result. The plotted curves show the half-cell potentials as a function of the DoL. Green lines represent the NCA cathode half cell, and blue lines represent the Si anode half cell. The black arrows together with labeled crosses show the utilization range of the respective half cell DoL for a full-cell SoC as previously identified in the balancing process. The anode DoL utilization between 0% and 100% SoC ranges between 0.0396 and 0.3068, respectively, and that of the cathode ranges between 0.7954 and 0.1396. The corresponding equilibrium potentials, averaged for charge and discharge, are 739.9 mV and 182.1 mV for the anode and 3.558 V and 4.378 V for the cathode. Table 3 lists these average values together with the individual values for charging and discharging.

Table 3. Degree of lithiation and corresponding equilibrium potential at 0% and 100% SoC from Figure 1. The values are given for the full-cell charge (CHA) and discharge (DIS) as well as the averaged (AVE) value at the respective SoC.

	Anode		Cathode		
SoC	**DoL in –**	E_{eq} **in V**	**DoL in –**	E_{eq} **in V**	
0%	$\chi_{0,Si} = 0.0396$	0.7531	$\chi_{0,NCA} = 0.7954$	3.5518	DIS
		0.7267		3.5635	CHA
		0.7399		3.5577	AVE
100%	$\chi_{1,Si} = 0.3068$	0.1932	$\chi_{1,NCA} = 0.1396$	4.3797	DIS
		0.1710		4.3764	CHA
		0.1821		4.3781	AVE

From these utilization ranges, it can be seen that Si is not entirely delithiated. This is due to silicon's significantly better aging behavior when not fully delithiated. Wetjen et al. [5] showed this for nanometer-sized silicon and Haufe et al. [13] showed this for the same micrometer-sized silicon that is also used in the present work. This results in a half cell balancing procedure that is slightly more complex for Si anodes than for graphite anodes because, for graphite anodes, the minimum DoL can usually be set close to zero, while for partially lithiated silicon, this is not recommended. If this effect were not taken into account, significant errors between simulation and measurement results would be expected at low full-cell SoC.

4.4. Fitting of Solid-Phase Concentrations

So far, electrodes' equilibrium potentials from the qOCP measurements have been balanced using a DVA-aided process and the resulting full-cell SoC has been transformed to a degree of lithiation unique to each electrode. The curves shown in Figure 1 are used as a model input and depend on the DoL from Equation (2). Thus, the value of $c_{s,max}$ directly influences the course of the DoL and the cell potential when a fixed intercalation rate is given. Initially, the values given by Equation (1) are used for the first simulation of a complete cell charge and discharge at C/50. An evaluation of the DoL revealed that the Si electrode was lithiated to a higher degree than that obtained from the transformation to the DoL ($\chi_{1,Si} = 0.3068$, see Figure 1).

This effect is also shown in Figure 4a. The blue dotted line shows the full-cell potential at C/50 (dis)charging using the initial parametrization. The data are plotted as a function of the (dis)charge capacity normalized to the capacity during a C/50 charge for both, the measurement (black) and model (colored), individually. Only CC phases are plotted, CV phases are not shown. Because the DoL overshoots during the simulation, the equilibrium potential is beyond the defined range and thus not valid. This becomes especially evident during the discharge at approx. 98% normalized capacity. At this point, the voltage curve shows a sharp bend. This is caused by the DoL entering the defined range again and the change in the shape of the anode equilibrium potential drastically impacts the full-cell potential. This bend is not comparable to that during the experiments.

To prevent this non-physical behavior, the maximum solid-phase concentration is fitted in such way that the DoL no longer overshoots. This fitting of the anode maximum solid-phase concentration does not alter the electrode potential at a given DoL, but it does alter the balancing between anode and cathode. To compensate for this overlithiation of the anode, the anode maximum solid-phase concentration was increased by a factor of 9.56%. Considering the volume expansion and its effect on the mass density in Equation (1), this is considered to be within the range of physically interpretable values.

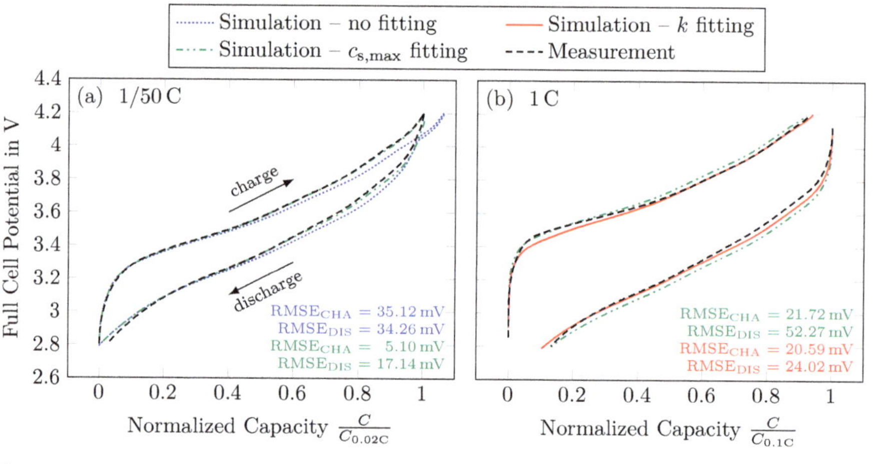

Figure 4. Full-cell potentials during the two different fitting procedures compared to measurement data. "No fitting" refers to the initial parametrization taken from literature, "$c_{s,max}$ fitting" is an intermediate step, and "k fitting" refers to the final parametrization. Data are plotted as a function of the (dis)charge capacity normalized to the capacity during a C/50 or C/10 charge for both, the measurement (black) and model (colored), individually. Only CC phases are plotted, CV phases are not shown. (**a**) shows the improvement due to the fitting of $c_{s,max}$ at C/50, and (**b**) shows the improvement due to the fitting of k at 1C. The RMSE is that between simulation and measurement and is color-coded according to the legend.

Subsequent to this individual fitting of the maximum solid phase concentration of the anode, the maximum solid-phase concentration of both electrodes was scaled equally. By scaling the solid phase concentrations by the same factor, the balancing between anode and cathode is retained while adjusting the absolute capacity of the model so that it matches the capacity of the measurements at C/10. Finally, the maximum solid-phase concentration of the anode is increased by 3.25%, already taking into account the anode-specific factor from above, and that of the cathode is decreased by 5.77%. The final results for the maximum solid-phase concentrations are summarized in Table A2. The green dash-dotted line in Figure 4a now shows a behavior similar to that during the experiments. Fitting of the maximum solid phase concentration resulted in an improvement of the RMSE by up to 30 mV in values.

4.5. Rate Capability

Up to this point, the model describes the experimental data satisfactorily at C-rates of C/50. To check the quality of the model's predictions at higher C-rates, measurements at different C-rates using a T-cell were conducted. The formation and stabilization procedure for T-cells and coin cells are identical, and the results are comparable. Hence, the model parametrization based on coin cells is expected to be valid for T-cells, too.

A comparison between measurement and simulation data (black dashed vs. green dash-dotted line) in Figure 4b shows that a certain error also remains at 1 C when using the current parametrization. This is due to the overpotential at higher C-rates which depends on the kinetic behavior. The literature reports a wide range for the exchange current density, and, as such, for the reaction rate constants k, the reaction rate constants were fitted so that the RMSE between simulation results and experimental data is minimized. Up to this point, the only quantity that was fitted is the maximum solid-phase concentration. The fitting of the reaction rates was done using the lsqnonlin optimizer in MATLAB® 2021b aiming to minimize the difference between simulation and measurement data for a 1 C constant current charge and discharge. All other C-rates were not part of the fitting procedure. They serve as validation data and can be seen in Figures 5 and 6.

The best agreement between the simulated and measured full-cell potentials was found with the reaction rate constants of $5.78 \times 10^{-12}\,\mathrm{m\,s^{-1}}$ for the anode and $1.64 \times 10^{-10}\,\mathrm{m\,s^{-1}}$ for the cathode. The values are also summarized in Table A2. Before further inspecting the full-cell potential, the optimized reaction rates are checked for plausibility regarding the range from the literature. To this end, the optimized reaction rates are re-transformed to exchange current densities. This re-transformation is analogous to the transformation of the initial exchange current densities to reaction rates, as described above. The results show a re-transformed and averaged exchange current density of $2.228\,\mathrm{A\,m^{-2}}$ for the Si anode and $9.108\,\mathrm{A\,m^{-2}}$ for the NCA cathode. These values deviate by a factor of about 2 to 9 from the initial value of $1\,\mathrm{A\,m^{-2}}$ for both anode and cathode, as taken from the literature. Considering a range of several orders of magnitude for the anode exchange current density, the fitting is considered to be within an acceptable range. For the NCA material, a value of approximately $9\,\mathrm{A\,m^{-2}}$ lies at the upper end of the range given in the literature.

Figure 4b allows for a further inspection of the full-cell potential and reveals the improvement due to the fitting of reaction rates. Here, the green dash-dotted line represents the simulation data after the fitting of the maximum solid phase concentrations, and the red solid line represents the subsequently optimized parametrization. Note that the optimized parametrization consists of a single reaction rate constant for each electrode that is optimized to match the charge and discharge at the same time. In other words: The reaction rate constant is not dependent on the sign of the cell current or the reaction direction. The RMSE given is that between the measurements and the optimized simulation results. The RMSE value during charging improves by only 1.13 mV and is almost unaffected by the optimization. This is because the regions where simulation and measurement differ have shifted from higher to lower SoC. The RMSE value during discharging, on the other hand, improved by approx. 28 mV, and the curves show an overall reduced deviation.

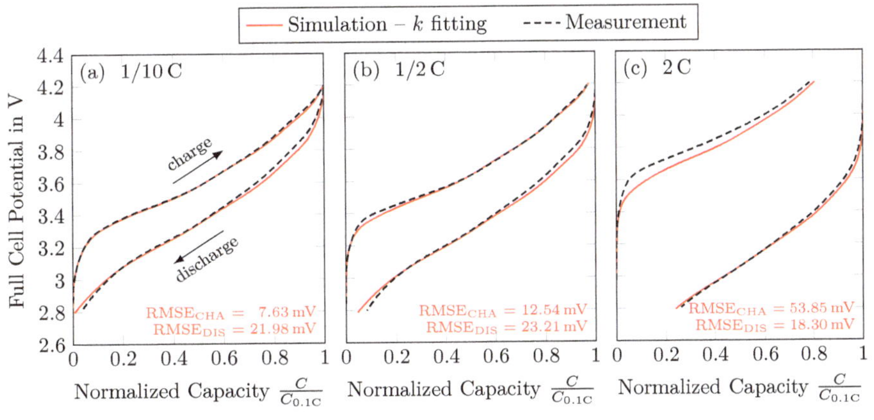

Figure 5. Full cell potentials as validation data for the final parametrization after fitting using C-rates of C/10 (**a**), C/2 (**b**), and 2C (**c**). Color-coding and representation identical to Figure 4.

Figure 5 is similar to Figure 4 but shows the validation data at different C-rates instead. Only the final parametrization including the optimized reaction rates is used for the simulation data. From both figures together it can be seen that the RMSE for discharging remains in the range between 18.30 mV and 24.02 mV for different C-rates. This indicates that the overpotentials during discharging are well described using the symmetric Butler–Volmer kinetics implemented here. The RMSE for charging, however, increases with increasing C-rate, revealing that overpotentials during charging at high C-rates are more pronounced than at low C-rates. This might indicate that kinetics are asymmetric. Another possible explanation for this observation is an SoC-dependent porosity. With increasing SoC, the porosity of the Si anode decreases due to intercalation-induced volume expansion. A decrease in porosity would first occur near the separator during charging and gradually propagate towards the current collector. Similarly, the porosity would first increase near the separator during discharging and then gradually propagate towards the current collector. This way, an increasing overpotential with increasing C-rate due to porosity changes is to be expected during charging, while during discharging, a decreasing overpotential with increasing C-rate is to be expected. Because the model does not account for such porosity changes, the behavior in Figure 4 is different. Overall, further investigation regarding the overpotentials during fast (dis)charging is needed in order to accurately describe the full-cell potential. This statement agrees with the statement deduced from Figure 6.

Figure 6 shows the CC discharge capacity as a function of the C-rate. The discharge capacity is normalized to the CC charge capacity at C/10, just as in Figures 4b and 5. Up to 2 C, the capacities predicted by the model are in good agreement with those obtained from the measurements. The maximum deviation between simulation and measurement data up to 2 C is <2.4% for discharging and <1.8% for charging. Especially for the discharge drawn with dotted lines, an almost constant offset between measurement and simulation can be seen. This could originate from the strong relaxation behavior of the qOCP measurement which is present even at small C-rates. The model does not account for these effects. From 3 C onwards, the model overestimates the capacity for charging and discharging when compared to the measurements. This indicates that the model underestimates the total polarization at high C-rates.

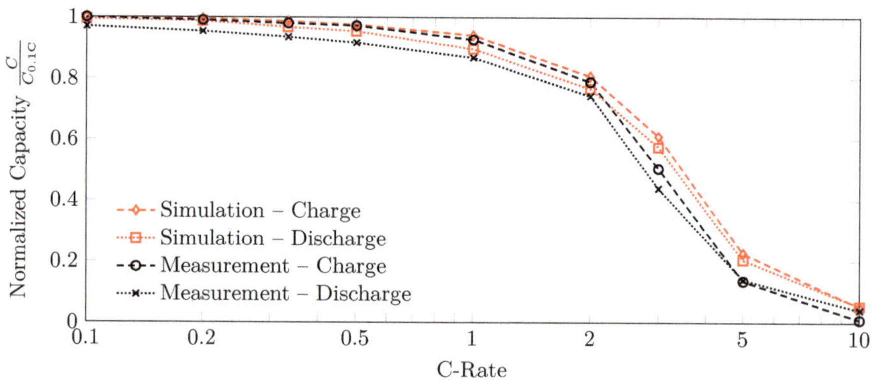

Figure 6. Rate capability of measurements using T-cells (black) and simulation results (red) for CC (dis)charge tests at C-rates of C/10, C/5, C/3, C/2, 1 C, 2 C, 3 C, 5 C, and 10 C. The (dis)charge capacity of the CC steps is normalized to the capacity during a C/10 CC charge for both, the measurements and the model, individually. The test was conducted between 2.8 V and 4.2 V at a constant temperature of 25 °C.

A possible explanation for the low capacity retention at elevated C-rates during the experiments are mass transport limitations. Xie and Yang [60] explained the interplay between transport in the solid and liquid phase using an electrochemical model including a graphite anode. Due to sluggish transport in the solid particles and preferred (de)intercalation near the separator-electrode interface, the particle surface will become saturated, resulting in a gradient of the surface concentration along the electrode thickness. This increases the diffusion path length in the electrolyte to allow for further reactions, increasing the total overpotential. The behavior described by Xie and Yang [60] is similar to that in Figure 6, suggesting that the rate capability as seen in the figure is not exclusive to silicon.

Despite the overall good representation of the real cell behavior, one potential aspect for optimization could lie in the model's porosity. For the model parametrization, the porosity calculated from data in the pristine state prior to cell formation is used. Further information on this topic can be found in Section S1 of the Supplementary Materials. However, because of the large volume expansion of silicon as well as the formation losses, it is likely that the active material volume fraction after formation is different from the value prior to formation. As a consequence, it is also likely that the increase in the active material volume fraction causes a decrease in porosity and thus poor rate performance at high C-rates as can be see in Figure 6. Furthermore, this decrease in porosity is likely to be a local effect linked to the gradient in the particle concentration mentioned above. Thus, an optimized and possibly local porosity value and as such an optimized MacMullin number would likely further increase the model accuracy. Lastly, the error between simulation and measurement in Figure 5 shows a clear trend with changing SoC, which might be due to an SoC-dependent porosity affecting the overpotentials. Accounting for this effect could enhance the model accuracy.

5. Conclusions and Outlook

Until now, the literature has not provided an electrochemical model for a silicon-dominant electrode. Hence, the goal of this study was to parametrize and validate a Newman-type model for a custom-built Si||NCA lithium-ion cell. Many parameters in the model presented here, such as the porosity, could be obtained from electrode manufacturing, while other parameters, such as the exchange current density, were taken from the literature and underwent a fitting procedure.

In a first step, the equilibrium potentials of the half cells and the full cell were measured. At C-rates as low as C/50, the Si half-cell potential still relaxes significantly, but the origin of this effect has not been identified yet. The behavior at C/50 is similar to that at C/10, indicating that more processes than just the charge transfer described by a single Butler–Volmer equation might be at play. Thus, the qOCP measurement at constant current have been used in the following, ignoring the model's equilibrium potential relaxation effect. Voltage hysteresis is accounted for by switching between lithiation and delithiation half-cell potentials.

In a second step, a balancing process of the measured half cell qOCP curves was performed. A proper superposition of the half-cell potentials then yields the reconstructed full-cell qOCP curve with a RMSE of 2.907 mV for the charge step and 5.432 mV for the discharge step. The measurements needed for this balancing process also allow for an estimation of the anode and cathode utilization, and as such allow for an estimation of the DoL.

In a third step, the maximum solid-phase concentrations were fitted in such way that the model's relative concentration is always bounded by the minimum and maximum DoL obtained in the previous step. Additionally, the concentrations were scaled so that the model's capacity meets that of the measurements.

In a fourth and final step, the model's reaction rate constants were fitted for a 1C constant current charge and discharge step. The fitting was performed in such a way that the full-cell potential during both charging and discharging is described by a single parameter set. Different C-rates between C/10 and 10 C were used to validate the model parametrization. The model can describe the full-cell potential and also meets the constant current (dis)charge capacity. This is true up to a C-rate of 2 C. At higher C-rates, the model underestimates the polarization when comparing the simulation results with measurements.

Overall, the model results are satisfying, with an RMSE between measurements and the simulations of 20.59 mV and 24.02 mV during 1 C charging and discharging, respectively. At the same time, the absolute error in the normalized capacities is 1.27% and 2.75%, respectively. Thus, the presented parametrization is considered valid to describe a silicon-dominant anode when a Newman-type model is used, under the premise that:

- Si is partially lithiated to a similar degree as in this study and as such does not undergo further amorphization or crystallization.
- Model equations have not been adjusted, for example with respect to complex electrode kinetics, solid state diffusion, or volumetric changes.
- Voltage hysteresis is accounted for via two equilibrium curves rather than a dedicated hysteresis model.

In future work, several aspects regarding silicon-dominant anodes can be further investigated. Optimized parameters such as DoL-dependencies for the solid-phase diffusivity or the electrical conductivity could further improve the model quality. Additionally, a dedicated voltage hysteresis model could prove valuable to better describe the voltage behavior during open-circuit periods. An overview, including suggestions for experimental studies, is given for each of the parameters that have been taken from literature. An understanding of the origin of the strong relaxation behavior, including an appropriate kinetics model, would benefit subsequent studies. Closely related to this is an analysis of polarization at high currents. Since only laboratory cells were used, isothermal conditions were assumed for all experiments and simulations. A scale-up to a large-format pouch cell, for example, would require a thermal model to capture temperature-related effects such as joule heating and entropic heating. Similarly, a suitable mechanical model, because of the large volumetric expansion, is needed to capture porosity changes and related effects on the mass transport. Eventually, a fully coupled physicochemical-mechanical-thermal model would be required to capture all mentioned effects properly.

Supplementary Materials: The following supporting information can be downloaded at: https://www.mdpi.com/article/10.3390/batteries9110558/s1, Figure S1: Quasi open-circuit potentials as shown in Figure 1 but with additional information.; Table S1: Electrode composition and mass densities for each cell component; Table S2: Characteristic properties of double-coated electrode samples; Table S3: Information on the cycling procedure; Table S4: Step-by-step overview for the data conversion from charge throughput to the degree of lithiation. References [61–63] are cited in the supplementary materials.

Author Contributions: Conceptualization, A.D.; methodology, A.D. and S.F.; software, A.D.; validation, A.D., S.F. and L.H.; formal analysis, A.D. and L.H.; investigation, A.D., S.F. and L.H.; resources, A.J.; data curation, A.D. and L.H.; writing—original draft preparation, A.D.; writing—review and editing, A.D., S.F., L.H. and A.J.; visualization, A.D. and L.H.; supervision, A.J.; project administration, A.D. and A.J.; funding acquisition, A.J. All authors have read and agreed to the published version of the manuscript.

Funding: This research is funded by the German Federal Ministry of Education and Research (BMBF) via the research project ExZellTUM III (grant number: 03XP0255). The project is overseen by Project Management Juelich (PtJ).

Data Availability Statement: The data presented in this study are available on request from the corresponding author. The data are not publicly available due to privacy policy.

Acknowledgments: The authors gratefully thank Philip Rapp for the support in T-cell assembly and Simon Helmer for measurements regarding of MacMullin numbers.

Conflicts of Interest: The authors declare no conflict of interest. The funders had no role in the design of the study; in the collection, analyses, or interpretation of data; in the writing of the manuscript; or in the decision to publish the results.

Abbreviations

The following abbreviations are used in this manuscript:

Acronyms	
AM	active material
CC	constant current
CV	constant voltage
DoL	degree of lithiation
DVA	differential voltage analysis
GITT	galvanostatic intermittent titration technique
Li	lithium
LIB	lithium-ion battery
NCA	nickel-cobalt-aluminum-oxide
PITT	potentiostatic intermittent titration technique
pOCP	pulsed open-circuit potential
qOCP	quasi open-circuit potential
RMSE	root-mean-squared error
SEI	solid-electrolyte interphase
Si	silicon
SoC	state of charge
Roman Symbols	
C	capacity, Ah
c	concentration, mol m^{-3}
D	diffusivity, m^2 s^{-1}
$\frac{d \ln f \pm}{d \ln c_1(x,t)}$	activity, no unit
E_{eq}	equilibrium potential, V
$\mathcal{F}$	Faraday's constant, 96,485 A s mol^{-1}
i	current density, A m^{-2}
i_0	exchange current density, A m^{-2}
j_n	pore-wall flux, mol m^{-2} s^{-1}
k	reaction rate constant, m s^{-1}
L	through-plane thickness, m
N_M	MacMullin number, no unit
r	r-axis or r-dimension (pseudo dimension), m
$\mathcal{R}$	universal gas constant, 8.314 J mol^{-1} K^{-1}
R_p	particle radius, m
T	temperature, K
t	time, s
t_0^+	transference number, no unit
x	x-axis or x-dimension, m
Greek Symbols	
α	charge transfer coefficient, no unit
χ	stoichiometry, no unit
ε	volume fraction, no unit
η	overpotential, V
κ	conductivity, S m^{-1}
φ	electrical potential, V
ρ	density, kg m^{-3}

Subscripts & Superscripts

a	anodic
app	applied
c	cathodic
eff	effective
gr	gravimetric
l	liquid phase
max	maximum
ref	reference
s	solid phase
surf	surface
th	theoretical
tot	total

Appendix A. Model Equations and Parameters

Table A1. Equations for the p2D model. Here, ∇_x is the spatial gradient in the real dimension, i.e., along the through-plane direction of the cell stack, and ∇_r is the spatial gradient in the pseudo dimesion, i.e., along the radius of the spherical active material particles.

Spatial gradients	$\nabla_x = \frac{\partial}{\partial x}, \quad \nabla_r = \frac{\partial}{\partial r}$ (r in spherical pseudo dimension)	(A1)
Mass balance	$\varepsilon_l \frac{\partial c_l(x,t)}{\partial t} = \nabla_x \left(D_l^{\text{eff}}(c_l(x,t)) \nabla_x c_l(x,t) + \frac{i_l(x,t)(1-t_+^0)}{\mathcal{F}} \right)$	(A2)
Mass balance	$\frac{\partial c_s(x,t,r)}{\partial t} = \nabla_r (D_s \nabla_r c_s(x,t,r)) = \frac{1}{r^2} \frac{\partial}{\partial r} \left(D_s r^2 \frac{\partial c_s(x,t,r)}{\partial r} \right)$	(A3)
Potentials	$\nabla_x \varphi_l(x,t) = -\frac{i_l(x,t)}{\kappa_l^{\text{eff}}} + \frac{2\mathcal{R}T}{\mathcal{F}} (1 - t_+^0) \left(1 + \frac{d \ln f_\pm}{d \ln c_l(x,t)}\right) \nabla_x \ln c_l(x,t)$	(A4)
	$\nabla_x \varphi_s(x,t) = -\frac{i_s(x,t)}{\kappa_s}$ with $i_{\text{app}}(t) = i_s(x,t) + i_l(x,t) \quad \forall\, x,t$	(A5)
Charge balance	$\nabla_x i_l(x,t) + \nabla_x i_s(x,t) = 0$ with $\nabla_x i_s(x,t) = -\frac{3\varepsilon_{\text{AM}}}{R_p} \mathcal{F} j_n(x,t)$	(A6)
Electrode kinetics	$j_n(x,t) = \frac{i_0(x,t)}{\mathcal{F}} \left(\exp\left(\frac{\alpha_a \mathcal{F} \eta(x,t)}{\mathcal{R}T}\right) - \exp\left(-\frac{\alpha_c \mathcal{F} \eta(x,t)}{\mathcal{R}T}\right) \right)$	(A7)
	$\eta(x,t) = \varphi_s(x,t) - \varphi_l(x,t) - E_{\text{eq}}(x,t)$	(A8)
	$i_0(x,t) = \mathcal{F} k_c^{\alpha_a} k_a^{\alpha_c} (c_{\text{s,max}} - c_{\text{s,surf}})^{\alpha_a} (c_{\text{s,surf}})^{\alpha_c} \left(\frac{c_l}{c_{\text{l,ref}}}\right)^{\alpha_a}$	(A9)
Effective transport parameters	$\kappa_s^{\text{eff}} = \frac{\kappa_s}{N_M}, \quad \kappa_l^{\text{eff}} = \frac{\kappa_l}{N_M}, \quad D_l^{\text{eff}} = \frac{D_l}{N_M}$	(A10)
Boundary conditions	$\nabla_x c_l(x,t)\big|_{x=0\, \wedge\, x=L_{\text{tot}}} = 0$	(A11)
	$\nabla_x \varphi_l(x,t)\big|_{x=0\, \wedge\, x=L_{\text{tot}}} = 0$	(A12)
	$\varphi_s(x,t)\big|_{x=0} = 0$	(A13)
	$\nabla_x \varphi_s(x,t)\big|_{x=L_{\text{tot}}} = -\frac{i_{\text{app}}}{\kappa_s^{\text{eff}}}$	(A14)
	$\nabla_r c_s(x,t,r)\big|_{r=0} = 0$	(A15)
	$\nabla_r c_s(x,t,r)\big|_{r=R_p} = -\frac{j_n}{D_s}$	(A16)

Table A2. Parametrization of the p2D model with NCA | | Si electrodes.

	Anode	Separator	Cathode
Geometry			
Thickness coating L	46 µm [m,c]	2×260 µm [d]	68 µm [m,c]
d_{50} particle radius R_p	2.25 µm [L1]		3.0755 µm [d]
Active material volume fraction ε_{AM}	0.32 [c]		0.61 [c]
Electrolyte volume fraction (porosity) ε_l	0.50 [c]	0.55 [L8]	0.32 [c]
MacMullin number N_M	5.385 [c]	1.29 [L9,c]	7.333 [c]
Thermodynamics			
Equilibrium potential E_{eq}	Figure 1 [m]		Figure 1 [m]
Stoichiometry χ 100% SoC	0.3068 [f]		0.1396 [f]
0% SoC	0.0396 [f]		0.7954 [f]
Max. concentration $c_{s,max}$	322,067 mol m^{-3} [c,f]		46,400 mol m^{-3} [c,f]
Transport			
Solid diffusivity D_s	2×10^{-15} m^2 s^{-1} [L3]		1×10^{-14} m^2 s^{-1} [L7]
Electric conductivity κ_s	33 S m^{-1} [L3]		1 S m^{-1} [L6]
Kinetics			
Reaction rate constant k	5.78×10^{-12} m s^{-1} [L2,f]		1.64×10^{-10} m s^{-1} [L4,f]
Anodic charge transfer coefficient α_a	0.5 [a]		0.5 [a]
Cathodic charge transfer coefficient α_c	0.5 [a]		0.5 [a]
Electrolyte *			
Salt diffusivity D_l in m^2 s^{-1} *, [L5]		$10^{-4} \cdot 10^{-4.43-\frac{54}{T-(229+5c_l)}-0.22c_l}$	
Ionic conductivity κ_l in S m^{-1} *, [L5,f]		$0.5358 \cdot 10^{-1} c_l(-10.5 + 0.668c_l + 0.494c_l^2 + 0.0740T - 0.0178c_l T - 8.86 \cdot 10^{-4} c_l^2 T - 6.96 \cdot 10^{-5} T^2 + 2.8 \cdot 10^{-5} c_l T^2)^2$	
Activity $\frac{d \ln f_\pm}{d \ln c_l(x,t)}$ (no unit) *, [L5]		$\frac{(0.601-0.24c_l^{0.5}+0.982(1-0.0052(T-293.15))c_l^{1.5})}{1-t_+^0} - 1$	
Transference number t_+^0 [L5]		0.38	
Reference concentration $c_{l,ref}$ [a]		1 mol m^{-3}	

* Analytic functions for the electrolyte are physically not interpretable. Input arguments are to be normalized (concentration c_l per mol L^{-1} and temperature T per K). The functions are only valid for $c_l \in [0\,\text{mol}\,\text{L}^{-1}; \leq 4\,\text{mol}\,\text{L}^{-1}]$. The output value is to be interpreted in the unit given in the table. [a] assumption; [c] calculated; [d] from supplier data sheet; [f] fitted, based on data sheet ([d,f]) or based on the literature ([L,f]); [L] from the literature; [L1] [13]; [L2] [20]; [L3] [37]; [L4] [40]; [L5] [46]; [L6] [49]; [L7] [53]; [L8] [64]; [L9] [65]; [m] measured.

Appendix B. Additional Measurement Results

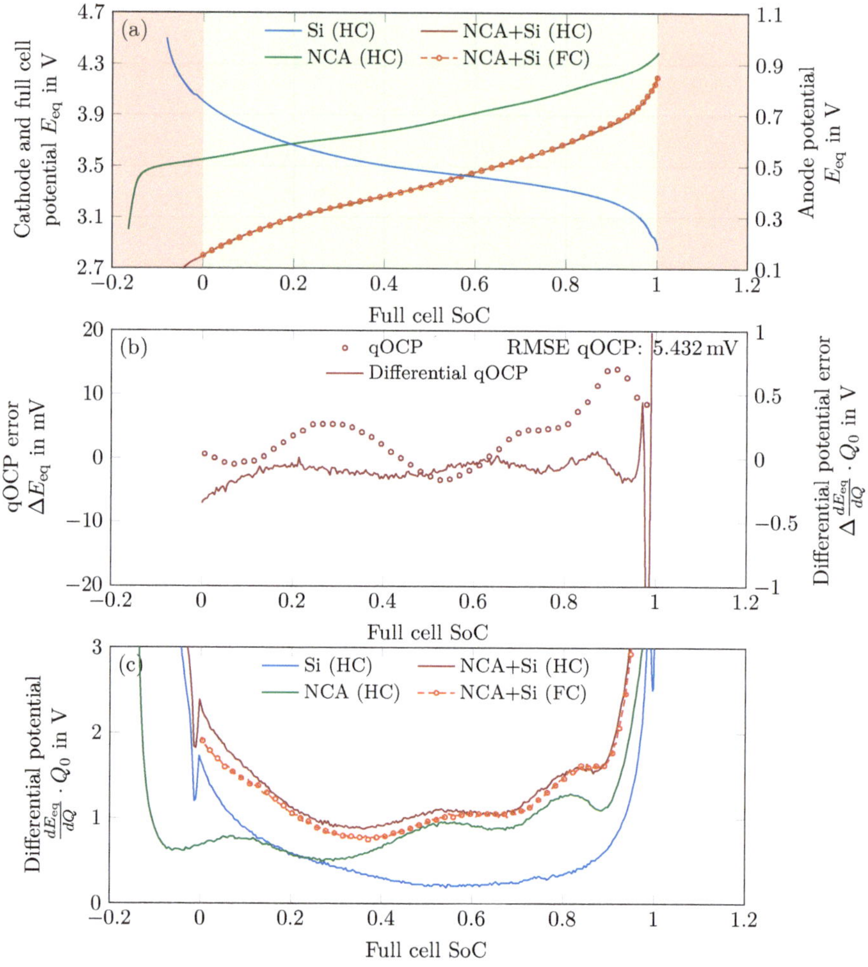

Figure A1. Approximation of the quasi open-circuit potential (qOCP) under constant current discharge at C/50 and 25 °C. Differential voltage analysis (**c**) was used to reconstruct the measured qOCP vs. full-cell SoC (**a**) under discharge of the full cell (see NCA+Si (FC)) via the (de)lithiation paths of the respective half cells (see NCA (HC) and Si (HC)). The approximation errors vs. SoC as well as the RMSE of the qOCP are shown in subplot (**b**)).

References

1. Scrosati, B. History of lithium batteries. *J. Solid State Electrochem.* **2011**, *15*, 1623–1630. [CrossRef]
2. Yoshino, A. The birth of the lithium-ion battery. *Angew. Chem. Int. Ed. Engl.* **2012**, *51*, 5798–5800. [CrossRef] [PubMed]
3. Zhang, J.G.; Wang, W.; Xiao, J.; Xu, W.; Graff, G.L.; Yang, G.; Choi, D.; Wang, D.; Li, X.; Liu, J. Silicon-Based Anodes for Li-ion Batteries. In *Encyclopedia of Sustainability Science and Technology*; Meyers, R.A., Ed.; Springer: New York, NY, USA, 2012; pp. 9293–9316. [CrossRef]
4. Wetjen, M.; Pritzl, D.; Jung, R.; Solchenbach, S.; Ghadimi, R.; Gasteiger, H.A. Differentiating the Degradation Phenomena in Silicon-Graphite Electrodes for Lithium-Ion Batteries. *J. Electrochem. Soc.* **2017**, *164*, A2840–A2852. [CrossRef]
5. Wetjen, M.; Solchenbach, S.; Pritzl, D.; Hou, J.; Tileli, V.; Gasteiger, H.A. Morphological Changes of Silicon Nanoparticles and the Influence of Cutoff Potentials in Silicon-Graphite Electrodes. *J. Electrochem. Soc.* **2018**, *165*, A1503–A1514. [CrossRef]
6. Xu, Q.; Wang, Q.; Chen, D.; Zhong, Y.; Wu, Z.; Song, Y.; Wang, G.; Liu, Y.; Zhong, B.; Guo, X. Silicon/graphite composite anode with constrained swelling and a stable solid electrolyte interphase enabled by spent graphite. *Green Chem.* **2021**, *23*, 4531–4539. [CrossRef]

7. Sturm, J.; Rheinfeld, A.; Zilberman, I.; Spingler, F.B.; Kosch, S.; Frie, F.; Jossen, A. Modeling and simulation of inhomogeneities in a 18650 nickel-rich, silicon-graphite lithium-ion cell during fast charging. *J. Power Sources* **2019**, *412*, 204–223. [CrossRef]
8. Schmitt, J.; Schindler, M.; Jossen, A. Change in the half-cell open-circuit potential curves of silicon–graphite and nickel-rich lithium nickel manganese cobalt oxide during cycle aging. *J. Power Sources* **2021**, *506*, 230240. [CrossRef]
9. Schindler, M.; Sturm, J.; Ludwig, S.; Durdel, A.; Jossen, A. Comprehensive Analysis of the Aging Behavior of Nickel-Rich, Silicon-Graphite Lithium-Ion Cells Subject to Varying Temperature and Charging Profiles. *J. Electrochem. Soc.* **2021**, *168*, 060522. [CrossRef]
10. Ko, M.; Chae, S.; Cho, J. Challenges in Accommodating Volume Change of Si Anodes for Li-Ion Batteries. *ChemElectroChem* **2015**, *2*, 1645–1651. [CrossRef]
11. Jantke, D.; Bernhard, R.; Hanelt, E.; Buhrmester, T.; Pfeiffer, J.; Haufe, S. Silicon-Dominant Anodes Based on Microscale Silicon Particles under Partial Lithiation with High Capacity and Cycle Stability. *J. Electrochem. Soc.* **2019**, *166*, A3881–A3885. [CrossRef]
12. Obrovac, M.N.; Krause, L.J. Reversible Cycling of Crystalline Silicon Powder. *J. Electrochem. Soc.* **2007**, *154*, A103. [CrossRef]
13. Haufe, S.; Bernhard, R.; Pfeiffer, J. Revealing the Failure Mechanism of Partially Lithiated Silicon-Dominant Anodes Based on Microscale Silicon Particles. *J. Electrochem. Soc.* **2021**, *168*, 080531. [CrossRef]
14. Doyle, M.; Fuller, T.F.; Newman, J. Modeling of Galvanostatic Charge and Discharge of the Lithium/Polymer/Insertion Cell. *J. Electrochem. Soc.* **1993**, *140*, 1526–1533. [CrossRef]
15. Fuller, T.F.; Doyle, M.; Newman, J. Relaxation Phenomena in Lithium–Ion–Insertion Cells. *J. Electrochem. Soc.* **1994**, *141*, 982–990. [CrossRef]
16. Rheinfeld, A.; Sturm, J.; Noel, A.; Wilhelm, J.; Kriston, A.; Pfrang, A.; Jossen, A. Quasi-Isothermal External Short Circuit Tests Applied to Lithium-Ion Cells: Part II. Modeling and Simulation. *J. Electrochem. Soc.* **2019**, *166*, A151–A177. [CrossRef]
17. Dees, D.W.; Battaglia, V.S.; Bélanger, A. Electrochemical modeling of lithium polymer batteries. *J. Power Sources* **2002**, *110*, 310–320. [CrossRef]
18. Jiang, Y.; Offer, G.; Jiang, J.; Marinescu, M.; Wang, H. Voltage Hysteresis Model for Silicon Electrodes for Lithium Ion Batteries, Including Multi-Step Phase Transformations, Crystallization and Amorphization. *J. Electrochem. Soc.* **2020**, *167*, 130533. [CrossRef]
19. Lory, P.F.; Mathieu, B.; Genies, S.; Reynier, Y.; Boulineau, A.; Hong, W.; Chandesris, M. Probing Silicon Lithiation in Silicon-Carbon Blended Anodes with a Multi-Scale Porous Electrode Model. *J. Electrochem. Soc.* **2020**, *167*, 120506. [CrossRef]
20. Swamy, T.; Chiang, Y.M. Electrochemical Charge Transfer Reaction Kinetics at the Silicon-Liquid Electrolyte Interface. *J. Electrochem. Soc.* **2015**, *162*, A7129–A7134. [CrossRef]
21. Wang, M.; Xiao, X.; Huang, X. Study of lithium diffusivity in amorphous silicon via finite element analysis. *J. Power Sources* **2016**, *307*, 77–85. [CrossRef]
22. Chandrasekaran, R.; Magasinski, A.; Yushin, G.; Fuller, T.F. Analysis of Lithium Insertion/Deinsertion in a Silicon Electrode Particle at Room Temperature. *J. Power Sources* **2010**, *157*, A1139. [CrossRef]
23. Chandrasekaran, R.; Fuller, T.F. Analysis of the Lithium-Ion Insertion Silicon Composite Electrode/Separator/Lithium Foil Cell. *J. Electrochem. Soc.* **2011**, *158*, A859. [CrossRef]
24. Golmon, S.; Maute, K.; Lee, S.H.; Dunn, M.L. Stress generation in silicon particles during lithium insertion. *Appl. Phys. Lett.* **2010**, *97*, 033111. [CrossRef]
25. Wang, M.; Xiao, X. Investigation of the chemo-mechanical coupling in lithiation/delithiation of amorphous Si through simulations of Si thin films and Si nanospheres. *J. Power Sources* **2016**, *326*, 365–376. [CrossRef]
26. Wang, M.; Xiao, X.; Huang, X. A multiphysics microstructure-resolved model for silicon anode lithium-ion batteries. *J. Power Sources* **2017**, *348*, 66–79. [CrossRef]
27. Dhillon, S.; Hernández, G.; Wagner, N.P.; Svensson, A.M.; Brandell, D. Modelling capacity fade in silicon-graphite composite electrodes for lithium-ion batteries. *Electrochim. Acta* **2021**, *377*, 138067. [CrossRef]
28. Dasari, H.; Eisenbraun, E. Predicting Capacity Fade in Silicon Anode-Based Li-Ion Batteries. *Energies* **2021**, *14*, 1448. [CrossRef]
29. Lu, B.; Song, Y.; Zhang, Q.; Pan, J.; Cheng, Y.T.; Zhang, J. Voltage hysteresis of lithium ion batteries caused by mechanical stress. *Phys. Chem. Chem. Phys. PCCP* **2016**, *18*, 4721–4727. [CrossRef]
30. Chevrier, V.L.; Dahn, J.R. First Principles Model of Amorphous Silicon Lithiation. *J. Electrochem. Soc.* **2009**, *156*, A454. [CrossRef]
31. McDowell, M.T.; Lee, S.W.; Harris, J.T.; Korgel, B.A.; Wang, C.; Nix, W.D.; Cui, Y. In situ TEM of two-phase lithiation of amorphous silicon nanospheres. *Nano Lett.* **2013**, *13*, 758–764. [CrossRef]
32. Bordes, A.; de Vito, E.; Haon, C.; Boulineau, A.; Montani, A.; Marcus, P. Multiscale Investigation of Silicon Anode Li Insertion Mechanisms by Time-of-Flight Secondary Ion Mass Spectrometer Imaging Performed on an In Situ Focused Ion Beam Cross Section. *Chem. Mater.* **2016**, *28*, 1566–1573. [CrossRef]
33. Chockla, A.M.; Harris, J.T.; Akhavan, V.A.; Bogart, T.D.; Holmberg, V.C.; Steinhagen, C.; Mullins, C.B.; Stevenson, K.J.; Korgel, B.A. Silicon nanowire fabric as a lithium ion battery electrode material. *J. Am. Chem. Soc.* **2011**, *133*, 20914–20921. [CrossRef]
34. McDowell, M.T.; Cui, Y. Single Nanostructure Electrochemical Devices for Studying Electronic Properties and Structural Changes in Lithiated Si Nanowires. *Adv. Energy Mater.* **2011**, *1*, 894–900. [CrossRef]
35. Pollak, E.; Salitra, G.; Baranchugov, V.; Aurbach, D. In Situ Conductivity, Impedance Spectroscopy, and Ex Situ Raman Spectra of Amorphous Silicon during the Insertion/Extraction of Lithium. *J. Phys. Chem. C* **2007**, *111*, 11437–11444. [CrossRef]

36. Kim, J.W.; Ryu, J.H.; Lee, K.T.; Oh, S.M. Improvement of silicon powder negative electrodes by copper electroless deposition for lithium secondary batteries. *J. Power Sources* **2005**, *147*, 227–233. [CrossRef]
37. Wang, M. Simulation of Amorphous Silicon Anode in Lithium-Ion Batteries. Ph.D. Thesis, Michigan State University, East Lansing, MI, USA, 2017.
38. Bucci, G.; Nadimpalli, S.P.; Sethuraman, V.A.; Bower, A.F.; Guduru, P.R. Measurement and modeling of the mechanical and electrochemical response of amorphous Si thin film electrodes during cyclic lithiation. *J. Mech. Phys. Solids* **2014**, *62*, 276–294. [CrossRef]
39. Sivonxay, E.; Aykol, M.; Persson, K.A. The lithiation process and Li diffusion in amorphous SiO_2 and Si from first-principles. *Electrochim. Acta* **2020**, *331*, 135344. [CrossRef]
40. Dees, D.W.; Gallagher, K.G.; Abraham, D.P.; Jansen, A.N. Electrochemical Modeling the Impedance of a Lithium-Ion Positive Electrode Single Particle. *J. Electrochem. Soc.* **2013**, *160*, A478–A486. [CrossRef]
41. Li, J.; Xiao, X.; Yang, F.; Verbrugge, M.W.; Cheng, Y.T. Potentiostatic Intermittent Titration Technique for Electrodes Governed by Diffusion and Interfacial Reaction. *J. Phys. Chem. C* **2012**, *116*, 1472–1478. [CrossRef]
42. Sethuraman, V.A.; Srinivasan, V.; Newman, J. Analysis of Electrochemical Lithiation and Delithiation Kinetics in Silicon. *J. Electrochem. Soc.* **2013**, *160*, A394–A403. [CrossRef]
43. Guidelli, R.; Compton, R.G.; Felio, J.M.; Gileadi, E.; Lipkowski, J.; Schmickler, W. Defining the transfer coefficient in electrochemistry: An assessment (IUPAC Technical Report). *Pure Appl. Chem.* **2014**, *86*, 245–258. [CrossRef]
44. Baggetto, L.; Niessen, R.A.H.; Roozeboom, F.; Notten, P.H.L. High Energy Density All-Solid-State Batteries: A Challenging Concept Towards 3D Integration. *Adv. Funct. Mater.* **2008**, *18*, 1057–1066. [CrossRef]
45. Tjandra, R.; Thanagasundram, S.; Tseng, K.J.; Jossen, A. Improved lithium-ion battery model with hysteresis effect. In Proceedings of the 2014 IEEE Transportation Electrification Conference and Expo (ITEC), Dearborn, MI, USA, 15–18 June 2014; IEEE: Piscataway, NJ, USA, 2014; pp. 1–8. [CrossRef]
46. Valøen, L.O.; Reimers, J.N. Transport Properties of $LiPF_6$-Based Li-Ion Battery Electrolytes. *J. Electrochem. Soc.* **2005**, *152*, A882–A891. [CrossRef]
47. Marinho, B.; Ghislandi, M.; Tkalya, E.; Koning, C.E.; de With, G. Electrical conductivity of compacts of graphene, multi-wall carbon nanotubes, carbon black, and graphite powder. *Powder Technol.* **2012**, *221*, 351–358. [CrossRef]
48. Amin, R.; Ravnsbæk, D.B.; Chiang, Y.M. Characterization of Electronic and Ionic Transport in $Li_{1-x}Ni_{0.8}Co_{0.15}Al_{0.05}O_2$ (NCA). *J. Electrochem. Soc.* **2015**, *162*, A1163–A1169. [CrossRef]
49. Ashwin, T.R.; McGordon, A.; Widanage, W.D.; Jennings, P.A. Modified electrochemical parameter estimation of NCR18650BD battery using implicit finite volume method. *J. Power Sources* **2017**, *341*, 387–395. [CrossRef]
50. Grundmann, M. *The Physics of Semiconductors*; Springer International Publishing: Cham, Switzerland, 2016. [CrossRef]
51. Pearson, G.L.; Bardeen, J. Electrical Properties of Pure Silicon and Silicon Alloys Containing Boron and Phosphorus. *Phys. Rev.* **1949**, *75*, 865–883. [CrossRef]
52. Fan, S.; Plascencia, G.; Utigard, T. High Temperature Electric Conductivity of Pure Silicon. *Can. Metall. Q.* **2008**, *47*, 509–512. [CrossRef]
53. Dees, D.W.; Kawauchi, S.; Abraham, D.P.; Prakash, J. Analysis of the Galvanostatic Intermittent Titration Technique (GITT) as applied to a lithium-ion porous electrode. *J. Power Sources* **2009**, *189*, 263–268. [CrossRef]
54. Park, J.H.; Yoon, H.; Cho, Y.; Yoo, C.Y. Investigation of Lithium Ion Diffusion of Graphite Anode by the Galvanostatic Intermittent Titration Technique. *Materials* **2021**, *14*, 4683. [CrossRef]
55. Tsai, P.C.; Wen, B.; Wolfman, M.; Choe, M.J.; Pan, M.S.; Su, L.; Thornton, K.; Cabana, J.; Chiang, Y.M. Single-particle measurements of electrochemical kinetics in NMC and NCA cathodes for Li-ion batteries. *Energy Environ. Sci.* **2018**, *11*, 860–871. [CrossRef]
56. Chan, C.K.; Peng, H.; Liu, G.; McIlwrath, K.; Zhang, X.F.; Huggins, R.A.; Cui, Y. High-performance lithium battery anodes using silicon nanowires. *Nat. Nanotechnol.* **2008**, *3*, 31–35. [CrossRef] [PubMed]
57. Pan, K.; Zou, F.; Canova, M.; Zhu, Y.; Kim, J.H. Systematic electrochemical characterizations of Si and SiO anodes for high-capacity Li-Ion batteries. *J. Power Sources* **2019**, *413*, 20–28. [CrossRef]
58. Smith, A.J.; Dahn, H.M.; Burns, J.C.; Dahn, J.R. Long-Term Low-Rate Cycling of $LiCoO_2$ /Graphite Li-Ion Cells at 55 °C. *J. Electrochem. Soc.* **2012**, *159*, A705–A710. [CrossRef]
59. Dahn, H.M.; Smith, A.J.; Burns, J.C.; Stevens, D.A.; Dahn, J.R. User-Friendly Differential Voltage Analysis Freeware for the Analysis of Degradation Mechanisms in Li-Ion Batteries. *J. Electrochem. Soc.* **2012**, *159*, A1405–A1409. [CrossRef]
60. Xie, W.; Yang, S. Charging Optimization of Lithium-Ion Batteries Based on Charge Transfer Limitation and Mass Transport Limitation. *J. Electrochem. Soc.* **2023**, *170*, 010506. [CrossRef]
61. Li, J.; Dahn, J.R. An In Situ X-Ray Diffraction Study of the Reaction of Li with Crystalline Si. *J. Power Sources* **2007**, *154*, A156. [CrossRef]
62. Graf, M.; Berg, C.; Bernhard, R.; Haufe, S.; Pfeiffer, J.; Gasteiger, H.A. Effect and Progress of the Amorphization Process for Microscale Silicon Particles under Partial Lithiation as Active Material in Lithium-Ion Batteries. *J. Electrochem. Soc.* **2022**, *169*, 020536. [CrossRef]
63. Nitta, N.; Wu, F.; Lee, J.T.; Yushin, G. Li-ion battery materials: Present and future. *Mater. Today* **2015**, *18*, 252–264. [CrossRef]

64. Landesfeind, J.; Hattendorff, J.; Ehrl, A.; Wall, W.A.; Gasteiger, H.A. Tortuosity Determination of Battery Electrodes and Separators by Impedance Spectroscopy. *J. Electrochem. Soc.* **2016**, *163*, A1372–A1387. [CrossRef]
65. Habedank, J.B.; Kraft, L.; Rheinfeld, A.; Krezdorn, C.; Jossen, A.; Zaeh, M.F. Increasing the Discharge Rate Capability of Lithium-Ion Cells with Laser-Structured Graphite Anodes: Modeling and Simulation. *J. Electrochem. Soc.* **2018**, *165*, A1563–A1573. [CrossRef]

nanomaterials

Article

Spray-Flame Synthesis of NASICON-Type Rhombohedral (α) $Li_{1+x}Y_xZr_{2-x}(PO_4)_3$ [x = 0–0.2] Solid Electrolytes

Md Yusuf Ali [1,†], Tianyu Chen [1,†], Hans Orthner [1] and Hartmut Wiggers [1,2,*]

1 Institute for Energy and Materials Processes—Reactive Fluids, University of Duisburg-Essen, 47057 Duisburg, Germany; yusuf.ali@uni-due.de (M.Y.A.); tianyu-chen@marubeni.com (T.C.); hans.orthner@uni-due.de (H.O.)
2 CENIDE, Center for Nanointegration Duisburg-Essen, 47057 Duisburg, Germany
* Correspondence: hartmut.wiggers@uni-due.de
† These authors contributed equally to this work.

Abstract: Since solid electrolytes have a broad electrochemical stability window, are exceptionally electrochemically stable against Li metal, and function as a physical separator to prevent dendrite growth, they are at the forefront of alternate possibilities, further increasing the stability and energy density of Li-ion batteries. NASICON-type electrolytes are a promising candidate due to their negligible moisture sensitivity, which results in outstanding stability and a lower probability of Li_2CO_3 passivity under the ambient atmosphere. However, one of the most promising representatives, $Li_{1+x}Y_xZr_{2-x}(PO_4)_3$ (LYZP), has multiple stable phases with significant variation in their corresponding Li-ion conductivity. In this paper, we have successfully synthesized the highly ionically conductive rhombohedral phase of LYZP via spray-flame synthesis. Two different solvent mixtures (e.g., 2-ethyl hexanoic acid/ethanol, propanol/propanoic acid) were chosen to explore the effect of precursor composition and combustion enthalpy on the phase composition of the nanoparticle. The as-synthesized nanoparticles from spray-flame synthesis consisted of the crystalline tetragonal zirconia (t-ZrO_2) phase, while lithium, yttrium, and phosphate were present on the nanoparticles' surface as amorphous phases. However, a short annealing step (1 h) was sufficient to obtain the NASICON phase. Moreover, we have shown the gradual phase conversion from orthorhombic β phase to rhombohedral α phase as the annealing temperature increased from 700 °C to 1300 °C (complete removal of β phase). In this context, Y^{3+} doping was also crucial, along with the appropriate solvent mixture and annealing temperature, for obtaining the much-desired rhombohedral α phase. Further, 0.2 at% Y^3+ doping was added to the solvent mixture of 2-ethyl hexanoic acid/ethanol, and annealing at 1300 °C for 1 h resulted in a high ionic conductivity of $1.14 \cdot 10^{-5}$ S cm^{-1}.

Keywords: spray-flame synthesis; solid-state electrolyte; NASICON type $Li_{1+x}Y_xZr_{2-x}(PO_4)_3$; nanoparticles; Li-ion batteries

Citation: Ali, M.Y.; Chen, T.; Orthner, H.; Wiggers, H. Spray-Flame Synthesis of NASICON-Type Rhombohedral (α) $Li_{1+x}Y_xZr_{2-x}(PO_4)_3$ [x = 0–0.2] Solid Electrolytes. *Nanomaterials* **2024**, *14*, 1278. https://doi.org/10.3390/nano14151278

Academic Editor: Jun-ho Yum

Received: 28 June 2024
Revised: 18 July 2024
Accepted: 24 July 2024
Published: 30 July 2024

1. Introduction

Electrical energy has become an indispensable part of our modern civilization. However, the significant origin of the same is still from nonrenewable sources. Moreover, the energy market will remain volatile in the future due to constant depletion and geopolitical instabilities [1,2]. Consequently, sustainable energy resources such as wind, solar, bioenergy, and hydropower come to the center of the stage [3–5]. Nevertheless, no matter how "low carbon" our market is, or furthermore, no matter how advanced and considerate our energy management system is [6], the fluctuation of energy usage plays a part in Achilles' heel. Amongst systems in the field of battery energy storage [7], solutions for serving energy support based on rechargeable batteries, mainly lithium-ion batteries (LIBs), have become the spotlight of this stage [7]. In this context, energy density, storage capacity, and thermal sustainability play an important role in establishing the next-best energy storage system. Since the term LIB has been available in the market since the late

1960s [8] and successfully commercialized in the 1990s [8,9], LIB storage systems have gained increasing interest in numerous application scenarios. However, employing high-energy-density Li metal anodes with conventional liquid or polymer electrolytes is not considered a 'termination' of lithium anodes for high-energy batteries. Regarding strict requirements of heat management [10–12], solid electrolytes (SEs) (e.g., inorganic [13–15], polymer [16], and hybrid composites [17], etc.) are believed to overcome obstacles due to their robust mechanical strength and high Li-ion transference number [18–21]. In addition, compared with liquid organic electrolytes, SEs have a wide electrochemical potential window of up to 5 V for battery cycling [22], which could empower the electric vehicle revolution and stationary demands. Employing solid composite electrolytes (SCE) could eliminate polymer-based mechanical separators, avoiding the risk of flammability and Li dendrite growth [23]. Moreover, the good wetting ability of solid electrolytes by the metallic lithium anode can reduce or even avoid the lithium dendrite formation in all-solid-state Li-metal batteries, resulting in high energy-density batteries [24]. Amongst many inorganic solid electrolytes, perovskite-type $Li_{3x}La_{2/3-x}TiO_3$ (LLTO) [25], garnet-type $Li_7La_3Zr_2O_{12}$ (LLZO) [26], NASICON-type $Li_{1+x}Al_xTi_{2-x}(PO_4)_3$ (LATP) [27], and anti-perovskite structure Li_2OHX (X = Cl, Br) [28] exhibit high Li-ion conductivity up to >10^{-4} S·cm^{-1} at room temperature. However, there are also challenges when using inorganic solid electrolytes. Y. Shimonishi et al. and Y. Li et al. [29,30] demonstrated the instability of garnet-type SEs in moist air (e.g., $Li_{7-x}La_3Zr_2O_{12-(1/2x)}$). So is the anti-perovskite structure of Li_3OB formation [31]. The chemical reaction between those SEs and moisture causes structural damage to solid electrolytes [32]. Consequently, the Li-ion conductivity of ceramics decreases, resulting in a significant increase in Li-ion interface impedance. On the other hand, NASICON-type and perovskite-type oxides containing Ti^{4+} have shown high stability in water [33].

NASICON is an acronym for sodium (Na) super-ionic conductor with the chemical formula $A_1B_2(PO_4)_3$, where the A site is occupied by a monovalent cation and the B site is either a single tetravalent, or a combination of tri, tetra, and pentavalent ions [34]. One potential Li-ion conducting candidate of the NASICON family is $LiZr_2(PO_4)_3$ (LZO), which is electrochemically stable in contact with lithium metal [35]. The ionic conduction of LZO is lower than that of its analogs, $LiGe_2(PO_4)_3$ (LGP) and $LiTi_2(PO_4)$ (LTP). It is a known fact that due to Ti^{4+} reduction in the presence of a Li anode, the use of LTP is very limited [36,37]. On the other hand, the cost of Ge precursors is generally relatively high, which makes affordable SEs obsolete [38]. However, Zr^{4+} demonstrates good chemical stability against a Li anode [39], as the energy of the Zr^{4+}/Zr^0 in $LiZr_2(PO_4)_3$ is above the Fermi level of Li metal [40]. An obstacle to NASICON-type LZO is its polymorphism with four crystalline forms (α and α', β and β' phase) [41–43]. Replacing tetravalent Zr^{4+} with trivalent metal ions (e.g., Y^{3+}) can influence the stability of lithium metal and increase the ionic conductivity above 10^{-5} S·cm^{-1} in accounting for mobile ions [44]. It has been shown that, depending upon the synthesis route and reaction temperature, NASICON-type LZO undergoes considerable phase transformations. According to Li et al. [44], the rhombohedral α phase (space group R-3c) transfers to the triclinic α' phase at temperatures below 60 °C. On the other hand, the orthorhombic β phase transfers to the monolithic β' phase at a higher temperature (~300 °C). The goal of this work is to synthesize the rhombohedral α phase, which is stable at room temperature with the highest ionic conductivity (above 10^{-5} S cm^{-1}) compared to its polymorph phases (e.g., triclinic α'), and to avoid the presence of the β phase.

Synthesis methods involving multiple pre- and post-processing steps are primary in determining SE characteristics. Nano-size materials have been shown to have size-induced high ionic conductivity and a high surface area with excellent mechanical properties [26,45]. The rising interest in nanosolid electrolytes is prompting the search for better synthesis technology. The commonly used high-temperature solid-state reaction of suitable starting materials requires an excellent mixture to form a homogeneous precipitation and, consequently, a time-consuming calcination and/or sintering process at over 1000 °C [46,47].

Nevertheless, it might result in undesired powder due to using course materials in the micrometer-size range [48,49]. In contrast, the sol-gel method based on dissolved precursors with subsequent annealing doesn't require that high energy; however, impurities in other phases seem inevitable [50]. Accordingly, the performance of the final product [51] is strongly influenced by the starting material [52,53]. Therefore, a highly versatile and flexible production process with high throughput is required that is also capable of providing materials with high purity. Hence, the choice of a suitable synthesis route is, therefore, decisive for reliable nanoparticle production that enables the manufacture of high-quality materials.

As an inexpensive manufacturer of ceramics, especially ceramics at the nanoscale, flame synthesis technology has shown its distinguishing characteristics [54]. Considering the homogeneity of particles and the predetermination of the composition of multinary metal oxides, we focus on gas-phase-based spray-flame synthesis (SFS). By applying a sprayer or an atomizer [55], the limited utilization of volatile and cost-effective vaporizable precursors such as organic metal salts is unlocked. Since the successful establishment of SFS in 2002 [56], this route has gained increasing interest. SFS has proven its versatile module application sharing routes with plasma, laser, and electrically heated wall reactors [57–59]. The merits of applying for SFS are self-explanatory. In some cases, the properties of materials in combination with a suitable thermal post-treatment, such as control over particle size (e.g., nm to μm), crystallinity, morphology and shape, and phase composition of products, can be well controlled [60]. Unlike batch processes, i.e., time-consuming sol-gel synthesis requiring pre- and post-processing steps such as gelation and drying, SFS based on spraying a solution containing dissolved precursors allows fast and continuous production.

In this work, our focus was to synthesize rhombohedral α phase NASICON $Li_{1+x}Y_xZr_{2-x}(PO_4)_3$ (LYZP) with the highest phase purity possible. Due to the benefits stated above, SFS was chosen as the synthesis route. As this is the very first (to the best of our knowledge) report to synthesize NASICON-type LYZP nanoparticles via SFS, based on literature [61,62], we investigated their synthesis using two different solvent mixtures that have been used in the past for spray-flame synthesis of metal oxides, i.e., propanol/propionic acid (PrOH/PA) and ethanol/2-ethyl hexanoic acid (EtOH/2-EHA). Furthermore, we also optimized Y^{3+} doping in LYZP to observe the effect of doping on phase purity and subsequent ionic conductivity.

2. Materials and Methods

In our prior experience [63], the Zr-based system does not directly produce a multicomponent final product by SFS. This is probably due to the first nucleation of ZrO_2 in the vapor phase rather than the rest of the elements. As a result, the remaining elements (Y, Li, P) are generally present in a synthesized sample as an amorphous phase on ZrO_2 nanoparticles. Furthermore, due to the similar phase transformation (cubic to monoclinic) of ZrO_2 at elevated temperatures, the successful stabilization of the cubic phase of ZrO_2 at room temperature was accomplished by Y^{3+} doping [64,65]. It has been widely studied that Y^{3+} ions replace Zr^{4+} at the cationic sublattice [65], resulting in oxygen vacancies. This directly affects the oxygen ion conductivity of ZrO_2 and makes it appropriate for applications such as solid oxide fuel cells and sensors [66]. The stated observations are one of the major factors in synthesizing Y^{3+} undoped (LZP) and doped LZP ($Li_{1+x}Y_xZr_{2-x}(PO_4)_3$ with x = 0.1 ($LY_{0.1}ZP$) and 0.2 ($LY_{0.2}ZP$), respectively), to not only effect the desired phase compositions but also alleviate the ionic conductivity. Like the conventional synthesis routes [41,44,67,68], SFS also has a substantial and reasonable combination of operating parameters. Regarding the following criteria: (i) solubility of precursors in solution; (ii) melting/boiling points of precursors and solvents; and (iii) costs of precursors and solvents, we decided on suitable precursors and solvents. $LiNO_3$ and $Y(NO_3)_3$ were applied as the sources of lithium and yttrium since they are both miscible in organic solvents and inexpensive. According to Lieber et al. [69], a higher metal precursor melting/decomposition temperature point relative to the solvent boiling point, i.e., T_{bp} (solvent)/$T_{d/mp}$(precursor) less than one generates inhomogeneous particles. Moreover, their ratio values between the boiling point/decomposition of solvents (e.g., T_{bp} of 2-EHA is 217 °C) and melting point of precursors (e.g., T_{mp} of $Li(NO_3)_3$ is 264 °C, T_{mp} of $Y(NO_3)_3$ is 52 °C, Table S1) satisfy the criterion to some extent for the synthesis of homogenous particles at the nanoscale [70].

Tributyl phosphate is favored due to its low cost and solubility in our preferred solvents. As for the Zirconium precursors, a list of suitable candidates concludes zirconium nitrate ($Zr(NO_3)_4$, abbreviated as "ZN"; decomposes at 100 °C), zirconium acetate ($Zr^{x+} \cdot xH_3CCOOH$, abbreviated as "ZA", 16 wt.% diluted in acetic acid), and zirconium(IV) propoxide ($Zr(OCH_2CH_2CH_3)_4$, abbreviated as "ZP", 70 wt.% in 1-propanol; T_{bp} = 208 °C). In this study, we chose ZP as a Zr precursor because, unlike ZN, can cause explosions [71] while ZP is relatively safer to use. Although metal acetate precursors are verified to result in small nanoparticles [72], ZA, only obtainable in dissolved and thus diluted form (15.0–17.0 wt% in acetic acid), has some drawbacks regarding the energy supply for combustion (low enthalpy of combustion of acetic acid with 15.32 kJ/mL). Considering all the above arguments, ZP has the highest potential regarding the above-mentioned criteria, while ZN and ZA were discarded.

As briefly stated above, depending upon the melting point/boiling point of the solvent and solute, the solvent mixture is expected to affect the synthesized particles (e.g., size, shape, surface characteristics, etc.). In this study, the chosen precursor solutions are denoted as $(LY_xZP)_{PA50}$ (x = 0.2) involving solvent mixture 'A', i.e., propanol and propionic acid (1:1 by volume), and $(LY_xZP)_{EA50}$ (x = 0.2) involving solvent mixture 'B', i.e., ethanol and 2-ethylhexanoic acid (1:1 by volume), as shown in Table 1. To study the maximum effect of Y doping in SFS and subsequent calcination, x = 0.2 (i.e., Y content at 20%) was chosen.

Table 1. Scheme of precursors and solvents and their corresponding nomenclature.

Nomenclature	Precursors				Solvents	
	Li	Y	Zr	P	(A)	(B)
	$LiNO_3$ (50% excess Li)	$Y(NO_3)_3 \cdot 6H_2O$	ZP	TBP	Propanol/Propionic Acid (PrOH/PA) *V*/*V*	Ethanol/2-Ethylhexanoic Acid (EtOH/2-EHA) *V*/*V*
$(LZP)_{EA50}$	✓	0	✓	✓	×	1:1
$(LY_{0.1}ZP)_{EA50}$	✓	0.1	✓	✓	×	1:1
$(LY_{0.2}ZP)_{PA50}$	✓	0.2	✓	✓	1:1	×
$(LY_{0.2}ZP)_{EA50}$	✓	0.2	✓	✓	×	1:1
$(LY_{0.2}ZP)_{EA70}$	✓	0.2	✓	✓	×	3:7

Precursors were used as supplied: $LiNO_3$ (VWR, ≥99.0% purity; Leuven, Belgium), $Y(NO_3)_3 \cdot 6H_2O$ (Aldrich, ≥99.8% purity; Darmstadt, Germany), Zirconium-tetra-propoxide (ZP, 70 wt% in 1-propanol, Sigma Aldrich; Darmstadt, Germany), and tributyl phosphate ($(C_4H_9)_3PO$, TBP, Sigma Aldrich, ≥99% purity; Darmstadt, Germany). Solvent mixtures (A) and (B) were prepared from 2-propanol (PrOH, BASF, ≥99.5% purity; Ludwigshafen, Germany) and propionic acid (PA, Acros Organics, ≥99% purity; Geel, Belgium), ethanol (EtOH, VWR, ≥99.9% purity; Leuven, Belgium), and 2-ethylhexanoic acid (2-EHA, Alfa Aesar, ≥99% purity; Haverhill, MA, USA), respectively. Moreover, to compensate for the loss of lithium during synthesis and subsequent annealing [73], lithium content was increased by 50 wt.% in all experiments. The total precursor concentration was 0.5 M in all cases.

The reactor used in this research work has been described in our previous work [63]. Details of gas flow parameters and reaction pressure are listed in Table 2. The yield of as-synthesized particles is around 1 g for each experiment. For further annealing, as-synthesized particles were pressed into pellets (diameter: 5 mm, pressure: 15 kN for 15 min). A horizontal tube furnace (Carbolite Tube Furnace, type: MTF 12/38/400) was used for annealing the pellets at 700 °C and 1000 °C for 1 h under oxygen flow

(6 slm) [74]. Furthermore, for annealing at higher temperatures (1150 °C and 1300 °C), a TGA (NETZSCH STA 449 F1 Jupiter) (heating rate: 10 K/min, dwelling time: 1 h, gas flow: O_2 [44]) was used.

Table 2. Spray-flame reactor operating conditions.

Operating Parameters						
Dispersion CH_4 [slm]	Dispersion O_2 [slm]	Pilot Flame CH_4 [slm]	Pilot Flame O_2 [slm]	Quench Gas Air [slm]	Coaxial Sheath Air [slm]	Reactor Pressure [mbar]
1	9	2	16	240	140	800–820

Powder diffraction XRD patterns were measured using an X-ray diffractometer (Malvern Empyrean diffractometer PANalytical with Cu Kα radiation). Transmission electron microscopy (TEM, Jeol JEM-2200FS; Japan) was used for particle morphology, size, and structure determination. The surface areas of the as-synthesized powders were measured using a Brunauer–Emmett–Teller (BET) device (Nova 2200, 3P Instruments GmbH; Odelzhausen, Germany). XPS spectra were recorded using a VersaProbe II (ULVAC-PHI, Chanhassen, MN, USA) equipped with Al Kα radiation. Simultaneous thermal analysis (STA) consisting of thermogravimetric analyses (TGA) and differential scanning calorimetry (DSC) was carried out with a Netsch STA 449 F1 Jupiter (Germany) under synthetic air with a heating rate of 10 k/min up to 1200 °C and combined with gas analysis during sample heating by quadrupole mass spectrometry (QMS 403 D, NETZSCH-Gerätebau GmbH, Selb, Germany).

The ionic conductivity of the powders was measured in the form of 5 mm diameter pellets (prepared as mentioned above) by an impedance analyzer, Solartron 1260, over a 1 Hz to 1 MHz frequency range. Before, thin conductive gold films were deposited on both sides of the sintered pellets by sputtering. To clarify, all impedance measurements were done at room temperature (~22 °C).

3. Results

First, we investigated the influence of the solvent mixture on nanoparticle production using the example of LYZP doped with 20% yttrium. The most suitable solvent mixture was then selected before discussing the synthesis and characteristics of materials with lower and no yttrium doping.

Characterization of $(LY_{0.2}ZP)_{PA50}$ and $(LY_{0.2}ZP)_{EA50}$ Nanoparticles

To determine the phase compositions of as-synthesized $(LY_{0.2}ZP)_{PA50}$ and $(LY_{0.2}ZP)_{EA50}$, nanoparticles were analyzed with X-ray diffraction. The XRD patterns presented in Figure 1a,b indicate that both as-synthesized powders consist of more than one phase. It was not possible to conclusively determine whether the primary phase (main signal at 30.1°) of as-prepared $(LY_{0.2}ZP)_{PA}$ and $(LY_{0.2}ZP)_{EA}$ samples is tetragonal ZrO_2 (t-ZrO_2, ICSD 66781, space group P 42/n m c Z) or cubic ZrO_2 (c-ZrO_2, ICSD 72955, space group F m -3 m) due to the peak broadening and overlapping [75]. In the case of $(LY_{0.2}ZP)_{PA50}$ (Figure 1a), a splitting of peaks at 35° (220) and 60° (311) (indicating t-ZrO_2) was observed; however, the splitting of peaks at 50° (indicating both c- and t-ZrO_2) is not clear enough to be evaluated [76]. Furthermore, traces of the monoclinic zirconia phase (ICSD 18190, space group P1 21/c 1) at 28° and 31° were observed as an additional phase. In addition, the presence of monoclinic Li_2CO_3 (ICSD 66941, space group C1 2/c 1) at 21.3° can be cross-validated with the results from other measurement methods such as TGA and XPS, as will be shown later. It must be concluded that the intended phase of α-LYZP was not observed in the pristine particles made from propanol/propionic acid solution. In contrast, additional peaks at 20° and 23° were observed for particles made from ethanol/2-ethylhexanoic acid $(LY_{0.2}ZP)_{EA50}$, as shown in Figure 1b. It indicates the appearance of a new phase compared to $(LY_{0.2}ZP)_{PA50}$, which can most probably be assigned to the composition of a Y-doped $LiZr_2(PO_4)_3$ (ICSD 191891). The peak broadening compared to (Figure 1a) is because the product from $(LY_{0.2}ZP)_{EA50}$ has a significantly lower crystallinity, which applies in particular to the signals for the new phase.

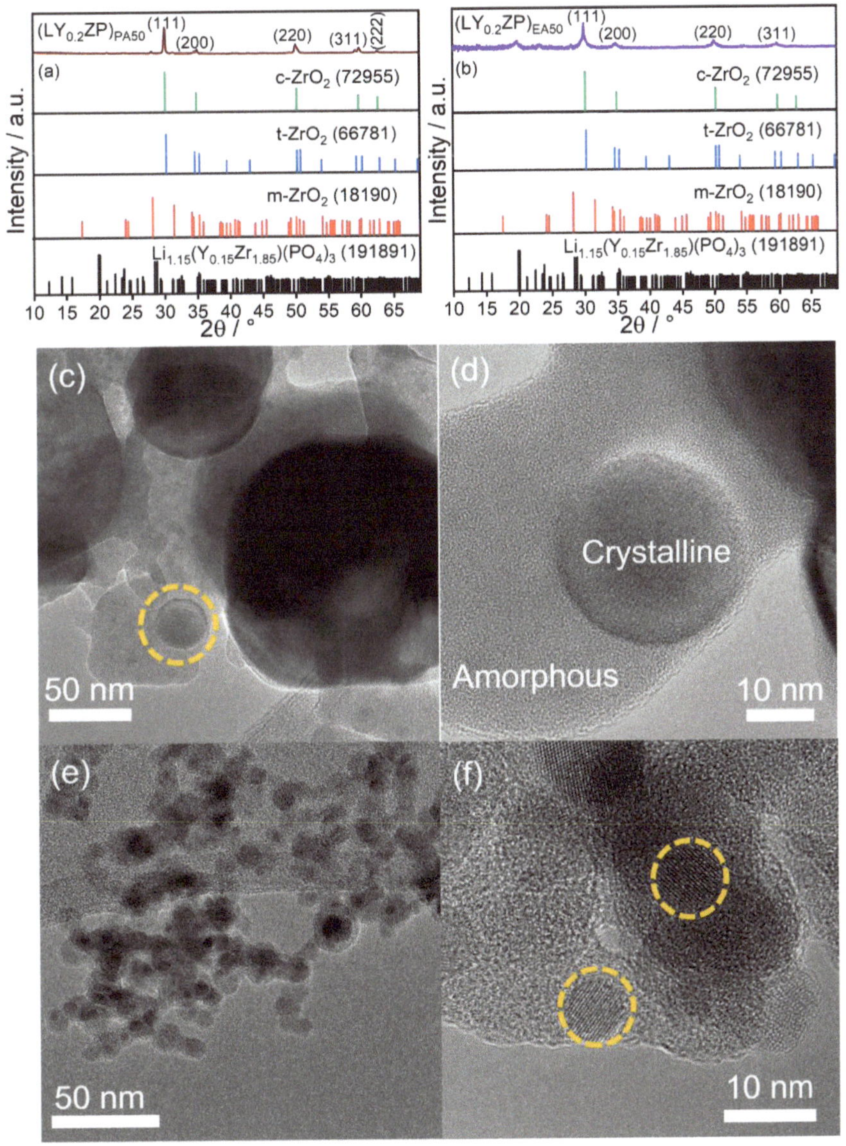

Figure 1. Phase composition and particle size of as-synthesized $(LY_{0.2}ZP)_{PA50}$ nanoparticles were determined by XRD (**a**), TEM (**c**), and HRTEM (**d**), respectively. Similarly, for $(LA_{0.2}ZP)_{EA50}$, the same was determined by XRD (**b**), TEM (**e**), and HRTEM (**f**), respectively. The yellowish circles highlight the single crystalline structures of as-synthesized samples.

Prior to further analysis, the particle morphology and size distribution of as-synthesized $(LY_{0.2}ZP)_{PA50}$ and $(LY_{0.2}ZP)_{EA50}$ particles were carried out using TEM (Figure 1c,e) and HRTEM (Figure 1d,f), respectively. Both nano- and sub-micron particles were observed for particles produced from PrOH/PA (Figure 1c). The yellowish circles (Figure 1c) highlight the single crystalline structures of as-synthesized samples. We observed that, in contrast to EtOH/2-EHA, the use of PrOH/PA as solvent $(LY_{0.2}ZP)_{PA50}$ results in products with a significantly broader range (Figure S1a) [75]. In terms of mass, the large particles are important, but due to an insufficient number, it was not possible to fit the particle size distribution of these large particles in a meaningful way. The obtained size distribution for the smaller mode was plotted and fitted with a lognormal function (Figure S1a). The count median diameter (CMD) of this smaller mode is 16.6 nm with a geometric standard deviation (σ_g) 1.9. Due to the large mass fraction of big particles, it is obvious that the fitted CMD for the small mode does not agree with the estimated BET-based mean particle size, assuming a monodisperse spherical particle size. Thus, the measured specific surface area (SSA) of 5.74 m^2/g corresponds to an average size of around 180 nm, mainly reflecting the big particle mode.

Srinivasan et al. have concluded that the particle sizes of t-ZrO_2 and m-ZrO_2 range from 2 to 20 nm and 20 to 70 nm, respectively [77]. Based on that, the size distribution of as-synthesized particles from $(LY_{0.2}ZP)_{PA}$ is assumed as a combined multinary model. To show the particle size distribution of as-synthesized particles from the experiment $(LY_{0.2}ZP)_{PA50}$ in high resolution, identical counted data from Figure 1c,d in the range from 2 nm to 50 nm are shown in Figure S1a,b. Figure 1e shows TEM images of as-synthesized materials from EtOH/2-EHA $(LY_{0.2}ZP)_{EA50}$ showing a significantly more homogeneous, monomodal particle size distribution, and no particles larger than 50 nm were observed (Figure 1e). The areas highlighted in Figure 1f show a single crystalline structure of zirconia and/or α-LYZP. Unlike the multimodal size distribution from $(LY_{0.2}ZP)_{PA50}$, the histogram of pristine $(LY_{0.2}ZP)_{EA50}$ shows a monomodal size distribution and could be fitted to a CMD of 7.8 nm with a geometric standard deviation σ_g of 1.54 (Figure S1b). This result is in very good agreement with the BET-based mean particle size of 8.9 nm, see Table S2.

Besides the crystalline areas, the TEM images in Figure 1d suggest huge amounts of amorphous materials surrounding the crystalline particles. For further inspection, the compositions of the pristine materials were analyzed by XPS. On the one hand, it was possible to determine that the elements used are present in the product and, on the other, to clarify their chemical composition. For $(LY_{0.2}ZP)_{PA50}$, Figure S2a–c indicates the presence of Li_2CO_3. Detailed analysis of the P 2p spectra (Figure S2d) shows that phosphorus is present in the form of phosphate (134 eV binding energy, indicating tetrahedral PO_4^{3-}). The presence of Yttrium as oxide was proven on the basis of the binding energy contributions at 160.4 eV corresponding to Y $3d_{3/2}$ and 158.3 eV corresponding to Y $3d_{5/2}$ (Figure S2e). Similarly, the XPS spectrum of zirconium in Figure S2f reveals the presence of zirconium core level at 185.8 eV in the form of Zr $3d_{3/2}$, and a higher binding energy contribution at 183.4 eV in the form of Zr $3d_{5/2}$, respectively. It is difficult to distinguish between the tetragonal and monoclinic phases of ZrO_2 via XPS. However, the higher binding energy position of the $3d_{5/2}$ spectra is well within the range of tetragonal ZrO_2 (i.e., monoclinic ZrO_2 generally appears at lower binding energy) [78]. Thus, this finding agrees with our XRD data. However, the shoulder interface spectra at ~181 eV are not observed for tetragonal ZrO_2 in this case [78]. For $(LY_{0.2}ZP)_{EA50}$, a significant peak at BE ~291 eV was detected in the spectrum of C 1s (Figure S3b), which indicates the presence of a carbonate bond. Since hints for crystalline Li_2CO_3 are almost missing in the XRD pattern, we assume that the pristine materials mainly contain amorphous Li_2CO_3 in addition to amorphous compounds of yttrium and phosphorous.

Further characterization of the as-synthesized $(LY_{0.2}ZP)_{PA50}$ and $(LY_{0.2}ZP)_{EA50}$ (Figure 2a,b) was performed with TGA/DSC/QMS analysis, with the bold line showing the TGA curve and the dotted line showing the DSC signal. The TGA curve demonstrates a total weight loss of around 17% and 8% for $(LY_{0.2}ZP)_{PA50}$ and $(LY_{0.2}ZP)_{EA50}$, respectively. The initial mass loss (~12% and ~3% for $(LY_{0.2}ZP)_{PA50}$ and $(LY_{0.2}ZP)_{EA50}$, respectively) between ambient temperature and <220 °C can be ascribed to the removal of physi- and chemisorbed H_2O since only H_2O was detected for both $(LY_{0.2}ZP)_{PA50}$ and $(LY_{0.2}ZP)_{EA50}$ via QMS (Figure 2c,d).

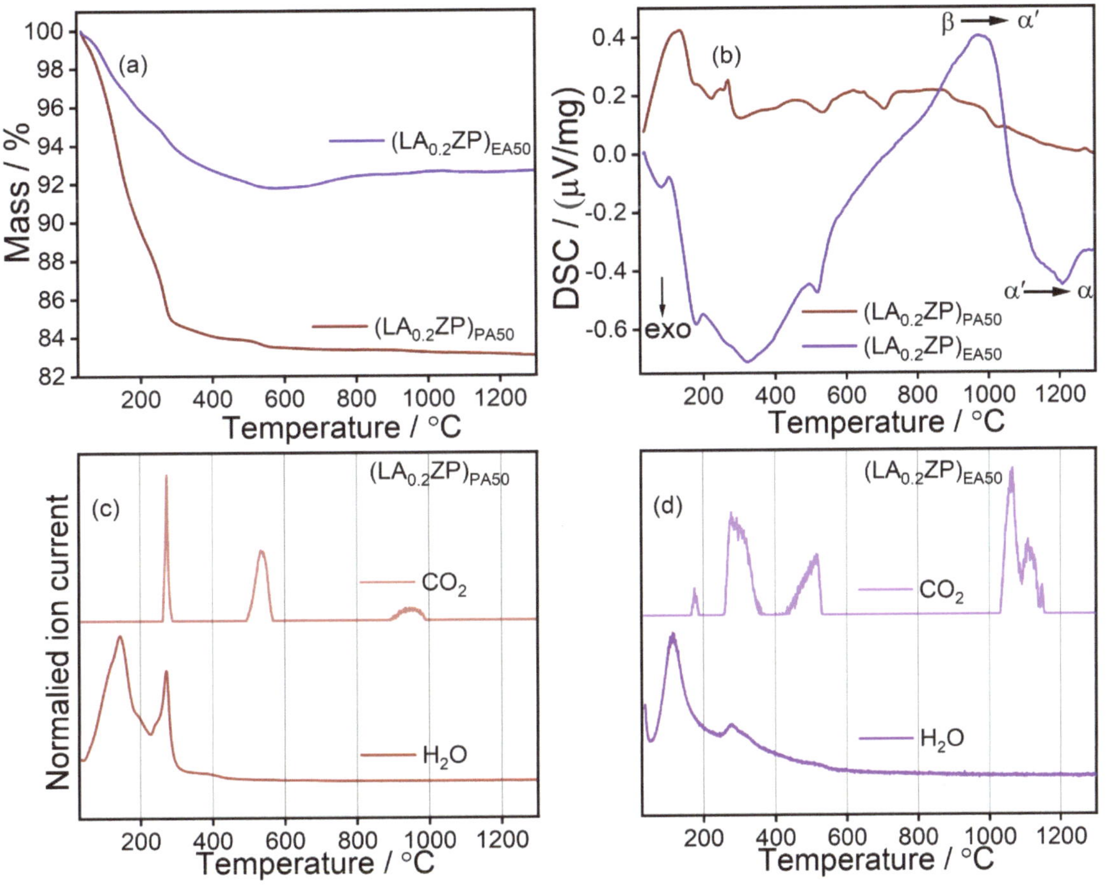

Figure 2. TGA and DSC of as-synthesized samples are shown in (**a**,**b**), respectively. (**c**,**d**) describe the CO_2 and H_2O QMS signal during TGA measurement of $(LY_{0.2}ZP)_{PA50}$ and $(LY_{0.2}ZP)_{EA50}$, respectively.

However, a slight amount of CO_2 signal was detected for $(LY_{0.2}ZP)_{EA50}$ (Figure 2d) in the same region, indicating the thermal decomposition of physisorbed hydrocarbons, i.e., the carboxylic acids. The subsequent mass loss (<5% and ~1% for $(LY_{0.2}ZP)_{PA50}$ and $(LY_{0.2}ZP)_{EA50}$, respectively) between 230 °C and 400 °C can be attributed to the oxidation of higher hydrocarbons, which is accompanied by the release of CO_2 and H_2O. According to Egger et al., the next mass loss of around 500 °C for both the as-synthesized samples can be attributed to the decomposition of the residual Zr-precursor [79].

From the XPS and XRD studies, it is clear that the as-synthesized materials contain all the elements used in the precursor solution; however, most (Zr, P, Y) are present as an amorphous phase, and the material requires thermal annealing to reach the desired LYZP phase. From the DSC measurements, it can be deduced that the endothermic signal at ~980 °C for $(LY_{0.2}ZP)_{EA50}$ (Figure 2b) is most probably caused by the formation of a new phase (Equation (1), [43,44,74]), presumably triggered by the melting of lithium carbonate [80]. Furthermore, the exothermic process at about 1200 °C indicates a phase transformation (Equation (2), [43,44,74]) after the thermal decomposition of leftover lithium carbonate, releasing CO_2. The slight increase in mass in the case of $(LA_{0.2}ZP)_{EA50}$ (<800 °C) may be due to a slight oxidation of the sample, which is supported by the DSC measurement showing an exothermic signal at about 550 °C.

$$\beta \rightarrow \alpha'(\sim 1000\ ^{\circ}\mathrm{C}) \quad (1)$$

$$\alpha' \rightarrow \alpha(\sim 1200\ ^{\circ}\mathrm{C}) \quad (2)$$

As discussed above, the phase change occurs for both the as-synthesized samples at relatively higher temperatures. Thus, the as-prepared materials were annealed at 700 and 1000 °C in a tube furnace under oxygen, with subsequent analysis of the crystal structure. Figure 3a,c reveals the effect of calcination on as-synthesized samples at different temperatures. When the pressed pellet from $(LY_{0.2}ZP)_{PA50}$ was heated at 700 °C for 1 h, XRD patterns showed a significant difference compared to the XRD pattern of pristine nanoparticles (Figure 3a). At this point, a significant decrease in the overall presence of t-ZrO_2 can be observed. Moreover, a sharp peak at 20° appears after annealing at 700 °C in the case of $(LY_{0.2}ZP)_{PA50}$, indicating the formation of the $LiZr_2(PO_4)_3$ phase (Figure 3a). Although the decrease in the t-ZrO_2 phase is welcomed, the intended rhombohedral phase has not been achieved. This further compels to anneal the samples at even higher temperatures.

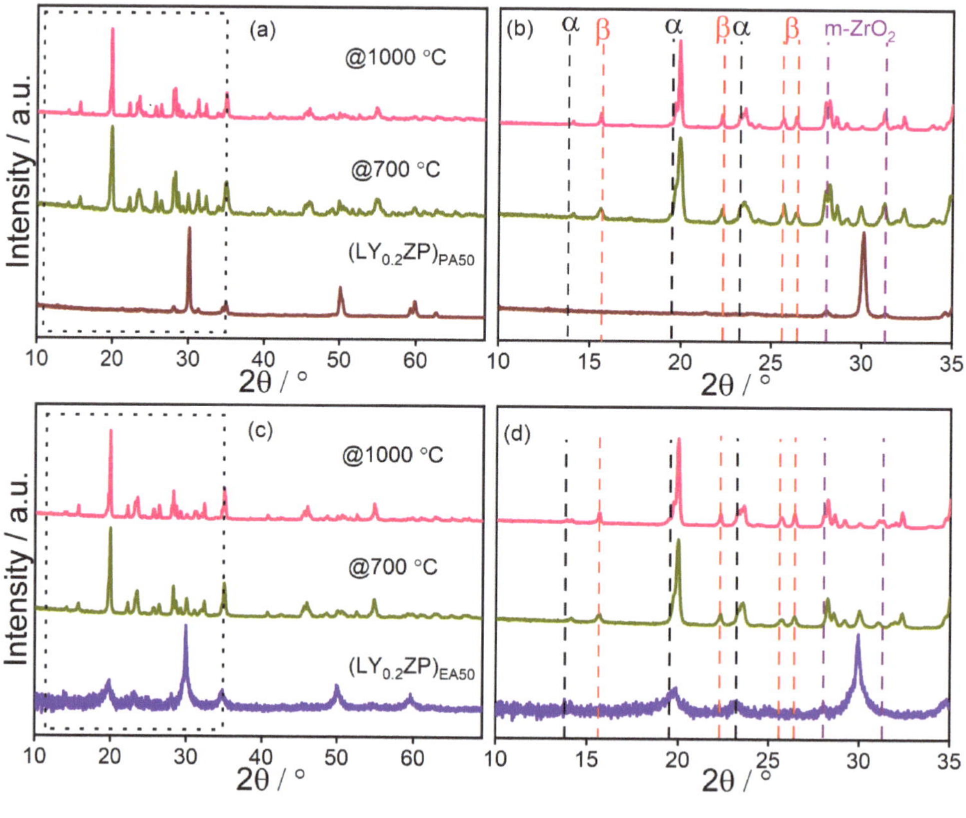

Figure 3. (**a**,**c**) XRD patterns comparison before and after annealing of $(LY_{0.2}ZP)_{PA50}$ and $(LY_{0.2}ZP)_{EA50}$, respectively, at different temperature conditions for 1 h under O_2. (**b**,**d**) corresponding detailed illustration in the range of 10° to 35° 2θ, α refers to rhombohedral phase $Li_{1+x}Y_xZr_{2-x}(PO_4)_3$ and β refers to orthorhombic phase $Li_{1+x}Y_xZr_{2-x}(PO_4)_3$.

The detailed illustration of the dotted rectangular in Figure 3a between 10° and 30° 2θ is shown in Figures 3b and 3d for $(LY_{0.2}ZP)_{PA50}$ and $(LY_{0.2}ZP)_{EA50}$, respectively. Based on that, it can be determined that β phase (orthorhombic) LYZP was obtained at 700 °C as the primary phase in both of the as-synthesized samples. As stated above, the t-ZrO_2 phase decreases significantly at 1000 °C. However, an increment in m-ZrO_2 has been observed (Figure 3b,d). Furthermore, t-ZrO_2 transforms into m-ZrO_2 after annealing at 1150 °C for 1 h (peak at 28°) (Figure S4a,b) [81]. When the annealing temperature further increased to 1150 °C, peaks belonging to the (rhombohedral) α-LYZP phase at 2θ 14°, 19.5°, and 23° start to appear. Ultimately, part of β phase transforms α phase under the heating condition of 1300 °C for 1 h. Nevertheless, α and β phases of LYZP still coexist even after annealing at high temperatures (1300 °C) (Figure S4a–d). Similarly, the smaller peaks between 2θ = 40–60° can be attributed to the (rhombohedral) α-LYZP phase (2θ: 42, 42.6, 45.3, 49, 54.8°) and the β phase of LYZP (2θ: 40.6, 46.1°). It can be concluded that, in addition to high temperatures over 1150 °C, the transition of transformation from the β phase of LYZP to the α phase needs a longer dwell time [74].

Raman spectroscopy (Figure S5) is used as a complementary to XRD to confirm the phase change during the calcination of the $(LY_{0.2}ZP)_{PA}$ as-synthesized sample. Here it is to be noted that as-synthesized samples from SFS are generally covered with unburned residual carbonaceous components (Figure 1d), which hinders the regular Raman measurements. Thus, almost no Raman signals could be observed for as-synthesized samples. However, after annealing, the absorption bands of m-ZrO_2 are easily recognized in Figure S5. Annealing at 1150 °C for 1 h results in extensive transformation of t-ZrO_2 to the m-ZrO_2 structure. All absorption bands in the range of 180 cm^{-1} to 616 cm^{-1} highlighted with dotted rectangular in Figure S5 refer to the monoclinic phase zirconia [82], while the main bands of t-ZrO_2 at 250 cm^{-1} and 640 cm^{-1} are absent [83,84]. Due to the phase transition, tetragonal zirconia as a minor impurity has not been detected by Raman after annealing. A band at 1028 cm^{-1} was assigned to P-O stretching vibrations [85], which accounted for the presence of phosphate, confirming LYZP phase formation.

To determine the phase composition of $(LY_{0.2}ZP)_{PA50}$ after annealing at 1300 °C for 1 h under O_2, the result of fitted data using Rietveld refinement is shown in Figure S4e. The XRD pattern from $(LY_{0.2}ZP)_{PA50}$ after annealing could be fitted well with the α-phase LYZP (ICSD 191891, space group R-3 c H). There is a substantial amount of β-$Li_{1.2}Y_{0.2}Zr_{1.8}(PO_4)_3$ (ICSD 91113, space group P b n a) in the pellet after heat treatment. m-ZrO_2 and t-ZrO_2 were also observed as impurities. A peak at 21° might be ascribed to the presence of $Y(PO_3)_3$ (ICSD 420121, space group C 1 c 1). Based on the result after refinement, the contents of the rhombohedral α-LYZP and β-LYZP are 31.5% and 40.3%, respectively (Table 3).

Table 3. Composition of phases before and after annealing of $(LY_{0.2}ZP)_{EA50}$ and $(LY_{0.2}ZP)_{PA50}$.

	Status	Content of Phase [%]			
		t-ZrO_2	**m-ZrO_2**	**α-LYZP**	**β-LYZP**
$(LY_{0.2}ZP)_{PA50}$	As-synthesized	34.0	39.9	/	/
	@1300 °C	0.6	27.6	31.5	40.3
$(LY_{0.2}ZP)_{EA50}$	As-synthesized	21.8	63.5	14.7	/
	@ 1300 °C	0.5	14.2	49.6	35.6

As seen above, the phase transformation progress indicates the co-existence of α and β phase LYZP after annealing at 1300 °C for 1 h, even for $(LY_{0.2}ZP)_{EA50}$. Nevertheless, the sharpest peak in Figures S4d and 3d after annealing at 1300 °C is at 19.8°, belonging to α-LYZP, while the sharpest peak for $(LY_{0.2}ZP)_{PA50}$ was at 28°, belonging to m-ZrO_2 (Figures S4b and 3b). However, m-ZrO_2 (shown in purple dotted line, Figure S4d) is still present as an impurity in the case of $(LY_{0.2}ZP)_{EA50}$ even after annealing at 1300 °C similar to $(LY_{0.2}ZP)_{PA50}$. The fitted XRD pattern in Figure S4f indicates that the primary phase of sintered material from $(LY_{0.2}ZP)_{EA50}$ is α-LYZP while the secondary phase is β-LYZP. Additionally, the composition of m-ZrO_2 as a main impurity phase decreases significantly from 27.6 wt.% (in the case of $(LY_{0.2}ZP)_{PA50}$, Table 3) to 14.2 wt.% (Table 3; in the case of $(LY_{0.2}ZP)_{PA50}$). Correspondingly, the composition of α-LYZP increases from 31.5 wt.% to 49.6 wt.%. In contrast to $(LY_{0.2}ZP)_{PA50}$ Figure S4e, the results shown in Figure S4f testify to the potential of solvent 'B' to form desirable homogeneous nanoparticles of α-LYZP with higher purity.

Based on further study of literature [72], and at this point to further increase α- phase, the volumetric concentration of 2-ethylhexanoic acid (2-EHA) (in solvent mixture 'B') was increased to 70 vol.% in the experiment denoted as $(LY_{0.2}ZP)_{EA70}$ to investigate its corresponding effect on the composition development after heat treatment.

The as-synthesized samples were named $(LY_{0.2}ZP)_{70}$, keeping the Y^{3+} doping as constant as the previous $(LY_{0.2}ZP)_{EA50}$. The XRD pattern of $(LY_{0.2}ZP)_{EA70}$ is shown in Figure 4, and it shows almost no visible difference from $(LY_{0.2}ZP)_{EA50}$ sample. A significant difference illustrated in Figure 4b is that β-LYZP was not detected in the XRD pattern after annealing at 1300 °C for 1 h under O_2. An increased concentration of 2-EHA may be beneficial to obtain α-LYZP with higher purity. According to Rietveld refinements of the XRD patterns shown in Figure 5, α-LYZP is present as the primary phase, while m-ZrO_2 is the secondary phase. Impurities such as t-ZrO_2 and $Y(PO_4)_3$ (ICSD 79754, space group I 41/a m d Z) were observed. The stoichiometric setup of the experiment might have an influence on the presence of impurity ZrO_2 and $Y(PO_4)_3$ after heat treatment at 1300 °C. The loss of lithium during high-temperature SFS and subsequent annealing might result in the presence of impurities such as $Y(PO_4)_3$ [63].

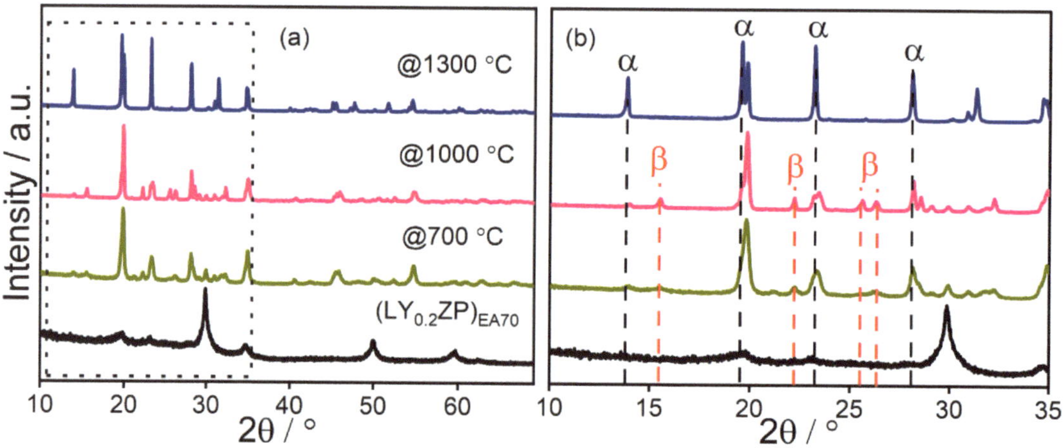

Figure 4. (**a**) XRD patterns of materials from $(LY_{0.2}ZP)_{EA70}$ before and after annealing at different temperature conditions for 1 h under O_2. (**b**) corresponding detailed illustration in the range of 10° to 35° 2θ. α and β refer to the rhombohedral phase and orthorhombic.

To summarize and show the effective manipulation of solvent mixture keeping Y^{3+} doping constant, a comparative XRD of $(LY_{0.2}ZP)_{PA50}$, $(LY_{0.2}ZP)_{EA50}$, $(LY_{0.2}ZP)_{EA70}$ graph is shown in Figure S7a. To obtain the α-$Li_{1.2}Y_{0.2}Zr_{1.8}(PO_4)_3$, heat treatment of pristine material above 1200 °C is inevitable in practice [44]. To illustrate the peak-intensive zone (highlighted with dotted rectangular in Figure S7a in the range of 10° to 35° 2theta more clearly), Figure S7b is shown. Compared to XRD patterns of materials after annealing from $(LY_{0.2}ZP)_{PA50}$ and $(LY_{0.2}ZP)_{EA50}$, a significant difference in XRD pattern after heat treatment from $(LY_{0.2}ZP)_{EA70}$ is the absence of β-LYZP. Regarding the improvement of the content of α-LYZP, the utilization of the solvent mixture ethanol/2-EHA (1:1 by volume) has a slight effect on composition after heat treatment at 1300 °C. On the other hand, an increased proportion of 2-EHA from 50 vol% to 70 vol% has a significant effect on composition after heat treatment at 1300 °C. To better understand this, the composition of materials after annealing at 1300 °C for 1 h under O_2 from cases $(LY_{0.2}ZP)_{PA50}$, $(LY_{0.2}ZP)_{EA50}$, and $(LY_{0.2}ZP)_{PA50}$ was shown in Table 4 via Rietveld refinement.

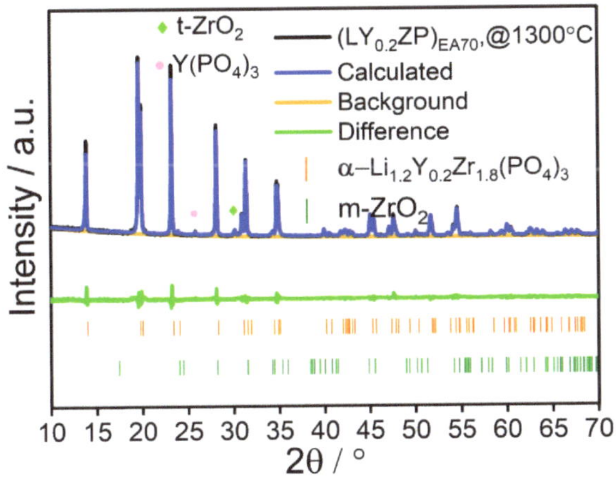

Figure 5. Fitted data of XRD pattern from $(LY_{0.2}ZP)_{EA70}$ after annealing at 1300 °C for 1 h under O_2.

Table 4. Composition development of materials after annealing at 1300 °C for 1 h under O_2.

Nomenclature	Solvent Mixture	Composition [wt%]		
	Propanol/Propionic Acid (1:1 by Volume)	α-LYZP	β-LYZP	m-ZrO_2
$(LY_{0.2}ZP)_{PA50}$	Ethanol/2-EHA (1:1 by volume)	31.5	40.3	27.6
$(LY_{0.2}ZP)_{EA50}$	Ethanol/2-EHA (3:7 by volume)	49.6	35.6	14.2
$(LY_{0.2}ZP)_{EA70}$	Propanol/propionic acid (1:1 by volume)	94.7	/	3.1

The most attractive observation is that the crystalline content of α-LYZP of material from an experiment $(LY_{0.2}ZP)_{EA70}$ after sintering is 94.7 wt.%, which indicates the beneficial potential of employed solvent mixture ethanol/2-EHA (3:7 by volume). Again, one issue that must be mentioned is that the calculation of crystalline content based on Rietveld refinement is used qualitatively.

Compared to solvent mixture 'A', solvent mixture 'B' has a higher boiling point (Table S1), which might lead to a gas-to-particle mechanism. Moreover, the esterification reaction was found in a solution containing metal nitrates, ethanol, and 2-EHA [72]. Moreover, this in turn results in a gas-to-particle formation pathway leading to smaller nanoparticles [86]. Furthermore, it has been shown that the breakup mechanism of microexplosion during spray remains the same as that of a single droplet [87]. Thus, it indicates that precursor chemistry (e.g., the boiling point of the final solvent mixture, melting point of the solutes, chemical stability of the precursors, etc.) affects not only the behavior of the micro explosion but also the final synthesized materials. The decomposition of zirconium(iv) propoxide above 600 °C may influence the composition in the high-temperature synthesis of α-LYZP above 1000 °C.

4. Discussion

In our synthesis conditions, a significant amount of zirconium propoxide was decomposed into product zirconia which was not incorporated into the LYZP phase. So, to summarize, solvent 'B' encourages the micro-explosion/droplet breakup/fragmentation, leading to a very small size of the final droplets. Due to the low melting point of Li, Y percussor leads to fast and foremost vaporization. On the other hand, having a high melting point of Zr, the formation of zirconia will be the first step on the route of condensation & nucleation from high spray-flame temperature. The smaller the final droplet, resulting in smaller the zirconia in size. At last, other species will nucleate surrounding zirconia Figure 1d forming an amorphous layer. The smaller the size higher the sintering activity leading to better phase composition. The same logic applies to why EHA/EtOH (7:3/*v*:*v*) produces a much better phase after annealing than solvent 'B'.

To investigate the effect of doped Y^{3+} concentration on $Li_{1+x}Y_xZr_{2-x}(PO_4)_3$ morphology involving solvent mixture 'B', experiments are denoted $(LY_{0.2}ZP)_{EA50}$, $(LY_{0.1}ZP)_{EA50}$, and $(LZP)_{EA50}$, respectively. In this section, the effect of doping (varying Y concentration) has been discussed. As shown in Figure 6a,b, the thermal properties of the LY_xZP are comparable to each other. The first mass loss was due to physisorbed unburned precursors and later may be due to Li_2CO_3 degradation [63]. Moreover, the exothermic change ~1200 °C might be due to phase change which is a similar property to $(LY_{0.2}ZP)_{PA50}$ (Figure 2).

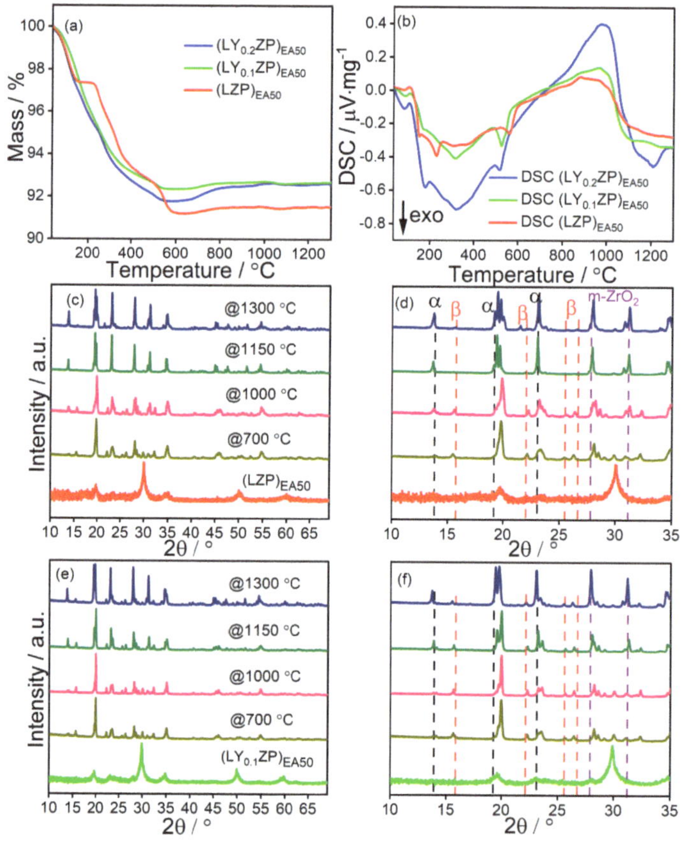

Figure 6. (**a**) TGA analysis of particles from $(LY_{0.2}ZP)_{EA50}$, $(LY_{0.1}ZP)_{EA50}$, and $(LZP)_{EA50}$. (**b**) corresponding DSC analysis. (**c**) XRD patterns of materials from $(LZP)_{EA50}$ before and after annealing at different conditions. (**d**) corresponding detailed illustration in the range of 10° to 35° 2θ. (**e**) XRD patterns of materials from $(LY_{0.1}ZP)_{EA50}$ before and after annealing at different temperature conditions for 1 h under O_2. (**f**) corresponding detailed illustration in the range of 10° to 30° 2θ. α and β refer to the rhombohedral and orthorhombic phases of $Li_{1+x}Y_xZr_{2-x}(PO_4)_3$, respectively.

The phase composition of as-prepared LYxZP particles involving the solvent mixture ethanol/2-EHA (1:1 by volume) was determined with XRD in an overview way (Figure S6a). Similarly, the primary phase of as-synthesized materials from three experiments is t-ZrO_2, and the secondary phase is m-ZrO_2. Nevertheless, in contrast to the as-synthesized material from $(LY_{0.2}ZP)_{PA50}$, the significant presence of peaks at 20 °C and 23 °C (marked with a point-point-dash line) was observed in Figure S7a. It indicates that the utilized solvent mixture of ethanol and 2-EHA results in the presence of a new phase, which may be the expected Y-doped $LiZr_2(PO_4)_3$ (ICSD 191891, Figure S7a).

At this point, it has been established that solvent mixture 'B' results in smaller nanoparticles with a narrower particle size distribution as well as a better phase composition than solvent mixture 'A'. However, it is unclear if Y incorporation has any subsequent effect on phase composition. XRD results of $(LZP)_{EA50}$ and $(LY_{0.1}ZP)_{EA50}$ are shown in Figure S6a and Figure 6b,e, respectively. The XRD patterns of materials from $(LY_{0.1}ZP)_{EA50}$ in Figure 6e show a similarity of the phase transformation as the case $LY_{0.2}ZP$ described in Figures 3c and 5c. No β′-LYZP was detected in XRD patterns after annealing at 1300 °C after 1 h under O_2. However, peak splitting at 20° 2θ in the XRD profile from $(LZP)_{EA50}$ (Figure 6c,d) at a higher annealing temperature (i.e., 1300 °C) was observed. A similar kind of peak splitting has not been observed in any other samples, even after annealing at 1300 °C. There might be another reaction when the annealing temperature reaches 1300 °C. Unfortunately, after an intensive investigation of the literature, reasonable assumptions and convincing explanations are not supportive of this phenomenon. Further investigation of the mentioned splitting peak needs to be done and is beyond this work.

It is clear from Figure 6c–f that the peak intensity of the α phase increases with increasing Y^{3+} doping. However, m-ZrO_2 impurity remains within the system with varying Y^{3+} doping. The high-resolution XPS results of P 2p, Zr 3d, and Y 3d elements are shown in Figure S6b–d, respectively. All three P 2p peaks from the experiment involving solvent mixture 'B' demonstrate a symmetric shape, which manifests the presence of one type of phosphorus (i.e., phosphate). The peaks related to the BE of ~183.2 eV and ~185.6 eV (Figure S6c), pointed with arrows, correspond to Zr 3d5/2 and Zr 3d3/2, while peaks related to the BE of ~158.1 eV and ~160.1 eV (Figure S6d, pointed with arrows) correspond to Y 3d5/2 and Y 3d3/2, respectively. XPS spectra of Y and Zr from experiments $(LY_{0.2}ZP)_{EA50}$, $(LY_{0.1}ZP)_{EA50}$, and $(LZP)_{EA50}$ in Figure S6c, d demonstrate the structure change of the bonds P-O-Zr and P-O-Y as a different compositions of aliovalent yttrium ion were incorporated.

After quantitative analysis via Rietveld refinement, there is no evidence that shows materials after annealing at 1300 °C from cases $LY_{0.2}ZP)_{EA50}$, $(LY_{0.1}ZP)_{EA50}$, and $(LZP)_{EA50}$ consist of any triclinic phase LYZP (α′-LYZP). Fitted data via Rietveld refinement of materials after annealing at 1300 °C from cases LZP is shown in (Figure S8). The impedance spectra of $(LY_{0.2}ZP)_{70}$ and $(LZP)_{EA50}$ particles after annealing at different temperatures are shown in (Figure 7) and (Figure S8), respectively. Both the as-synthesized samples show the highest resistance at room temperature compared to annealed samples.

The impedance of the as-synthesized $(LY_{0.2}ZP)_{70}$ is reduced by almost three orders of magnitude by annealing at 1300 °C, although the shape of the spectra remains almost the same (Figure 7a–c). As known from the literature [88], the impedance spectrum of NASICON-type structures exhibits relaxation processes for the bulk and grain boundary conductivity, which can be seen in two more or less pronounced semicircles in the higher-frequency part of the spectrum. In the low-frequency part, the spectrum is characterized by the capacitive properties of the ionic conductor, which manifest themselves in a straight line whose imaginary part of the impedance increases with decreasing frequency. This can be seen particularly well for the material tempered at 1000 °C (Figure 7c) and 1300 °C (Figure 7a). This low-frequency behavior is also characteristic of pure ionic conductors. We attribute this change in impedance to the fact that t-ZrO_2 transforms into the NASICON structure with increasing annealing temperature. This, combined with significant grain growth, leads to a highly increased ionic conductivity compared to the as-synthesized samples. And true to our XRD analysis (Figure 4), Figure 7a indicates that after annealing at

1300 °C, $(LY_{0.2}ZP)_{EA}$ shows the best conductivity (1.14×10^{-5} S cm^{-1}), which is comparable to the reported value [44]. In comparison, the undoped sample shows poor performance at similar conditions (Figure S8, 1.88×10^{-6} S cm^{-1}).

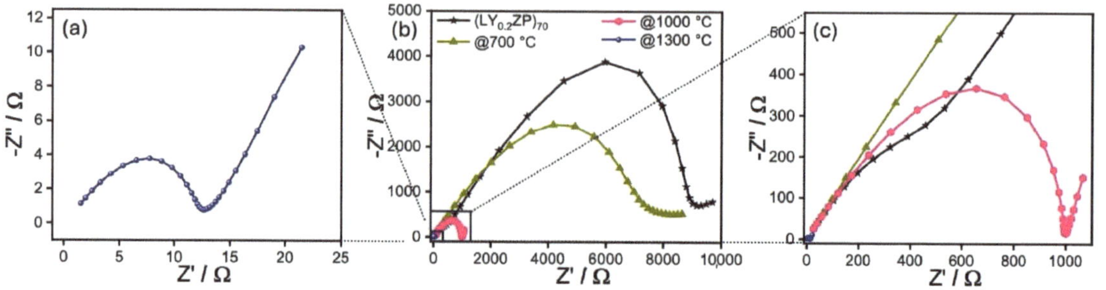

Figure 7. Impedance spectra of as-synthesized $(LY_{0.2}ZP)_{EA70}$ and after annealing at different temperatures. Figure (**b**) shows an overview of all spectra, (**a**,**c**) are enlarged sections of the measurements on samples after annealing at 1300 and 1000 °C, respectively.

5. Conclusions

Here, we have successfully synthesized nanocrystalline rhombohedral (α) lithium yttrium zirconium phosphate (LYZP) based on a spray-flame process followed by a short annealing step. In this case, the addition of Y^{3+} as a dopant is a mandatory step to stabilize the α phase. We have witnessed that direct synthesis of LYZP is not possible due to the shorter residual time and rapid nucleation of tetragonal zirconia (t-ZrO_2) in contrast to species containing Li, Y, and phosphate species during gas-phase synthesis. This reaction mechanism is universal with different solvent mixtures (e.g., 2-ethyl hexanoic acid/ethanol, propanol/propanoic acid). It is shown that the 2-ethyl hexanoic acid/ethanol solvent mixture ('B') is suitable to synthesize nanoparticles with narrow size distribution with higher specific surface area compared to the nanoparticles synthesized by solvent mixture propanol/propanoic acid ('A'). All products, in the case of A, annealed at <1300 °C exhibited the undesirable, poorly ion-conducting β phase with a mixture of a smaller percentage α phase. However, the particularly good mixing of t-ZrO_2 and Li, Y, phosphate precursor, as occurs in spray flame synthesis, with appropriate Y^{3+} doping and solvent mixture 'B' shows advantages. To the best of our knowledge, this is the only bottom-up approach for α-LYZP synthesis.

Supplementary Materials: The following supporting information can be downloaded at: https://www.mdpi.com/article/10.3390/nano14151278/s1, Table S1: Physical properties of solvents and precursors; Figure S1: Particle size distribution of as-synthesized particles from case $(LY_{0.2}ZP)_{PA50}$ (a) and $(LY_{0.2}ZP)_{EA50}$ (b) respectively; Figure S2: Results of XPS measurements of (a) Li 1s. (b) C 1s. (c) O 1s. (d) P 2p. (e) Y 3d. (f) Zr 3d of as-synthesized $(LY_{0.2}ZP)_{PA50}$; Figure S3: Results of XPS measurements of (a) Li 1s. (b) C 1s. (c) O 1s of as-synthesized $(LY_{0.2}ZP)_{EA50}$; Figure S4: (a,c) XRD patterns comparison before and after annealing at 1150 & 1300 °C of $(LY_{0.2}ZP)_{PA}$ and $(LY_{0.2}ZP)_{EA}$ respectively, for 1 h under O2. (b,d) corresponding detailed illustration in the range of 10° to 35° 2θ, α refers to rhombohedral phase $Li_{1+x}Y_xZr_{2-x}(PO_4)_3$ and β refers to orthorhombic phase $Li_{1+x}Y_xZr_{2-x}(PO_4)_3$. (e,f) phase composition of material from $(LY_{0.2}ZP)_{PA}$ and $(LY_{0.2}ZP)_{EA}$ respectively using Rietveld refinement after annealing at 1300 °C for 1 h under O_2; Figure S5: Raman spectroscopy of materials from $(LY_{0.2}ZP)_{PA50}$ after sintering at different temperature conditions for 1 h under O_2. All absorptions bands in dotted rectangular indicates the presence of m-ZrO_2, while the dotted line is attributed to P-O stretching; Figure S6: (a) Compositions of as-synthesized materials from experiments involving solvent mixture ethanol/2-EHA (1:1 by volume). Point-point-dash lines refer to new present peaks compared to the compositions of as-synthesized particles from case $(LY0_{.2}ZP)_{PA50}$. XPS results from (b) P 2p. (c) Zr 3d. (d) Y 3d from experiments involving solvent mixture 'B' with varying Y^{3+} doping; Table S2. Comparison of as-synthesized average particle size; Figure S7: (a) XRD patterns of materials from $(LY_{0.2}ZP)_{PA50}$, $(LY_{0.2}ZP)_{EA50}$ and $(LY_{0.2}ZP)_{EA70}$ (from bottom to top) after annealing at different temperature conditions for 1h under O_2. (b) corresponding detailed illustration in the range of 10° to 35° 2θ. α and β refers to rhombohedral phase and orthorhombic phase of $Li_{1+x}Y_xZr_{2-x}(PO_4)_3$ respectively; Figure S8: Impedance spectra of (LZP)EA50 particles after annealing at different temperatures (a). And (b) represents the zoomed image of the square in (a); Figure S9: (a) TEM (b) HRTEM of particles from $(LY_{0.1}ZP)_{EA50}$ (c) corresponding particle size distribution and fitted lognormal curve. (d) TEM (e) HRTEM of particles from $(LZP)_{EA50}$ (f) corresponding particle size distribution and fitted lognormal curve.

Author Contributions: Initial literature investigation, M.Y.A.; conceptualization, M.Y.A., H.O. and H.W.; experiments, T.C. and M.Y.A.; data analysis, T.C. and M.Y.A.; writing—original draft preparation, T.C. and M.Y.A.; writing—review and editing, M.Y.A., H.O. and H.W.; supervision, project administration, funding acquisition, H.W. All authors have read and agreed to the published version of the manuscript.

Funding: This work was supported by the Federal Ministry of Education and Research of Germany, project 03XPO273B (FestBatt) and FestBatt2 (FB-2 Hybrid) 03XPO428B.

Data Availability Statement: Data can be made available upon request.

Acknowledgments: The authors would like to thank Steven Angel for many meaningful discussions. Special thanks are to AG Lupascu (University of Duisburg Essen) and Mohammed-Ali Sheikh for XRD measurements and the ICAN, Markus Heidelmann for TEM, and Ulrich Hagemann for XPS measurements. The authors are grateful for financial support from the Federal Ministry of Education and Research of Germany, project 03XPO273B (FestBatt) and FestBatt2 (03XP0428D). M.Y.A acknowledges support from the IMPRS–SurMat.

Conflicts of Interest: The authors declare no conflicts of interest.

References

1. Gaies, B.; Guesmi, K.; Porcher, T.; Boroumand, R. Financial instability and oil price fluctuations: Evidence from oil exporting developing countries. *Eur. J. Comp. Econ.* **2020**, *17*, 55–71. [CrossRef]
2. Brand, K. Umstrittene Ostsee-Pipeline: Grünes Licht für Nord Stream 2. 2021. Available online: https://www.tagesschau.de/ausland/amerika/nord-stream-2-einigung-103.html (accessed on 2 February 2022).
3. Kougias, I. *Hydropower—Technology Development Report 2020*; Publications Office of the European Union: Luxembourg, 2021.
4. Taylor, N.; Jager-Waldau, A. *Photovoltaics: Technology Development Report 2020*; Publications Office of the European Union: Luxembourg, 2021; p. 30504.
5. Telsnig, T. *Wind Energy Technology Development Report 2020*; Publications Office of the European Union: Luxembourg, 2021.
6. Yoo, J.; Park, B.; An, K.; Al-Ammar, E.; Khan, Y.; Hur, K.; Kim, J. Look-Ahead Energy Management of a Grid-Connected Residential PV System with Energy Storage under Time-Based Rate Programs. *Energies* **2012**, *5*, 1116–1134. [CrossRef]
7. Hesse, H.C.; Schimpe, M.; Kucevic, D.; Jossen, A. Lithium-Ion Battery Storage for the Grid—A Review of Stationary Battery Storage System Design Tailored for Applications in Modern Power Grids. *Energies* **2017**, *10*, 2107. [CrossRef]
8. Meyers, W.F.; Simmons, J.W. Electric Current-Producing Cell with Anhydrous Organic Liquid Electrolyte. U.S. Patent 3,423,242, 21 January 1969.
9. Brodd, R.J.; Tagawa, K. Lithium-Ion Cell Production Processes. In *Advances in Lithium-Ion Batteries*; van Schalkwijk, W.A., Scrosati, B., Eds.; Springer: Boston, MA, USA, 2002; pp. 267–288.
10. Ravdel, B.; Abraham, K.M.; Gitzendanner, R.; DiCarlo, J.; Lucht, B.; Campion, C. Thermal stability of lithium-ion battery electrolytes. *J. Power Sources* **2003**, *119–121*, 805–810. [CrossRef]
11. Abada, S.; Marlair, G.; Lecocq, A.; Petit, M.; Sauvant-Moynot, V.; Huet, F. Safety focused modeling of lithium-ion batteries: A review. *J. Power Sources* **2016**, *306*, 178–192. [CrossRef]
12. Arbizzani, C.; Gabrielli, G.; Mastragostino, M. Thermal stability and flammability of electrolytes for lithium-ion batteries. *J. Power Sources* **2011**, *196*, 4801–4805. [CrossRef]
13. Takada, K.; Nakano, S.; Inada, T.; Kajiyama, A.; Kouguchi, M.; Sasaki, H.; Kondo, S.; Watanabe, M.; Murayama, M.; Kanno, R. Solid-State Lithium Batteries with Sulfide-Based Solid Electrolytes. *Solid State Ion.* **2004**, *172*, 25–30. [CrossRef]
14. Wang, J.; Huang, J.; Huang, S.; Komine, Y.; Notohara, H.; Urita, K.; Moriguchi, I.; Wei, M. Regulating the effects of SnS shrinkage in all-solid-state lithium-ion batteries with excellent electrochemical performance. *Chem. Eng. J.* **2022**, *429*, 132424. [CrossRef]
15. Wang, J.; Okabe, J.; Komine, Y.; Notohara, H.; Urita, K.; Moriguchi, I.; Wei, M. The optimized interface engineering of VS2 as cathodes for high performance all-solid-state lithium-ion battery. *Sci. China Technol. Sci.* **2022**, *65*, 1859–1866. [CrossRef]
16. Judez, X.; Zhang, H.; Li, C.; Eshetu, G.G.; González-Marcos, J.A.; Armand, M.; Rodriguez-Martinez, L.M. Review—Solid Electrolytes for Safe and High Energy Density Lithium-Sulfur Batteries: Promises and Challenges. *J. Electrochem. Soc.* **2018**, *165*, A6008. [CrossRef]

17. Overhoff, G.M.; Ali, M.Y.; Brinkmann, J.-P.; Lennartz, P.; Orthner, H.; Hammad, M.; Wiggers, H.; Winter, M.; Brunklaus, G. Ceramic-in-Polymer Hybrid Electrolytes with Enhanced Electrochemical Performance. *ACS Appl. Mater. Interfaces* **2022**, *14*, 53636–53647. [CrossRef] [PubMed]
18. Janek, J.; Zeier, W.G. A solid future for battery development. *Nat. Energy* **2016**, *1*, 16141. [CrossRef]
19. Monroe, C.; Newman, J. The impact of elastic deformation on deposition kinetics at lithium/polymer interfaces. *J. Electrochem. Soc.* **2005**, *152*, A396–A404. [CrossRef]
20. Brissot, C.; Rosso, M.; Chazalviel, J.N.; Lascaud, S. Dendritic growth mechanisms in lithium/polymer cells. *J. Power Sources* **1999**, *81*, 925–929. [CrossRef]
21. Kerman, K.; Luntz, A.; Viswanathan, V.; Chiang, Y.M.; Chen, Z.B. Review-Practical Challenges Hindering the Development of Solid State Li Ion Batteries. *J. Electrochem. Soc.* **2017**, *164*, A1731–A1744. [CrossRef]
22. Duan, H.; Fan, M.; Chen, W.-P.; Li, J.-Y.; Wang, P.-F.; Wang, W.-P.; Shi, J.-L.; Yin, Y.-X.; Wan, L.-J.; Guo, Y.-G. Extended Electrochemical Window of Solid Electrolytes via Heterogeneous Multilayered Structure for High-Voltage Lithium Metal Batteries. *Adv. Mater.* **2019**, *31*, 1807789. [CrossRef] [PubMed]
23. Wu, B.; Wang, S.; Lochala, J.; Desrochers, D.; Liu, B.; Zhang, W.; Yang, J.; Xiao, J. The role of the solid electrolyte interphase layer in preventing Li dendrite growth in solid-state batteries. *Energy Environ. Sci.* **2018**, *11*, 1803–1810. [CrossRef]
24. Luo, W.; Gong, Y.; Zhu, Y.; Fu, K.K.; Dai, J.; Lacey, S.D.; Wang, C.; Liu, B.; Han, X.; Mo, Y.; et al. Transition from Superlithiophobicity to Superlithiophilicity of Garnet Solid-State Electrolyte. *J. Am. Chem. Soc.* **2016**, *138*, 12258–12262. [CrossRef]
25. Lu, J.Y.; Li, Y. Perovskite-type Li-ion solid electrolytes: A review. *J. Mater. Sci.-Mater. Electron.* **2021**, *32*, 9736–9754. [CrossRef]
26. Wang, C.; Fu, K.; Kammampata, S.P.; McOwen, D.W.; Samson, A.J.; Zhang, L.; Hitz, G.T.; Nolan, A.M.; Wachsman, E.D.; Mo, Y.; et al. Garnet-Type Solid-State Electrolytes: Materials, Interfaces, and Batteries. *Chem. Rev.* **2020**, *120*, 4257–4300. [CrossRef] [PubMed]
27. DeWees, R.; Wang, H. Synthesis and Properties of NaSICON-type LATP and LAGP Solid Electrolytes. *Chemsuschem* **2019**, *12*, 3713–3725. [CrossRef]
28. Hood, Z.D.; Wang, H.; Pandian, A.S.; Keum, J.K.; Liang, C.D. Li_2OHCl Crystalline Electrolyte for Stable Metallic Lithium Anodes. *J. Am. Chem. Soc.* **2016**, *138*, 1768–1771. [CrossRef] [PubMed]
29. Li, Y.; Han, J.; Vogel, S.; Wang, C.-A. The reaction of $Li_{6.5}La_3Zr_{1.5}Ta_{0.5}O_{12}$ with water. *Solid. State. Ion.* **2014**, *269*, 57–61. [CrossRef]
30. Shimonishi, Y.; Toda, A.; Zhang, T.; Hirano, A.; Imanishi, N.; Yamamoto, O.; Takeda, Y. Synthesis of Garnet-Type $Li_7{-}xLa_3Zr_2O_{12}{-}1/2x$ and Its Stability in Aqueous Solutions. *Solid State Ion.* **2011**, *183*, 48–53. [CrossRef]
31. Schroeder, D.J.; Hubaud, A.A.; Vaughey, J.T. Stability of the solid electrolyte Li_3OBr to common battery solvents. *Mater. Res. Bull.* **2014**, *49*, 614–617. [CrossRef]
32. Dashjav, E.; Ma, Q.; Xu, Q.; Tsai, C.-L.; Giarola, M.; Mariotto, G.; Tietz, F. The influence of water on the electrical conductivity of aluminum-substituted lithium titanium phosphates. *Solid State Ion.* **2018**, *321*, 83–90. [CrossRef]
33. Zhao, Y.; Ding, Y.; Li, Y.; Peng, L.; Byon, H.R.; Goodenough, J.B.; Yu, G. A chemistry and material perspective on lithium redox flow batteries towards high-density electrical energy storage. *Chem. Soc. Rev.* **2015**, *44*, 7968–7996. [CrossRef]
34. Bronowski, J. CHAPTER 6—D.C. CONDUCTIVITY. In *Structural Chemistry of Glasses*; Rao, K.J., Ed.; Elsevier Science Ltd.: Oxford, UK, 2002; pp. 203–261.
35. Li, Y.; Liu, M.; Liu, K.; Wang, C.-A. High Li+ conduction in NASICON-type $Li_1{+}xYxZr_2{-}x(PO_4)_3$ at room temperature. *J. Power Sources* **2013**, *240*, 50–53. [CrossRef]
36. Guo, Z.; Qin, X.; Xie, Y.; Lei, C.; Wei, T.; Zhang, Y. Advanced NASICON-type LiTi2(PO4)3 as electrode materials for lithium-ion batteries. *Chem. Phys. Lett.* **2022**, *806*, 140010. [CrossRef]
37. Khatua, S.; Rao, Y.B.; Achary, K.R.; Patro, L.N. Li-ion transport studies of NASICON-type $LiZr_2(PO_4)_3$ solid electrolyte crystallizing in rhombohedral structure at room temperature. *Surf. Interfaces* **2023**, *41*, 103212. [CrossRef]
38. Hou, M.; Liang, F.; Chen, K.; Dai, Y.; Xue, D. Challenges and perspectives of NASICON-type solid electrolytes for all-solid-state lithium batteries. *Nanotechnology* **2020**, *31*, 132003. [CrossRef] [PubMed]
39. Zhang, Y.; Chen, K.; Shen, Y.; Lin, Y.; Nan, C.-W. Enhanced lithium-ion conductivity in a LiZr2(PO4)3 solid electrolyte by Al doping. *Ceram. Int.* **2017**, *43*, S598–S602. [CrossRef]
40. Zhou, Q.; Xu, B.; Chien, P.-H.; Li, Y.; Huang, B.; Wu, N.; Xu, H.; Grundish, N.S.; Hu, Y.-Y.; Goodenough, J.B. NASICON $Li_{1.2}Mg_{0.1}Zr_{1.9}(PO_4)_3$ Solid Electrolyte for an All-Solid-State Li-Metal Battery. *Small Methods* **2020**, *4*, 2000764. [CrossRef]
41. Catti, M.; Comotti, A.; Di Blas, S. High-Temperature Lithium Mobility in α-$LiZr_2(PO_4)_3$ NASICON by Neutron Diffraction. *Chem. Mater.* **2003**, *15*, 1628–1632. [CrossRef]
42. Nomura, K.; Ikeda, S.; Ito, K.; Einaga, H. Ionic conduction behavior in zirconium phosphate framework. *Solid State Ion.* **1993**, *61*, 293–301. [CrossRef]
43. Arbi, K.; Ayadi-Trabelsi, M.; Sanz, J. Li mobility in triclinic and rhombohedral phases of the Nasicon-type compound $LiZr_2(PO_4)_3$ as deduced from NMR spectroscopy. *J. Mater. Chem.* **2002**, *12*, 2985–2990. [CrossRef]
44. Li, Q.H.; Xu, C.; Huang, B.; Yin, X. Rhombohedral $Li_{1+x}Y_xZr_{2-x}(PO_4)_3$ Solid Electrolyte Prepared by Hot-Pressing for All-Solid-State Li-Metal Batteries. *Materials* **2020**, *13*, 035930. [CrossRef] [PubMed]
45. Zhang, Y.-H.; Wu, L.-R.; Ma, J.; Cui, G. Nanotechnology in solid state batteries, what's next? *Next Nanotechnol.* **2023**, *2*, 100011. [CrossRef]

46. Fuentes, R.O.; Figueiredo, F.; Marques, F.; Franco, J. Influence of microstructure on the electrical properties of NASICON materials. *Solid State Ion.* **2001**, *140*, 173–179. [CrossRef]
47. Fuentes, R.; Figueiredo, F.; Marques, F.; Franco, J. Processing and Electrical Properties of NASICON Prepared from Yttria-Doped Zirconia Precursors. *J. Eur. Ceram. Soc.* **2001**, *21*, 737–743. [CrossRef]
48. Buscaglia, M.T.; Bassoli, M.; Buscaglia, V. Solid-state synthesis of nanocrystalline BaTiO3: Reaction kinetics and powder properties. *J. Am. Ceram. Soc.* **2008**, *91*, 2862–2869. [CrossRef]
49. Clabel, J.L.; Awan, I.T.; Pinto, A.H.; Nogueira, I.C.; Bezzon, V.D.N.; Leite, E.R.; Balogh, D.T.; Mastelaro, V.R.; Ferreira, S.O.; Marega, E. Insights on the mechanism of solid state reaction between TiO_2 and $BaCO_3$ to produce BaTiO3 powders: The role of calcination, milling, and mixing solvent. *Ceram. Int.* **2020**, *46*, 2987–3001. [CrossRef]
50. Kotobuki, M.; Koishi, M. Preparation of $Li_{1.5}Al_{0.5}Ti_{1.5}(PO_4)(_3)$ solid electrolyte via a sol-gel route using various Al sources. *Ceram. Int.* **2013**, *39*, 4645–4649. [CrossRef]
51. Ugemuge, N.; Parauha, Y.R.; Dhoble, S.J. Chapter 15—Synthesis and luminescence study of silicate-based phosphors for energy-saving light-emitting diodes. In *Energy Materials*; Dhoble, S.J., Kalyani, N.T., Vengadaesvaran, B., Kariem Arof, A., Eds.; Elsevier: Amsterdam, The Netherlands, 2021; pp. 445–480.
52. Takada, K.; Fujimoto, K.; Inada, T.; Kajiyama, A.L.; Kouguchi, M.; Kondo, S.; Watanabe, M. Sol-gel preparation of Li+ ion conductive thin film. *Appl. Surf. Sci.* **2002**, *189*, 300–306. [CrossRef]
53. Xu, X.X.; Wen, Z.Y.; Wu, J.G.; Yang, X.L. Preparation and electrical properties of NASICON-type structured $Li_{1.4}Al_{0.4}Ti_{1.6}(PO_4)(_3)$ glass-ceramics by the citric acid-assisted sol-gel method. *Solid State Ion.* **2007**, *178*, 29–34. [CrossRef]
54. Ulrich, G.D. Flame Synthesis of Fine Particles. *Chem. Eng. News* **1984**, *62*, 22–29. [CrossRef]
55. Guo, J.Z.; Goodings, J.M.; Hayhurst, A.N.; Taylor, S.G. A simple method for measuring positive ion concentrations in flames and the calibration of a nebulizer/atomizer. *Combust. Flame* **2003**, *133*, 335–343. [CrossRef]
56. Madler, L.; Kammler, H.K.; Mueller, R.; Pratsinis, S.E. Controlled synthesis of nanostructured particles by flame spray pyrolysis. *J. Aerosol Sci.* **2002**, *33*, 369–389. [CrossRef]
57. Dasgupta, M.; Fortugno, P.; Wiggers, H. Plasma-assisted gas-phase synthesis and in-line coating of silicon nanoparticles. *Plasma Process Polym.* **2020**, *17*, 1900245. [CrossRef]
58. Chrystie, R.S.M.; Ebertz, F.L.; Dreier, T.; Schulz, C. Absolute SiO concentration imaging in low-pressure nanoparticle-synthesis flames via laser-induced fluorescence. *Appl. Phys. B-Lasers Opt.* **2019**, *125*, 29. [CrossRef]
59. Abdali, A.; Moritz, B.; Gupta, A.; Wiggers, H.; Schulz, C. Hybrid microwave-plasma hot-wall reactor for synthesis of silica nanoparticles under well-controlled conditions. *J. Optoelectron. Adv. Mater.* **2010**, *12*, 440–444.
60. Schneider, F.; Suleiman, S.; Menser, J.; Borukhovich, E.; Wlokas, I.; Kempf, A.; Wiggers, H.; Schulz, C. SpraySyn-A standardized burner configuration for nanoparticle synthesis in spray flames. *Rev. Sci. Instrum.* **2019**, *90*, 085108. [CrossRef] [PubMed]
61. Angel, S.; Tapia, J.D.; Gallego, J.; Hagemann, U.; Wiggers, H. Spray-Flame Synthesis of $LaMnO_{3+\delta}$ Nanoparticles for Selective CO Oxidation (SELOX). *Energy Fuels* **2021**, *35*, 4367–4376. [CrossRef]
62. Alkan, B.; Cychy, S.; Varhade, S.; Muhler, M.; Schulz, C.; Schuhmann, W.; Wiggers, H.; Andronescu, C. Spray-Flame-Synthesized $LaCo_{1-x}Fe_xO_3$ Perovskite Nanoparticles as Electrocatalysts for Water and Ethanol Oxidation. *ChemElectroChem* **2019**, *6*, 4266–4274. [CrossRef]
63. Ali, M.Y.; Orthner, H.; Wiggers, H. Spray Flame Synthesis (SFS) of Lithium Lanthanum Zirconate (LLZO) Solid Electrolyte. *Materials* **2021**, *14*, 3472. [CrossRef] [PubMed]
64. Yanagida, H.; Kōmoto, K.; Miyayama, M.; Yamada, H. *The Chemistry of Ceramics*; Wiley: Chichester, UK, 1996.
65. Hund, F. Anomale Mischkristalle im System $ZrO_2Y_2O_3$ Kristallbau der Nernst-Stifte. *Z. Elektrochem. Und Angew. Phys. Chem.* **1951**, *55*, 363–366. [CrossRef]
66. Yamada, T.; Kubota, Y.; Makinose, Y.; Suzuki, N.; Nakata, K.; Terashima, C.; Matsushita, N.; Okada, K.; Fujishima, A.; Katsumata, K.-i. Single Crystal ZrO2 Nanosheets Formed by Thermal Transformation for Solid Oxide Fuel Cells and Oxygen Sensors. *ACS Appl. Nano Mater.* **2019**, *2*, 6866–6873. [CrossRef]
67. El-Shinawi, H.; Greaves, C.; Janek, J. Sol–gel synthesis and room-temperature properties of α-$LiZr_2(PO_4)_3$. *RSC Adv.* **2015**, *5*, 17054–17059. [CrossRef]
68. Lai, Y.; Sun, Z.; Jiang, L.; Hao, X.; Jia, M.; Wang, L.; Liu, F. Rapid sintering of ceramic solid electrolytes $LiZr_2(PO_4)_3$ and $Li_{1.2}Ca_{0.1}Zr_{1.9}(PO_4)_3$ using a microwave sintering process at low temperatures. *Ceram. Int.* **2019**, *45*, 11068–11072. [CrossRef]
69. Lieber, C.M.; Wang, Z.L. Functional Nanowires. *MRS Bull.* **2007**, *32*, 99–108. [CrossRef]
70. Teoh, W.Y.; Amal, R.; Madler, L. Flame spray pyrolysis: An enabling technology for nanoparticles design and fabrication. *Nanoscale* **2010**, *2*, 1324–1347. [CrossRef] [PubMed]
71. *PubChem Compound Summary for CID 26251, Zirconium Nitrate*; National Library Medicine: Bethesda, MD, USA, 2021.
72. Angel, S.; Neises, J.; Dreyer, M.; Friedel Ortega, K.; Behrens, M.; Wang, Y.; Arandiyan, H.; Schulz, C.; Wiggers, H. Spray-flame synthesis of La(Fe, Co)O_3 nano-perovskites from metal nitrates. *AIChE J.* **2020**, *66*, e16748. [CrossRef]
73. Wang, D.; Kou, R.; Ren, Y.; Sun, C.-J.; Zhao, H.; Zhang, M.-J.; Li, Y.; Huq, A.; Ko, J.Y.P.; Pan, F.; et al. Synthetic Control of Kinetic Reaction Pathway and Cationic Ordering in High-Ni Layered Oxide Cathodes. *Adv. Mater.* **2017**, *29*, 1606715. [CrossRef] [PubMed]
74. Xu, H.; Wang, S.; Wilson, H.; Zhao, F.; Manthiram, A. Y-Doped NASICON-type $LiZr_2(PO_4)_3$ Solid Electrolytes for Lithium-Metal Batteries. *Chem. Mater.* **2017**, *29*, 7206–7212. [CrossRef]

75. Chraska, T.; King, A.H.; Berndt, C.C.; Karthikeyan, J. Phase Transformation as a Function of Particle Size in Nanocrystalline Zirconia. *MRS Online Proc. Libr.* **1997**, *481*, 613–617. [CrossRef]
76. Ramirez, L.; Mecartney, M.L.; Krumdieck, S.P. Nanocrystalline ZrO_2 thin films on silicon fabricated by pulsed-pressure metalorganic chemical vapor deposition (PP-MOCVD). *J. Mater. Res.* **2008**, *23*, 2202–2211. [CrossRef]
77. Srinivasan, R.; Rice, L.; Davis, B. Critical Particle Size and Phase Transformation in Zirconia: Transmission Electron Microscopy and X-Ray Diffraction Studies. *J. Am. Ceram. Soc.* **2005**, *73*, 3528–3530. [CrossRef]
78. Lackner, P.; Zou, Z.; Mayr, S.; Diebold, U.; Schmid, M. Using photoelectron spectroscopy to observe oxygen spillover to zirconia. *Phys. Chem. Chem. Phys.* **2019**, *21*, 17613–17620. [CrossRef] [PubMed]
79. Egger, P.; Dirè, S.; Ischia, M.; Campostrini, R. Pyrolysis study of sol-gel derived zirconia by TG-GC-MS. *J. Therm. Anal. Calorim.* **2005**, *81*, 407–415. [CrossRef]
80. Shi, L.; Qu, T.; Liu, D.; Deng, Y.; Yang, B.; Dai, Y. Process of Thermal Decomposition of Lithium Carbonate. In Proceedings of the Materials Processing Fundamentals 2020, Cham, Switzerland, 9 January 2020; pp. 107–116.
81. Geiculescu, A.C.; Spencer, H.G. Thermal Decomposition and Crystallization of Aqueous Sol-Gel Derived Zirconium Acetate Gels: Effects of the Additive Anions. *J. Sol-Gel Sci. Technol.* **2000**, *17*, 25–35. [CrossRef]
82. Efaw, C.M.; Vandegrift, J.L.; Reynolds, M.; McMurdie, S.; Jaques, B.J.; Hu, H.; Xiong, H.; Hurley, M.F. Characterization of zirconium oxides part I: Raman mapping and spectral feature analysis. *Nucl. Mater. Energy* **2019**, *21*, 100707. [CrossRef]
83. Kim, D.-J.; Jung, H.-J.; Yang, I.-S. Raman Spectroscopy of Tetragonal Zirconia Solid Solutions. *J. Am. Ceram. Soc.* **1993**, *76*, 2106–2108. [CrossRef]
84. Colbea, C.; Avram, D.; Cojocaru, B.; Negrea, R.; Ghica, C.; Kessler, V.G.; Seisenbaeva, G.A.; Parvulescu, V.; Tiseanu, C. Full Tetragonal Phase Stabilization in ZrO_2 Nanoparticles Using Wet Impregnation: Interplay of Host Structure, Dopant Concentration and Sensitivity of Characterization Technique. *Nanomaterials* **2018**, *8*, 988. [CrossRef] [PubMed]
85. Long, D.A. Infrared and Raman characteristic group frequencies. Tables and charts George Socrates John Wiley and Sons, Ltd., Chichester, Third Edition, 2001. Price £135. *J. Raman Spectrosc.* **2004**, *35*, 905. [CrossRef]
86. Strobel, R.; Pratsinis, S.E. Effect of solvent composition on oxide morphology during flame spray pyrolysis of metal nitrates. *Phys. Chem. Chem. Phys.* **2011**, *13*, 92469252. [CrossRef] [PubMed]
87. Stodt, M.F.B.; Groeneveld Jan, D.; Mädler, L.; Kiefer, J.; Fritsching, U. Microexplosions of multicomponent drops in spray flames. *Combust. Flame* **2022**, *240*, 112043. [CrossRef]
88. Wang, J.; He, T.; Yang, X.; Cai, Z.; Wang, Y.; Lacivita, V.; Kim, H.; Ouyang, B.; Ceder, G. Design principles for NASICON super-ionic conductors. *Nat. Commun.* **2023**, *14*, 5210. [CrossRef] [PubMed]

energies

MDPI

Article

Surface Modification of Ga-Doped-LLZO ($Li_7La_3Zr_2O_{12}$) by the Addition of Polyacrylonitrile for the Electrochemical Stability of Composite Solid Electrolytes

Hyewoo Noh [1,2,†], Daeil Kim [1,†], Wooyoung Lee [1], Boyun Jang [1,*], Jeong Sook Ha [2,*] and Ji Haeng Yu [1,*]

1 Korea Institute of Energy Research, Daejeon 34129, Republic of Korea; suk04092@naver.com (H.N.); daeilkim@kier.re.kr (D.K.); leewyo225@kier.re.kr (W.L.)
2 Department of Chemical and Biological Engineering, Korea University, Seoul 02841, Republic of Korea
* Correspondence: byjang@kier.re.kr (B.J.); jeongsha@korea.ac.kr (J.S.H.); jhyu@kier.re.kr (J.H.Y.)
† These authors contributed equally to this work.

Abstract: Composite solid electrolytes (CSEs), often incorporating succinonitrile (SCN), offer promi I confirm sing solutions for improving the performance of all-solid-state batteries. These electrolytes are typically made of ceramics such as $Li_7La_3Zr_2O_{12}$ (LLZO) and polymers such as poly(vinylidene fluoride-co-hexafluoropropylene) (PVDF-HFP). Garnet-applied polymer–ceramic electrolyte (g-PCE) is composed of PVDF-HFP, SCN, and LLZO. However, the interface between SCN and LLZO is reportedly unstable owing to the polymerization of SCN. This polymerization could cause two serious problems: (1) gelation during the mixing of LLZO and SCN and (2) degradation of ionic performance during charge and discharge. To prevent this catalytic reaction, polyacrylonitrile (PAN) can be added to the g-PCE (g-PPCE). PAN blocks the polymerization of SCN through a cyclization process involving La ions which occurs more rapidly than SCN polymerization. In this study, the enhanced chemical stability of the garnet-applied PAN-added polymer ceramic electrolyte (g-PPCE) was achieved by using an impregnation process which added SCN with 5 wt.% of PAN. The resulting CSE has an ionic conductivity of ~10^{-4} S/cm at room temperature. Coin-type cells assembled with LFP ($LiFePO_4$) and LNCM ($LiNi_{0.6}Co_{0.2}Mn_{0.2}O_2$) cathodes with Li-metal anodes show specific discharge capacities of 150 and 167 mAh/g at 0.1 C, respectively, and stable cycle performance. Additionally, a pouch-type cell with a discharge capacity of 5 mAh also exhibits potential electrochemical performance.

Keywords: composite solid electrolyte; garnet oxide; polymerization; cyclization; solid-state Li-metal battery

Citation: Noh, H.; Kim, D.; Lee, W.; Jang, B.; Ha, J.S.; Yu, J.H. Surface Modification of Ga-Doped-LLZO ($Li_7La_3Zr_2O_{12}$) by the Addition of Polyacrylonitrile for the Electrochemical Stability of Composite Solid Electrolytes. *Energies* **2023**, *16*, 7695. https://doi.org/10.3390/en16237695

Academic Editors: Huang Zhang and Yuan Ma

Received: 24 October 2023
Revised: 11 November 2023
Accepted: 15 November 2023
Published: 21 November 2023

1. Introduction

As the global consumption of electricity increases, electricity storage, not merely generation, has become increasingly important because of fluctuations in levels of electricity generation from renewable sources [1–3]. Recently, all-solid-state lithium batteries have been widely studied owing to their potential to offer higher energy density, extended cycle life, and enhanced safety compared to conventional lithium-ion batteries [2–5]. Among the solid-state electrolytes, oxide-based solid-state electrolytes have attracted attention owing to their high ionic conductivity and thermal and mechanical stability, along with sulfide electrolytes [6–10]. Oxide-based solid electrolytes are primarily employed in the form of pellets or sheets, which results in various limitations, such as excessive electrolyte thickness or susceptibility to external impacts [11]. To overcome these limitations, composite solid electrolytes (CSEs) based on oxide-based and polymer electrolytes have been actively studied [12–17].

CSEs containing $Li_7La_3Zr_2O_{12}$ (LLZO), an oxide-based electrolyte, along with succinonitrile (SCN) serving as a plasticizer in a polymer electrolyte, have been proposed. Vanita et al. prepared a ceramic-rich (50 wt.%) composite electrolyte with $Li_{6.4}La_3Zr_{1.4}Ta_{0.6}O_{12}$

(LLZTO), polyethylene oxide (PEO), lithium bis(trifluoromethanesulfonyl)imide (LiTFSI), and SCN using solvent-free cryomilling. The mixture of SCN and LiTFSI (20:1), called plastic-crystal electrolyte (PCE), exhibits high room-temperature conductivity (5.26×10^{-3} S/cm at 298 K). When it was mixed with LLZTO, it was observed that the mixture promoted lithium-ion transport at the grain boundary. Moreover, the addition of PEO at 5 wt.% reduces the brittleness/hardness of the CSE, thus overcoming the electrode–electrolyte interfacial resistance [18]. Kumlachew et al. fabricated a tri-layer composite polymer electrolyte (Tri-CPE) by uniformly dispersing Ga/F-doped LLZO into a poly(vinyl fluoride) (PVDF)/polyacrylonitrile (PAN)/LiTFSI/SCN matrix using a simple solution-casting method. The top and bottom membranes did not contain LLZO, and the Tri-CPE exhibited an ionic conductivity at room temperature of 4.5×10^{-4} S/cm, with a high Li^+ ion transfer number (0.84). The cell with the LMO@T-LNCM811-based composite cathode showed a Coulombic efficiency of 99.4% and capacity retention of 89.8% at 1 C over 300 cycles [19]. Despite the good performance shown in the above studies, problems such as the coordination or dehydrogenation of SCN with transition metals in ceramic electrolytes have been reported [20]. The primary issue lies in the gelation of the electrolytes during the mixing of SCN and $Li_{6.25}Ga_{0.25}La_3Zr_2O_{12}$ (LGLZO), rendering the fabrication of CSEs challenging.

Various methods have been developed to address above issues. Cheng et al. fabricated a flexible Al-LLZO sheet with a thickness of approximately 75 μm through a tape-casting method and impregnated the sheet with an ionic liquid, Li(G4)FSI, which consists of a lithium bis(fluorosulfonyl)-imide (LiFSI) and tetraglyme (G4) in an equimolar complex. [21]. The impregnated ionic liquid yielded positive results by bridging the gap between the LLZO particles, thereby creating an effective Li-ion pathway. Consequently, the impregnated electrolyte film exhibited a high ionic conductivity (approximately 0.1 mS/cm) and good cycling performance, owing to the excellent interfacial stability with the electrode. Because the electrolyte is fabricated through a tape-casting method, mass production is possible, which considerably improves processability. Shen et al. fabricated LLZO scaffolds with a porosity of 75% using an aqueous freeze tape-casting method [22]. Using this method, LLZO scaffolds with constant pore size and shape can be fabricated. A porous LLZO scaffold impregnated with PEO/LiTFSI exhibited excellent ionic conductivity. However, these studies did not consider the side reactions of the impregnated SCN.

Recently, a sharp increase in resistance during the charge and discharge of cells with these composite electrolytes was reported, which was attributed to the polymerization of SCN during the side reaction of SCN with La in LGLZO [23,24]. Yang et al. confirmed the coordination between the nitrile groups of SCN and the La atoms of LLZTO [23]. More specifically, Fourier-transform infrared spectroscopy (FTIR) and x-ray photoelectron spectroscopy (XPS) analyses indicated that the -C≡N groups of SCN transform into C=N groups, which suggests that SCN is polymerized. This phenomenon increases CSE resistance and decreases cell performance. To address this issue, they dissolved 20 wt.% of PAN in the PCE and applied the solution as a coating on the prepared LLZTO substrate (~800 μm). The cell with the PAN-modified electrolyte exhibited a conductivity of approximately 10^{-4} S/cm as well as excellent Coulombic efficiency (95.6%) and capacity retention (99% at 0.1 C, 25 °C) after 250 cycles [23]. In addition, Zhang et al. reported the reaction of the nitrile group of SCN with LLZTO to decrease the ionic conductivity of SCN. They applied a PAN coating on LLZTO and subsequently dissolved it in SCN to obtain an electrolyte with a thickness of approximately 13 μm using the tape-casting method. The SCN/PAN-coated LLZTO exhibited a conductivity of ~10^{-4} S/cm at room temperature. In cell applications, it demonstrated a capacity of 163 mAh/g and an 87% capacity retention after 500 cycles at 0.2 C [24]. However, these studies primarily focus on the interface between the LLZTO substrate and the SCN layer, while investigations into the interfacial behavior of SCN and LLZO particles during the impregnation process remain insufficient.

To overcome these problems, we designed a process, electrolyte materials, and a structure that can solve the gelation problem during the mixing process as well as the difficulties incurred by the polymerization of SCN during charging and discharging. By using a simple tape-casting and impregnation process, three-layer g-PCE consisting of a PVDF-HFP-based SPE-LGLZO was successfully fabricated without gelation during the process. As mentioned above, however, the SCN-LLZO interface poses a risk of unstable electrochemical performance due to the polymerization of SCN during the charge/discharge process. Also, the interface area between SCN and LGLZO in this study is much higher than that reported previously [25]. SCN's polymerization is much more severe and critical to electrochemical performance. Therefore, the effect of added PAN on SCN's polymerization and the electrochemical performance of a solid-state Li-metal battery was carefully investigated. At first, polymerization of SCN with LGLZO and the role of PAN were described with diagrams. To visualize the polymerization, polymerization behaviors of SCN in a composite of LGLZO and SCN were studied by a thermal aging experiment. Additionally, the optimization of the amount of added PAN was conducted by measuring basic ionic-conducting properties. Finally, electrochemical performance levels with LFP||Li-metal and NCM||Li-metal batteries were evaluated to confirm the role of PAN by comparing those with and without PAN in CSE.

2. Materials and Methods

2.1. Synthesis of $Li_{6.25}Ga_{0.25}La_3Zr_2O_{12}$ Powders

The LGLZO was synthesized via a solid-state reaction. Li_2CO_3 (Sigma-Aldrich, Saint Louis, MO, USA, 99%), La_2O_3 (Sigma-Aldrich, 99.99%), ZrO_2 (Terio Co., Qingdao, China, 99.9%) and Ga_2O_3 (Sigma-Aldrich, >99.99%) were weighed stoichiometrically. La_2O_3 was dried at 900 °C for 6 h before weighing to remove moisture. A 10 mol% excess of Li_2CO_3 was added to compensate for the Li loss during the subsequent heat treatment of LGLZO. The mixture was ball-milled for 24 h with ethanol and first-calcined at 900 °C for 6 h in an alumina crucible under ambient atmosphere to obtain an LGLZO powder. The heating rate was 5 °C/min. After the first calcination stage, the LGLZO powder was ball-milled in a zirconia jar for 24 h. The powder was subsequently sintered at 1200 °C for 12 h with a heating rate of 5 °C/min.

2.2. Preparation of LGLZO Sheets

The tape-casting method was used to produce LGLZO green sheets. The LGLZO powder was heated at 900 °C for 1 h in dry air before use. The average particle size of the LGLZO powder was 5–7 μm. To create a slurry, we added 3 wt.% of BYK 103 dispersant (BYK), 8 wt.% of dioctyl phthalate plasticizer (DoP, Sigma Aldrich), and 11 wt.% of ethyl cellulose binder (10 cP, Sigma Aldrich). Ethanol and propyl acetate (38 wt.%) were used as solvents. All concentrations were relative to the weighed LGLZO powder. The tape slurry was ball-milled for 48 h at room temperature. After that, the slurry was casted on the polyester (PET) film with a comma coater. The gap of the coater was 150 μm and the temperature of the plate was 40 °C. The speed of the coater was 3.6 mm/s. Lastly, the sheet was dried at 50 °C for 1 h after casting.

2.3. Preparation of the SPE

A PVDF-HFP-based polymer electrolyte was prepared from PVDF-HFP (Mw = 400, 000, Sigma-Aldrich), SCN (Sigma-Aldrich, 99%), LiTFSI (Sigma-Aldrich), and 4-fluoro ethylenecarbonate (FEC) (Sigma-Aldrich). First, a PCE, which is essential for Li ion conductivity, was prepared by mixing SCN and LiTFSI in a molar ratio of 19:1. The solution was stirred at 50 °C for 12 h to ensure homogeneous dispersion. For the SPE solution, 1.5 g of PVDF-HFP, 2.5 g of PCE, 0.12 g of FEC and 6 g of acetone (Daejung, Nonsan, Republic of Korea, 99.5%) were mixed at 80 °C for 12 h. The incorporation of 3 wt.% of FEC served as a beneficial electrolyte additive, effectively causing the formation of LiF and enhancing the stability of the solid electrolyte interphase (SEI) on the lithium metal anodes [26]. The solution was cast onto a glass plate using the doctor blade method to form a film with a thickness of 15 μm.

2.4. *Synthesis of g-PCE and g-PPCE*

2.4.1. g-PCE (Garnet-Applied Polymer–Ceramic Electrolyte)

The g-PCE is a garnet-applied polymer–ceramic electrolyte. It is composed of PVDF-HFP, SCN, LiTFSI, and LGLZO. To prepare the g-PCE, 3 ml of the prepared SPE solution was cast onto a glass plate using the doctor blade method and dried at 50 °C for 10 min in a glove box. The cast SPE films were punched into disks 20 and 18 mm in diameter. Disks of 20 mm diameter were used at the bottom of the electrolyte. Then, 10 μL of liquid-phase PCE was pipetted onto the SPE disks. An LGLZO sheet (16 mm in diameter) was placed at the center of the SPE disk. Subsequently, 10 μL of PCE was pipetted onto the LGLZO green sheet. Finally, an 18 mm diameter SPE disk was placed on top of the second PCE aliquot. The prepared g-PCE was covered with PET film to prevent SCN volatilization and heated on a hot plate at 50 °C for 5 h to uniformly impregnate the LGLZO green sheet with the PCE.

2.4.2. g-PPCE (Garnet-Applied PAN Added Polymer–Ceramic Electrolyte)

The g-PPCE is a garnet-applied PAN-added polymer–ceramic electrolyte. It is composed of PVDF-HFP, SCN, LiTFSI, LGLZO and PAN. PPCE was prepared by adding 5 wt.% PAN to the PCE and stirring at 120 °C for 24 h to obtain a homogeneous solution. This solution was impregnated onto a green sheet using the same process as described above in Section 2.4.1.

2.5. *Material Characterizations*

Field emission scanning electron microscopy (FE-SEM, Hitachi S-4800, Tokyo, Japan) and environmental scanning electron microscopy (ESEM, QuattroS, London, UK) were used to analyze the microstructures of the LGLZO green sheet and g-PPCE. Additionally, an energy-dispersive spectrometer (EDS, QuattroS, London, UK) was used to determine the elemental distribution of the electrolyte. X-ray diffraction (XRD) patterns were measured from 20° to 50° using a Rigaku Smart Lab (Houston, TX, USA) to reveal the crystal structure of the electrolyte. The thermal stability of each electrolyte in the g-PPCE was analyzed by thermogravimetry analysis (TGA, Labsys Evo TG-DTA, Pennsauken, NJ, USA) in an air atmosphere, with a ramp-up condition of 10 °C/min and a temperature range of 25–800 °C. The chemical structure of the electrolyte was characterized, and the mechanism for the competitive coordination protection of SCN with PAN was analyzed using Fourier Transform Infrared Spectroscopy (FTIR spectroscopy, Bruker Alpha II., Billerica, MA, USA)

2.6. *Preparation of Cells*

LFP/CSE/Li cells were used to evaluate the charge and discharge performance levels. The cathode was made of LFP/carbon black/PVDF (80:10:10). The cathode was punched into circular pieces with diameters of 14 mm. The active substance was loaded at a density of approximately 2.4 mg/cm^2. Additionally, NCM/CSE/Li cells were fabricated, with the cathode consisting of NCM/carbon black/PVDF (94:3:3). The cathode was punched into the circular pieces with a diameter of 14 mm. The active substance was loaded at a density of approximately 7 mg/cm^2.

Moreover, the electrochemical properties of CSE were determined using a 3450 sized pouch cell. First, the cathode LFP with a density of 3.3 mg/cm^2 was cut into a 3 × 4 cm^2 size. Li metal with 100 μm thickness was also cut into the same dimensions and hot-pressed with copper foil. The g-PCE and g-PPCE were prepared with 3.4 × 4.4 cm^2-sized SPE and a 3.2 × 4.2 cm^2-sized LGLZO sheet, and the electrodes and electrolytes were stacked carefully. The pouch cells with g-PCE and g-PPCE were prepared with cutting and sealing processes. The pouch cells received uniform pressure using the cold isostatic press (CIP) after production (100 bar, 5 min). During the process, stacking three layers of electrolytes accurately was not easy, but the items were successfully prepared. Although the solid electrolyte was investigated for solid-state battery, an additional solid electrolyte, also known as catholyte, was needed to gain a stable interface inside of the cathode. Because this research was only focused on controlling interfaces in a solid electrolyte, the effect of a cathode–electrolyte interface in cathode should be eliminated. A small amount of gel electrolyte was added to form a gel state, but only on the inside of the electrode. Specifically, 10 μL of PVA-CN-based gel electrolyte (3 wt.% of PVA-CN-added liquid electrolyte; the liquid electrolyte was 1 M LiPF6 in EC:DEC(1:1 vol.%)) was dropped on the cathode's surface and aged at 50 °C for 12 hrs for gelation after cell assembly. All of the incorporation of the gel electrolyte for cathode preparation was performed according to previously reported research [26–28].

2.7. Electrochemical Performance Measurements

Electrochemical impedance spectroscopy (EIS, WonATech) was performed on SS | g-PCE | SS and SS | g-PPCE | SS shaped in a sandwich structure. The measurement conditions were fixed in the frequency range between 1 MHz and approximately 0.1 mHz at 25 °C. The ionic conductivity (σ) was calculated by Equation (1).

$$\sigma = \frac{L}{RS} \tag{1}$$

where R is the bulk resistance, S is the area of the electrode, and L is the thickness of the electrolyte. Linear sweep voltammetry (LSV) was performed using a cell with the SS/g-PCE/Li and SS/g-PPCE/Li structure. The voltage range was 3–5.3 V, at a scan rate of 1 mV/s at 25 °C. Cyclic voltammetry (CV) was performed using a cell with the SS/g-PCE/Li structure. The voltage range was 1–6 V, at a scan rate of 1 mV/s at 25 °C. SS/g-PCE/Li and SS/g-PPCE/Li cells were required to measure the Li^+ transference number. The Li^+ transference number is calculated using Equation (2), where ΔV is 10 mV; I_o and I_{ss} represent the initial and steady-state currents, respectively; and R_o and R_{ss} are the initial and steady-state interfacial resistances between the Li metal and the electrolyte.

$$t_{Li^+} = \frac{I_{SS}(\Delta V - I_o R_o)}{I_o(\Delta V - I_{SS} R_{SS})} \tag{2}$$

The activation energy E_a for the conduction of Li^+ can be obtained from the Nernst equation (Equation (3)).

$$\ln \sigma \cdot T = \ln (N \cdot q \cdot e \cdot D_0) - (E_a/k \cdot T) \tag{3}$$

where σ is the conductivity, N is the charge carrier density, q is the electric charge, e is the elementary-charge constant, D_0 is temperature-independent pre-exponential constant, k is the Boltzmann constant, and T is the temperature. The electrochemical properties of the g-PPCE were analyzed using a CR2032 coin cell. LFP/g-PPCE/Li cells were tested at 25 °C with a voltage range of 3–4 V and a C-rate of 0.1 C. Additionally, NCM/g-PCE/Li and NCM/g-PPCE/Li cells were tested at 25 °C with a voltage range of 3–4.25 V and a C-rate of 0.1 C. Moreover, the electrochemical properties of g-PPCE as shown were determined using 3450 sized pouch cells. The LFP/g-PPCE/Li cell was tested at 25 °C with a voltage range of 3–4 V and a C-rate of 0.1 C.

3. Results

Figure 1 shows a schematic of the SPE-LGLZO CSE and the chemical reaction between SCN and PAN on the LGLZO surface with and without PAN. Specifically, the structure on the left shows a three-layer CSE using a ceramic middle layer and two polymer top/bottom layers. The ceramic layer is brittle and has a rough surface due to its high ceramic content (~80 wt.%). There are two main weaknesses of this application as a solid electrolyte. First, it is difficult to handle during cell preparation. Secondly, large interface resistance levels with the electrodes are unavoidable. Therefore, we have designed a three-layered ceramic–polymer composite electrolyte. The ceramic layer was an LGLZO sheet impregnated with PCE. Two solid polymer electrolyte (SPE) layers consisted of poly(vinylidene fluoride-co-hexafluoropropylene) (PVDF-HFP) and PCE positioned on the top and bottom, and a PCE-impregnated LGLZO sheet was positioned in the middle. SPEs on the top and bottom were sufficiently flexible and strong to form a good interface with the electrode and exhibited a high ionic conductivity (~1 mS/cm) owing to the inclusion of PCE [29]. In Figure S1, the Nyquist impedance plots of PCEs with various ratios of LiTFSI and SCN are shown. The PCE with a specific ratio of LiTFSI:SCN = 1:19 mol% was selected due to its solid-state nature at that particular ratio and its excellent conductivity. The LGLZO sheet in the middle had an LGLZO content of approximately 80%, which provided the main ionic conduction paths through the ceramics. In addition, the impregnation of the LGLZO sheet with PCE increased the mobility of Li-ions and reduced the interfaces between the LGLZO powders.

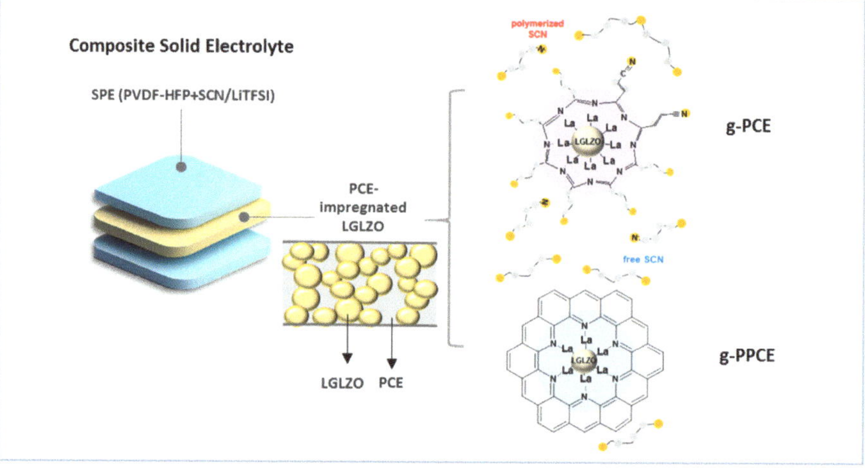

Figure 1. Schematic illustration of SPE-LGLZO composite solid electrolyte and the chemical reactions on the LGLZO particles of g-PCE and g-PPCE.

During the preparation of SPE-LGLZO CSEs, the La of LGLZO and the nitrile groups of SCN reacted chemically, resulting in the polymerization of SCN. The structure at the top right of Figure 1 shows the chemical structures of the species involved in SCN polymerization, which entailed the reaction between SCN and LGLZO in g-PCE. The coordination between La on the surface of LGLZO and the nitrile group of SCN leads to the conversion of the triple bond in -C≡N into a double bond, forming a -C=N- group. This was observed in the chemical structure of the resulting polymer. Additionally, the polymerization of the surrounding SCN was observed. This reaction occurred due to the π-back donation effect, in which the nitrile group donated an electron to the unoccupied d-orbital of the La ion, while the inner orbital electrons of the La ion transferred to the nitrile group. Eventually, an overall charge transfer occurred from LGLZO to the nitrile group. As a result, electrons accumulated on the nitrile groups of SCN, increasing the charge density and creating a Lewis alkaline environment, thereby causing the polymerization of the nitrile group. The polymerized nitrile groups manifested as a continuous decrease in ionic conductivity due to loss of its original gauche-trans conformations [24]. A more serious problem is that the Lewis alkaline environment causes polymerization of not only the SCN on the LGLZO surface, but also the surrounding SCN away from the surface, which keeps reducing the electrochemical functionality of the SCN during the charge/discharge process [23]. The bottom-right structure in Figure 1 represents the reaction between La and PAN on the surface of LGLZOs in the g-PPCE, which is a CSE containing PAN. The introduction of PAN, which has a strong polarity, caused La on the surface of LGLZO to react with PAN more readily than with SCN. During the coordination of the La in LGLZO and PAN, the conversion of the -C≡N in PAN into -C=N- induced PAN cyclization. [30,31] Owing to the strong polarity of cyclized PAN, free SCN in the vicinity remained unpolymerized. Therefore, PAN prevented the polymerization of SCN, both on the LGLZO surface and in the surroundings, which retained its electrochemical function during cell cycles. It is certain that, because the cyclized PAN cannot cover LGLZO particle perfectly, there were still Li-ion conducting paths from the plastic crystal electrolyte (LiTFSI-added SCN) into the bulk LGLZO, as the simply drawn images in Figure S2 depict.

As to the gelation during the mixing of LGLZO and SPE, photo images of the SPE/LGLZO and SPE/PAN/LGLZO slurry before and after mixing were obtained, as shown in Figure S3a,b. Without PAN, the slurry gelled, along with a severe color change. On the other hand, slurry with PAN exhibited no gelation, but still evinced a color change. The photo images of fabricated sheets are also given. With slurry without PAN, it was impossible to gain a film due to the gelation, while with the slurry with PAN, it was possible to fabricate a sheet. Even though there was no gelation, the color change still indicated a part-polymerization of SCN. Therefore, we have developed a different fabrication, one with tape-casting of a ceramic sheet and impregnation of PCE, as in the following description. Figure 2 shows a schematic of the fabrication of SPE, LGLZO sheets and CSE, along with SEM images of each electrolyte. The SEM image of the SPE fabricated through blade coating reveals small pores distributed on the surface. The cross-sectional SEM image of the LGLZO sheet prepared using the tape-casting method reveals a constant thickness, suggesting that a constant interface can be achieved by placing the SPEs at the top and bottom of the LGLZO sheet. Also, Figure S4 shows a cross-sectional SEM image of g-PPCE along with the corresponding EDS mapping results. Notably, the nitrogen (N) content originating from SCN exhibits an even distribution within both the SPE and the LGLZO sheets. Similarly, the fluorine (F) content from PVDF-HFP and LiTFSI is uniformly distributed, with a higher concentration found in the SPE. Finally, lanthanum (La), a constituent of LGLZO, is primarily concentrated within, and uniformly distributed across, the LGLZO sheet. The SEM image and the XRD pattern of the LGLZO particles are shown in Figure S5. The SEM image reveals the constant particle size of the LGLZO, whereas XRD demonstrates its absolute cubic phase. The average particle size of the LGLZO is 5–7 μm. Numerous research studies focusing on polymer–ceramic composite solid electrolytes employ nano-scale ceramic particles to increase specific surface area, thereby preventing polymer crystallization and consequently improving ionic conductivities [32–34]. These studies suppose an increase in the particle volumetric distribution. However, as the ceramic content in the composite solid electrolyte increases, aggregation of ceramic particles occurs [35]. In our research, we employ micro-sized ceramic materials to mitigate particle aggregation, a common issue when using nano-sized particles. This approach allows us to substantially enhance the ceramic content, creating a predominant pathway for lithium ions within the ceramic structure. [36,37]. Cubic-phase LGLZO has an ionic conductivity that is approximately 10 times higher than that of the tetragonal phase [7]. The XRD patterns of the LGLZO green sheet, SPE, and CSE are shown in Figure S6. The XRD pattern of the CSE also showed the same cubic phase of LGLZO. In addition, the existence of pores in the cross-sectional image demonstrates an effective PCE impregnation. The surface of the three-layer CSE closely resembles that of the SPE. Furthermore, the cross-sectional image evidenced an electrolyte thickness of approximately 75 μm in total. Additionally, a smooth and constant connection between the SPE and LGLZO was demonstrated. It should be mentioned that there were no differences in the microstructures between samples with and without PAN, because only a small amount of PAN, lower than 10 wt.%, was substituted for SCN. Moreover, to prove the role of PAN, electrochemical performance levels as well as the chemical change of g-PCE was precisely observed, as described in the following.

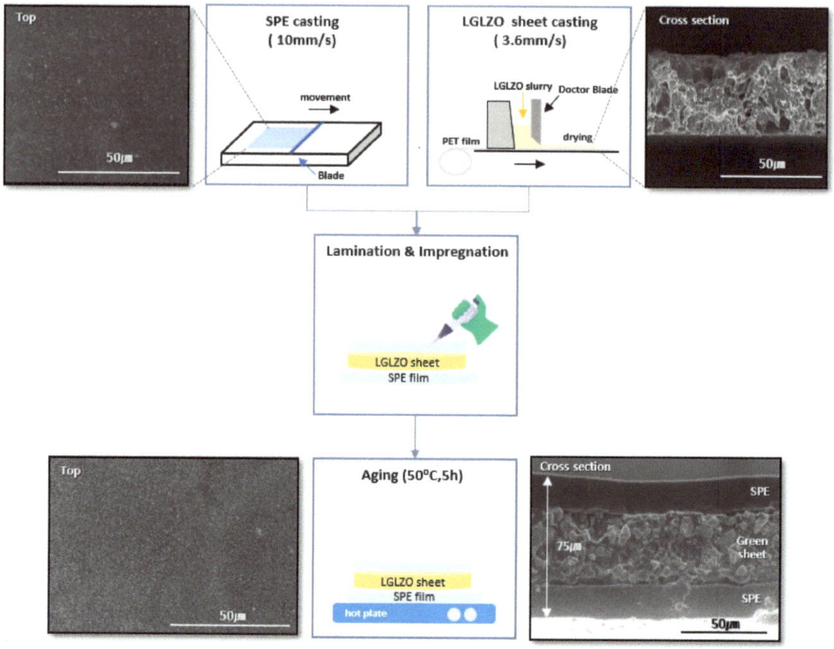

Figure 2. Schematic diagram of preparing g-PPCE and SEM images of the SPE, LGLZO sheet, and g-PPCE membrane.

It is well-known that the color of PCE continuously changes from ivory to brown, and finally to dark brown, with the proceeding of polymerization of SCN [38–40]. To effectively illustrate the browning phenomenon, wherein the slurry turns brown, a simple thermal aging experiment was conducted with beaker cells. To quantify the alterations in ionic conductivity resulting from the reaction between SCN and LGLZO and the influence of PAN, EIS measurements were conducted on the PCE/LGLZO and PPCE/LGLZO slurries. The left inset image of Figure 3a shows the beaker cell. After aging at 80 °C for 1 h, the color of the slurry changed from ivory to dark brown, as shown in the right inset image, which clearly indicates polymerization of SCN. As shown in EIS results of the PCE/LGLZO slurry before and after aging, its resistance increased from 425 Ω to 1395 Ω. This can be attributed to the reaction between SCN with LGLZO, which led to the polymerization of SCN, resulting in the browning and gelling of the slurry.

Figure 3b shows the images and EIS measurements of the PPCE/LGLZO slurry before and after aging. After aging at 80 °C for 1 h, the resistance value increased from 503 Ω to 601 Ω. The pristine PPCE/LGLZO slurry shows a higher series resistance than that of PCE/LGLZO slurry due to the effect of PAN, which does not aid the dissociation of Li-salts compared to SCN. The browning progressed less, compared to the case without PAN. In addition, no additional color change was detected after the initial change into bright brown. This color change is surely attributed to cyclization of PAN on the LGLZO surface [41]. Nevertheless, the inclusion of PAN prevented the severe reaction between SCN and LGLZO. The chemical instability of SCN toward LGLZO was corroborated by FTIR analysis to validate the alteration in the chemical structure of SCN with adding PAN.

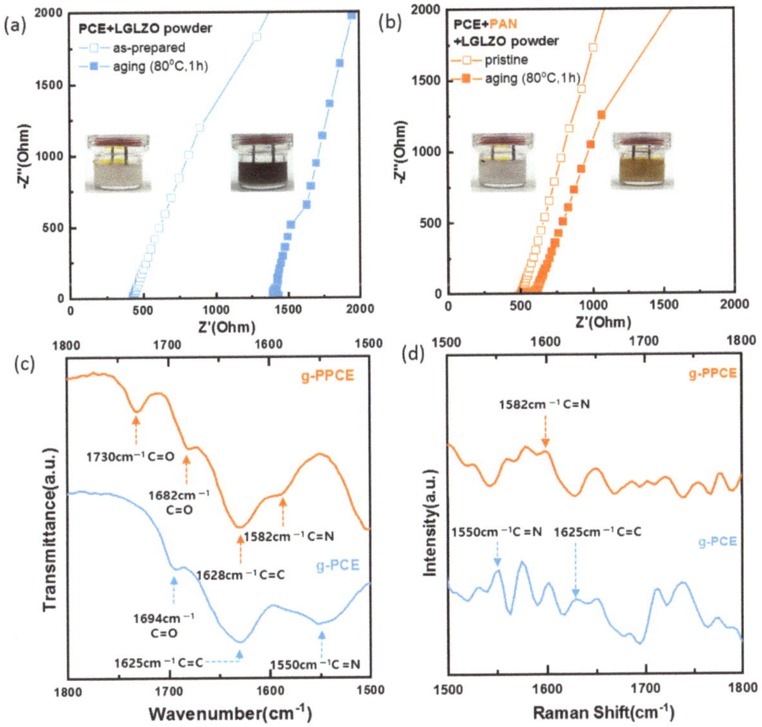

Figure 3. (**a**) Nyquist impedance spectra and the images (inset) of the beaker cell of PCE/LGLZO slurry before and after aging at 80 °C, 1 h; (**b**) Nyquist impedance spectra and the images (inset) of the beaker cell of PPCE/LGLZO slurry before and after aging at 80 °C, 1 h; (**c**) FTIR images of g-PCE and g-PPCE samples; and (**d**) Raman spectroscopy of g-PCE and g-PPCE samples.

In Figure 3c, FTIR spectra of the g-PCE and g-PPCE samples aged at 80 °C for 1 h are shown. FTIR analysis was performed to confirm the reaction with SCN on the surface of LGLZO. Peaks at 1694, 1625, and 1550 cm^{-1} were observed in the FTIR spectrum of g-PCE. The peak at 1694 cm^{-1} represents the vibration mode of the carbonyl group (C=O) [42], the peak at 1625 cm^{-1} represents C=N bonds, and the peak at 1550 cm^{-1} represents C=C bonds [23,24]. In the case of g-PCE, as it contains SCN, C=N and C=C bonds should not be observed. However, peaks corresponding to these bonds were observed in the FTIR spectrum of g-PCE, indicating that -C≡N changed to C=N and C=C. This phenomenon is attributed to the nitrile group of SCN reacting with LGLZO and the resulting polymerization. As shown in bottom graph of Figure 3c, however, the FTIR spectrum of g-PPCE did not show polymerization peaks at 1550 and 1625 cm^{-1}. The peaks at 1730 cm^{-1} and 1682 cm^{-1} represent the stretching vibration absorption of the free and hydrogen-bonded C=O [43,44]. These peaks originated from the end of the chemical structure of PAN. The peak at 1628 cm^{-1} represents the vibration mode of the C=C bond of the cyclized PAN, and the peak at 1582 cm^{-1} represents the cyclic C=N group [30,45]. As PAN also contains a C≡N bond, the FTIR results of g-PPCE should not present a C=N or C=C bond. This suggests that the triple bond of PAN breaks and transforms into double bonds, including C=C and C=N. Consequently, PAN reacts with LGLZO more readily than SCN in the g-PPCE, effectively preventing its polymerization. The full range of FTIR data is presented in Figure S7. To compare the thermal properties, TGA graphs of the LGLZO green sheet, SPE, PPCE-impregnated green sheet, and g-PPCE are shown in Figure S8. After heat treatment at 800 °C, the LGLZO remains about 26 wt.% of g-PPCE.

To further verify the results of FTIR analysis, Raman analyses of g-PCE and g-PPCE were performed (Figure 3d). The Raman measurements exhibit peaks at 1550 and 1625 cm^{-1} in g-PCE, representing the C=N and C=C bonds, respectively. This indicates that the triple bonds of C and N in the nitrile groups of SCN changed to double bonds as the SCN polymerization progressed. However, in the case of g-PPCE, no peaks at 1550 and 1625 cm^{-1}, which would have indicated polymerization, appear. Conversely, a peak at 1582 cm^{-1} appears, demonstrating the presence of a cyclic C=N bond. This confirms the cyclization phenomenon of PAN and demonstrates that SCN polymerization is prevented because PAN readily reacts with LGLZO.

In order to optimize the amount of added PAN and evaluate its effects, we measured basic ionic conducting properties. Figure 4a displays the Nyquist plots of the measured EIS spectra for the three-layer CSE samples with different ratios of wt.% of PAN. The inset shows a magnified view of the high-frequency region. The Nyquist plot of each electrolyte consisted of a semicircle representing the interface resistance and a Warburg plot representing Li-ion diffusion. The series resistance among the electrolytes is similar, approximately 10 Ω, indicating that the addition of PAN does not change the resistance of the electrolyte itself. The interface resistance of g-PPCE with 0, 5, and 10 wt.% PAN is 27, 31, and 104 Ω, respectively. Figure S9 shows that the interface resistance levels of g-PPCE with 2.5 and 7.5 wt.% PAN are 28 Ω and 41 Ω, respectively. Except for the electrolyte containing 10 wt.% PAN, the resistance was 15–20 times higher than that of the liquid electrolyte [46,47]. The diameter of the semicircle increased as the wt.% of PAN increased. Even if there is no PAN addition, there is still a semicircle observed in this Nyquist plot. In addition, no semicircles were observed in beaker cell experiment, both with and without PAN, as described in Figure 3. From the above results, this presence of a semicircle in the Nyquist plot is attributed to the interface between LGLZO and PCE (or PPCE). Specifically, applying a ceramic sheet for impregnation of PCE resulted in insufficient contact between ceramic powders and PCE (or PPCE). Another clue is that there is no semicircle detected in three-layer CSE fabricated by blade coating of mixture of LGLZO and PCE solution, even though the result is not shown in this research. Adding PAN might make for more difficulty in formation of contacts between LGLZO and PCE by increasing the viscosity of PCE. Consequently, PAN is not critical to ionic conductivity of PCE itself, but the impregnation condition should be carefully applied to ensure sufficient contact between LGLZO and PCE.

Figure 4b illustrates the ionic conductivities of the g-PPCEs with different ratings of wt.% of PAN, including error bars. In addition to the CSE, Figure S10 shows a Nyquist plot of an LGLZO pellet. The inset of the graph shows a photo image of an LGLZO pellet. The conductivity of the LGLZO pellet was 1.4 mS/cm. A Nyquist plot of the SPE is shown in Figure S11. The SPE exhibited a conductivity of 0.7 mS/cm. With PAN contents of 0, 2.5, and 5 wt.%, the g-PPCE exhibited similar conductivities (0.23 mS/cm, 0.22 mS/cm, and 0.20 mS/cm, respectively). However, the resistance of the g-PPCE with a 7.5 wt.% PAN content increased to 41 Ω and exhibited an ionic conductivity of 0.15 mS/cm. The resistance of the g-PPCE with 10 wt.% PAN increased significantly to 104 Ω and exhibited an ionic conductivity of 0.06 mS/cm. In addition, the error rate increased by more than 10 wt.% of PAN content. This suggests that PAN, as a minor conductor, did not significantly affect the resistance of the CSE up to 5 wt.% and reacted first with LGLZO to prevent the polymerization of the nitrile group of SCN. However, when g-PPCE contained more than 5 wt.% of PAN, the resistance significantly increased because of the reduced proportion of SCN, resulting in a reduction in Li^+ conductivity [24]. As a result, g-PPCE containing 5 wt.% PAN demonstrates that PAN blocks the polymerization of SCN while having no significant effect on conductivity. Thus, all the g-PPCE have 5 wt.% of PAN content for the following evaluations of electrochemical properties.

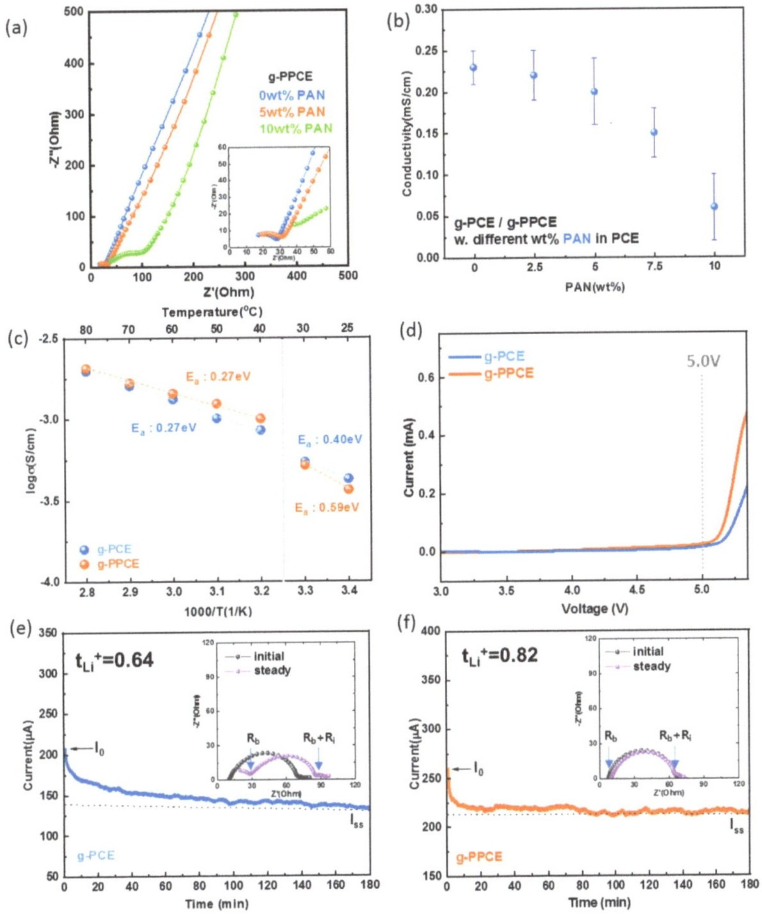

Figure 4. Electrochemical properties of the g-PCE and g-PPCE samples. (**a–c**) Nyquist impedance spectra, ionic conductivities and Arrhenius plots of the SS| |g-PCE| |SS and SS| |g-PPCE| |SS samples; (**d**) LSV curves of the SS| |g-PCE| |Li and SS| |g-PPCE| |Li samples; (**e**,**f**) polarization curves (inset) and steady-state impedance diagram of the Li| |g-PCE| |Li and Li| |g-PPCE| |Li samples, respectively.

Figure 4c depicts the Arrhenius curves of the temperature-dependent conductivities of g-PCE and g-PPCE, both exhibiting two zones with distinct slopes. The turning point indicating a change in slope in the Arrhenius curves corresponds to the melting point of PCE. The melting point of SCN is 58 °C; With the addition of LiTFSI, however, its melting point was reduced, enabling it to partially exist as a liquid at 40 °C [48]. This leads to a change in the slope at 40 °C. Consequently, the E_a values for g-PCE and g-PPCE were 0.40 and 0.59 eV, respectively, below 40 °C. However, the E_a decreased to 0.27 eV for both electrolytes above 40 °C, indicating an enhancement of the ion transport capacity of the CSE during the solid-to-liquid transition. This is attributed to SCN reducing the crystallinity of the polymer, serving as a Li-ion carrier and promoting Li-ion transfer. The activation energy of an LGLZO pellet is 0.30 eV [49,50], and that of PCE is 0.23 eV below 40 °C and 0.16 eV above 40 °C [51]. From above results, the activation energy levels of g-PCE and g-PPCE approximated those of LGLZO pellets (0.30 eV) rather than that of PCE. This suggests that LGLZO serves as the primary pathway for Li-ion conduction in the g-PCE and g-PPCE.

A wide potential window for the electrolyte is essential to drive batteries at high voltages. The graph in Figure 4d shows the LSV (linear sweep voltammetry) curves of g-PCE and g-PPCE. As the voltage changed, both g-PCE and g-PPCE exhibited virtually no increase in current, up to 5.0 V. However, g-PPCE exhibited a slightly faster current increase above 5.0 V, reaching 0.2 mA for g-PCE and 0.46 mA for g-PPCE at 5.3 V. Consequently, there was no obvious oxidative degradation of either g-PCE or g-PPCE over Li^+/Li, at least within operating voltages of up to 5.0 V.

For a deeper understanding of the Li-ion conduction, the polarization curves and steady-state impedance diagrams (insets) of g-PCE and g-PPCE are shown in Figure 4e and 4f, respectively. First, the polarization curve of g-PCE shows that the initial current (I_0) was 207 μA and reached a stable region after approximately 180 min, with a steady-state current (I_{ss}) of 133 μA. The inset shows the steady-state impedance diagram of g-PCE, where R_b represents the bulk resistance, R_i is the interfacial resistance, and R_b+R_i shows the charge-transfer resistance. In the steady state, g-PCE exhibited a R_b of 28 Ω; R_b+R_i was 84 Ω. The charge-transfer resistance increased after the polarization test. The calculated Li^+ transfer number of g-PCE was 0.64, which is higher than the 0.5 of conventional polymer electrolytes [52,53]. With the same polarization procedure, the calculated Li^+ transfer number of g-PPCE was 0.82, which is higher than that of g-PCE. Better Li-ion transport ability in g-PPCE containing PAN is obtained because PAN reacts with LGLZO first and prevents the severe polymerization of SCN. Moreover, the higher t_{Li+} values of g-PCE and g-PPCE, both higher than 0.5, demonstrate that g-PCE and g-PPCE are single-ion conducting materials and that LGLZO in the electrolyte dominates the Li-ion conduction process [54,55]. Consequently, g-PPCE is a promising CSE, in spite of its relatively low ionic conductivity due to its higher Li^+ transference number.

The electrochemical compatibility of g-PCE and g-PPCE with the metallic Li anode was assessed through the use of Li/g-PCE/Li and Li/g-PPCE/Li symmetric cells. As shown in Figure 5, galvanostatic cycling curves of the symmetric cells of g-PCE and g-PPCE are gained with current density and capacity of 0.1 mA/cm^2 and 0.1 mAh/cm^2, respectively. The voltage of the g-PCE cell gradually increased after 350 h due to the formation of dead Li. Furthermore, the sudden short circuit was detected after 500 h, which was obviously due to Li-dendrite. However, no indication of a short circuit event was detected in the cycle of g-PPCE within 500 h cycling; nevertheless, the voltage was gradually increased. Cyclic voltammetry of g-PPCE (as shown in Figure S12) also shows no noticeable peaks up to 5 V, which also indicates voltage stability of the solid electrolyte. All of the results confirm that the g-PPCE has stable interface properties which are superior to those of the g-PCE with Li-metal. Consequently, the stable electrochemical performance of solid electrolyte due to PAN addition also guarantees a stable interface with the Li-metal anode, which is critical to the electrochemical performance levels of the battery.

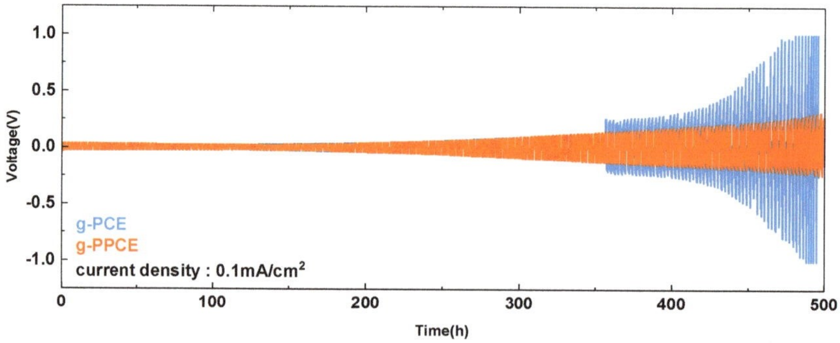

Figure 5. Galvanostatic cycling curves of the Li symmetric cells of g-PCE and g-PPCE at 0.1 mA /cm^2.

To compare the electrochemical performance levels of g-PCE and g-PPCE in Li-metal batteries, confirming the effect of PAN, LFP | | g-PCE | | Li and LFP | | g-PPCE | | Li cells were fabricated. First of all, excellent rate capability is an essential feature of all-solid-state batteries. The rate-performances of g-PCE and g-PPCE are shown in Figure 6a. The specific capacities of g-PCE were 147, 130, 115, 104, and 90 mAh/g at 0.1, 0.3, 0.5, 0.7, and 1 C, respectively. However, the specific capacities of g-PPCE were 150, 141, 138, 134, and 127 mAh/g under the same conditions, higher than those of g-PCE. At a current density of 0.1 C, no significant difference existed between the capacities with and without PAN. However, as the C-rate increased, g-PPCE maintained a good discharge capacity 1.5 times higher than that exhibited by g-PCE at 1 C. This suggests that the SCN in g-PCE reacted with LGLZO and was polymerized, leading to the gelation of the electrolyte and an increase in resistance. Conversely, in the g-PPCE, PAN prevented the polymerization of the nitrile groups of SCN, preventing battery degradation and maintaining stability at a high current density. The graph also illustrates a gradual decrease in current density. Upon restoring the current density to 0.1 C, the reversible capacities of both g-PCE and g-PPCE remained stable. The Coulombic efficiencies of g-PCE and g-PPCE were high (~99%). However, changes in C-rate caused instability at some points. This was considered an instrument operating error in the measurement process, as the CE values were stable after the changes in current density.

The discharge curves of the LFP | | g-PCE | | Li and LFP | | g-PPCE | | Li cells at different C rates are shown in Figure 6b,c. The g-PPCE curves showed clear potential plateaus, indicating a reversible cycling process [56,57]. Contrastingly, the discharge curve exhibited by g-PCE was non-linear, with a leftward slant at current densities exceeding 0.5 C. This variation in voltage plateau is generally related to battery degradation, indicating changes in the structure and composition of the electrode materials [58]. The curves with unstable potential plateaus also suggest that the polymerization of the nitrile groups of SCN in the g-PCE may have caused some degradation reactions at high cycle rates, leading to stability issues between the electrolyte and electrode.

Secondarily, cyclic performance levels of g-PCE and g-PPCE were compared. Figure 7a displays the charge and discharge curves of the LFP | | g-PCE | | Li cell at a rate of 1 C and 25 °C. The initial polarization voltage of the g-PCE cell was 0.2 V and the capacities at cycles number 1, 2, 50 and 100 were 118, 118, 95, and 85 mAh/g, respectively. The discharge graph sloped slightly after 50 cycles, indicating a significant decrease in capacity. Figure 7b illustrates the charge and discharge curves of the LFP | | g-PPCE | | Li cell at a rate of 1 C and 25 °C. The initial polarization voltage of the g-PPCE cell was 0.19 V, and the capacities at cycles number 1, 2, 50 and 100 were 132, 132, 131 and 116 mAh/g, respectively. The discharge capacity was 130 mAh/g up to the 50th cycle and reduced subsequently. The charge/discharge curves of the g-PPCE show flat and reversible plateaus, indicating minimal polarization. Moreover, the Nyquist plots of LFP | | g-PPCE | | Li cell before and after galvanostatic cycling at 1 C and 25 °C are shown in Figure S9. These Nyquist plots were subsequently analyzed by fitting them to the equivalent circuits presented in Figure S10. Following 100 cycles, a stable formation of the solid electrolyte interface (SEI) with the increased charge transfer resistance is observed, showing the appearance of two semicircles. Figure 7c shows the cycle performance levels of LFP | | g-PCE | | Li and LFP | | g-PPCE | | Li cells (1 C, 25 °C). The capacity of g-PCE decreases continuously, resulting in a Coulombic efficiency of 98% and specific capacity of 92 mAh/g at 100 cycles, corresponding to 69% of the initial capacity. Conversely, the capacity of g-PPCE remained constant; the CE was 99% and the specific capacity was 117 mAh/g at 100 cycles, 89% of the initial capacity. In terms of performance over 100 cycles, g-PPCE exhibited a higher capacity retention rate than g-PCE. This indicates that the unstable cycling performance of g-PCE at high current densities is due to the coordination of SCN and LGLZO in g-PCE, leading to gelation of the electrolyte and increased resistance. In contrast, the stable cycling performance of g-PPCE can be attributed to PAN reacting with LGLZO more readily than SCN, preventing SCN polymerization and, consequently, performance degradation.

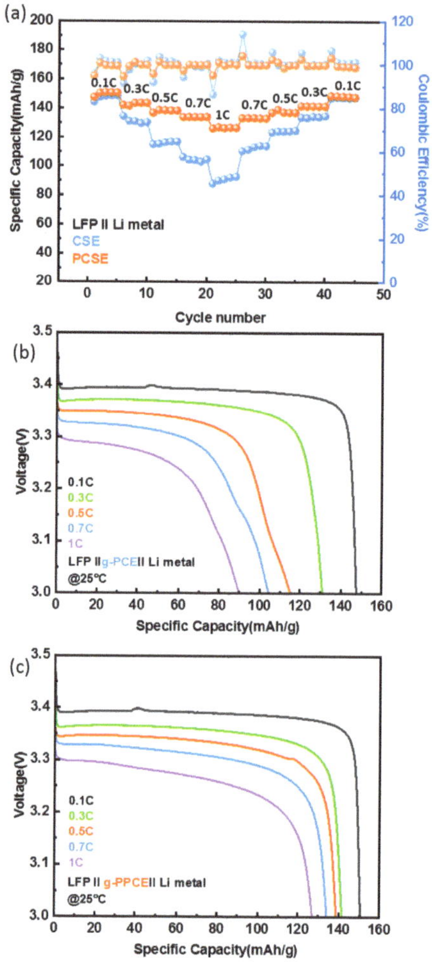

Figure 6. (**a**) The rate performance of the LFP| |g-PCE| |Li and LFP| |g-PPCE| |Li batteries at room temperature. (**b**,**c**) The discharge curves of the LFP| |g-PCE| |Li and LFP| |g-PPCE| |Li batteries at different rates at 25 °C.

For a higher voltage operation of solid-state Li-metal battery, NCM was applied as a cathode active material. Figure 8a displays the charge and discharge curves of the NCM| |g-PCE| |Li cell at 0.1 C and 25 °C. The capacities at cycles number 1, 2, and 20 were 151, 149, and 119 mAh/g, respectively, with overcharging occurring at the 28th cycle. This suggests that cells with g-PCE undergo micro-short-circuits as the Li dendrites grow. SCN polymerization appears to induce gelation of the electrolyte and locally change its resistance distribution, leading to the growth of Li dendrites. Figure 8b depicts the cycling performance of the NCM| |g-PCE| |Li cell. The Coulombic efficiency at the 27th cycle (before overcharging) was 94%, and the discharge capacity was 94 mAh/g, indicating a 62% capacity retention. The unstable data at the 15th cycle is considered to be due to instrument errors. In addition, the charge and discharge curves of the NCM| |g-PPCE| |Li cell at 0.1 C and 25 °C are displayed in Figure 8c. The capacities at cycles number 1, 2, 20, and 50 were 167, 167, 164, and 155 mAh/g, respectively. Figure 8d shows the cycle performance of the same cell. The Coulombic efficiency at the 50th cycle was 97%, and the discharge capacity was 92% of the initial one. The Coulombic efficiencies around the 12th, 35th, and

42nd cycles exhibited fluctuations. This was attributed to operational errors in the device and recovery in later cycles. The g-PPCE exhibited stable cycle performance with no side reactions at high voltages.

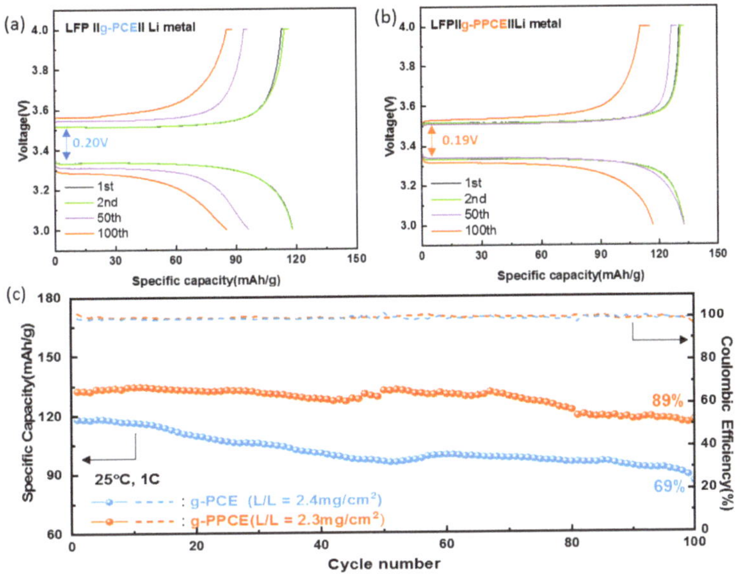

Figure 7. (**a**,**b**) The charge and discharge curves of the LFP||g-PCE||Li and LFP||g-PPCE||Li batteries at 1 C and 25 °C; (**c**) The cycle performance levels of the LFP||g-PCE||Li and LFP||g-PPCE||Li batteries at 1 C and 25 °C.

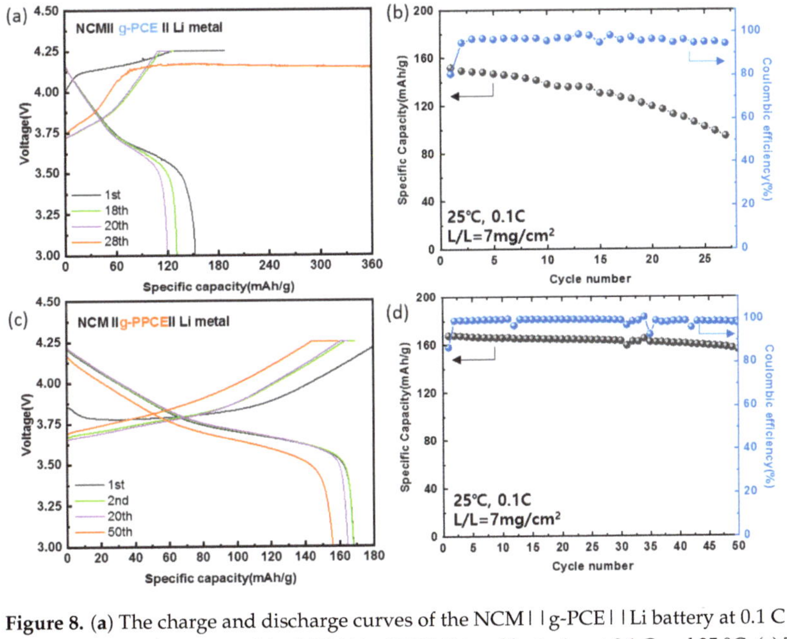

Figure 8. (**a**) The charge and discharge curves of the NCM||g-PCE||Li battery at 0.1 C and 25 °C. (**b**) The cycle performance of the NCM||g-PCE||Li and batteries at 0.1 C and 25 °C. (**c**) The charge and discharge curves of the NCM||g-PPCE||Li battery at 0.1 C and 25 °C. (**d**) The cycle performance of the NCM||g-PPCE||Li and batteries at 0.1 C and 25 °C.

Finally, a pouch cell was fabricated with a g-PPCE and cell tests were conducted to verify the feasibility of large-area battery processing. Figure 9a illustrates the charge/discharge curves of a 3450 size-pouch cell assembled with an LFP| |g-PPCE| |Li structure. The cell exhibited an initial capacity of 5 mAh at 0.1 C and 25 °C, and the capacities at the 2nd, 5th, and 10th cycles were 5.0, 4.9, and 4.8 mAh/g, respectively. Figure 9b depicts the cycling performance of the pouch cell. After 10 cycles, the Coulombic efficiency was 98% and the capacity was 4.9 mAh, with a 98% capacity retention. This indicated that the electrolyte/electrode interface was stable even in the pouch cell configuration.

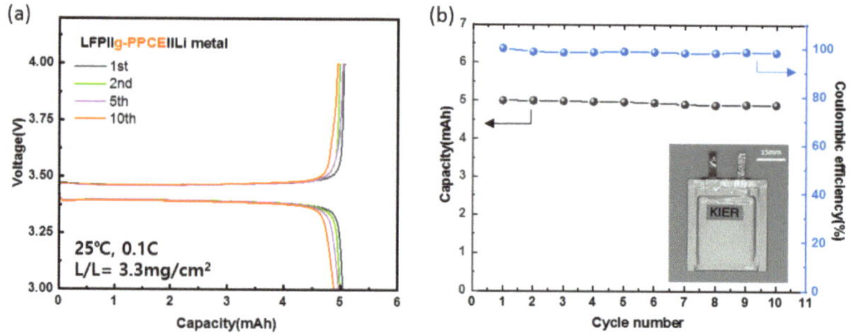

Figure 9. (**a**) The charge and discharge curves of the LFP| |g-PPCE| |Li battery at 0.1 C and 25 °C (**b**) The cycle performance of 3450 sized solid state-pouch cell with g-PPCE at 0.1 C and 25 °C and photographic image of the pouch cell with LFP| |g-PPCE| |Li.

4. Conclusions

CSE with SCN are known to have two serious problems with LLZO ceramics, in spite of SCN's superior performance as a solid plasticizer: (1) gelation during the mixing process, and (2) polymerization during the charge/discharge process. By using a tape-casting and impregnation process, we successfully fabricated a PVDF-HFP-based SPE-LGLZO CSE without gelation. To solve the second problem, we simply added PAN to SCN prior to the impregnation process. By an aging test with a beaker cell, it was proved that the addition of PAN to SCN provided a protection mechanism for the coordination between the nitrile groups of SCN and La in LGLZO. Even though the interface area is high, PAN effectively prevents SCN polymerization via the rapid coordination with LGLZO due to its higher dielectric constant. The g-PPCE with 5 wt.% of PAN shows an ionic conductivity of 2×10^{-4} S/cm at 25 °C with a high Li+ transference number of 0.85. The Li-metal cell (LFP| |Li) assembled using the g-PPCE exhibited a high discharge capacity of 113 mAh/g at 1 C and cycle stability over 100 cycles. With a high-loading (~7 mg/cm^2) NCM cathode, Li-metal cell exhibited stable cycle performance, with a capacity of 167 mAh/g. Moreover, the g-PPCE was more durable than the g-PCE at high current densities. The preparation of a pouch cell with a g-PPCE and a cell capacity of 5 mAh demonstrated the potential for large-scale production. Consequently, adding PAN is key to enhancing the electrochemical performance levels of solid-state Li-metal battery with CSE containing LLZO and SCN. As for future work, a more effective impregnation process will be developed to overcome the increased interface resistance due to the higher viscosity of PPCE.

Supplementary Materials: The following supporting information can be downloaded from: https://www.mdpi.com/article/10.3390/en16237695/s1, Figure S1. Nyquist impedance plots of LiTFSI/SCN(PCE) with various ratios of concentration. Figure S2. The schematic image and the real image of LGLZO particle that is protected by the cyclized PAN. Figure S3. Images of SPE/LGLZO and SPE/PAN/LGLZO slurry before and after 80 °C, 2 h mixing and the image of aged slurry after fabrication. Figure S4. The cross-section SEM image of g-PPCE and the corresponding EDS mappings of N, F and La. Figure S5. SEM image of LGLZO and XRD pattern of LGLZO. Figure S6. XRD graphs of LGLZO green sheet, SPE, impregnated green sheet and g-PPCE samples. Figure S7. FT-IR analysis. Figure S8. TGA graphs of LGLZO green sheet, SPE, g-PPCE and g-PPCE samples. Figure S9. Nyquist impedance plots of SSllg-PPCEllSS with different percent of PAN. Figure S10. Nyquist impedance plot and the image of the LGLZO pellet(inset), and the conductivity of LGLZO pellet. Figure S11. Nyquist impedance plot of SPE. Figure S12. Cyclic voltammetry curves of g-PPCE. Figure S13. Nyquist impedance plots of a LFP | | g-PPCE | | Li-metal cell before and after galvanostatic cycling at 1C and 25 °C. Figure S14. The equivalent circuits of Nyquist impedance plots of a LFP | | g-PPCE | | Li-metal cell before and after galvanostatic cycling at 1C and 25 °C.

Author Contributions: Conceptualization, J.S.H.; Methodology, J.H.Y.; Investigation, W.L.; Writing—original draft, H.N. and D.K.; Writing—review & editing, B.J. All authors have read and agreed to the published version of the manuscript.

Funding: This work was partly supported by the research and development program of the Korea Institute of Energy Research (C3-2405) and the Technological innovation R&D program of MSS [S3143178].

Data Availability Statement: Data are contained within the article and Supplementary Materials.

Conflicts of Interest: The authors declare no conflict of interest.

References

1. Li, C.; Wang, Z.-y.; He, Z.-j.; Li, Y.-j.; Mao, J.; Dai, K.-h.; Yan, C.; Zheng, J.-c. An Advance Review of Solid-State Battery: Challenges, Progress and Prospects. *Sustain. Mater. Technol.* **2021**, *29*, e00297. [CrossRef]
2. Xiao, Y.; Wang, Y.; Bo, S.H.; Kim, J.C.; Miara, L.J.; Ceder, G. Understanding Interface Stability in Solid-State Batteries. *Nat. Rev. Mater.* **2020**, *5*, 105–126.
3. Manthiram, A.; Yu, X.; Wang, S. Lithium Battery Chemistries Enabled by Solid-State Electrolytes. *Nat. Rev. Mater.* **2017**, *2*, 16103.
4. Park, K.; Yu, B.C.; Jung, J.W.; Li, Y.; Zhou, W.; Gao, H.; Son, S.; Goodenough, J.B. Electrochemical Nature of the Cathode Interface for a Solid-State Lithium-Ion Battery: Interface between $LiCoO_2$ and Garnet-$Li_7La_3Zr_2O_{12}$. *Chem. Mater.* **2016**, *28*, 8051–8059. [CrossRef]
5. Janek, J.; Zeier, W.G. Challenges in Speeding up Solid-State Battery Development. *Nat. Energy* **2023**, *8*, 230–240.
6. Kim, D.H.; Kim, M.Y.; Yang, S.H.; Ryu, H.M.; Jung, H.Y.; Ban, H.J.; Park, S.J.; Lim, J.S.; Kim, H.S. Fabrication and Electrochemical Characteristics of NCM-Based All-Solid Lithium Batteries Using Nano-Grade Garnet Al-LLZO Powder. *J. Ind. Eng. Chem.* **2019**, *71*, 445–451. [CrossRef]
7. Yang, S.H.; Kim, M.Y.; Kim, D.H.; Jung, H.Y.; Ryu, H.M.; Han, J.H.; Lee, M.S.; Kim, H.S. Ionic Conductivity of Ga-Doped LLZO Prepared Using Couette–Taylor Reactor for All-Solid Lithium Batteries. *J. Ind. Eng. Chem.* **2017**, *56*, 422–427. [CrossRef]
8. Delluva, A.A.; Kulberg-Savercool, J.; Holewinski, A. Decomposition of Trace Li_2CO_3 during Charging Leads to Cathode Interface Degradation with the Solid Electrolyte LLZO. *Adv. Funct. Mater.* **2021**, *31*, 2103716. [CrossRef]
9. Krauskopf, T.; Dippel, R.; Hartmann, H.; Peppler, K.; Mogwitz, B.; Richter, F.H.; Zeier, W.G.; Janek, J. Lithium-Metal Growth Kinetics on LLZO Garnet-Type Solid Electrolytes. *Joule* **2019**, *3*, 2030–2049. [CrossRef]
10. Ohta, S.; Komagata, S.; Seki, J.; Saeki, T.; Morishita, S.; Asaoka, T. Short Communication All-Solid-State Lithium Ion Battery Using Garnet-Type Oxide and Li_3BO_3 Solid Electrolytes Fabricated by Screen-Printing. *J. Power Sources* **2013**, *238*, 53–56. [CrossRef]
11. Costa, C.M.; Lee, Y.H.; Kim, J.H.; Lee, S.Y.; Lanceros-Méndez, S. Recent Advances on Separator Membranes for Lithium-Ion Battery Applications: From Porous Membranes to Solid Electrolytes. *Energy Storage Mater.* **2019**, *22*, 346–375.
12. Wang, W.; Yi, E.; Fici, A.J.; Laine, R.M.; Kieffer, J. Lithium Ion Conducting Poly(Ethylene Oxide)-Based Solid Electrolytes Containing Active or Passive Ceramic Nanoparticles. *J. Phys. Chem. C* **2017**, *121*, 2563–2573. [CrossRef]
13. He, K.; Chen, C.; Fan, R.; Liu, C.; Liao, C.; Xu, Y.; Tang, J.; Li, R.K.Y. Polyethylene Oxide/Garnet-Type $Li_{6.4}La_3Zr_{1.4}Nb_{0.6}O_{12}$ Composite Electrolytes with Improved Electrochemical Performance for Solid State Lithium Rechargeable Batteries. *Compos. Sci. Technol.* **2019**, *175*, 28–34. [CrossRef]
14. Chen, W.P.; Duan, H.; Shi, J.L.; Qian, Y.; Wan, J.; Zhang, X.D.; Sheng, H.; Guan, B.; Wen, R.; Yin, Y.X.; et al. Bridging Interparticle Li+Conduction in a Soft Ceramic Oxide Electrolyte. *J. Am. Chem. Soc.* **2021**, *143*, 5717–5726. [CrossRef]
15. Nguyen, V.A.; Kuss, C. Review—Conducting Polymer-Based Binders for Lithium-Ion Batteries and Beyond. *J. Electrochem. Soc.* **2020**, *167*, 065501. [CrossRef]
16. Kumar, B.; Scanlon, L.G. POWEA Polymer-Ceramic Composite Electrolytes. *J. Power Sources* **1994**, *52*, 261–268. [CrossRef]
17. Merrill, L.C.; Chen, X.C.; Zhang, Y.; Ford, H.O.; Lou, K.; Zhang, Y.; Yang, G.; Wang, Y.; Wang, Y.; Schaefer, J.L.; et al. Polymer-Ceramic Composite Electrolytes for Lithium Batteries: A Comparison between the Single-Ion-Conducting Polymer Matrix and Its Counterpart. *ACS Appl. Energy Mater.* **2020**, *3*, 8871–8881. [CrossRef]
18. Vanita, V.; Waidha, A.I.; Yadav, S.; Schneider, J.J.; Clemens, O. Conductivity Enhancement within Garnet-Rich Polymer Composite Electrolytes via the Addition of Succinonitrile. *Int. J. Appl. Ceram. Technol.* **2023**, *20*, 236–250. [CrossRef]
19. Walle, K.Z.; Musuvadhi Babulal, L.; Wu, S.H.; Chien, W.C.; Jose, R.; Lue, S.J.; Chang, J.K.; Yang, C.C. Electrochemical Characteristics of a Polymer/Garnet Trilayer Composite Electrolyte for Solid-State Lithium-Metal Batteries. *ACS Appl. Mater. Interfaces* **2021**, *13*, 2507–2520. [CrossRef]

20. Fu, C.; Ma, Y.; Zuo, P.; Zhao, W.; Tang, W.; Yin, G.; Wang, J.; Gao, Y. In-Situ Thermal Polymerization Boosts Succinonitrile-Based Composite Solid-State Electrolyte for High Performance Li-Metal Battery. *J. Power Sources* **2021**, *496*, 229861. [CrossRef]
21. Cheng, E.J.; Kimura, T.; Shoji, M.; Ueda, H.; Munakata, H.; Kanamura, K. Ceramic-Based Flexible Sheet Electrolyte for Li Batteries. *ACS Appl. Mater. Interfaces* **2020**, *12*, 10382–10388. [CrossRef]
22. Shen, H.; Yi, E.; Heywood, S.; Parkinson, D.Y.; Chen, G.; Tamura, N.; Sofie, S.; Chen, K.; Doeff, M.M. Scalable Freeze-Tape-Casting Fabrication and Pore Structure Analysis of 3D LLZO Solid-State Electrolytes. *ACS Appl. Mater. Interfaces* **2020**, *12*, 3494–3501. [CrossRef]
23. Yang, Y.N.; Jiang, F.L.; Li, Y.Q.; Wang, Z.X.; Zhang, T. A Surface Coordination Interphase Stabilizes a Solid-State Battery. *Angew. Chem.-Int. Ed.* **2021**, *60*, 24162–24170. [CrossRef]
24. Gao, Y.; Sun, Z.; Cui, C.; Wang, H.; Cao, W.; Hou, Z.; Zhu, D.; Yang, Y.; Zhang, T. An Ultrathin, Flexible Solid Electrolyte with High Ionic Conductivity Enhanced by a Mutual Promotion Mechanism. *ACS Appl. Mater. Interfaces* **2022**, *14*, 45373–45381. [CrossRef]
25. Tran, H.K.; Truong, B.T.; Zhang, B.R.; Jose, R.; Chang, J.K.; Yang, C.C. Sandwich-Structured Composite Polymer Electrolyte Based on PVDF-HFP/PPC/Al-Doped LLZO for High-Voltage Solid-State Lithium Batteries. *ACS Appl. Energy Mater.* **2023**, *6*, 1475–1487. [CrossRef]
26. Hou, T.; Yang, G.; Rajput, N.N.; Self, J.; Park, S.W.; Nanda, J.; Persson, K.A. The Influence of FEC on the Solvation Structure and Reduction Reaction of LiPF6/EC Electrolytes and Its Implication for Solid Electrolyte Interphase Formation. *Nano Energy* **2019**, *64*, 103881. [CrossRef]
27. Kim, Y.S.; Cho, Y.G.; Odkhuu, D.; Park, N.; Song, H.K. A Physical Organogel Electrolyte: Characterized by in Situ Thermo-Irreversible Gelation and Single-Ion-Predominent Conduction. *Sci. Rep.* **2013**, *3*, 1917. [CrossRef]
28. Zhou, D.; He, Y.B.; Liu, R.; Liu, M.; Du, H.; Li, B.; Cai, Q.; Yang, Q.H.; Kang, F. In Situ Synthesis of a Hierarchical All-Solid-State Electrolyte Based on Nitrile Materials for High-Performance Lithium-Ion Batteries. *Adv. Energy Mater.* **2015**, *5*, 1500353. [CrossRef]
29. Xu, K.; Xu, C.; Jiang, Y.; Cai, J.; Ni, J.; Lai, C. Sandwich Structured PVDF-HFP-Based Composite Solid Electrolytes for Solid-State Lithium Metal Batteries. *Ionics* **2022**, *28*, 3243–3253. [CrossRef]
30. Zhang, X.; Xu, B.Q.; Lin, Y.H.; Shen, Y.; Li, L.; Nan, C.W. Effects of $Li_{6.75}La_3Zr_{1.75}Ta_{0.25}O_{12}$ on Chemical and Electrochemical Properties of Polyacrylonitrile-Based Solid Electrolytes. *Solid State Ion.* **2018**, *327*, 32–38. [CrossRef]
31. Xu, H.; Zhang, X.; Jiang, J.; Li, M.; Shen, Y. Ultrathin $Li_7La_3Zr_2O_{12}$@PAN Composite Polymer Electrolyte with High Conductivity for All-Solid-State Lithium-Ion Battery. *Solid State Ion.* **2020**, *347*, 115227. [CrossRef]
32. Tong, R.A.; Chen, L.; Fan, B.; Shao, G.; Liu, R.; Wang, C.A. Solvent-Free Process for Blended PVDF-HFP/PEO and LLZTO Composite Solid Electrolytes with Enhanced Mechanical and Electrochemical Properties for Lithium Metal Batteries. *ACS Appl. Energy Mater.* **2021**, *4*, 11802–11812. [CrossRef]
33. Li, Z.; Huang, H.-M.; Zhu, J.-K.; Wu, J.-F.; Yang, H.; Wei, L.; Guo, X. Supporting Information Ionic Conduction in Composite Polymer Electrolytes: Case of PEO:Ga-LLZO Composites. *ACS Appl. Mater. Interfaces* **2019**, *11*, 784–791. [CrossRef]
34. Li, S.; Zhang, S.Q.; Shen, L.; Liu, Q.; Ma, J.B.; Lv, W.; He, Y.B.; Yang, Q.H. Progress and Perspective of Ceramic/Polymer Composite Solid Electrolytes for Lithium Batteries. *Adv. Sci.* **2020**, *7*, 1903088. [CrossRef]
35. Chen, L.; Li, Y.; Li, S.P.; Fan, L.Z.; Nan, C.W.; Goodenough, J.B. PEO/Garnet Composite Electrolytes for Solid-State Lithium Batteries: From "Ceramic-in-Polymer" to "Polymer-in-Ceramic". *Nano Energy* **2018**, *46*, 176–184. [CrossRef]
36. Bonilla, M.R.; García Daza, F.A.; Ranque, P.; Aguesse, F.; Carrasco, J.; Akhmatskaya, E. Unveiling Interfacial Li-Ion Dynamics in $Li_7La_3Zr_2O_{12}$/PEO(LiTFSI) Composite Polymer-Ceramic Solid Electrolytes for All-Solid-State Lithium Batteries. *ACS Appl. Mater. Interfaces* **2021**, *13*, 30653–30667. [CrossRef]
37. Isaac, J.A.; Devaux, D.; Bouchet, R. Dense Inorganic Electrolyte Particles as a Lever to Promote Composite Electrolyte Conductivity. *Nat. Mater.* **2022**, *21*, 1412–1418. [CrossRef]
38. Zhu, D.M.; Bai, F.; Zhang, Y.N.; Yang, Y.N.; Sun, Z.; Deng, J.W.; Zhang, T. A Solid-State Lithium-Oxygen Battery Operating at Ambient Temperature and Full Charge-Discharge. *J. Energy Storage* **2022**, *56*, 105790. [CrossRef]
39. Zhang, X.; Fu, C.; Cheng, S.; Zhang, C.; Zhang, L.; Jiang, M.; Wang, J.; Ma, Y.; Zuo, P.; Du, C.; et al. Novel PEO-Based Composite Electrolyte for Low-Temperature All-Solid-State Lithium Metal Batteries Enabled by Interfacial Cation-Assistance. *Energy Storage Mater.* **2023**, *56*, 121–131. [CrossRef]
40. Hara, K. High pressure and high temperature reactions in the organic solid state: Polymerization of Nitrites. *Rev. Phys. Chem. Jpn.* **1970**, *40*, 73–92.
41. Jin, S.Y.; Kim, M.H.; Jeong, Y.G.; Yoon, Y.I.; Park, W.H. Effect of Alkaline Hydrolysis on Cyclization Reaction of PAN Nanofibers. *Mater. Des.* **2017**, *124*, 69–77. [CrossRef]
42. Haque, N.; Biswas, S.; Basu, P.; Haque Biswas, I.; Khatun, R.; Khan, A.; Islam, S.M. Triazinetriamine-Derived Porous Organic Polymer-Supported Copper Nanoparticles (Cu-NPs@TzTa-POP): An Efficient Catalyst for the Synthesis of: N -Methylated Products via CO2fixation and Primary Carbamates from Alcohols and Urea. *New J. Chem.* **2020**, *44*, 15446–15458. [CrossRef]
43. Liu, W.; Lu, G.; Xiao, B.; Xie, C. Potassium Iodide-Polyethylene Glycol Catalyzed Cycloaddition Reaction of Epoxidized Soybean Oil Fatty Acid Methyl Esters with CO_2. *RSC Adv.* **2018**, *8*, 30860–30867. [CrossRef]
44. Lu, F.; Song, B.; He, P.; Wang, Z.; Wang, J. Electrochemical Impedance Spectroscopy (EIS) Study on the Degradation of Acrylic Polyurethane Coatings. *RSC Adv.* **2017**, *7*, 13742–13748. [CrossRef]
45. Dang, W.; Liu, J.; Wang, X.; Yan, K.; Zhang, A.; Yang, J.; Chen, L.; Liang, J. Structural Transformation of Polyacrylonitrile (PAN) Fibers during Rapid Thermal Pretreatment in Nitrogen Atmosphere. *Polymers* **2020**, *12*, 63. [CrossRef]

46. Lundgren, H.; Behm, M.; Lindbergh, G. Electrochemical Characterization and Temperature Dependency of Mass-Transport Properties of LiPF 6 in EC:DEC. *J. Electrochem. Soc.* **2015**, *162*, A413–A420. [CrossRef]
47. Zhang, S.S.; Xu, K.; Jow, T.R. EIS Study on the Formation of Solid Electrolyte Interface in Li-Ion Battery. *Electrochim. Acta* **2006**, *51*, 1636–1640. [CrossRef]
48. Echeverri, M.; Kim, N.; Kyu, T. Ionic Conductivity in Relation to Ternary Phase Diagram of Poly(Ethylene Oxide), Succinonitrile, and Lithium Bis(Trifluoromethane)Sulfonimide Blends. *Macromolecules* **2012**, *45*, 6068–6077. [CrossRef]
49. Pesci, F.M.; Bertei, A.; Brugge, R.H.; Emge, S.P.; Hekselman, A.K.O.; Marbella, L.E.; Grey, C.P.; Aguadero, A. Establishing Ultralow Activation Energies for Lithium Transport in Garnet Electrolytes. *ACS Appl. Mater. Interfaces* **2020**, *12*, 32806–32816. [CrossRef]
50. Zhang, Y.; Chen, F.; Li, J.; Zhang, L.; Gu, J.; Zhang, D.; Saito, K.; Guo, Q.; Luo, P.; Dong, S. Regulation Mechanism of Bottleneck Size on Li+ Migration Activation Energy in Garnet-Type $Li_7La_3Zr_2O_{12}$. *Electrochim. Acta* **2018**, *261*, 137–142. [CrossRef]
51. Fan, L.Z.; Hu, Y.S.; Bhattacharyya, A.J.; Maier, J. Succinonitrile as a Versatile Additive for Polymer Electrolytes. *Adv. Funct. Mater.* **2007**, *17*, 2800–2807. [CrossRef]
52. Pożyczka, K.; Marzantowicz, M.; Dygas, J.R.; Krok, F. Ionic conductivity and lithium transference number of poly(ethylene oxide):LiTFSI system. *Electrochim. Acta* **2017**, *227*, 127–135. [CrossRef]
53. Wang, Y.; Huang, K.; Zhang, P.; Li, H.; Mi, H. PVDF-HFP Based Polymer Electrolytes with High Li+ Transference Number Enhancing the Cycling Performance and Rate Capability of Lithium Metal Batteries. *Appl. Surf. Sci.* **2022**, *574*, 151593. [CrossRef]
54. Buschmann, H.; Dölle, J.; Berendts, S.; Kuhn, A.; Bottke, P.; Wilkening, M.; Heitjans, P.; Senyshyn, A.; Ehrenberg, H.; Lotnyk, A.; et al. Structure and Dynamics of the Fast Lithium Ion Conductor "$Li_7La_3Zr_2O_{12}$". *Phys. Chem. Chem. Phys.* **2011**, *13*, 19378–19392. [CrossRef]
55. Jia, M.; Khurram Tufail, M.; Guo, X. Insight into the Key Factors in High Li+ Transference Number Composite Electrolytes for Solid Lithium Batteries. *ChemSusChem* **2023**, *16*, e202201801. [CrossRef]
56. Xu, D.; He, Y.B.; Chu, X.; Ding, Z.; Li, B.; He, J.; Du, H.; Qin, X.; Kang, F. Synthesis of Lithium Iron Phosphate/Carbon Microspheres by Using Polyacrylic Acid Coated Iron Phosphate Nanoparticles Derived from Iron(III) Acrylate. *ChemSusChem* **2015**, *8*, 1009–1016. [CrossRef]
57. Zhou, D.; Liu, R.; He, Y.B.; Li, F.; Liu, M.; Li, B.; Yang, Q.H.; Cai, Q.; Kang, F. SiO_2 Hollow Nanosphere-Based Composite Solid Electrolyte for Lithium Metal Batteries to Suppress Lithium Dendrite Growth and Enhance Cycle Life. *Adv. Energy Mater.* **2016**, *6*, 1502214. [CrossRef]
58. Yao, X.; Olsson, E.; Zhao, J.; Feng, W.; Luo, W.; Tan, S.; Huang, M.; Zhao, Y.; Huang, J.; Cai, Q.; et al. Voltage Plateau Variation in a Bismuth-Potassium Battery. *J. Mater. Chem. A Mater. Energy Sustain.* **2022**, *10*, 2917–2923. [CrossRef]

energies

MDPI

Article

On the Thermal Stability of Selected Electrode Materials and Electrolytes for Na-Ion Batteries

Ruslan R. Samigullin, Zoya V. Bobyleva, Maxim V. Zakharkin, Emiliya V. Zharikova, Marina G. Rozova, Oleg A. Drozhzhin * and Evgeny V. Antipov

Department of Chemistry, Lomonosov Moscow State University, Leninskie Gory 1-3, Moscow 119991, Russia; ruslan.samigullin.rinatovich@yandex.ru (R.R.S.); evgeny.antipov@gmail.com (E.V.A.)
* Correspondence: drozhzhin@elch.chem.msu.ru

Abstract: Sodium-ion batteries are a technology rapidly approaching widespread adoption, so studying the thermal stability and safety of their components is a pressing issue. In this work, we employed differential scanning calorimetry (DSC) and ex situ powder X-ray diffraction to study the thermal stability of several types of sodium-ion electrolytes ($NaClO_4$ and $NaPF_6$ solutions in PC, EC, DEC, and their mixtures) and various cathode and anode materials ($Na_3V_2(PO_4)_3$, $Na_3(VO)_2(PO_4)_2F$, β-$NaVP_2O_7$, and hard carbon) in combination with electrolytes. The obtained results indicate, first, the satisfactory thermal stability of liquid Na-ion electrolytes, which start to decompose only at 270~300 °C. Second, we observed that charged vanadium-based polyanionic cathodes, which appear to be very stable in the "dry" state, demonstrate an increase in decomposition enthalpy and a shift of the DSC peaks to lower temperatures when in contact with 1 M $NaPF_6$ in the EC:DEC solution. However, the greatest thermal effect from the "electrode–electrolyte" interaction is demonstrated by the anode material: the heat of decomposition of the soaked electrode in the charged state is almost 40% higher than the sum of the decomposition enthalpies of the electrolyte and dry electrode separately.

Keywords: Na-ion batteries; differential scanning calorimetry; thermal stability; electrode material; electrolyte; cathode; anode

Citation: Samigullin, R.R.; Bobyleva, Z.V.; Zakharkin, M.V.; Zharikova, E.V.; Rozova, M.G.; Drozhzhin, O.A.; Antipov, E.V. On the Thermal Stability of Selected Electrode Materials and Electrolytes for Na-Ion Batteries. *Energies* **2024**, *17*, 3970. https://doi.org/10.3390/en17163970

Academic Editors: Huang Zhang and Yuan Ma

Received: 11 July 2024
Revised: 6 August 2024
Accepted: 8 August 2024
Published: 10 August 2024

1. Introduction

Lithium-ion batteries (LIBs) are the most efficient electricity storage solution available today, and their applications have steadily expanded over the past thirty years. Yet, the safety of LIBs is an extremely important issue, which has become the subject of many studies during the last decades [1–13]. It mainly concerns the thermal stability of the charged electrode materials, electrolytes, and their combinations. In general, the accumulated experimental experience consists of the following theses: (a) oxide cathode materials are less stable than phosphate ones, although the release of oxygen when heating charged cathodes occurs in both cases; (b) in combination with an electrolyte, decomposition processes, as a rule, occur more intensely and at lower temperatures; (c) anode materials (primarily graphite) demonstrate enthalpies of decomposition that are several times higher than cathode ones. Sodium-ion batteries (SIBs), which are considered the most likely successors to LIBs in many areas of technology, may also cause certain operational safety issues, and, in recent years, there has been active research devoted to the thermal stability of their components [14–23]. Thus, the analysis of the thermal behavior of Prussian blue analogs revealed the generation of toxic cyanides and their exothermic reactions with electrolytes, indicating that PBA cathodes may bring safety concerns to nonaqueous Na-ion batteries [18,19]. Zhang et al. studied the thermal stability of hard carbon in the presence of an electrolyte at different states of charge and showed that the thermal effect of decomposition varies from 1.9 kJ/g for an initial electrode (2.5 V vs. Na/Na^+) to 3.0 kJ/g for an electrode at a potential of 0 V vs. Na/Na^+, which is close to the values for a metallic sodium anode (3.6 kJ/g). [20]. Mohsin et al. demonstrated an increase in enthalpy and a decrease in the temperature of decomposition peaks for the $Na_3V_2(PO_4)_3$ cathode and hard carbon anode materials with increasing their degree of charge [21]. Samigullin et al. conducted a systematic study of the thermal stability of several charged materials for LIBs and SIBs in the absence of an electrolyte, demonstrating the advantages of polyanionic vanadium compounds over oxides and carbon materials [22,23]. A natural question arises regarding the stability of such systems in the presence of an electrolyte, as well as the stability of the liquid electrolytes themselves. A significant difference in the thermal stability of LIB electrode materials between dry and electrolyte-soaked states, due to certain chemical interactions, was previously observed [3,6,24]. It was shown that polyanionic cathodes, such as $LiFePO_4$ (LFP) or $LiFe_{1-x}Mn_xPO_4$ (LFMP), that begin to decompose at 300–400 °C in the absence of an electrolyte, demonstrate a strong decrease in the onset temperature of the exothermic process, i.e., down to 200–250 °C [3,24,25]. Most authors consider PF_5, produced from the decomposition of $LiPF_6$ during the initial stage of heating, to be the most reactive component of the reaction media. On the other hand, $NaPF_6$ salt has a better thermal stability with a higher decomposition temperature [11,26,27]. Therefore, the behavior of the "electrode-electrolyte" composition in the Na-ion system seems quite ambiguous and worth studying. In the recent paper, Gan et al. conducted a systematic study of the thermal stability of the $NaNi_{1/3}Fe_{1/3}Mn_{1/3}O_2$ (NaNFM) cathode material in different Na-ion electrolytes and revealed its better compatibility with $NaPF_6$- and NaTFSI-based solutions; the combination of ethylene carbonate and diethyl carbonate demonstrated better thermal behavior then propylene-carbonate-based electrolytes [28].

In this work, we investigated the thermal stability of several types of electrolytes ($NaClO_4$ and $NaPF_6$ solutions in propylene carbonate (PC), ethylene carbonate (EC), diethyl carbonate (DEC), and their mixtures), as well as cathode and anode materials ($Na_3V_2(PO_4)_3$ (NVP), $Na_3(VO)_2(PO_4)_2F$ (NVOPF), β-$NaVP_2O_7$ (NVPO), and hard carbon) in combination with electrolytes utilizing differential scanning calorimetry (DSC) and ex-situ powder X-ray diffraction (PXRD).

2. Materials and Methods

2.1. Material Preparation

All electrode materials were synthesized according to previously published routes. All the initial reagents were purchased from Merck (formerly Sigma Aldrich, Rahway, NJ, USA). Briefly, the synthesis procedure is as follows.

2.1.1. $Na_3V_2(PO_4)_3$ (NVP)

The $Na_3V_2(PO_4)_3$ sample was prepared by a two-step synthesis. In the first step, vanadium oxide and citric acid were dissolved in distilled water in a molar ratio of 1:3, followed by stirring for 40 min at 70 °C. Then, a solution of sodium dihydrogen phosphate in distilled water was added dropwise and stirred for another 40 min, after which the solution was evaporated overnight at 95 °C. The solid residue was ground in a mortar and annealed at 350 °C for 3 h under argon, after which it was ground in a SPEX-8000M vibration mill for 5 min and annealed at 750 °C for 8 h under argon [29].

2.1.2. $Na_3(VO)_2(PO_4)_2F$ (NVOPF)

V_2O_5, NaF, $NH_4H_2PO_4$, and oxalic acid were mixed in 1:3:2:2 molar ratio. Then, the initial precursors were dissolved in deionized water, and the solution was magnetically stirred at 65–70 °C for about 30 min while adding 1 mL of ammonia solution (25 mass%). Next, 5 mL of the obtained homogeneous solution was transferred to a 10 mL glass vessel, placed into a microwave hydrothermal reactor Anton Paar 400 and treated at 180 °C for 30 min. The final solid products were centrifuged and washed with deionized water several times and dried in air. To obtain carbon coating, the material was mixed with glucose

(8 wt.%) in a mortar to receive homogeneous mixture. Then, the mixture was annealed at 600 °C for 1 h under Ar flow [30].

2.1.3. β-$NaVP_2O_7$ (NVPO)

The sample was synthesized in two stages. At the first stage, the synthesis was carried out using the hydrothermal method. At the second stage, the obtained precipitate was annealed in an inert atmosphere. The hydrothermal stage was carried out according to the following scheme. The initial reagents—$VOSO_4$, NaOH, H_3PO_4, and $N_2H_6SO_4$—were dissolved in distilled water at room temperature with constant stirring. Then, the solution was poured into a Teflon beaker, loaded into a steel autoclave (50 mL). The autoclave was heated with stirring and maintained at 230 °C for 24 h. After that, the autoclave was cooled in air or subjected to rapid cooling under a stream of cold water. The resulting powder was centrifuged (3000 rpm, 3 min) with repeated washing with distilled water and dried in air. At the second stage, the obtained samples were annealed in a furnace at 650 °C in an argon flow (gas flow rate of 10 mL/min, heating time of 3 h) with preliminary argon purging for 30 min [31].

2.1.4. Hard Carbon (HC)

First, D-glucose powder was caramelized in air at 200 °C for 24 h. Then, obtained precursor was powdered in an agate mortar. The second step of the synthesis was annealing at 1300 °C under argon flow for 1 h. The obtained HC material was ground at 400 rpm in a ball mill (Fritsch Pulverisette 5 classic line) for 1 h [32].

2.1.5. Electrodes

For the preparation of the electrodes, all powders were mixed with super P carbon black (Timcal, Willebroek, Belgium) and polyvinylidene fluoride (PVdF, Kureha Chemical, Tokyo, Japan) in a weight ratio of 80:10:10 with N-methyl-2-pyrrolidone (NMP) as the solvent. The homogenous slurry was cast on an aluminum foil by the Doctor Blade technique. The cast sheet was dried at 80 °C to remove NMP, roll-pressed, and punched into 10 or 16 mm diameter electrodes. Subsequently, the punched electrodes were dried at 120 °C under vacuum, and then taken for cell assembly.

2.2. *Characterization*

The electrochemical measurements were carried out using two-electrode half-cells with hard carbon, NVP, NVOPF, and NVPO as the working electrode; sodium metal as the counter electrode; and glass fiber (Schleicher & Schuell MicroScience, Dassel, Germany) as the separator. Electrolytes were prepared by dissolving $NaPF_6$ or $NaClO_4$ in the corresponding solvents: PC, EC, and DEC, pre-dried with molecular sieves. All electrolyte components were supplied by Merck (formerly Sigma Aldrich, Rahway, NJ, USA) and Kishida Chemicals. Electrochemical cells were assembled in an Ar-filled glove box (MBraun, Garching, Germany). Galvanostatic experiments were carried out using an Elins P-20X8 potentiostat-galvanostat (ES8 software, v. 4.192).

Thermal stability of the electrode materials was investigated by differential scanning calorimetry (DSC) using a Netzsch DSC 204 F1 Phoenix instrument (Selb, Germany) within the temperature range 50–450 °C (5 °C·min^{-1} heating rate) in an argon atmosphere. Electrodes from fully charged "half-cells" were extracted, washed with DMC, and dried under vacuum, and then electrode materials were scrapped from the current collector. The resulting powders (about 6 mg), either separately or in the presence of an electrolyte (about 5 μL), were placed in sealed crucibles made of high-pressure Cr–Ni stainless steel in an Ar-filled glove box. The electrode powder/electrolyte mass ratio was 1:1 to ensure an excess of the electrolyte during the experiment. To calculate the magnitude of the thermal effect, the area under the peak on the DSC curve was considered with subtraction of the background (if any) using the Netzsch Proteus 6.0 software.

Morphology of the samples was investigated using a JEOL JSM-6490LV (Tokyo, Japan) scanning electron microscope (SEM). The phase composition of the electrodes (ex situ experiments) was investigated by powder X-ray diffraction (PXRD) data assembled from a Huber Guinier camera G670 (Image Plate detector, curved Ge (111) monochromator, and CuKα_1 radiation λ = 1.54051 Å) (Rimsting, Germany).

3. Results

First, we performed the DSC experiments in several types of liquid Na-ion electrolytes often used in research works (Figure 1). PC, EC, and DEC were chosen as co-solvents.

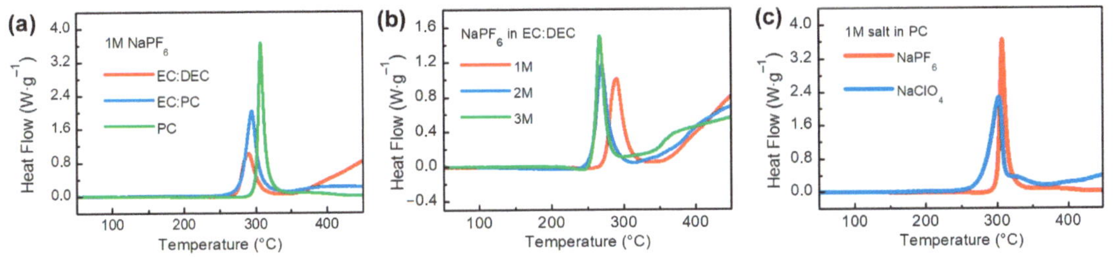

Figure 1. DSC curves for different electrolyte solutions: (**a**) 1M $NaPF_6$ in EC:DEC, EC:PC, and PC; (**b**) 1M, 2M, and 3M $NaPF_6$ in EC:DEC; and (**c**) 1M $NaPF_6$ and $NaClO_4$ in PC.

The results of the studies on 1M $NaPF_6$ solutions in various solvents do not clearly indicate a definitive advantage for any particular type of electrolyte. Although a PC-based electrolyte has a higher temperature of the onset and peak of decomposition (Table 1), the heat released is 1.7 times greater than in the case of 1M $NaPF_6$ in the EC:DEC solution. Similar results were published earlier when comparing solutions of sodium salts in PC and a mixture of EC and dimethyl carbonate (DMC) [10].

Table 1. DSC data for different electrolyte solutions.

Electrolyte	Onset Temperature, °C	Peak Temperature, °C	Enthalpy, J·g^{-1}
1M $NaPF_6$ in EC:DEC	274	290	268
2M $NaPF_6$ in EC:DEC	258	270	281
3M $NaPF_6$ in EC:DEC	257	267	246
1M $NaPF_6$ in PC	301	307	476
1M $NaPF_6$ in PC:EC	279	295	456
1M $NaClO_4$ in PC	279	302	629

The study of electrolytes with various $NaPF_6$ concentrations indicates that the thermal stability decreases with an increasing amount of salt in the solution. This observation may seem to contradict the common point of view regarding the better safety of concentrated electrolytes in a lithium-ion system. However, early works devoted to the unique properties of concentrated solutions were concerned rather with the lower volatility of the organic solvent in such systems, whereas later studies using DSC and accelerated rate calorimetry (ARC) revealed a decrease in the decomposition onset temperature with the increasing concentration of lithium salts in different solvents [33–35]. Hence, our data for the concentrated sodium electrolytes are in good agreement with the results previously published for lithium-ion systems, as well as with our experiments. As for the salts, according to our data, the decomposition enthalpy in the case of the $NaClO_4$-based solution is approx. 1.5 times higher than for the $NaPF_6$ solution (Figure 1c, Table 1).

In general, the better stability of sodium-ion electrolytes compared to lithium-ion ones is worth noting. Research shows that solutions of $LiPF_6$ in alkyl carbonates have an S-type decomposition curve, with an endothermic peak of $LiPF_6$ decomposition followed by an exothermic reaction of the interaction between PF_5 (a strong Lewis acid) and solvent components [3,6,7]. The position of the decomposition peak of 1M $LiPF_6$ solutions varies slightly among different studies but is generally within the range of 230–260 °C. In contrast, the Na-ion electrolyte demonstrates a higher temperature of the exothermic peak without any signs of endothermic salt decomposition (this effect is slightly noticeable only in a 3M $NaPF_6$ solution; Figure 1b), which suggests a safety advantage of SIBs.

The obtained electrode materials (their morphology is shown in Figure 2a) demonstrated typical electrochemical properties, with discharge capacities (Figure 2b) in the range of 83–95% of the theoretical ones. Detailed information about the crystal structure, phase composition, and electrochemical properties of the studied electrode materials is provided in previous publications [29–32].

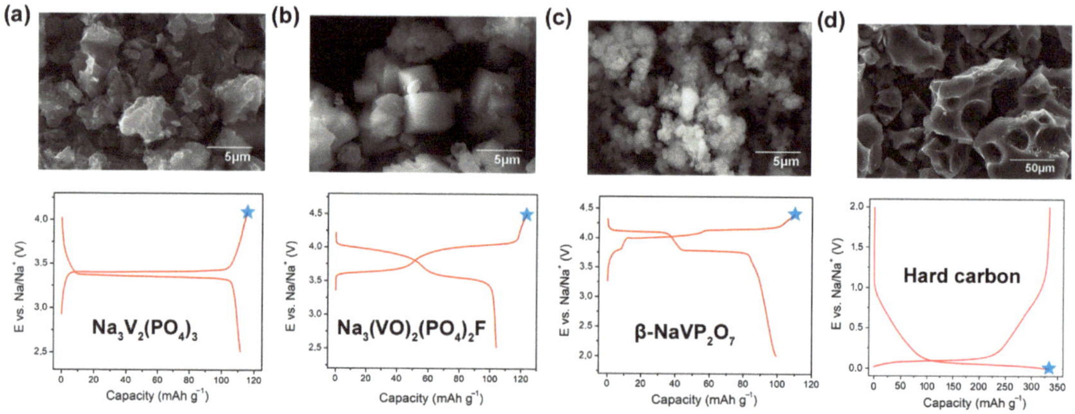

Figure 2. SEM images and charge–discharge curves for the electrode samples: $Na_3V_2(PO_4)_3$ (**a**), $Na_3(VO)_2(PO_4)_2F$ (**b**), β–$NaVP_2O_7$ (**c**), and hard carbon (**d**). Points on the curve where the electrodes were tested are marked with an asterisk.

The study of Na-ion cathode materials reveals the following pattern. Desodiated vanadium-based polyanionic cathodes—$Na_3V_2(PO_4)_3$, $Na_3(VO)_2(PO_4)_2F$, and β–$NaVP_2O_7$—demonstrate a significant increase in the decomposition enthalpy in the presence of an electrolyte compared to their "dry" state (Figures 3a–c and 4). Here and below, the thermal effect of the mixture is normalized to the mass of the electrode material. In the absence of chemical interaction, this effect is defined as:

$$\Delta H_{electrode+electrolyte} = \Delta H_{electrode} + \Delta H_{electrolyte}.$$

Moreover, the decomposition onset temperature of vanadium-based polyanionic cathodes soaked in electrolytes is greatly reduced compared to their dried state, and even lower than that of the liquid electrolyte. This can be a consequence of the increased reactivity of vanadium in high oxidation states with the electrolyte components. This effect was observed by S. Whittingham's group in delithiated $LiVOPO_4$. The authors suggested that the interaction between the charged material ($VOPO_4$) and the solvent in the electrolyte is accelerated by $LiPF_6$, and they proposed 11 possible chemical reactions to describe this process [3]. In the present work, we have obtained similar results for all three studied vanadium-containing cathode materials in the presence of an electrolyte, i.e., a strong decrease in the onset temperature of exothermic processes compared to "dry" materials and electrolytes. Notably, in the case of charged $Na_3(VO)_2(PO_4)_2F$ and β–$NaVP_2O_7$ ($Na(VO)_2(PO_4)_2F$ and β–VP_2O_7, correspondingly), the first decomposition peak appears at 170~180 °C. However, this temperature is higher for $Na_1V_2(PO_4)_3$, reaching ~225 °C (Figure 4).

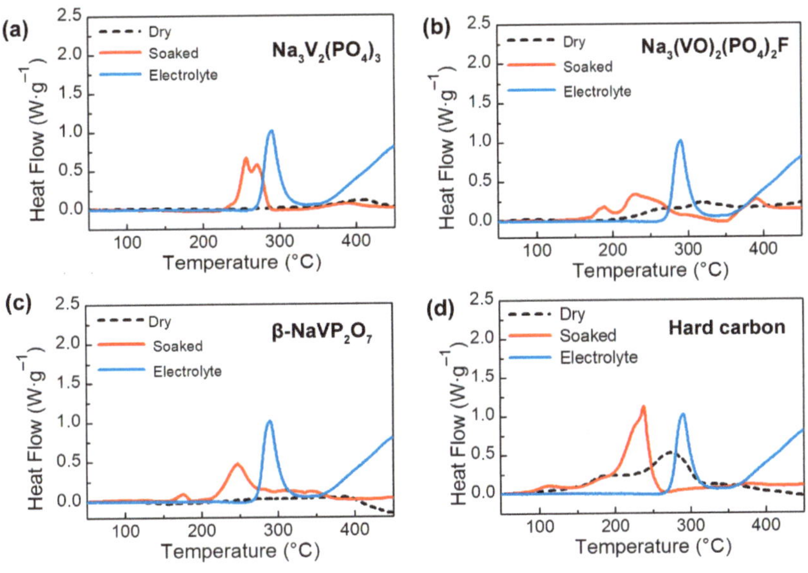

Figure 3. DSC curves (red line) for the charged electrode materials soaked in electrolyte: $Na_3V_2(PO_4)_3$ (**a**), $Na_3(VO)_2(PO_4)_2F$ (**b**), β–$NaVP_2O_7$ (**c**), and hard carbon (**d**). For comparison, DSC curves for the dried electrodes (dot line) and 1M $NaPF_6$ in EC:DEC electrolyte (blue line) are also plotted.

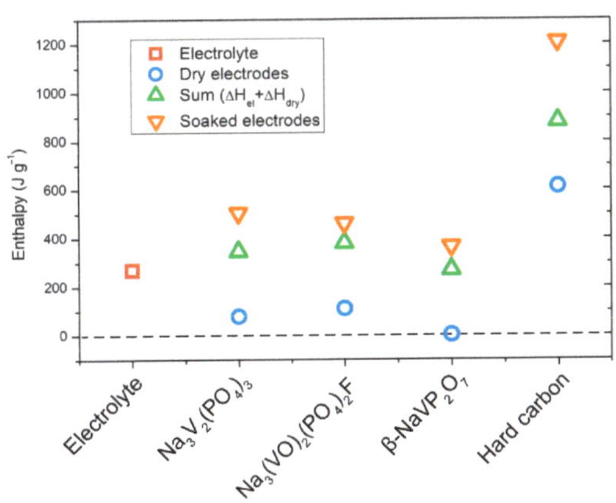

Figure 4. Schematic representation of the data on the enthalpy of dried [20] and soaked electrodes.

Recently, Pablos et al. studied the thermal stability of $Na_{3-x}V_2(PO_4)_2F_3$ and $Na_{3-x}V_2(PO_4)_2F_{3-y}O_y$ at different states of charge and observed a similar low-temperature decomposition peak at a high desodiation state in all samples (x = 2) [36]. According to our findings and literature data, the onset of interaction between the charged vanadium-containing material and the electrolyte during heating is largely determined by the degree of oxidation of vanadium cations, i.e., the desodiation potential. Additionally, an increase in the charging potential leads to a change in the quantity and composition of electrolyte oxidation products at the cathode–electrolyte interface, potentially decreasing the temperature at which the material starts to decompose. Thus, in the work of Samigullin and co-authors, an increase in the enthalpy and a decrease in the onset temperature of the decomposition process for $Na_3V_2(PO_4)_3$ charged at a higher potential (4.5 versus 3.8 V vs. Na/Na^+) were found [22]. Considering that, at 3.8 V, the material is already completely charged, it was concluded that the decrease in thermal stability was associated precisely with an increase in the amount and reactivity of the CEI (cathode–electrolyte interface) components.

In terms of decomposition enthalpy, β–$NaVP_2O_7$ remains the most stable among the studied cathode materials despite exhibiting the highest charging potential and the lowest onset temperature (Table 2). This is likely due to the well-known stability of the pyrophosphate group, as well as the fact that the main product of the electrode–electrolyte reaction, α–$NaVP_2O_7$, is quite similar to β–VP_2O_7 in terms of its composition and structure (Figure 5) [31].

Table 2. DSC data and phase composition of the charged electrodes after the DSC experiment in the presence of electrolyte (ex situ PXRD).

Initial Sample	Onset Temperature, °C	Peak Temperature, °C	Enthalpy, Dry *, $J\cdot g^{-1}$	Enthalpy, Soaked, $J\cdot g^{-1}$	Main Crystalline Phases by Ex Situ PXRD
$Na_3V_2(PO_4)_3$	243	256, 270	78	503	α–$NaVP_2O_7$
$Na_3(VO)_2(PO_4)_2F$	170	188, 229, 391	111	460	$Na_5V_3O_3F_{11}$
β–$NaVP_2O_7$	150	175, 247	–	364	α–$NaVP_2O_7$
Hard carbon	85	114, 238	610	1205	NaF
Electrolyte (1M $NaPF_6$ in EC:DEC)	274	290		268	

* Data for the charged, washed, and dried electrode materials obtained in our previous study [23]. All calculations are normalized to the weight of the "dry" electrode material.

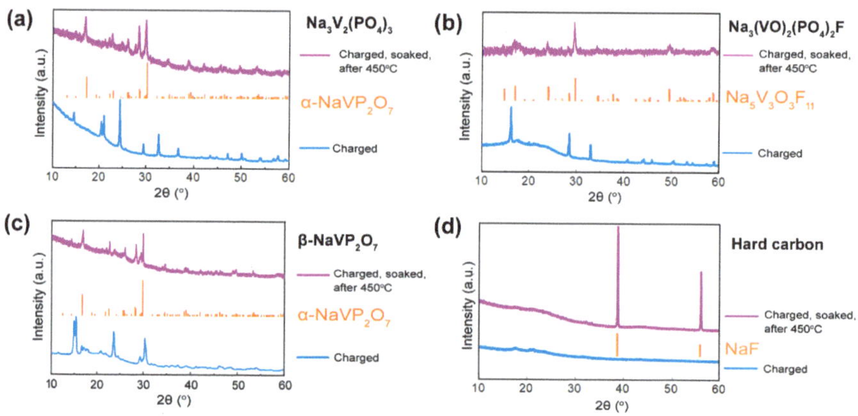

Figure 5. Results of ex situ PXRD for the charged electrodes soaked in 1M $NaPF_6$ in EC:DEC electrolyte before and after DSC experiments: $Na_3V_2(PO_4)_3$ (**a**), $Na_3(VO)_2(PO_4)_2F$ (**b**), β–$NaVP_2O_7$ (**c**), and hard carbon (**d**).

According to the X-ray diffraction data (Figure 5), the decomposition products of $NaV_2(PO_4)_3$ and $Na(VO)_2(PO_4)_2F$ in the presence of an electrolyte differ significantly from those of the dry electrodes described in literature [23]. This correlates well with the elevated values of thermal effects: while charged dry electrodes mainly produce desodiated phases after the DSC experiment, the presence of electrolytes results in the formation of sodium-containing pyrophosphate and oxofluoride (Figure 5, Table 2).

In the case of the anode material, the presence of electrolytes leads to similar results: the thermal effect of the decomposition increases while the onset and peak temperatures decrease (Figure 3, Table 2). The total enthalpy released during heating by the mixture of the charged hard carbon and electrolyte is 1205 $J\cdot g^{-1}$. Hard carbon shows the highest increase in enthalpy compared to the "electrode+electrolyte" sum ($\approx$38%) among all the studied samples (Figure 4). It is likely that the chemical interaction in the mixture during heating occurs in a manner similar to the formation of the solid–electrolyte interface (SEI). This process is well-known to proceed more intensely at the anode (i.e., reduction in the electrolyte) than at the cathode (oxidation of the electrolyte) side [37–40].

4. Discussion

The choice of a polyanionic cathode material for the development of sodium-ion battery technology is an extremely ambiguous task, since scientists and developers know a large family of different cathodes, each with its own advantages and disadvantages. In this paper, we studied such a key parameter of a sodium-ion battery as the thermal stability of the "electrode–electrolyte" combination for several vanadium-based cathode materials. Particularly, two of them are well-known and intensively studied—$Na_3V_2(PO_4)_3$ and $Na_3(VO)_2(PO_4)_2F$—and a third—β–$NaVP_2O_7$—was discovered much later than the others, but is gaining popularity, at least in laboratory studies. The inclusion of this material in this paper is also dictated by the fact that pyrophosphates are much more thermally stable than phosphates or fluoride phosphates, which means that the use of β–$NaVP_2O_7$ may be in demand in those areas where battery safety is more important than its weight and size characteristics [4]. Indeed, a comparison of the theoretical specific energy capacity of these cathodes does not speak in favor of $Na_3V_2(PO_4)_3$: $\approx$415 Wh kg^{-1} for $Na_3V_2(PO_4)_3$, $\approx$420 for β–$NaVP_2O_7$, and $\approx$480 for $Na_3(VO)_2(PO_4)_2F$. However, as the data presented in the article show, it exhibits less chemical activity during the thermal decomposition of the electrode wetted with an electrolyte: the increase in heat for β–$NaVP_2O_7$ (compared to the value for liquid electrolyte) was 36% versus 87% for $Na_3V_2(PO_4)_3$ and 71% for $Na_3(VO)_2(PO_4)_2F$.

For a deeper understanding of the results, it is necessary to compare the obtained data with the lithium-ion system, more precisely, with the most popular polyanionic material—$LiFePO_4$. There are some difficulties with this, since the literature data devoted to the study of the thermal stability of delithiated $LiFePO_4$ in "dry" form and in the presence of electrolytes are partly contradictory. Thus, Xiang et al., who studied the interaction of liquid electrolyte with various cathode materials, found that $LiFePO_4$ suppresses the decomposition of the electrolyte and reduces the magnitude of the thermal effect [24]. The authors explained this by the fact that some active groups on the surface (e.g., PO_4^{3-}) of the material interact with Lewis acid PF_5, which is a catalyst for the thermal decomposition of the electrolyte. On the other hand, Joachin et al., as well as Huang et al. and Yi et al., found a significant increase in the enthalpy of decomposition of $FePO_4$ in the presence of an electrolyte [3,5,41]. Joachin et al. estimated the thermal effect of the reaction at 145 J·g^{-1} (however, the authors did not indicate what exactly the thermal effect value was normalized to), Huang et al. at 251 J·g^{-1} (normalized to the mass of the electrode + the mass of the electrolyte, which was from 25 to 35% in the mixture), and Yi et al. 501 J·g^{-1} (normalized to the mass of the electrode material, with an electrode:electrolyte mass ratio = 1:2). Yamada et al. did not provide data for a "dry" electrode, but the one wetted with electrolytes showed thermal effect values similar to Joachin's—147 J·g^{-1} [42]. It is worth noting that, in this case, too, the authors did not indicate what exactly the thermal effect was normalized to. Thus, even among those researches who report an increase in the enthalpy of decomposition in contact with an electrolyte, there is no consensus on the values of thermal effects; this is probably largely due to different approaches to conducting the experiment. We performed our studies with hydrothermally synthesized $LiFePO_4$ and commercially available 1M $LiPF_6$ in EC:DEC:DMC = 1:1:1 (Figure 6a) and obtained a value of 276 J·g^{-1} (50 J·g^{-1} for an electrode without electrolyte), normalized to the mass of the electrode material (or 138 J·g^{-1}, if normalized to the sum of the masses of the electrode and electrolyte—i.e., using the approach of Huang et al.).

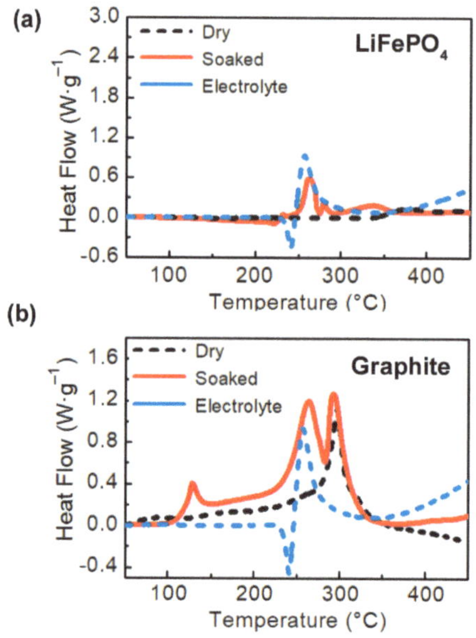

Figure 6. DSC curves (red line) for the charged Li-ion electrode materials soaked in electrolyte: $LiFePO_4$ (**a**) and graphite (**b**). For comparison, DSC curves for the dried electrodes (dot line) and 1M $LiPF_6$ in EC:DEC:DMC electrolyte (blue line) are also plotted.

The increase in heat release compared to a pure electrolyte in our experiment was 22%, which is less than for all studied sodium cathodes. Thus, even despite the better stability of the sodium-ion liquid electrolyte compared to the lithium-ion one, sodium-containing vanadium-based cathodes in a charged state demonstrate a greater increase in the enthalpy of decomposition in the presence of an electrolyte than $LiFePO_4$.

As for the anode material, the situation is rather the opposite. Wang et al. studied charged graphite in dry form and in the presence of an electrolyte and obtained the values of the thermal effect of decomposition of 1341 and 2253 $J{\cdot}g^{-1}$, respectively (normalized to the mass of the electrode material; the mass ratio electrode:electrolyte = 1:1) [43]. In our experiment, under the same conditions, we obtained 449 and 2334 $J{\cdot}g^{-1}$ with the same normalization (Figure 6b). Thus, the thermal effect of the reaction of charged graphite with an electrolyte is higher than that of hard carbon (610 and 1205 $J{\cdot}g^{-1}$, respectively).

5. Conclusions

To summarize the results, we would highlight the following observations.

Sodium-ion electrolytes, depending on their composition, begin to decompose when heated to 250~300 °C, which is 20–30 °C higher than the published temperatures for lithium electrolytes.

Vanadium-based polyanionic cathodes in contact with 1M $NaPF_6$ in the EC:DEC electrolyte demonstrate reduced decomposition temperatures and a more intense heat release in comparison with the dry state. The onset temperature depends on the desodiation potential; this effect may be a consequence of both a more reactive cathode–electrolyte interface and the previously described chemical interaction of vanadium cations with electrolyte components. However, in terms of decomposition enthalpy, β-$NaVP_2O_7$ demonstrates a better behavior compared to $Na_3V_2(PO_4)_3$ and $Na_3(VO)_2(PO_4)_2F$ despite its high charging potential.

The least stable part of the sodium-ion electrochemical cell is the anode material. In the presence of electrolytes, it demonstrates almost a 40% increase in decomposition enthalpy and a shift of the main peak maxima from 273 to 238 °C. However, both the increase in the heat of the decomposition of hard carbon in the presence of electrolytes and its absolute value are less than similar parameters for graphite in a lithium-ion system, in accordance with the literature data and the results of our own experiments.

Overall, the significant increase in the thermal effect of the decomposition of "charged material–electrolyte" mixtures compared to "dry" electrodes confirms the widespread belief that the liquid electrolyte is the main source of safety problems in metal-ion batteries. To eliminate these shortcomings, it is necessary to develop solid-state secondary current sources using polymeric and/or ceramic electrolyte materials.

Author Contributions: Conceptualization, R.R.S.; methodology, Z.V.B., M.V.Z., E.V.Z. and M.G.R.; investigation, Z.V.B., M.V.Z., E.V.Z., M.G.R. and R.R.S.; writing—original draft preparation, R.R.S.; writing—review and editing, Z.V.B., M.V.Z., E.V.Z., M.G.R., O.A.D. and E.V.A.; supervision, O.A.D. and E.V.A.; project administration, E.V.A.; funding acquisition, E.V.A. and O.A.D. All authors have read and agreed to the published version of the manuscript.

Funding: This work was supported by the Russian Science Foundation (grant No. 24-13-00107).

Data Availability Statement: The raw data supporting the conclusions of this article will be made available by the authors on request.

Acknowledgments: This work was supported in part by the Lomonosov Moscow State University Program of Development. The authors thank Alina Safiullina and Tatyana Perfilyeva for their help with preparing electrolyte solutions.

Conflicts of Interest: The authors declare no conflicts of interest.

References

1. Liu, B.; Jia, Y.; Yuan, C.; Wang, L.; Gao, X.; Yin, S.; Xu, J. Safety Issues and Mechanisms of Lithium-Ion Battery Cell upon Mechanical Abusive Loading: A Review. *Energy Storage Mater.* **2020**, *24*, 85–112. [CrossRef]
2. Martha, S.K.; Haik, O.; Zinigrad, E.; Exnar, I.; Drezen, T.; Miners, J.H.; Aurbach, D. On the Thermal Stability of Olivine Cathode Materials for Lithium-Ion Batteries. *J. Electrochem. Soc.* **2011**, *158*, A1115. [CrossRef]
3. Huang, Y.; Lin, Y.C.; Jenkins, D.M.; Chernova, N.A.; Chung, Y.; Radhakrishnan, B.; Chu, I.H.; Fang, J.; Wang, Q.; Omenya, F.; et al. Thermal Stability and Reactivity of Cathode Materials for Li-Ion Batteries. *ACS Appl. Mater. Interfaces* **2016**, *8*, 7013–7021. [CrossRef]
4. Tamaru, M.; Chung, S.C.; Shimizu, D.; Nishimura, S.I.; Yamada, A. Pyrophosphate Chemistry toward Safe Rechargeable Batteries. *Chem. Mater.* **2013**, *25*, 2538–2543. [CrossRef]
5. Yi, J.; Wang, C.; Xia, Y. Comparison of Thermal Stability between Micro- and Nano-Sized Materials for Lithium-Ion Batteries. *Electrochem. Commun.* **2013**, *33*, 115–118. [CrossRef]
6. Lee, Y.; Kim, S.O.; Mun, J.; Park, M.S.; Kim, K.J.; Lee, K.Y.; Choi, W. Influence of Salt, Solvents, and Additives on the Thermal Stability of Delithiated Cathodes in Lithium-Ion Batteries. *J. Electroanal. Chem.* **2017**, *807*, 174–180. [CrossRef]
7. Gnanaraj, J.S.; Zinigrad, E.; Asraf, L.; Gottlieb, H.E.; Sprecher, M.; Schmidt, M.; Geissler, W.; Aurbach, D. A Detailed Investigation of the Thermal Reactions of LiPF6 Solution in Organic Carbonates Using ARC and DSC. *J. Electrochem. Soc.* **2003**, *150*, A1533. [CrossRef]
8. Golubkov, A.W.; Fuchs, D.; Wagner, J.; Wiltsche, H.; Stangl, C.; Fauler, G.; Voitice, G.; Thalera, A.; Hacker, V. Thermal-runaway experiments on consumer Li-ion batteries with metal-oxide and olivin-type cathodes. *RSC Adv.* **2014**, *4*, 3633–3642. [CrossRef]
9. Velumani, D.; Bansal, A. Thermal Behavior of Lithium- and Sodium-Ion Batteries: A Review on Heat Generation, Battery Degradation, Thermal Runway—Perspective and Future Directions. *Energy Fuels* **2022**, *36*, 14000–14029. [CrossRef]
10. Bang, H.; Kim, D.-H.; Bae, Y.C.; Prakash, J.; Sun, Y.-K. Effects of Metal Ions on the Structural and Thermal Stabilities of $Li[Ni_1{-}x{-}yCoxMny]O_2$ (x + y $\leq$ 0.5) Studied by In Situ High Temperature XRD. *J. Electrochem. Soc.* **2008**, *155*, A952–A958. [CrossRef]
11. Sun, Y.K.; Myung, S.T.; Park, B.C.; Prakash, J.; Belharouak, I.; Amine, K. High-energy cathode material for long-life and safe lithium batteries. *Nat. Mater.* **2009**, *8*, 320–324. [CrossRef] [PubMed]
12. Bak, S.-M.; Hu, E.; Zhou, Y.; Yu, X.; Senanayake, S.D.; Cho, S.-J.; Kim, K.-B.; Chung, K.Y.; Yang, X.-Q.; Nam, K.-W. Structural Changes and Thermal Stability of Charged $LiNixMnyCozO_2$ Cathode Materials Studied by Combined In Situ Time-Resolved XRD and Mass Spectroscopy. *ACS Appl. Mater. Interfaces* **2014**, *6*, 22594–22601. [CrossRef] [PubMed]
13. Song, Y.; Cui, Y.; Li, B.; Geng, L.; Yan, J.; Zhu, D.; Zhou, P.; Zhou, J.; Yan, Z.; Xue, Q.; et al. Revealing the origin of high-thermal-stability of single-crystal Ni-rich cathodes toward higher-safety batteries. *Nano Energy* **2023**, *116*, 108846. [CrossRef]
14. Zhao, J.; Xu, J.; Lee, D.H.; Dimov, N.; Meng, Y.S.; Okada, S. Electrochemical and Thermal Properties of P2-Type $Na_2/_3Fe$ $_1/_3Mn_2/_3O_2$ for Na-Ion Batteries. *J. Power Sources* **2014**, *264*, 235–239. [CrossRef]

15. Zuo, W.; Liu, R.; Ortiz, G.F.; Rubio, S.; Chyrka, T.; Lavela, P.; Zheng, S.; Tirado, J.L.; Wang, D.; Yang, Y. Sodium Storage Behavior of $Na_{0.66}Ni_{0.33-x}Zn_xMn_{0.67}O_2$ (x = 0, 0.07 and 0.14) Positive Materials in Diglyme-Based Electrolytes. *J. Power Sources* **2018**, *400*, 317–324. [CrossRef]
16. Zhao, J.; Zhao, L.; Chihara, K.; Okada, S.; Yamaki, J.I.; Matsumoto, S.; Kuze, S.; Nakane, K. Electrochemical and Thermal Properties of Hard Carbon-Type Anodes for Na-Ion Batteries. *J. Power Sources* **2013**, *244*, 752–757. [CrossRef]
17. Ponrouch, A.; Marchante, E.; Courty, M.; Tarascon, J.M.; Palacín, M.R. In Search of an Optimized Electrolyte for Na-Ion Batteries. *Energy Environ. Sci.* **2012**, *5*, 8572–8583. [CrossRef]
18. Li, Z.; Dadsetan, M.; Gao, J.; Zhang, S.; Cai, L.; Naseri, A.; Jimenez-Castaneda, M.E.; Filley, T.; Miller, J.T.; Thomson, M.J.; et al. Revealing the Thermal Safety of Prussian Blue Cathode for Safer Nonaqueous Batteries. *Adv. Energy Mater.* **2021**, *11*, 2101764. [CrossRef]
19. Ge, L.; Song, Y.; Niu, P.; Li, B.; Zhou, L.; Feng, W.; Ma, C.; Li, X.; Kong, D.; Yan, Z.; et al. Elaborating the Crystal Water of Prussian Blue for Outstanding Performance of Sodium Ion Batteries. *ACS Nano* **2024**, *18*, 3542–3552. [CrossRef]
20. Zhang, X.; Dong, X.; Qiu, X.; Cao, Y.; Wang, C.; Wang, Y.; Xia, Y. Extended low-voltage plateau capacity of hard carbon spheres anode for sodium ion batteries, Extended low-voltage plateau capacity of hard carbon spheres anode for sodium ion batteries. *J. Power Sources* **2020**, *476*, 228550. [CrossRef]
21. Mohsin, I.U.; Hofmann, A.; Ziebert, C. Exploring the reactivity of $Na_3V_2(PO_4)_3$/C and hard carbon electrodes in sodium-ion batteries at various charge states. *Electrochim. Acta* **2024**, *487*, 144197. [CrossRef]
22. Samigullin, R.R.; Zakharkin, M.V.; Drozhzhin, O.A.; Antipov, E.V. Thermal Stability of NASICON-Type $Na_3V_2(PO_4)_3$ and $Na_4VMn(PO_4)_3$ as Cathode Materials for Sodium-ion Batteries. *Energies* **2023**, *16*, 3051. [CrossRef]
23. Samigullin, R.R.; Drozhzhin, O.A.; Antipov, E.V. Comparative Study of the Thermal Stability of Electrode Materials for Li-Ion and Na-Ion Batteries. *ACS Appl. Energy Mater.* **2022**, *5*, 14–19. [CrossRef]
24. Xiang, H.F.; Wang, H.; Chen, C.H.; Ge, X.W.; Guo, S.; Sun, J.H.; Hu, W.Q. Thermal stability of LiPF6-based electrolyte and effect of contact with various delithiated cathodes of Li-ion batteries. *J. Power Sources* **2009**, *191*, 575–581. [CrossRef]
25. Chen, G.; Richardson, T.J. Thermal instability of Olivine-type $LiMnPO_4$ cathodes. *J. Power Sources* **2010**, *195*, 1221–1224. [CrossRef]
26. He, X.; Ping, P.; Kong, D.; Wang, G.; Wang, D. Comparison study of electrochemical and thermal stability of $Na_3V_2(PO_4)_3$ in different electrolytes under room and elevated temperature. *Int. J. Energy Res.* **2022**, *46*, 23173–23194. [CrossRef]
27. Eshetu, G.G.; Grugeon, S.; Kim, H.; Jeong, S.; Wu, L.; Gachot, G.; Laruelle, S.; Armand, M.; Passerini, S. Comprehensive Insights into the Reactivity of Electrolytes Based on Sodium Ions. *ChemSusChem* **2016**, *9*, 462. [CrossRef]
28. Gan, Y.; Ping, P.; Wang, J.; Song, Y.; Gao, W. Comparative analysis of thermal stability and electrochemical performance of $NaNi_{1/3}Fe_{1/3}Mn_{1/3}O_2$ cathode in different electrolytes for sodium ion batteries. *J. Power Sources* **2024**, *594*, 234008. [CrossRef]
29. Zakharkin, M.V.; Drozhzhin, O.A.; Ryazantsev, S.V.; Chernyshov, D.; Kirsanova, M.A.; Mikheev, I.V.; Pazhetnov, E.M.; Antipov, E.V.; Stevenson, K.J. Electrochemical properties and evolution of the phase transformation behavior in the NASICON-type $Na_{3+x}Mn_xV_{2-x}(PO_4)_3$ ($0 \leq x \leq 1$) cathodes for Na-ion batteries. *J. Power Sources* **2020**, *470*, 228231. [CrossRef]
30. Burova, D.; Shakhova, I.; Morozova, P.; Iarchuk, A.; Drozhzhin, O.A.; Rozova, M.G.; Praneetha, S.; Murugan, V.; Tarascon, J.M.; Abakumov, A.M. The Rapid Microwave-Assisted Hydrothermal Synthesis of NASICON-Structured $Na_3V_2O_2$: $X(PO_4)_2F_{3-2x}$ ($0 < x \leq 1$) Cathode Materials for Na-Ion Batteries. *RSC Adv.* **2019**, *9*, 19429–19440. [CrossRef]
31. Drozhzhin, O.A.; Tertov, I.V.; Alekseeva, A.M.; Aksyonov, D.A.; Stevenson, K.J.; Abakumov, A.M.; Antipov, E.V. β-NaVP2O7 as a Superior Electrode Material for Na-Ion Batteries. *Chem. Mater.* **2019**, *31*, 7463–7469. [CrossRef]
32. Bobyleva, Z.V.; Drozhzhin, O.A.; Alekseeva, A.M.; Dosaev, K.A.; Peters, G.S.; Lakienko, G.P.; Perfilyeva, T.I.; Sobolev, N.A.; Maslakov, K.I.; Savilov, S.V.; et al. Caramelization as a Key Stage for the Preparation of Monolithic Hard Carbon with Advanced Performance in Sodium-Ion Batteries. *ACS Appl. Energy Mater.* **2023**, *6*, 181–190. [CrossRef]
33. Wang, J.; Yamada, Y.; Sodeyama, K.; Chiang, C.H.; Tateyama, Y. Superconcentrated electrolytes for a high-voltage lithium-ion battery. *Nat. Commun.* **2016**, *7*, 12032. [CrossRef] [PubMed]
34. Hou, J.; Lu, L.; Wang, L.; Ohma, A.; Ren, D.; Feng, X.; Li, Y.; Li, Y.; Ootani, I.; Han, X.; et al. Thermal runaway of Lithium-ion batteries employing $LiN(SO_2F)_2$-based concentrated electrolytes. *Nat. Commun.* **2020**, *11*, 5100. [CrossRef] [PubMed]
35. Zhao, L.; Inoishi, A.; Okada, S. Thermal risk evaluation of concentrated electrolytes for Li-ion batteries. *J. Power Sources Adv.* **2021**, *12*, 100079. [CrossRef]
36. Pablos, C.; Olchowka, J.; Petit, E.; Minart, G.; Duttine, M.; Weill, F.; Masquelier, C.; Carlier, D.; Croguennec, L. Thermal Stability of $Na_{3-x}V_2(PO_4)_2F_{3-y}O_y$: Influence of F− for O_2− Substitution and Degradation Mechanisms. *Chem. Mater.* **2023**, *35*, 4078–4088. [CrossRef]
37. Gauthier, M.; Carney, T.J.; Grimaud, A.; Giordano, L.; Pour, N.; Chang, H.-H.; Fenning, D.P.; Lux, S.F.; Paschos, O.; Bauer, C.; et al. Electrode–Electrolyte Interface in Li-Ion Batteries: Current Understanding and New Insights. *J. Phys. Chem. Lett.* **2015**, *6*, 4653–4672. [CrossRef] [PubMed]
38. Meng, J.; Jia, G.; Yang, H.; Wang, M. Recent advances for SEI of hard carbon anode in sodium-ion batteries: A mini review. *Front. Chem.* **2022**, *10*, 986541. [CrossRef] [PubMed]
39. Daboss, S.; Philipp, T.; Palanisamy, K.; Flowers, J.; Stein, H.S.; Kranz, C. Characterization of the solid/electrolyte interphase at hard carbon anodes via scanning (electrochemical) probe microscopy. *Electrochim. Acta* **2023**, *453*, 142345. [CrossRef]
40. Pan, Y.; Zhang, Y.; Parimalam, B.S.; Nguyen, C.C.; Wang, G.; Lucht, B.L. Investigation of the solid electrolyte interphase on hard carbon electrode for sodium ion batteries. *J. Electroanal. Chem.* **2017**, *799*, 181–186. [CrossRef]

41. Joachin, H.; Kaun, T.D.; Zaghib, K.; Prakash, J. Electrochemical and Thermal Studies of Carbon-Coated $LiFePO_4$ Cathode. *J. Electrochem. Soc.* **2009**, *156*, A401–A406. [CrossRef]
42. Yamada, A.; Chung, S.C.; Hinokuma, K. Optimized $LiFePO_4$ for Lithium Battery Cathodes. *J. Electrochem. Soc.* **2001**, *148*, A224–A229. [CrossRef]
43. Wang, Q.; Sun, J.; Yao, X.; Chen, C. Thermal Behavior of Lithiated Graphite with Electrolyte in Lithium-Ion Batteries. *J. Electrochem. Soc.* **2006**, *153*, A329–A333. [CrossRef]

materials MDPI

Article

Co/Al Co-Substituted Layered Manganese-Based Oxide Cathode for Stable and High-Rate Potassium-Ion Batteries

Junxian Li [1,2], Wenli Shu [1,2], Guangwan Zhang [1,2], Jiashen Meng [1], Chunhua Han [1,2,3], Xiujuan Wei [4,*] and Xuanpeng Wang [1,2,3,5,*]

1 School of Materials Science and Engineering, Wuhan University of Technology, Wuhan 430070, China
2 Sanya Science and Education Innovation Park, Wuhan University of Technology, Sanya 572000, China
3 Hubei Longzhong Laboratory, Wuhan University of Technology (Xiangyang Demonstration Zone), Xiangyang 441000, China
4 Institute for Sustainable Transformation, School of Chemical Engineering and Light Industry, Guangdong University of Technology, Guangzhou 510006, China
5 Department of Physical Science & Technology, School of Science, Wuhan University of Technology, Wuhan 430070, China
* Correspondence: xjwei@gdut.edu.cn (X.W.); wxp122525691@whut.edu.cn (X.W.); Tel.: +86-135-4528-4506 (Xuanpeng Wang)

Abstract: Manganese-based layered oxides are promising cathode materials for potassium-ion batteries (PIBs) due to their low cost and high theoretical energy density. However, the Jahn-Teller effect of Mn^{3+} and sluggish diffusion kinetics lead to rapid electrode deterioration and a poor rate performance, greatly limiting their practical application. Here, we report a Co/Al co-substitution strategy to construct a P3-type $K_{0.45}Mn_{0.7}Co_{0.2}Al_{0.1}O_2$ cathode material, where Co^{3+} and Al^{3+} ions occupy Mn^{3+} sites. This effectively suppresses the Jahn-Teller distortion and alleviates the severe phase transition during K^+ intercalation/de-intercalation processes. In addition, the Co element contributes to K^+ diffusion, while Al stabilizes the layer structure through strong Al-O bonds. As a result, the $K_{0.45}Mn_{0.7}Co_{0.2}Al_{0.1}O_2$ cathode exhibits high capacities of 111 mAh g^{-1} and 81 mAh g^{-1} at 0.05 A g^{-1} and 1 A g^{-1}, respectively. It also demonstrates a capacity retention of 71.6% after 500 cycles at 1 A g^{-1}. Compared to the pristine $K_{0.45}MnO_2$, the $K_{0.45}Mn_{0.7}Co_{0.2}Al_{0.1}O_2$ significantly alleviates severe phase transition, providing a more stable and effective pathway for K^+ transport, as investigated by in situ X-ray diffraction. The synergistic effect of Co/Al co-substitution significantly enhances the structural stability and electrochemical performance, contributing to the development of new Mn-based cathode materials for PIBs.

Keywords: potassium-ion batteries; Mn-based layered oxide cathodes; synergy effect; Jahn-Teller effect; potassium storage mechanism

Citation: Li, J.; Shu, W.; Zhang, G.; Meng, J.; Han, C.; Wei, X.; Wang, X. Co/Al Co-Substituted Layered Manganese-Based Oxide Cathode for Stable and High-Rate Potassium-Ion Batteries. *Materials* **2024**, *17*, 1277. https://doi.org/10.3390/ma17061277

Academic Editor: Alessandro Dell'Era

Received: 21 February 2024
Revised: 7 March 2024
Accepted: 8 March 2024
Published: 10 March 2024

1. Introduction

In the current era of rapidly depleting fossil fuels and the growing demand for renewable energy, lithium-ion batteries (LIBs) have been widely used in various electronic devices due to their high energy density and long cycle life [1–3]. However, the huge consumption, rapidly increasing price and uneven distribution of lithium resources pose significant challenges to the sustainable development of LIBs. Consequently, there is a pressing need to identify suitable alternatives that are cost-effective and suitable for large-scale energy storage. In recent years, potassium-ion batteries (PIBs) have become highly promising candidates because of abundant resources and low cost [4–7]. PIBs have a similar operating mechanism to LIBs. And the low redox potential of K^+/K (−2.93 V vs. SHE, second only to Li^+/Li) gives it a high theoretical energy density. In addition, K^+ exhibits fast diffusion kinetics in organic electrolytes because of its weak Lewis acidity [8,9]. However, the inherent property of the large K^+ ionic radius (1.38 Å) greatly limits K^+ transport in the electrode material and the electrode material undergoes severe volume changes during electrochemical cycling, leading to rapid capacity loss [10,11]. Therefore, the development of suitable electrode materials plays a crucial role in advancing PIB technology.

Currently, the main cathode materials being investigated for PIBs are Prussian blue analogues [12,13], polyanionic compounds [14,15] and layered transition metal oxides [16–18]. Among these, layered transition metal oxides are one of the most promising cathode materials thanks to the ease of synthesis and high theoretical capacity. Specifically, Mn-based layered oxides have garnered significant attention in PIBs owing to their low cost and environmental friendliness [19]. However, the Jahn-Teller distortion of Mn^{3+} leads to an unbalanced lengthening of the Mn-O bond and reduces the symmetry of the molecular structure, thus exacerbating the overall structural instability [20,21]. In addition, slow K^+ transport, limited K^+ storage sites and severe phase transitions during cycling cause their poor performance in practical applications [22]. Kim et al. [23] reported layered P3-$K_{0.5}MnO_2$ with a complex phase transition during cycling, leading to rapid capacity loss and retaining only 70% of the initial discharge capacity after 50 cycles at 20 mA g^{-1}. Due to the poor diffusion of K^+, this electrode shows a low specific capacity of only 38 mAh g^{-1} at 300 mA g^{-1}, indicating a poor rate performance. Studies have shown that metal-ion substitution is a powerful measure to improve the above defects in K_xMnO_2. For example, Ni [24], Co [25], Fe [26], Mg [27] and Ni-Ti [28] substitutions have demonstrated the ability to improve their structural stability and electrochemical properties. Zhang et al. [29] reported a $K_{0.3}Mn_{0.95}Co_{0.05}O_2$ cathode. The Co doping suppressed Jahn-Teller distortion, allowing for more isotropic migration pathways for K^+ in the interlayer. This enhanced the ionic diffusion and, consequently, the rate capability. Zhong et al. [30] found that Co-Fe co-substitution could expand the interlayer distance and effectively suppressed the interlayer-gliding to achieve highly reversible phase evolution, thus improving the rate performance and cycle life of $K_{0.5}MnO_2$.

Herein, we have modified $K_{0.45}MnO_2$ (KMO) with two selected metal elements (Co and Al), and have innovatively synthesized the co-substituted P3-$K_{0.45}Mn_{0.7}Co_{0.2}Al_{0.1}O_2$ (KMCAO). Through the careful consideration of chemical bonding, Al^{3+} partially replaces the Mn^{3+} sites, forming stronger Al-O bonds compared to the Mn-O bonds. This substitution serves to stabilize the layer structure [31]. Additionally, the inclusion of Co^{3+} ions facilitated electrochemical reactions, providing active sites for K^+ storage. Moreover, the Co^{3+}/Al^{3+} co-substitution of Mn^{3+} sites effectively inhibited Jahn-Teller distortion, mitigating severe crystal structural transformations. Through in situ XRD, spherical aberration scanning transmission electron microscopy and first-principles simulation calculations, the KMCAO electrode exhibits a highly reversible phase structure transition during the cycling process. Meanwhile, Co/Al co-substitution widens the K layer spacing of the material and lowers the barrier energy for K^+ migration. Compared to the pristine KMO, the KMCAO cathode exhibits superior cycling stability and rate performance. The excellent electrochemical performance of the KMCAO || soft carbon full cell also demonstrates its potential for practical applications. These findings highlight the effectiveness of our modification strategy and provide new insights into the development of PIB cathode materials.

2. Materials and Methods

All reactants are analytical grade. P3-type $K_{0.45}Mn_{0.7}Co_{0.2}Al_{0.1}O_2$ and $K_{0.45}MnO_2$ samples are synthesized using a sol–gel method. Firstly, add 2 g of polyvinylpyrrolidone (PVP K90, Mw = 1,300,000) to 20 mL of deionized water and stir continuously to form a solution. KNO_3, $Mn(CH_3COO)_2{\cdot}4H_2O$, $Co(CH_3COO)_2{\cdot}4H_2O$ and $Al(NO_3)_3{\cdot}9H_2O$ are dissolved into the solution by stoichiometric amounts. The mixed solution is dried at 80 °C for 15 h and then pre-sinters at 350 °C in air for 3 h to obtain a black solid. Finally, the black solid powder is calcined at 800 °C in air for 12 h to obtain $K_{0.45}Mn_{0.7}Co_{0.2}Al_{0.1}O_2$. After slowly cooling down to 150 °C, the products are transferred promptly and stored in an Ar-filled glove box. The synthesis procedure of $K_{0.45}MnO_2$ is the same as that of $K_{0.45}Mn_{0.7}Co_{0.2}Al_{0.1}O_2$, except that $Co(CH_3COO)_2{\cdot}4H_2O$ and $Al(NO_3)_3{\cdot}9H_2O$ are not added to the mixed solution.

The soft carbon is synthesized from 3,4,9,10-perylenetetracarboxylic dianhydride (PTCDA) after pyrolysis at 900 °C for 10 h under a flowing Ar atmosphere [32].

3. Results and Discussion

3.1. Structure and Morphology

The ICP test results of KMCAO and KMO are shown in Table S1. The chemical composition of the elements is consistent with the nominal composition. The electrical conductivities of KMCAO and KMO are measured using the four-probe method. The conductivity values obtained are 9.73×10^{-6} and 6.21×10^{-6} S cm^{-1}, respectively (Table S2). Higher conductivity will favor the diffusion of ions, thus improving electrochemical performance. Figures 1a and S1a show the Rietveld refinement XRD patterns of KMCAO and KMO. These patterns indicate that both samples exhibit a P3-type layered structure and belong to the $R3m$ space group. Figure 1b shows the crystal structure of the P3 phase, where O ion layers are arranged in parallel in ABBCCA order, TM (Mn, Co, Al) ions occupy octahedral sites, and K ions occupy prismatic sites. The Rietveld refinement XRD reports of KMCAO and KMO are shown in Tables S3 and S4. The decreasing value of a (KMCAO: a = 2.8681 Å, KMO: a = 2.8745 Å) indicates that Co/Al co-substitution shrinks the TM layers, which helps stabilize the TM layers. The analysis of the cell parameters along the c-axis shows an enlargement of the K layers' spacing, as evidenced by the increase in c value (KMCAO: c = 21.1069 Å, KMO: c = 20.8983 Å). This expansion is supported by the observed shift of the diffraction peak of the KMCAO (003) plane to a smaller angle in the XRD pattern (Figure S1b). The expansion of the interlayer space facilitates the diffusion of K ions and reduces the electrostatic repulsion between O ions in the neighboring TM layers. These effects contribute to the mitigation of phase transitions and interlayer sliding [25].

A spherical-aberration-corrected transmission electron microscope (AC-TEM) is used to obtain detailed atomic-scale crystal structure information of KMCAO through annular bright-field (ABF) and high-angle annular dark-field (HAADF) imaging. In the ABF-STEM images, the O and K layers are represented by bright grey dotted contrasts, while the TM layer appears as dark dotted contrasts. Conversely, the HAADF-STEM images display a reversal of contrast. In the ABF-STEM images, the arrangement of the K and TM layers in an alternating manner, as well as the stacking of the O layer according to the ABBCCA sequence along the [010] zone axis, can be observed (Figure 1c). This is a typical P3-phase structure. The HAADF-STEM image confirms that the distance between the neighboring layers is about 0.7 nm (Figure 1d), which corresponds to the Rietveld refinement XRD results (c/3). In addition, the ABF-STEM image reveals a hexagonal symmetry in the arrangement of the TM atoms along the [001] zone axis (Figure 1e). In the HAADF-STEM image, the measured distance between adjacent TM atoms is about 0.28 nm, which is consistent with the cell parameter a in the Rietveld refinement XRD results (Figure 1f). The detailed crystal structure of KMO can also be observed in the ABF-STEM and HAADF-STEM images (Figure S2). The scanning electron microscopy (SEM) and transmission electron microscopy (TEM) images (Figure S3a,b,d,e) reveal that both the KMCAO and KMO samples exhibit irregular polygonal shapes in terms of their particle morphology. The average particle diameter is approximately 1 μm. The high-resolution TEM (HRTEM) images show clear lattice stripes, indicating their high crystallinity. The lattice spacing is measured to be approximately 0.24 nm, corresponding to the (012) plane of the P3-type layered structure (Figure S3c,f). And both inset Selected Area Electron Diffraction (SAED) patterns show the hexagonal structure. Further, the energy-dispersive spectroscopy (EDS) mapping images demonstrate a uniform distribution of K, Mn, (Co, Al) and O elements throughout all the particles in the samples (Figures 1g and S3g).

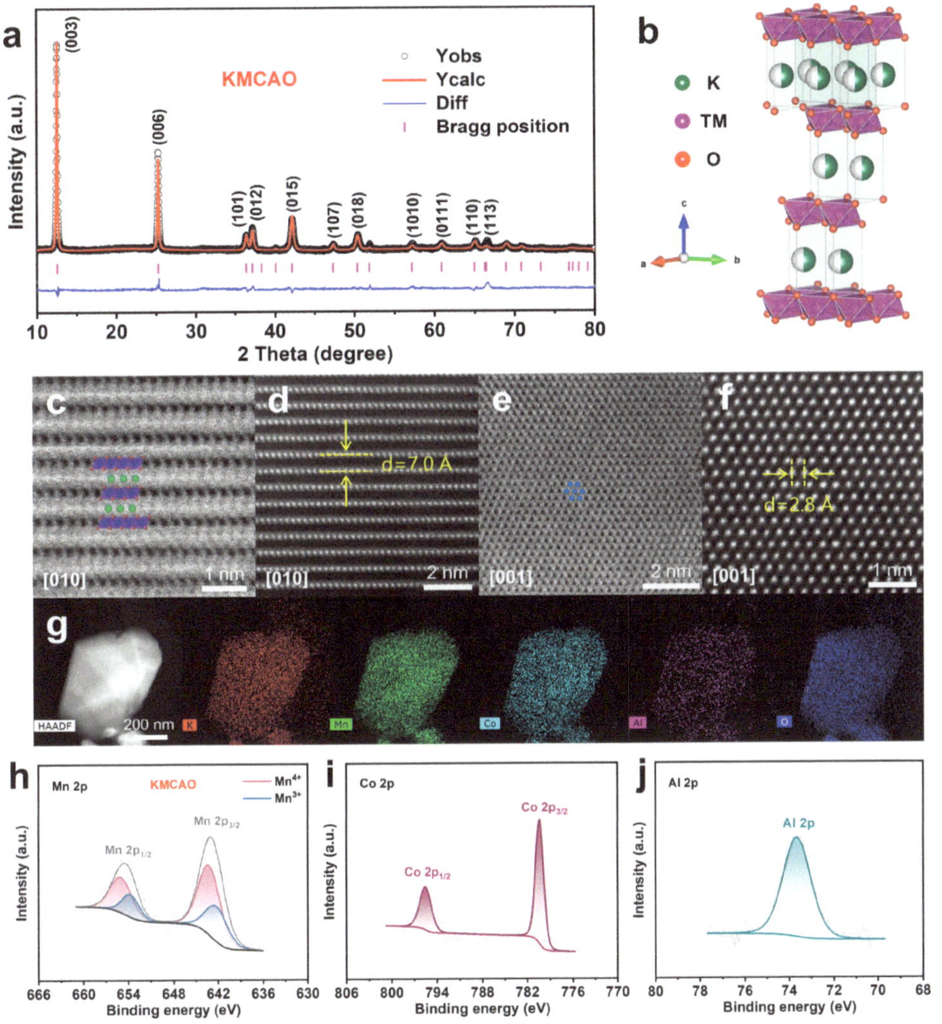

Figure 1. Structure characterizations of the KMCAO and KMO. (**a**) XRD Rietveld refinement of KMCAO. (**b**) P3 type structure schematic. (**c**) ABF-STEM and (**d**) HAADF-STEM images of KMCAO along the [010] zone axis. (**e**) ABF-STEM and (**f**) HAADF-STEM images of KMCAO along the [001] zone axis. (**g**) HAADF-STEM image of KMCAO and the corresponding EDS mappings for K, Mn, Co, Al and O elements. The XPS spectra of (**h**) Mn 2p, (**i**) Co 2p and (**j**) Al 2p of KMCAO.

X-ray photoelectron spectroscopy (XPS) is used to characterize the surface chemical compositions and valences of the corresponding elements in KMCAO and KMO. The full spectrum clearly shows signals of K, Mn, Co, Al and O elements, confirming the successful co-substitution of Co and Al elements (Figure S4a). In the Co 2p spectrum of KMCAO, a double peak feature is observed at 780.1 eV and 795.2 eV, corresponding to Co $2p_{3/2}$ and Co $2p_{1/2}$, respectively, indicating the trivalent state of Co ions in the sample (Figure 1i) [33]. The Al 2p spectrum of the KMCAO exhibits a characteristic peak at 73.3 eV, corresponding to Al^{3+} (Figure 1j) [34]. The Mn 2p spectrum exhibits two main peaks that can be deconvoluted into four characteristic peaks (641.9 eV and 642.9 eV, 653.4 eV and 654.7 eV), which are attributed to Mn $2p_{3/2}$ and Mn $2p_{1/2}$, respectively. This indicates the presence of both Mn^{3+} and Mn^{4+} in the sample (Figures 1h and S4b) [35]. It is worth noting that the area of the Mn^{4+} characteristic peaks is significantly larger in KMCAO compared to KMO, which effectively raises the average valence of Mn. The X-ray absorption near edge structure (XANES) tests are performed to further investigate the average oxidation state of Mn in KMCAO and KMO (Figure S4c). The comparison of the Mn_2O_3 and MnO_2 reference spectra shows that the average oxidation state of Mn is an intermediate value between +3 and +4. The photon energy of the Mn K-edge of KMCAO shifts to a higher energy, indicating an increased oxidation state of Mn after Co/Al co-substitution. This finding is also consistent with the results obtained from theoretical chemical component valence calculations (Table S5). This co-substitution strategy could potentially mitigate the structural degradation caused by the Jahn-Teller effect of Mn^{3+}, thereby enhancing the overall structural stability of the electrode materials.

3.2. Electrochemical Performance

The K storage performance of the prepared KMCAO and KMO cathodes was evaluated using cyclic voltammetry (CV) and constant current charge/discharge tests within the voltage range of 1.5~3.9 V. Figure 2a,b show the typical CV curves for KMCAO and KMO cathodes, respectively, at a scan rate of 0.2 mV s^{-1}. KMCAO shows five pairs of redox peaks. It is generally believed that the pair of redox peaks at 1.78/1.50 V may be related to the transformation of K^+/vacancy order/disorder due to the mixing of transition metal ions [36]. And the two pairs of redox peaks at 2.08/1.87 V and 2.50/2.28 V are attributed to the Mn^{3+}/Mn^{4+} redox pair, while the other two pairs of redox peaks at 2.84/3.14 V and 3.58/3.81 V are associated with Mn^{3+}/Mn^{4+} and Co^{3+}/Co^{4+} contributions [37,38]. The redox peaks in KMO are all attributed to Mn^{3+}/Mn^{4+} [18]. Compared to the pristine KMO, the Co/Al co-substitution significantly reduces the potential interval between the oxidation and reduction peaks, and the reduction in polarization will be beneficial for its practical application. Table S6 shows the calculation of voltage polarization. In addition, the better overlap of the CV curves indicates that KMCAO has remarkable reversibility during the electrochemical reaction process. The participation of active Co elements in the electrochemical reaction contributes to additional capacity, while the electrochemically inactive Al element stabilizes the layer structure through robust Al-O bonding [39]. Figures 2c and S5 show the charge/discharge curves of KMCAO and KMO cathodes for different cycles at a rate of 0.1 A g^{-1}. The voltage plateaus correspond to the redox peaks. Co/Al co-substitution in KMO can effectively smooth the charge–discharge profiles and increase the reversible capacity. At 0.1 A g^{-1}, KMCAO exhibits a high reversible discharge specific capacity of 102 mAh g^{-1}, with a capacity retention of 83.3% after 150 cycles. In contrast, the KMO cathode shows a rapid capacity decrease from 91 mAh g^{-1} to 48 mAh g^{-1} after 150 cycles, corresponding to a capacity retention of only 52.7% (Figure 2d). The SEM and TEM images after cycling for KMCAO and KMO are presented in Figure S6. The particle morphology of both KMCAO and KMO are preserved well after cycling. In comparison, KMCAO has a better integrity of particle morphology. In addition, the KMCAO cathode has an excellent rate performance with average discharge specific capacities of 111, 104, 96, 87, 77 and 67 mAh g^{-1} at 0.05, 0.1, 0.2, 0.5, 1 and 2 A g^{-1}, respectively. When the current density is reset to 0.05 A g^{-1}, a discharge specific capacity of 108 mAh g^{-1} can be obtained, which is close to the initial value (Figure 2e). The corresponding charge/discharge curves demonstrate the rapid K^+ storage property and low polarization characteristics of the KMCAO cathode (Figures 2f and S7). The KMCAO cathode also exhibits a long cycle lifespan. After 500 cycles at 1 A g^{-1}, the discharge specific capacity is 58 mAh g^{-1}, corresponding to a capacity retention of 71.6%. In comparison, the KMO cathode retains only 49.2% of its capacity (Figure 2g). This improved cycling performance can be attributed to the successful regulation of the Mn average valence through Co/Al co-substitution, which suppresses the Jahn-Teller effect. Notably, the KMCAO cathode demonstrates competitive K^+ storage performance compared to previously reported layered oxide cathodes for PIBs (Table S7). The electrochemical performance of the high mass loadings of KMCAO is investigated (Figure S8). It also performs well. After 100 cycles at 0.1 A g^{-1}, the KMCAO with a high mass loading of 7.11 mg cm^{-2} maintains a high reversible capacity of 64 mAh g^{-1} and capacity retention of 77.1%. Also, the KMCAO demonstrates exceptional rate performance even with a high mass loading of 6.34 mg cm^{-2}. The average discharge capacities are 88, 81, 74, 65 and 54 mAh g^{-1} at 0.1, 0.2, 0.5, 1 and 2 A g^{-1}, respectively. Further, galvanostatic intermittent titration technique (GITT) and electrochemical impedance spectroscopy (EIS) tests are performed to investigate the K^+ diffusion dynamics of the KMCAO and KMO cathodes. Based on the GITT test results (Figure S9a,b), the K^+ diffusion coefficient of KMCAO is calculated to be 10^{-9}–10^{-11} cm^2 s^{-1} during the first discharge cycle, which is generally higher than that of the pristine KMO cathode (Figure S9c). The EIS fitting results show that the KMCAO cathode has a lower charge transfer resistance and faster K^+ diffusion dynamics compared to the KMO cathode (Figure S10 and Table S8).

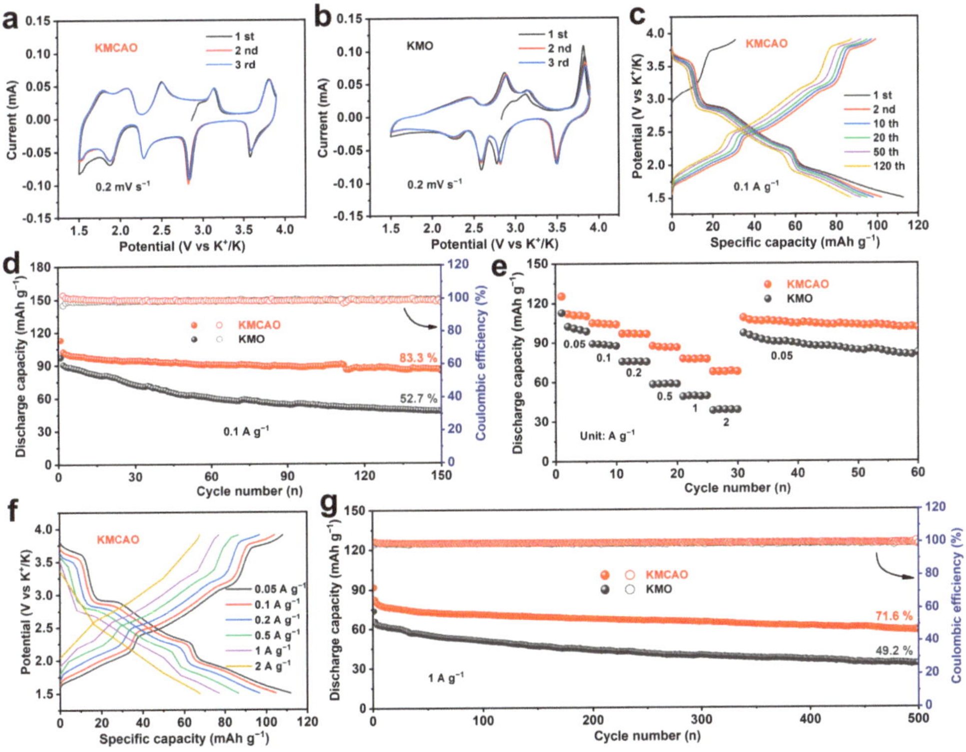

Figure 2. Electrochemical performance of KMCAO and KMO in 1.5~3.9 V (vs. K^+/K). CV curves of the first three cycles at 0.2 mV s^{-1} of (**a**) KMCAO and (**b**) KMO. (**c**) Charge/discharge curves of KMCAO at 0.1 A g^{-1}. (**d**) Cycling performance at 0.1 A g^{-1}. (**e**) Rate performance at 0.05, 0.1, 0.2, 0.5, 1 and 2 A g^{-1}. (**f**) The corresponding charge/discharge curves of KMCAO at different rates. (**g**) Long-term cycling performance at 1 A g^{-1}.

3.3. Potassium Storage Mechanism

In order to investigate the detailed crystal structure evolution during the K^+ intercalation/de-intercalation process, in situ XRD experiments are conducted to monitor the charging/discharging process of both the KMCAO and KMO cathodes (Figure 3). During the charging process, the (006) plane of KMCAO gradually shifts to a lower angle, while the (101), (012) and (015) planes shift to higher angles. These observations indicate the detachment of K^+. K^+ acts as an electrostatic shield between the O layers. With K^+ extraction, the shielding effect weakens and the electrostatic repulsion between O ions dominates. This leads to an expansion of the *c*-axis and a contraction of the *a*-*b* plane [40]. At higher voltages, the (015) peak disappears and a new diffraction peak appears at about 40.5°, corresponding to the (104) plane of the O3 phase. During discharge, the (015) peak reappears as the voltage decreases. At this stage, the electrostatic repulsion between the O layers decreases with the re-insertion of K^+. The (006) plane shifts back to a higher angle, and the (101), (012) and (015) planes shift to lower angles, restoring the initial P3 phase state. These shift behaviors of typical planes are the same in the first two charge/discharge cycles, showing the highly reversible K^+ intercalation/de-intercalation process (Figure 3a,b). In the case of KMO, the phase transition between P3 and O3 also occurs in the high-potential zone. However, a notable difference arises as the (006) plane experiences a sudden jump during the phase transition, corresponding to an abrupt change in the lattice parameter c value, which shows a drastic lattice distortion in KMO. On the other hand, the c value of KMCAO changes smoothly throughout the charging/discharging process (Figure S11). The (104) plane also exhibits a jump during the process (Figure 3c,d). In addition to the phase transition in the high-potential zone, the intensity of the KMO (006) plane diminishes in the low-potential zone (discharging to ~2.5 V and charging to ~2.8 V), indicating a weakening of the P3 phase feature. Furthermore, indistinguishable diffraction peaks appear near the (101) and (012) planes, while the (015) peak shows jumping displacements and disconnections. Therefore, we suggest that these complex and severe phase transitions may accelerate the further degradation of the lattice structure of the KMO cathode during charging/discharging, leading to the poor cycle stability observed during K^+ intercalation/de-intercalation [41].

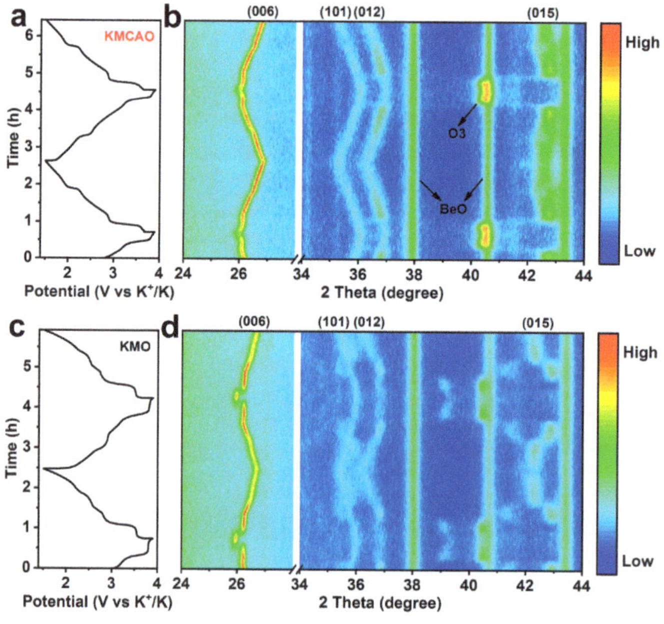

Figure 3. In situ XRD characterization of (**a**,**b**) KMCAO and (**c**,**d**) KMO at 50 mA g^{-1} in the range of 1.5~3.9 V.

3.4. *First-Principles Calculations*

First-principles calculations are conducted to further investigate the reasons for the superior electrochemical performance of the KMCAO cathode. First, based on the crystal structures of KMCAO (Figure 4a,b) and KMO (Figure S12a,b), the transition state structures are constructed using the Climbing Image Nudged Elastic Band (CINEB) method with linear interpolation points [42], illustrating the path for K^+ migration (the green balls in the figures). Subsequently, the migration energy barrier of K^+ in the KMCAO lattice is calculated to be about 0.61 eV (Figure 4c). The lower migration energy barriers favor K^+

intercalation and de-intercalation, thereby facilitating highly reversible redox reactions. In addition, the density of states (DOS) show that KMCAO exhibits a continuous distribution of the density of states near the Fermi energy level, with a significantly higher intensity of the density of states compared to KMO, which has a clear band gap near the Fermi energy level. This confirms the enhanced electrical conductivity of the KMCAO material. These results provide conclusive evidence for the excellent electrochemical performance of the KMCAO cathode. Overall, the Co/Al co-substitution offers more active channels for K^+ diffusion, which helps alleviate the irreversible phase transition during the reaction, thereby ensuring the maintenance of the structural stability of the crystals.

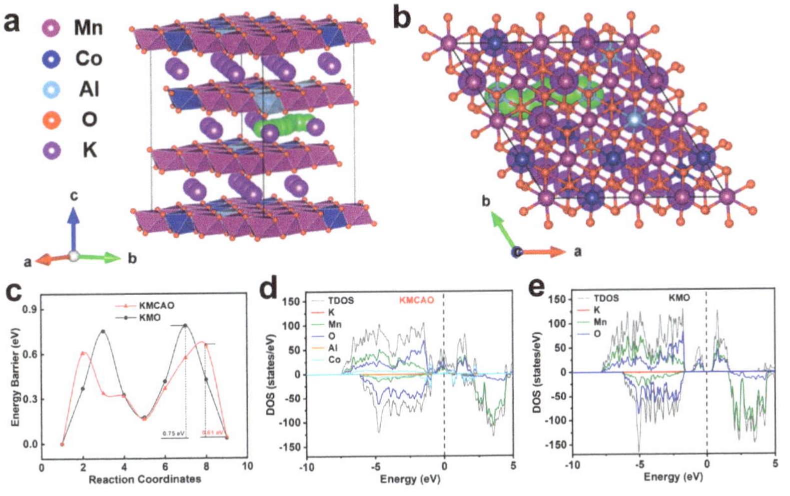

Figure 4. First-principles calculations. Schematic of the KMCAO crystal structure showing the K^+ migration pathways (indicated by the green balls) from (**a**) the side and (**b**) the top views. (**c**) The corresponding migration energy barriers in KMCAO and KMO. Density of states of (**d**) KMCAO and (**e**) KMO.

3.5. Full Cell Demonstration

To assess the practical application potential of the KMCAO cathode, performance tests are conducted on a potassium-ion full cell, paired with a soft carbon anode (Figure 5). The structural and morphological characterizations of the soft carbon are shown in Figure S13. And Figure S14 shows its electrochemical performances. Before assembling the full cell, the soft carbon anode was pre-cycled at 0.01~1.5 V (vs. K^+/K) to activate the material. The anode/cathode capacity ratio is adjusted to 1.2 to eliminate irreversibility. The configuration and operational mechanism of typical full PIBs are shown in Figure 5a. Figure 5b shows the normalized charge/discharge curves of KMCAO and soft carbon electrodes in half/full PIBs. The full cell maintains a high specific capacity of 82 mAh g^{-1} after 100 cycles at 0.1 A g^{-1} (Figure 5c). It also exhibits a good rate performance, with average discharge specific capacities of 85, 76, 73, 71 and 69 mAh g^{-1} at 0.1, 0.2, 0.3, 0.4 and 0.5 A g^{-1}, respectively (Figure 5d). When tested at 0.3 A g^{-1}, the initial discharge specific capacity of the full cell was 76 mAh g^{-1} and the initial coulombic efficiency was close to 99%. And the capacity remains at 80.2% after 300 cycles. (Figure 5e). These results highlight that KMCAO cathode has a great deal of potential for practical applications in PIBs.

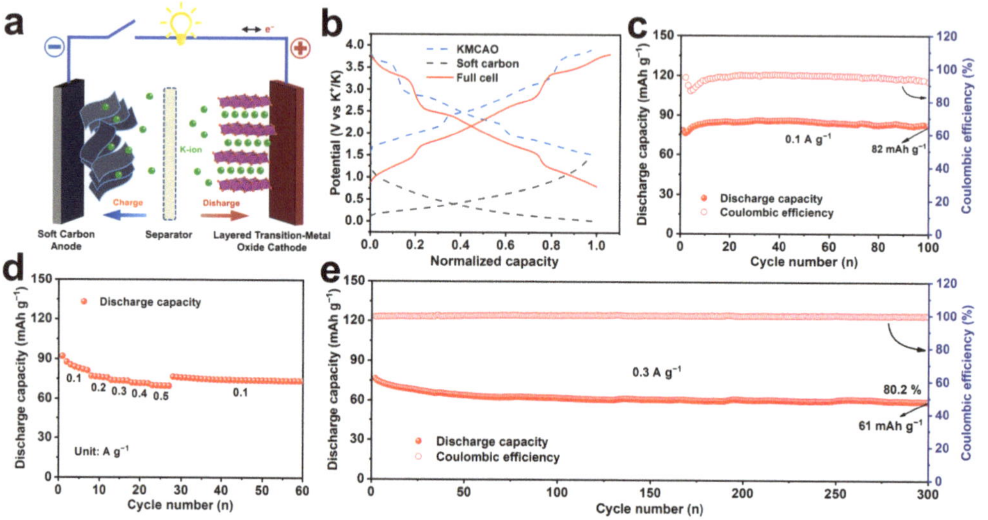

Figure 5. Electrochemical performance of potassium-ion full cell based on KMCAO/soft carbon in 0.8~3.8 V. (**a**) Schematic illustration of the cell configuration and operational mechanism of the full PIBs. (**b**) Normalized charge/discharge curves of the half and full PIBs. (**c**) Cycling performance at 0.1 A g^{-1}. (**d**) Rate performance at 0.1, 0.2, 0.3, 0.4 and 0.5 A g^{-1}. (**e**) Long-term cycling capability at 0.3 A g^{-1}.

4. Conclusions

In summary, the successful synthesis of a Co/Al co-substituted P3-type layered KMCAO cathode for PIBs has been achieved. The incorporation of Co^{3+}/Al^{3+} into the Mn^{3+} sites effectively suppresses the Jahn-Teller distortion. Additionally, Co contributes to the electrochemical reaction, enhancing capacity, while Al forms stable bonds with oxygen, further stabilizing the layer structure. The KMCAO cathode exhibits a specific discharge capacity of 111 mAh g^{-1} at 0.05 A g^{-1}. It also shows an excellent rate performance with a specific capacity of 81 mAh g^{-1} at 1 A g^{-1} and retains 71.6% of its capacity after 500 cycles. The Co/Al co-substitution also results in a wider spacing between the K layers, reduced K^+ diffusion barriers and faster K^+ diffusion dynamics compared to the pristine KMO cathode. Furthermore, the in situ XRD results show that the KMCAO cathode exhibits a milder phase transition during electrode cycling. The Co/Al co-substitution strategy employed in this work provides an effective approach for designing high-performance cathode materials in PIBs.

Supplementary Materials: The following supporting information can be downloaded at: https://www.mdpi.com/article/10.3390/ma17061277/s1. Figure S1. (a) XRD Rietveld refinement of KMO. (b) Zoomed-in image of XRD patterns of KMCAO and KMO for (003) peaks. Figure S2. (a) ABF-STEM and (b) HAADF-STEM images of KMO along the [010] zone axis. (c) ABF-STEM and (d) HAADF-STEM images of KMO along the [001] zone axis. Figure S3. (a) SEM, (b) TEM and (c) HRTEM images of KMCAO. (d) SEM, (e) TEM and (f) HRTEM images of KMO (inset: SEAD pattern). (g) HAADF-STEM image of KMO and the corresponding EDS mappings for K, Mn and O elements. Figure S4. (a) The XPS spectra of KMCAO and KMO. (b) Mn 2p of KMO. (c) XANES spectra of Mn K-edge. Figure S5. Charge/discharge curves of KMO at 0.1 A g^{-1}. Figure S6. SEM images after 150 cycles at 0.1 A g^{-1} for (a) KMCAO and (b) KMO. TEM images after 150 cycles at 0.1 A g^{-1} for (c) KMCAO and (d) KMO. Figure S7. The corresponding charge/discharge curves of KMO at different rates. Figure S8. Cycling and rate performances of KMCAO at the high mass loadings. (a) Charge/discharge curves at 0.1 A g^{-1} (b) Cycling performance with coulombic efficiencies measured at 0.1 A g^{-1}. (c) Rate performance conducted at 0.05, 0.1, 0.2, 0.5, 1 and 2 A g^{-1}. Figure S9. Potential response of (a) KMCAO and (b) KMO during GITT measurements. (c) The calculated diffusion coefficient of K^+ for KMCAO and KMO. Figure S10. Nyquist plots and the equivalent circuit model. Figure S11. Lattice parameter *c* variation of (a) KMCAO and (b) KMO during the second charge/discharge process. Figure S12. Schematic crystal structure of KMO showing the K^+ migration pathways (described by the green balls) from (a) the side and (b) the top views. Figure S13. Structural and morphological characterizations of soft carbon. (a) XRD pattern. (b) SEM image. (c) Raman spectrum. Figure S14. Electrochemical performances in the potential range of 0.01–1.5 V of soft carbon. (a) Charge/discharge curves at 0.1 A g^{-1}. (b) Cycling performance at 0.1 A g^{-1}. (c) Rate performance at 0.1, 0.2, 0.5, 1 and 2 A g^{-1}. Table S1. ICP measurement results of KMCAO and KMO. Table S2. The resulting electrical conductivities of KMCAO and KMO measured by the four-point probe method. Table S3. Structural parameters and atomic position of KMCAO from Rietveld refinement. Table S4. Structural parameters and atomic position of KMO from Rietveld refinement. Table S5. Mn average valence calculation results of $K_{0.45}Mn_{0.7}Co_{0.2}Al_{0.1}O_2$ and $K_{0.45}MnO_2$. Table S6. The voltage polarization calculation results of $K_{0.45}Mn_{0.7}Co_{0.2}Al_{0.1}O_2$ and $K_{0.45}MnO_2$. Table S7. Electrochemical performance of $K_{0.45}Mn_{0.7}Co_{0.2}Al_{0.1}O_2$ compared with other layered oxide cathodes in PIBs. Table S8. The obtained resistance values for $K_{0.45}MnO_2$ and $K_{0.45}Mn_{0.7}Co_{0.2}Al_{0.1}O_2$ through fitting the EIS spectra along the equivalent circuit. Refs. [28,40,43–51] are cited in the Supplementary Materials.

Author Contributions: Conceptualization, J.L. and X.W. (Xuanpeng Wang); Methodology, J.L.; Software, W.S.; Validation, W.S.; Formal analysis, G.Z.; Investigation, G.Z.; Resources, C.H.; Data curation, C.H.; Writing—original draft, J.L.; Writing—review & editing, J.M. and X.W. (Xuanpeng Wang); Visualization, X.W. (Xiujuan Wei); Supervision, X.W. (Xiujuan Wei); Project administration, X.W. (Xiujuan Wei); Funding acquisition, X.W. (Xuanpeng Wang). All authors have read and agreed to the published version of the manuscript.

Funding: This research was funded by the National Natural Science Foundation of China (52373306, 52102225), the Natural Science Foundation of Hubei Province (2023AFA053), and the Hainan Provincial Joint Project of Sanya Yazhou Bay Science and Technology City (2021CXLH0007). And the APC was funded by the Hainan Provincial Joint Project of Sanya Yazhou Bay Science and Technology City (2021CXLH0007).

Data Availability Statement: Data are contained within the article and Supplementary Materials.

Conflicts of Interest: The authors declare no conflict of interest.

References

1. Camargos, P.H.; dos Santos, P.H.J.; dos Santos, I.R.; Ribeiro, G.S.; Caetano, R.E. Perspectives on Li-ion battery categories for electric vehicle applications: A review of state of the art. *Int. J. Energy Res.* **2022**, *46*, 19258–19268. [CrossRef]
2. Kainat, S.; Anwer, J.; Hamid, A.; Gull, N.; Khan, S.M. Electrolytes in Lithium-Ion Batteries: Advancements in the Era of Twenties (2020's). *Mater. Chem. Phys.* **2024**, *313*, 128796. [CrossRef]
3. Kobayashi, T.; Ohnishi, T.; Osawa, T.; Pratt, A.; Tear, S.; Shimoda, S.; Baba, H.; Laitinen, M.; Sajavaara, T. In-Operando Lithium-Ion Transport Tracking in an All-Solid-State Battery. *Small* **2022**, *18*, 2204455. [CrossRef]
4. Tatara, R.; Ishihara, K.; Kosugi, M.; Aoki, K.; Takei, Y.; Matsui, T.; Takayama, T.; Komaba, S. Application of Potassium Ion Conducting $KTiOPO_4$ as Effective Inner Solid-Contact Layer in All-Solid-State Potassium Ion-Selective Electrode. *J. Electrochem. Soc.* **2023**, *170*, 027507. [CrossRef]
5. Woo, D.; Ban, M.; Lee, J.S.; Park, C.Y.; Kim, J.; Kim, S.; Lee, J. Anisotropic lens-shaped mesoporous carbon from interfacially perpendicular self-assembly for potassium-ion batteries. *Chem. Commun.* **2024**, *60*, 590–593. [CrossRef]
6. Wrogemann, J.M.; Fromm, O.; Deckwirth, F.; Beltrop, K.; Heckmann, A.; Winter, M.; Placke, T. Impact of Degree of Graphitization, Surface Properties and Particle Size Distribution on Electrochemical Performance of Carbon Anodes for Potassium-Ion Batteries. *Batter. Supercaps* **2022**, *5*, e202200206. [CrossRef]
7. Sada, K.; Darga, J.; Manthiram, A. Challenges and Prospects of Sodium-Ion and Potassium-Ion Batteries for Mass Production. *Adv. Energy Mater.* **2023**, *13*, 2302321. [CrossRef]
8. Dhir, S.; Wheeler, S.; Capone, I.; Pasta, M. Outlook on K-Ion Batteries. *Chem* **2020**, *6*, 2442–2460. [CrossRef]
9. Li, J.; Hu, Y.; Xie, H.; Peng, J.; Fan, L.; Zhou, J.; Lu, B. Weak Cation–Solvent Interactions in Ether-Based Electrolytes Stabilizing Potassium-ion Batteries. *Angew. Chem. Int. Ed.* **2022**, *61*, e202208291. [CrossRef]
10. Sohn, W.; Chae, J.S.; Lim, G.H.; Roh, K.C. Ion-exchange-assisted $Li_{0.27}K_{0.72}Ni_{0.6}Co_{0.2}Mn_{0.2}O_2$ cathode in potassium-ion batteries. *J. Alloys Compd.* **2022**, *898*, 162904. [CrossRef]
11. Ko, W.; Kim, J.; Kang, J.; Park, H.; Lee, Y.; Ahn, J.; Ku, B.; Choi, M.; Ahn, H.; Oh, G.; et al. Development of P3-type $K_{0.70}[Cr_{0.86}Sb_{0.14}]O_2$ cathode for high-performance K-ion batteries. *Mater. Today Energy* **2023**, *36*, 101356. [CrossRef]

12. Le Pham, P.N.; Wernert, R.; Cahu, M.; Sougrati, M.T.; Aquilanti, G.; Johansson, P.; Monconduit, L.; Stievano, L. Prussian blue analogues for potassium-ion batteries: Insights into the electrochemical mechanisms. *J. Mater. Chem. A* **2023**, *11*, 3091–3104. [CrossRef]
13. Shu, W.; Huang, M.; Geng, L.; Qiao, F.; Wang, X. Highly Crystalline Prussian Blue for Kinetics Enhanced Potassium Storage. *Small* **2023**, *19*, 2207080. [CrossRef]
14. Park, W.B.; Han, S.C.; Park, C.; Hong, S.U.; Han, U.; Singh, S.P.; Jung, Y.H.; Ahn, D.; Sohn, K.-S.; Pyo, M. KVP_2O_7 as a Robust High-Energy Cathode for Potassium-Ion Batteries: Pinpointed by a Full Screening of the Inorganic Registry under Specific Search Conditions. *Adv. Energy Mater.* **2018**, *8*, 1703099. [CrossRef]
15. Fedotov, S.S.; Luchinin, N.D.; Aksyonov, D.A.; Morozov, A.V.; Ryazantsev, S.V.; Gaboardi, M.; Plaisier, J.R.; Stevenson, K.J.; Abakumov, A.M.; Antipov, E.V. Titanium-based potassium-ion battery positive electrode with extraordinarily high redox potential. *Nat. Commun.* **2020**, *11*, 1484. [CrossRef]
16. Park, S.; Park, S.; Park, Y.; Alfaruqi, M.H.; Hwang, J.-Y.; Kim, J. A new material discovery platform of stable layered oxide cathodes for K-ion batteries. *Energy Environ. Sci.* **2021**, *14*, 5864–5874. [CrossRef]
17. Jha, P.K.; Totade, S.N.; Barpanda, P.; Sai Gautam, G. Evaluation of P3-Type Layered Oxides as K-Ion Battery Cathodes. *Inorg. Chem.* **2023**, *62*, 14971–14979. [CrossRef] [PubMed]
18. Kim, Y.; Oh, G.; Lee, J.; Kang, H.; Kim, H.; Park, J.; Kansara, S.; Hwang, J.-Y.; Park, Y.; Lestari, K.R.; et al. Stabilization of layered-type potassium manganese oxide cathode with fluorine treatment for high-performance K-ion batteries. *J. Power Sources* **2023**, *588*, 233729. [CrossRef]
19. Zhu, X.; Meng, F.; Zhang, Q.; Xue, L.; Zhu, H.; Lan, S.; Liu, Q.; Zhao, J.; Zhuang, Y.; Guo, Q.; et al. $LiMnO_2$ cathode stabilized by interfacial orbital ordering for sustainable lithium-ion batteries. *Nat. Sustain.* **2021**, *4*, 392–401. [CrossRef]
20. Zhao, Z.; Xu, T.; Yu, X. Unlock the Potassium Storage Behavior of Single-Phased Tungsten Selenide Nanorods via Large Cation Insertion. *Adv. Mater.* **2023**, *35*, 2208096. [CrossRef]
21. Li, X.; Ma, X.; Su, D.; Liu, L.; Chisnell, R.; Ong, S.P.; Chen, H.; Toumar, A.; Idrobo, J.-C.; Lei, Y.; et al. Direct visualization of the Jahn–Teller effect coupled to Na ordering in $Na_{5/8}MnO_2$. *Nat. Mater.* **2014**, *13*, 586–592. [CrossRef]
22. Hwang, J.-Y.; Kim, J.; Yu, T.-Y.; Jung, H.-G.; Kim, J.; Kim, K.-H.; Sun, Y.-K. A new P2-type layered oxide cathode with superior full-cell performances for K-ion batteries. *J. Mater. Chem. A* **2019**, *7*, 21362–21370. [CrossRef]
23. Kim, H.; Seo, D.-H.; Kim, J.C.; Bo, S.-H.; Liu, L.; Shi, T.; Ceder, G. Investigation of Potassium Storage in Layered P3-Type $K_{0.5}MnO_2$ Cathode. *Adv. Mater.* **2017**, *29*, 1702480. [CrossRef]
24. Cho, M.K.; Jo, J.H.; Choi, J.U.; Myung, S.-T. Cycling Stability of Layered Potassium Manganese Oxide in Nonaqueous Potassium Cells. *ACS Appl. Mater. Interfaces* **2019**, *11*, 27770–27779. [CrossRef]
25. Choi, J.U.; Kim, J.; Hwang, J.-Y.; Jo, J.H.; Sun, Y.-K.; Myung, S.-T. $K_{0.54}[Co_{0.5}Mn_{0.5}]O_2$: New cathode with high power capability for potassium-ion batteries. *Nano Energy* **2019**, *61*, 284–294. [CrossRef]
26. Liu, C.-L.; Luo, S.-H.; Huang, H.-B.; Liu, X.; Zhai, Y.-C.; Wang, Z.-W. Fe-doped layered P3-type $K_{0.45}Mn_{1-x}Fe_xO_2$ ($x \leq 0.5$) as cathode materials for low-cost potassium-ion batteries. *Chem. Eng. J.* **2019**, *378*, 122167. [CrossRef]
27. Weng, J.; Duan, J.; Sun, C.; Liu, P.; Li, A.; Zhou, P.; Zhou, J. Construction of hierarchical $K_{0.7}Mn_{0.7}Mg_{0.3}O_2$ microparticles as high capacity & long cycle life cathode materials for low-cost potassium-ion batteries. *Chem. Eng. J.* **2020**, *392*, 123649.
28. Xu, Y.-S.; Zhou, Y.-N.; Zhang, Q.-H.; Qi, M.-Y.; Guo, S.-J.; Luo, J.-M.; Sun, Y.-G.; Gu, L.; Cao, A.-M.; Wan, L.-J. Layered oxides with solid-solution reaction for high voltage potassium-ion batteries cathode. *Chem. Eng. J.* **2021**, *412*, 128735. [CrossRef]
29. Zhang, Q.; Didier, C.; Pang, W.K.; Liu, Y.; Wang, Z.; Li, S.; Peterson, V.K.; Mao, J.; Guo, Z. Structural Insight into Layer Gliding and Lattice Distortion in Layered Manganese Oxide Electrodes for Potassium-Ion Batteries. *Adv. Energy Mater.* **2019**, *9*, 1900568. [CrossRef]
30. Zhong, W.; Liu, X.; Cheng, Q.; Tan, T.; Huang, Q.; Deng, Q.; Hu, J.; Yang, C. Suppressing the interlayer-gliding of layered P3-type $K_{0.5}Mn_{0.7}Co_{0.2}Fe_{0.1}O_2$ cathode materials on electrochemical potassium-ion storage. *Appl. Phys. Rev.* **2021**, *8*, 031412. [CrossRef]
31. Kim, U.-H.; Myung, S.-T.; Yoon, C.S.; Sun, Y.-K. Extending the Battery Life Using an Al-Doped $Li[Ni_{0.76}Co_{0.09}Mn_{0.15}]O_2$ Cathode with Concentration Gradients for Lithium Ion Batteries. *ACS Energy Lett.* **2017**, *2*, 1848–1854. [CrossRef]
32. Wang, X.; Han, K.; Qin, D.; Li, Q.; Wang, C.; Niu, C.; Mai, L. Polycrystalline soft carbon semi-hollow microrods as anode for advanced K-ion full batteries. *Nanoscale* **2017**, *9*, 18216–18222. [CrossRef]
33. Gogula, S.K.; Gangadharappa, V.A.; Jayaraman, V.K.; Prakash, A.S. High-Voltage Layered $NaNi_{0.5}Co_{0.1}Ti_{0.3}Sb_{0.1}O_2$ Cathode for Sodium-Ion Batteries. *Energy Fuels* **2023**, *37*, 4143–4149. [CrossRef]
34. Chae, M.S.; Kim, H.J.; Lyoo, J.; Attias, R.; Gofer, Y.; Hong, S.-T.; Aurbach, D. Anomalous Sodium Storage Behavior in Al/F Dual-Doped P2-Type Sodium Manganese Oxide Cathode for Sodium-Ion Batteries. *Adv. Energy Mater.* **2020**, *10*, 2002205. [CrossRef]
35. Eom, S.; Jeong, S.H.; Lee, S.J.; Jung, Y.H.; Kim, J.-H. Mitigating the P2–O2 phase transition of Ni–Co–Mn based layered oxide for improved sodium-ion batteries via interlayered structural modulation. *Mater. Today Energy* **2023**, *38*, 101449. [CrossRef]
36. Liang, J.; Lin, C.; Meng, X.; Liang, M.; Lai, J.; Zheng, X.; Huang, Q.; Liu, L.; Shi, Z. P3-Type $K_{0.45}Co_{1/12}Mg_{1/12}Mn_{5/6}O_2$ as a superior cathode material for potassium-ion batteries with high structural reversibility ensured by Co–Mg Co-substitution. *J. Mater. Chem. A* **2021**, *9*, 17261–17269. [CrossRef]
37. Hao, J.; Xiong, K.; Zhou, J.; Rao, A.M.; Wang, X.; Liu, H.; Lu, B. Yolk–Shell P3-Type $K_{0.5}[Mn_{0.85}Ni_{0.1}Co_{0.05}]O_2$: A Low-Cost Cathode for Potassium-Ion Batteries. *Energy Environ. Mater.* **2022**, *5*, 261–269. [CrossRef]

38. Sada, K.; Barpanda, P. P3-type layered $K_{0.48}Mn_{0.4}Co_{0.6}O_2$: A novel cathode material for potassium-ion batteries. *Chem. Commun.* **2020**, *56*, 2272–2275. [CrossRef] [PubMed]
39. Singh, A.N.; Kim, M.-H.; Meena, A.; Wi, T.-U.; Lee, H.-W.; Kim, K.S. Na/Al Codoped Layered Cathode with Defects as Bifunctional Electrocatalyst for High-Performance Li-Ion Battery and Oxygen Evolution Reaction. *Small* **2021**, *17*, 2005605. [CrossRef] [PubMed]
40. Xiao, Z.; Xia, F.; Xu, L.; Wang, X.; Meng, J.; Wang, H.; Zhang, X.; Geng, L.; Wu, J.; Mai, L. Suppressing the Jahn–Teller Effect in Mn-Based Layered Oxide Cathode toward Long-Life Potassium-Ion Batteries. *Adv. Funct. Mater.* **2022**, *32*, 2108244. [CrossRef]
41. Caixiang, Z.; Hao, J.; Zhou, J.; Yu, X.; Lu, B. Interlayer-Engineering and Surface-Substituting Manganese-Based Self-Evolution for High-Performance Potassium Cathode. *Adv. Energy Mater.* **2023**, *13*, 2203126. [CrossRef]
42. Henkelman, G.; Jónsson, H. Improved tangent estimate in the nudged elastic band method for finding minimum energy paths and saddle points. *J. Chem. Phys.* **2000**, *113*, 9978–9985. [CrossRef]
43. Kresse, G.; Furthmüller, J. Efficiency of ab-initio total energy calculations for metals and semiconductors using a plane-wave basis set. *Comput. Mater. Sci.* **1996**, *6*, 15–50. [CrossRef]
44. Kresse, G.; Joubert, D. From ultrasoft pseudopotentials to the projector augmented-wave method. *Phys. Rev. B* **1999**, *59*, 1758–1775. [CrossRef]
45. Perdew, J.P.; Burke, K.; Ernzerhof, M. Generalized Gradient Approximation Made Simple. *Phys. Rev. Lett.* **1996**, *77*, 3865–3868. [CrossRef] [PubMed]
46. Monkhorst, H.J.; Pack, J.D. Special points for Brillouin-zone integrations. *Phys. Rev. B* **1976**, *13*, 5188–5192. [CrossRef]
47. Henkelman, G.; Uberuaga, B.P.; Jónsson, H. A climbing image nudged elastic band method for finding saddle points and minimum energy paths. *J. Chem. Phys.* **2000**, *113*, 9901–9904. [CrossRef]
48. Zhang, Z.; Hu, Q.; Liao, J.; Xu, Y.; Duan, L.; Tian, R.; Du, Y.; Shen, J.; Zhou, X. Uniform P2-$K_{0.6}CoO_2$ Microcubes as a High-Energy Cathode Material for Potassium-Ion Batteries. *Nano Lett.* **2023**, *23*, 694–700. [CrossRef]
49. Luo, R.-J.; Li, X.-L.; Ding, J.-Y.; Bao, J.; Ma, C.; Du, C.-Y.; Cai, X.-Y.; Wu, X.-J.; Zhou, Y.-N. Suppressing Jahn-Teller distortion and phase transition of $K_{0.5}MnO_2$ by K-site Mg substitution for potassium-ion batteries. *Energy Storage Mater.* **2022**, *47*, 408–414. [CrossRef]
50. Xu, Y.-S.; Zhang, Q.-H.; Wang, D.; Gao, J.-C.; Tao, X.-S.; Liu, Y.; Sun, Y.-G.; Gu, L.; Chang, B.-B.; Liu, C.-T.; et al. Enabling reversible phase transition on $K_{5/9}Mn_{7/9}Ti_{2/9}O_2$ for high-performance potassium-ion batteries cathodes. *Energy Storage Mater.* **2020**, *31*, 20–26. [CrossRef]
51. Jo, J.H.; Choi, J.U.; Park, Y.J.; Jung, Y.H.; Ahn, D.; Jeon, T.-Y.; Kim, H.; Kim, J.; Myung, S.-T. P2-$K_{0.75}[Ni_{1/3}Mn_{2/3}]O_2$ Cathode Material for High Power and Long Life Potassium-Ion Batteries. *Adv. Energy Mater.* **2020**, *10*, 1903605. [CrossRef]

materials

MDPI

Article

An Artificial $MnWO_4$ Cathode Electrolyte Interphase Enabling Enhanced Electrochemical Performance of δ-MnO_2 Cathode for Aqueous Zinc Ion Battery

Hao Tian, Huanlin Zhang, You Zuo, Lei Ling, Tengfei Meng, Hang Zhang, Xiaohong Sun and Shu Cai *

Key Laboratory of Advanced Ceramics and Machining Technology of Ministry of Education, School of Materials Science and Engineering, Tianjin University, Tianjin 300072, China
* Correspondence: caishu@tju.edu.cn

Abstract: The dissolution of active material in aqueous batteries can lead to a rapid deterioration in capacity, and the presence of free water can also accelerate the dissolution and trigger some side reactions that affect the service life of aqueous batteries. In this study, a $MnWO_4$ cathode electrolyte interphase (CEI) layer is constructed on a δ-MnO_2 cathode by cyclic voltammetry, which is effective in inhibiting the dissolution of Mn and improving the reaction kinetics. As a result, the CEI layer enables the δ-MnO_2 cathode to produce a better cycling performance, with the capacity maintained at 98.2% (vs. activated capacity at 500 cycles) after 2000 cycles at 10 A g^{-1}. In comparison, the capacity retention rate is merely 33.4% for pristine samples in the same state, indicating that this $MnWO_4$ CEI layer constructed by using a simple and general electrochemical method can promote the development of MnO_2 cathodes for aqueous zinc ion batteries.

Keywords: cathode electrolyte interphase; Mn dissolution; δ-MnO_2 cathode; electrochemical performance; aqueous zinc ion battery

Citation: Tian, H.; Zhang, H.; Zuo, Y.; Ling, L.; Meng, T.; Zhang, H.; Sun, X.; Cai, S. An Artificial $MnWO_4$ Cathode Electrolyte Interphase Enabling Enhanced Electrochemical Performance of δ-MnO_2 Cathode for Aqueous Zinc Ion Battery. *Materials* **2023**, *16*, 3228. https://doi.org/10.3390/ma16083228

Academic Editors: Huang Zhang and Yuan Ma

Received: 23 March 2023
Revised: 12 April 2023
Accepted: 15 April 2023
Published: 19 April 2023

1. Introduction

At present, aqueous zinc ion batteries (AZIBs) have been widely studied for energy storage due to various advantages such as low cost, environmental benignity, and high safety performance [1–10]. So far, there have been various materials reported to be suitable as cathodes for AZIBs, including manganese-based materials [11–15], vanadium-based materials [16–20], and Prussian blue analogs [21–24]. Among them, manganese oxide has attracted widespread attention because of its abundance, low toxicity, high energy density, and structural diversity [25]. However, Mn dissolution issues still limit the cycling stability of manganese-based materials for AZIBs [26].

In the past decades, there has been some progress made in the research on how to improve the performance of electrode materials [27] and electrolytes [28] in resisting the dissolution of cathodes. Apart from the existing strategies, researchers have also discovered that the construction of an electrode-electrolyte interphase (EEI), including anode electrolyte interphase and cathode electrolyte interphase, can also help improve the energy density, cycling performance, and power density of batteries [29]. Therefore, building the cathode electrolyte interphase (CEI) surface protection layer is considered a feasible solution to the dissolution of cathodes. However, the research on CEI is still limited now due to the complexity of the cathode energy storage mechanism for AZIBs and the difficulty in characterizing various interfacial reactions [30].

Typically, the methods of CEI construction are divided into two categories: in situ formation and artificial synthesis. In recent years, some studies have reportedly been conducted on the in situ formation of CEI. Liang et al. proposed to perform electrochemical synthesis of the $CaSO_4{\cdot}2H_2O$ layer in situ on a Ca_2MnO_4 cathode, which significantly improved the stability and service life of the battery [31]. Cao et al. also built an in situ CEI layer of $BaSO_4$ on the $Ba_{0.26}V_2O_5{\cdot}0.92H_2O$ cathode of AZIBs, which reduced the dissolution of cathodes, thus leading to an excellent cycling performance [32]. Compared with the synthesis of CEI in situ, the artificial construction of CEI is much easier to manipulate. Xiong et al. synthesized a reduced graphene oxide (rGO) layer coated with α-MnO_2 powder, which improved both rate performance and cycling stability [33]. For different coating materials, powder coating may cause hindrances to ion transport to some extent. Unlike the powder coating as described above, the artificial CEI is more similar to cathode coating, where a layer is formed between the cathode and the electrolyte. Guo et al. reported a H_fO_2 layer formed on a $Zn_3V_2O_7(OH)_2{\cdot}2H_2O$ electrode as an artificial solid electrolyte interphase. The H_fO_2 layer was built by means of atomic layer deposition and is capable of isolating the electrode from the electrolyte, thereby reducing the dissolution of the cathode in the electrolyte and inhibiting the formation of insulated by-products. As a result, the capacity retention rate was considerably improved from 45% to 90% after 100 cycles at 0.1 A g^{-1} [34]. Based on the construction of the CEI layer and its effect on cycling performance, there have been many studies carried out. For example, Paraffin [35], $(Zn(OH)_2)_3(ZnSO_4)(H_2O)_5$ [36], and $SrCO_3$ [37] have been reported as a kind of CEI layer. All these results demonstrate that the construction of CEI is effective in significantly enhancing the electrochemical performance of AZIBs.

For the manganese-based material used in AZIBs, the CEI layer can help prevent the cathode from direct exposure to the electrolyte, which suppresses Mn dissolution, thus maintaining high cycling performance and capacity. However, the strategy of CEI layer construction on manganese-based materials is not universal, and the economic benefits are unsatisfactory. In this study, a CEI layer of $MnWO_4$ is constructed on a δ-MnO_2 cathode through a facile electrochemical method (cyclic voltammetry), which not only inhibits the Mn dissolution but also improves the reaction kinetics. The key to this economical and efficient strategy lies in the dissolution of the Mn-based cathode in solutions that do not co-exist with Mn^{2+} ions. The prepared $MnWO_4$-coated δ-MnO_2 (denoted as W-MnO_2) shows an outstanding cycling performance (98.2% capacity retention vs. activated capacity at 500 cycles, after 2000 cycles at 10 A g^{-1}), indicating the effectiveness of the CEI construction strategy. In addition, the low-cost strategy of the CEI layer can be applied to other manganese-based AZIBs.

2. Experimental Section

2.1. Materials Preparation

The synthesis of the cubic $MnCO_3$ precursor was performed in the way as reported by others [38]. Firstly, 25 mL of 0.8 M NH_4HCO_3 aqueous solution, 25 mL of n-butanol, and 500 mL of cyclohexane were thoroughly mixed. Then, 20 g of cetyltrimethylammonium bromide (CTAB) was added into the mixed solution and stirred until it became clear. Next, 25 mL of 0.4 M $MnSO_4$ aqueous solution was added dropwise into the solution, which led to a white precipitate. Afterwards, the precipitate was collected through centrifugation, washed clean with alcohol and distilled water, respectively, and dried under vacuum at 100 °C to obtain white $MnCO_3$. To further oxidize $MnCO_3$ into MnO_2 [39,40], 1 g of the synthesized $MnCO_3$ precursor was added into 0.032 M of $KMnO_4$ aqueous solution. Then, the mixed solution was subjected to ultrasonic treatment for 30 min and stirred for 1 day. The δ-MnO_2 precursor was collected by centrifugation, washed (three times) with alcohol and distilled water, and finally dried at 75 °C.

The δ-MnO_2 cathode was produced by using N-methylpirpiridone (NMP) as a solvent to disperse the precursor powder (δ-MnO_2, 70 wt%), conductive additive (Super P, 20 wt%), and binder (PVDF, 10 wt%), and was coated on carbon fiber paper. The loading density of the cathode was set to about 1.5 mg cm^{-2}.

The $MnWO_4$-coated δ-MnO_2 (W-MnO_2) was constructed in a conventional three-electrode configuration by cyclic voltammetry (CV) at a scan rate of 50 mV/s (negative scan from −0.6 to 0.6 V for 100 segments). The electrolyte was 0.1 mol/L Na_2WO_4 solution, and the pH value was adjusted to 7 by using H_2SO_4. The pristine δ-MnO_2 cathode was treated as the working electrode of the three-electrode system, while Ag/AgCl and graphite were taken as the reference and counter electrodes, respectively. Finally, W-MnO_2 was obtained by washing it thoroughly with distilled water (three times) and drying it at 75 °C. The loading density of the W-MnO_2 cathode was approximately 0.5–1% higher than the pristine cathode.

2.2. *Materials Characterization*

X-ray diffraction (XRD, D8 Advance, Bruker, Cu Kα) data were collected at a scan range of 5–70° (2θ) and a step size of 0.02°. Both SEM (Sigma 300, Zeiss, operating voltage 5 kV) and HR-TEM (JEM-2100F, JEOL) were employed to examine the morphology and microstructure of the samples. X-ray photoelectron spectroscopy (XPS, PHI-1600, PerkinElmer) was performed to record the valence states of the samples loaded with Cu. The C 1s peak with a binding energy of 284.8 eV was used to calibrate all XPS spectra. Nitrogen adsorption measurements for Brunauer–Emmett–Teller (BET) analysis were tested at 77 K using an ASAP 2460.

2.3. *Electrochemical Measurements*

The 2032-type coin cells were assembled with prepared W-MnO_2 as the cathode, a Zn foil as the counter electrode, and an aqueous $ZnSO_4$ (3 M) solution with a $MnSO_4$ additive (0.2 M) as the electrolyte. The CV and EIS (100 kHz to 10 mHz) were measured on a CHI 660E electrochemical workstation. To conduct the CV tests at different scan rates, the peak current (i) and scan rate (v) were determined through Equation (1) [41]:

$$i = a\nu^b, \quad (1)$$

where a and b represent variable parameters, and the b-value is obtained through the slope of log(i) vs. log(ν). Furthermore, the current contribution is divided into capacitive and diffusion contributions according to Equation (2) [42]:

$$i = k_1\mathcal{V} + k_2\mathcal{V}^{\frac{1}{2}}, \quad (2)$$

where k_1 and k_2 refer to the coefficients of proportionality for capacitive and diffusion contributions, respectively. The GCD curves, cycling performance, and GITT measurements were achieved by using the LAND CT2001A battery test system at room temperature. Moreover, the pause and rest time of GITT at 0.2 A g^{-1} lasted 10 min and 180 min, respectively. The diffusion coefficient can be determined through the following equation [43]:

$$D = \frac{4l^2}{\pi\tau}, \quad (3)$$

where D represents the diffusion coefficient, l indicates the diffusion length (cm) of active materials, and τ refers to the duration of the current pause (s). ΔE_s and ΔE_t represent the voltage difference by the current pulse and the voltage difference during the constant current pulse, respectively.

3. Results and Discussion

The $MnWO_4$-coated δ-MnO_2 (W-MnO_2) was obtained by means of the electrochemical treatment (cyclic voltammetry) conducted in a three-electrode system, as shown in Figure 1a. The working electrode was the δ-MnO_2 cathode. Figures S1 and S2 show the XRD and BET results of δ-MnO_2 powder, respectively. According to the N_2 adsorption isotherm, the specific area of δ-MnO_2 is 20 m^2 g^{-1}. To confirm the chemical composition of the CEI layer on the δ-MnO_2 cathode, XRD was performed for the W-MnO_2 cathode, as shown in Figure 1b. In addition to the weak characteristic peaks of δ-MnO_2, a peak appears at 18° corresponding to $MnWO_4$ (JCPDS No. 72-0478) after the CV process. Moreover, there are some other characteristic peaks of $MnWO_4$ observed at around 37° and 52°, indicating the presence of the $MnWO_4$ after the CV process. It is suspected that the absence of the δ-MnO_2 characteristic peaks may result from the limited crystallinity of δ-MnO_2 and the strong diffraction peak of the carbon fiber paper. XPS was performed to determine the Mn valence during the CV process. As shown in Figure 1c, the splitting magnitude of two splitting components for the Mn 3s peak increases to 6.13 eV from 4.87 eV after the electrochemical treatment. In general, the Mn 3s peak consists of two multiple splitting components [11,44], with the oxidation state of Mn determined by the splitting magnitude ΔE, which is 6.0 eV and 4.7 eV for Mn^{2+} and MnO_2 (Mn^{4+}), respectively [45]. It can be found that the valence state of Mn shifted from +4 in δ-MnO_2 to +2 in W-MnO_2, indicating the formation of the $MnWO_4$ on the δ-MnO_2 cathode. To determine the effect of $MnWO_4$ formation on the morphology of δ-MnO_2 and the area of $MnWO_4$ distribution, SEM and TEM tests were performed. As shown in Figure 1d, the size of δ-MnO_2 cubes is approximately 500 nm, and the morphology of the δ-MnO_2 cubes is barely changed during the formation of $MnWO_4$ (Figure 1e). The EDS element mapping of W-MnO_2 (Figure 1f) shows a uniform distribution of element W on the surface of the MnO_2 cube, indicating that the $MnWO_4$ is formed uniformly on the surface of the δ-MnO_2 cathode. Moreover, the HRTEM (Figure 1g) images of the W-MnO_2 surface show that lattice fringes are 0.22 nm and 0.249 nm, which correspond to the (121) and (002) crystal planes of $MnWO_4$, respectively. Meanwhile, the (121), (002), and (−113) crystal planes of $MnWO_4$ are also observable in the results of selected area electron diffraction (SAED). Judging from the image of TEM (Figure 1f), it can be concluded that $MnWO_4$ was formed on the surface of δ-MnO_2 cathodes as a CEI layer. Thus, it can be inferred that during the CV process, the δ-MnO_2 surface is partially dissolved and rapidly reacts with $WO_4{}^{2-}$ to form $MnWO_4$ during the CV process. Finally, the $MnWO_4$ CEI layer is successfully constructed on the δ-MnO_2 cathode.

To examine the effect of the $MnWO_4$ CEI layer on the electrochemical performance of the δ-MnO_2 cathode, a number of coin cells were assembled with 3 M $ZnSO_4$ + 0.2 M $MnSO_4$ as the electrolyte and zinc foil as the anode. Figure 2a presents the CV curves drawn for the W-MnO_2 cathode in the initial five CV cycles. The peak of the CV curves almost overlap, and their intensity increases at a slow pace after the second cycle, indicating that the W-MnO_2 cathode maintains excellent performance in electrochemical activity and reversibility after the construction of the CEI layer. For the W-MnO_2 cathode, the two cathodic peaks at 1.2 V and 1.4 V correspond to different stages of charge carrier insertion [15,46]. By drawing a comparison with the CV curves of the δ-MnO_2 cathode (Figure S3), the increased intensity of cathodic peak shown by W-MnO_2 near 1.4 V is suspected to result from the improvement of reaction kinetics by the $MnWO_4$ CEI layer. The galvanostatic charge and discharge (GCD) curve of the W-MnO_2 cathode at 0.2 A g^{-1} (Figure 2b) shows a slow-paced improvement of capacity during cycling, suggesting the activation of the W-MnO_2 cathode. Afterwards, the W-MnO_2 cathode exhibits two-stage charge carrier intercalation, which is coherent with the CV results. In comparison with the GCD curves of the δ-MnO_2 cathode (Figure S4) and W-MnO_2 cathode, there is almost no difference found between them, indicating that the CEI layer did not change the characteristics of the two-stage charge carrier intercalation. As confirmed by the cycling test conducted at 0.2 A g^{-1} (Figure 2c), the W-MnO_2 cathode is slowly activated by the $MnWO_4$ CEI layer. In the first 100 cycles, the capacity of the W-MnO_2 cathode improves slowly and stabilizes gradually at around 301.2 mAh g^{-1}, which is close to the initial capacity of the δ-MnO_2 cathode. However, the capacity of the δ-MnO_2 cathode declines continuously, which indicates that the $MnWO_4$ CEI layer improves the cycling stability significantly. As shown in Figure 2d, the rate capability of the W-MnO_2 cathode was evaluated after the activation process. To be specific, the W-MnO_2 cathode achieves a specific discharge capacity of 295.2, 260.5, 237.4, 210.3, 158.1, and 105.5 mAh g^{-1} at the current density of 0.2, 0.5, 1, 2, 5, and 10 A g^{-1}, respectively. The corresponding GCD curves of W-MnO_2 and δ-MnO_2 cathodes at various current densities are presented in Figures S5 and S6, respectively. Compared with the corresponding values of the δ-MnO_2 cathode that vary from 0.2 to 10 A g^{-1}, the capacity rate of W-MnO_2 and δ-MnO_2 cathodes reaches 35.7% and 26.9%, respectively. It implies that the $MnWO_4$ CEI layer is conducive to improving rate performance. Notably, the W-MnO_2 cathode achieves an outstanding performance in cycling stability at 10 A g^{-1}, as shown in Figure 2e, from which it can be seen that the capacity of the W-MnO_2 cathode slowly increases to 98.6 mAh g^{-1} during activation (initial 500 cycles). After 2000 cycles, the W-MnO_2 cathode maintains a capacity retention rate of 98.2% (vs. the activated capacity at 500 cycles). However, the cycling capacity of the δ-MnO_2 cathode without the $MnWO_4$ CEI layer decreases rapidly after the initial 100 cycles. Subsequently, the capacity is gradually reduced to 35 mAh g^{-1} after 2000 cycles. At this point, the capacity retention rate is merely 33.4%. This result confirms that the electrochemical cycling performance can be improved by the $MnWO_4$ CEI layer on the δ-MnO_2 cathode. Without any significant change in the structure and morphology of the cathode material, a thin layer constructed on the cathode surface is sufficient to improve the electrochemical performance significantly. The construction of the CEI layer is more universal than the adjustment for electrodes [12] and electrolytes [13]. In comparison with other reported CEI or SEI layers (Table S1, Supporting Information) [31–37], the $MnWO_4$ CEI layer, as constructed in this paper, leads to a significant improvement of high current cycle performance.

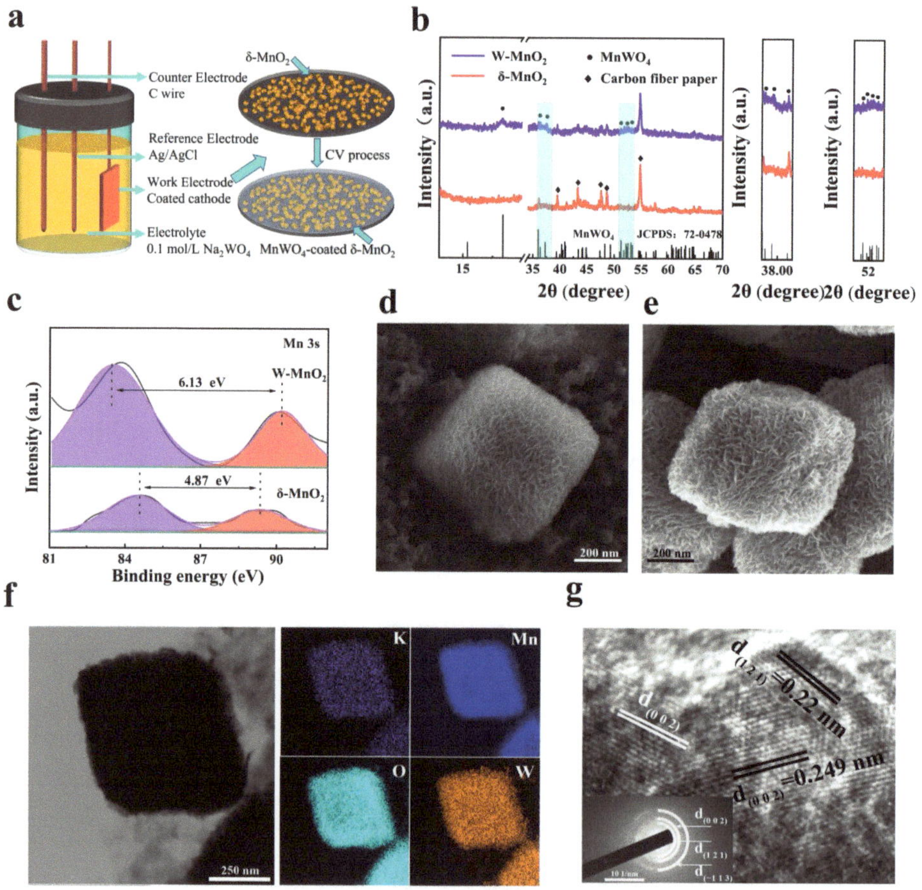

Figure 1. (**a**) Schematic illustration of the preparation process of $MnWO_4$ CEI layer. (**b**) XRD patterns of δ-MnO_2 and W-MnO_2. (**c**) High-resolution XPS spectra of Mn 3s region in δ-MnO_2 and W-MnO_2. (**d**,**e**) The morphology of (**d**) δ-MnO_2 and (**e**) W-MnO_2. (**f**) TEM image and the elemental mappings of W-MnO_2. (**g**) HRTEM image of $MnWO_4$ on the surface of W-MnO_2 and the corresponding SAED pattern (inset).

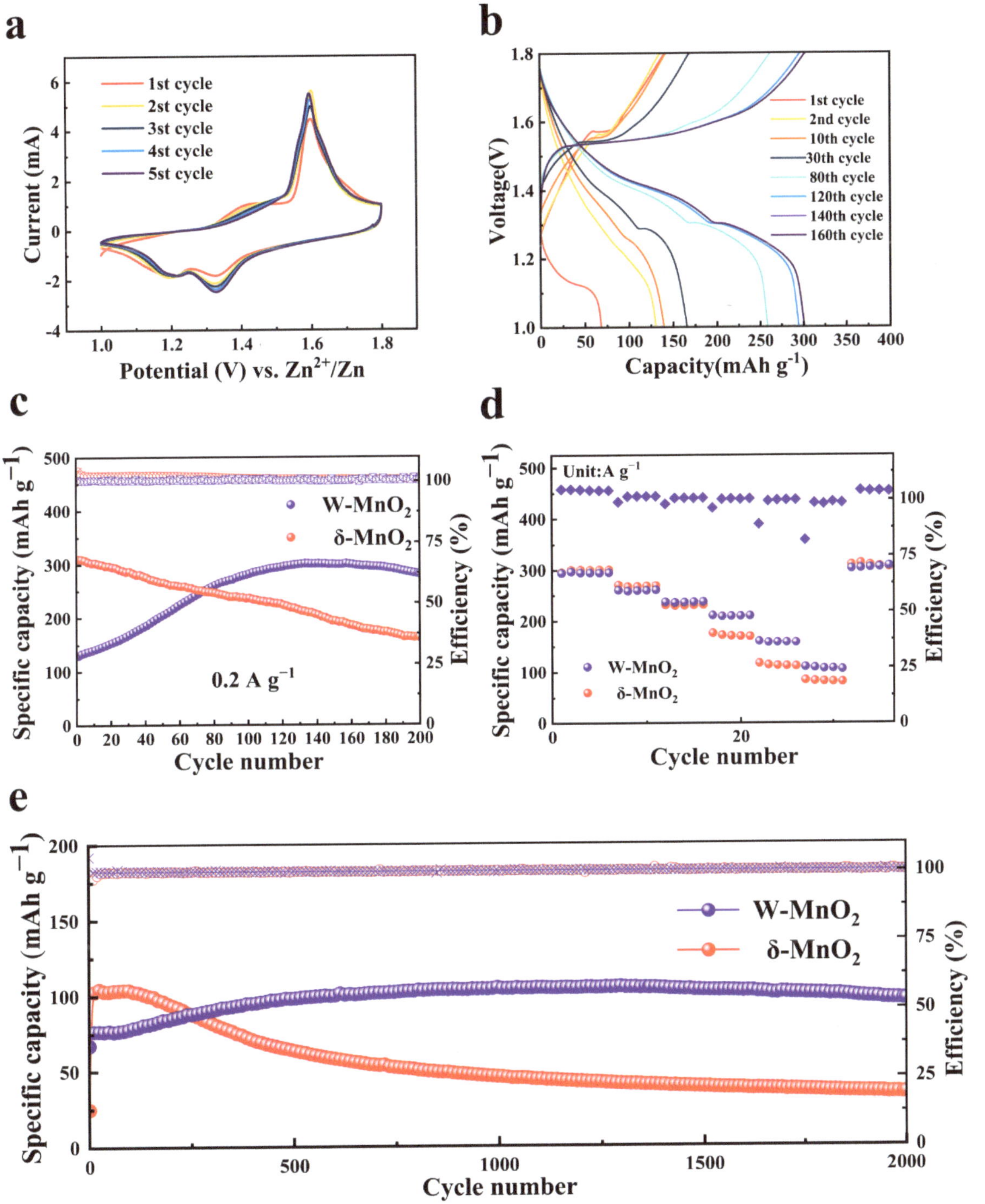

Figure 2. Electrochemical performance of δ-MnO_2 and W-MnO_2 cathodes. (**a**) CV curves of W-MnO_2 cathode at 1 mV s^{-1}. (**b**,**c**) Galvanostatic charge and discharge curves of the W-MnO_2 cathode and corresponding cycling performance at 0.2 A g^{-1}. (**d**) Rate capacity of δ-MnO_2 and W-MnO_2 cathodes with current density from 0.2 to 10 A g^{-1}. (**e**) Cycling performance at 10 A g^{-1}.

To explore the effect of the CEI layer on the mechanism of energy storage, the structure evolution of the W-MnO_2 cathode was analyzed by means of ex situ XRD, XPS, and TEM. Taking into account the BET result of δ-MnO_2 and the morphology change of the W-MnO_2 cathode, the storage mechanism was analyzed through bulk diffusion rather than surface adsorption. The ex situ XRD of the W-MnO_2 cathode (Figure 3a) reveals the incremental increase of characteristic peaks (around 10° and 33°) corresponding to $Zn_4SO_4(OH)_6{\cdot}xH_2O$ upon the entire discharge process, which evidences the occurrence of H^+ insertion. This is consistent with the findings of previous research [47–50]. In addition, the formation of $Zn_4SO_4(OH)_6{\cdot}xH_2O$ nanosheet at the discharge stage is revealed by ex situ TEM (Figure S7). The results of TEM mapping show the presence of S, Zn, and O elements. Moreover, it can be seen from the SAED pattern (Figure S7) that the nanosheet is $Zn_4SO_4(OH)_6{\cdot}xH_2O$. In contrast, H^+ is gradually released from the W-MnO_2 cathode during the subsequent charge to 1.8 V, which is accompanied by the disappearance of $Zn_4SO_4(OH)_6{\cdot}xH_2O$, as shown in Figure 3c. That is to say, reversible (de)insertion occurs to H^+ throughout the storage process. The reversible storage of Zn^{2+} in the W-MnO_2 host is confirmed by the ex situ XPS performed on the acid-washed cathodes (Figure 3b). When the cathode is discharged to 1.3 V, there are two strong peaks emerging at 1045.8 and 1022.7 eV, which can be considered evidence of Zn^{2+} intercalation [51,52]. The peak strength of the Zn 2p further increases when the cathode is fully discharged (1 V), indicating the occurrence of Zn^{2+} intercalation throughout the discharge process. In addition, the stability of the $MnWO_4$ CEI film during the cycle process is indicated by TEM, HRTEM, and corresponding SAED (Figure S8) in full charge and discharge states. Therefore, H^+/Zn^{2+} co-insertion is confirmed as the storage mechanism of the W-MnO_2 cathode during the discharge process.

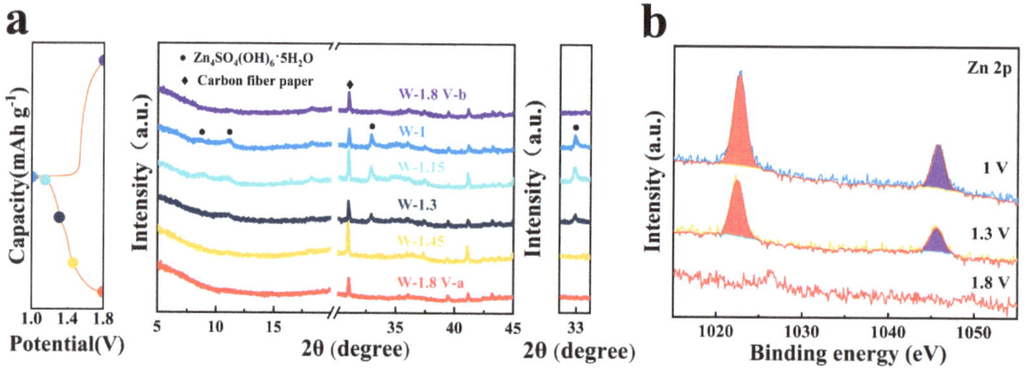

Figure 3. (**a**) Ex situ XRD analysis of W-MnO_2 cathode at various voltages. (**b**) XPS spectra of Zn 2p at different discharge stages.

When the mechanism of energy storage is investigated, the reversible formation of the by-product ($Zn_4SO_4(OH)_6{\cdot}xH_2O$) on the cathode is worth noting. During the discharge process, the formation of $Zn_4SO_4(OH)_6{\cdot}xH_2O$ nanosheets could inhibit the electrochemical reaction in the cathode to some extent, as reported in other studies [34]. To demonstrate the impact of the by-product ($Zn_4SO_4(OH)_6{\cdot}xH_2O$) on the charge transfer resistance, the EIS test was performed during the discharge process. The EIS data (Figure S9) of the W-MnO_2 cathode were fitted with the equivalent circuit template, as indicated by two semicircles in the medium and high-frequency regions. The semicircle at a high frequency is considered as the constructed $MnWO_4$ CEI layer and the $Zn_4SO_4(OH)_6{\cdot}xH_2O$ formed during the discharge process, while that at a medium frequency is attributed to the charge transfer resistance (R_{ct}). Figure 4a shows the variation and comparison of the R_{ct} during different stages of discharge for both W-MnO_2 and δ-MnO_2 cathodes. Apparently, the R_{ct} of δ-MnO_2 cathode increases rapidly (from 12.62 to 201 Ω), which suggests that the existence of $Zn_4SO_4(OH)_6{\cdot}xH_2O$ nanosheet plays a part in insulating the active material, which impedes electron transport and increases internal resistance. For the W-MnO_2 cathode, the incremental of R_{ct} is more significant compared to the δ-MnO_2 cathode, indicating that the impact of $Zn_4SO_4(OH)_6{\cdot}xH_2O$ is mitigated by the presence of the $MnWO_4$ CEI layer. Moreover, the R_{ct} of the W-MnO_2 cathode is higher than that of the δ-MnO_2 cathode in the initial state, which is due to the relatively low conductivity of the $MnWO_4$ CEI layer.

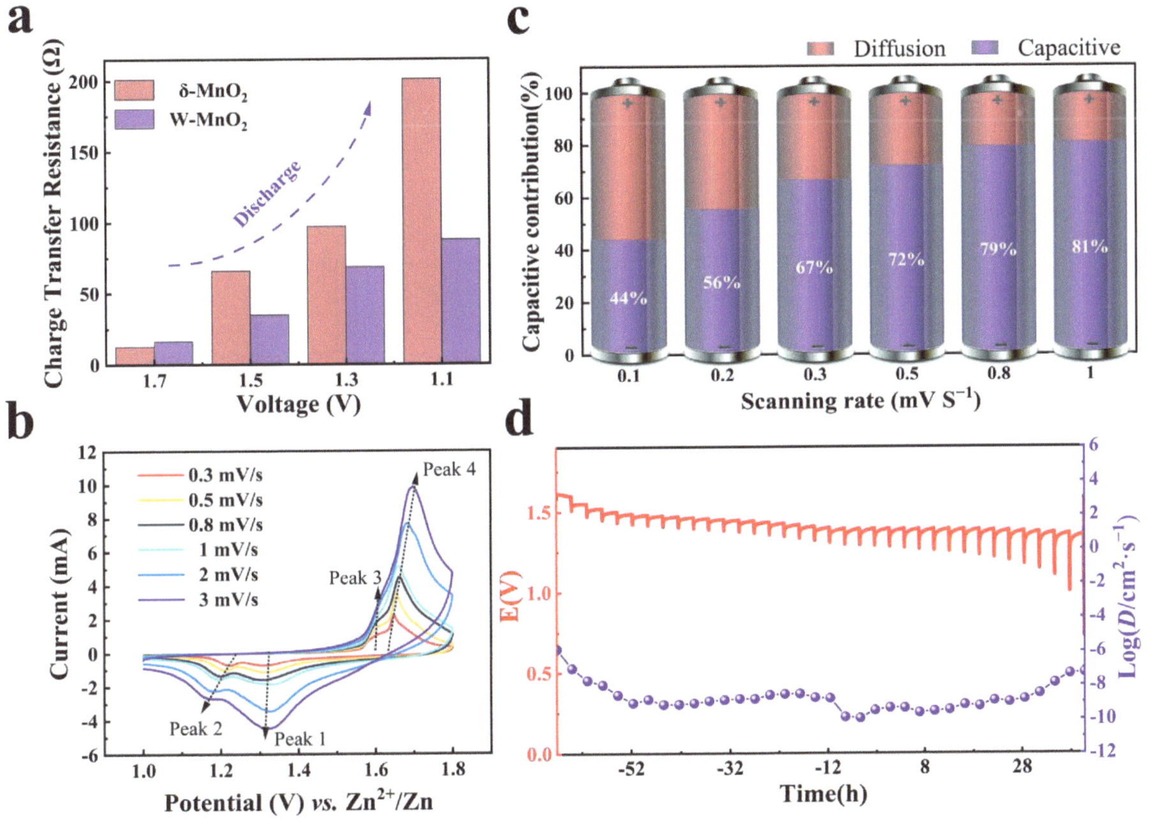

Figure 4. (**a**) The charge transfer resistance of W-MnO_2 and δ-MnO_2 cathodes at the discharge stage. (**b**) CV tests at various scan rates from 0.3 to 3 mV s^{-1}. (**c**) The proportion of capacitive contributions at different scan rates for W-MnO_2. (**d**) GITT curves of the W-MnO_2 cathode after activation during the discharge process.

To reveal the effect of the $MnWO_4$ CEI layer and $Zn_4SO_4(OH)_6{\cdot}xH_2O$ intermediate on the reaction kinetics of the W-MnO_2 cathode, the kinetics behaviors were analyzed by carrying out CV (cyclic voltammetric curve) tests at varying scan rates (0.3 to 3.0 mV s^{-1}), as shown in Figure 4b. For the W-MnO_2 cathode, the b-value (Figure S11a) of the three different peaks is calculated to be 0.80, 0.58, and 0.62 for peaks 1, 2, and 4, respectively. As for δ-MnO_2 cathode (Figures S10 and S11b), the b-value of peaks 1, 2, and 4 is 0.6, 0.41, and 0.53, respectively. The rise in the b-value of the W-MnO_2 cathode indicates that the improvement of reaction kinetics contributes to an excellent rate performance [53]. Furthermore, the capacitive-controlled contribution for the W-MnO_2 cathode is calculated to be 44.1%, 55.5%, 66.5%, 71.9%, 79.2%, and 80.7% at a scan rate of 0.1, 0.2, 0.3, 0.5, 0.8, and 1 mV s^{-1}, respectively (Figure 4c). The proportion of capacitive contribution to the whole capacity for the W-MnO_2 cathode at 1 mV s^{-1} is 80.7%, suggesting that the pseudocapacitive behavior dominates the storage mechanism. Compared with the δ-MnO_2 cathode, the capacitive contributions of the δ-MnO_2 cathode (Figure S12) is less significant at different scan rates, which reaffirms the improvement of reaction kinetics by the construction of the $MnWO_4$ CEI layer. Finally, to gain an insight into the diffusion dynamics, the galvanostatic intermittent titration technique (GITT) was applied to calculate the diffusion coefficient (D) at different stages of discharge (Figures 4d and S13). It can be found that the W-MnO_2 and δ-MnO_2 cathodes experience two stages of discharge according to the D value. In the first one, D is between 10^{-8} and 10^{-9}. In the second one, D decreases to the range of 10^{-9}–10^{-10}. It is noteworthy that the $MnWO_4$ CEI layer causes the diffusion coefficient of the W-MnO_2 cathode to be relatively more stable, which is always above 10^{-10}. To sum up, the $MnWO_4$ CEI layer of δ-MnO_2 can mitigate the impact of $Zn_4SO_4(OH)_6{\cdot}xH_2O$ on the cathode and ensure sufficient reaction kinetics, which explains the better electrochemical performance.

4. Conclusions

In the present study, a $MnWO_4$ CEI layer was constructed on the δ-MnO_2 surface by following a facile cyclic voltammetry method, which significantly reduced the impact of by-product ($Zn_4SO_4(OH)_6{\cdot}xH_2O$) on the cathode and improved the reaction kinetics during the process of H^+/Zn^{2+} co-intercalation, thus enhancing the rate performance (295.2 mA at 0.1 A g^{-1} and 105.5 mA at 10 A g^{-1}). More importantly, the dissolution of Mn was inhibited in the AZIBs by the $MnWO_4$ CEI layer, thus ensuring its long cycling lifespan. Compared to the activated capacity at 500 cycles, the capacity retention rate at 10 A g^{-1} was maintained at 98.2% after 2000 cycles, which is much higher than the retention rate of 33.4% for the pristine MnO_2 cathode. This CEI construction strategy could contribute to exploring the stable Mn-based cathode of AZIBS.

Supplementary Materials: The following supporting information can be downloaded at: https://www.mdpi.com/article/10.3390/ma16083228/s1, Figure S1: XRD pattern of synthetic δ-MnO_2 pristine powder; Figure S2: N_2 adsorption isotherms of δ-MnO_2; Figure S3: CV curves of the δ-MnO_2 electrode at 1 mV/s; Figure S4: Galvanostatic charge and discharge curves of the δ-MnO_2 cathode at 0.2 A g^{-1}; Figure S5: Galvanostatic charge and discharge curves of W-MnO_2 cathode at various current densities ranging from 0.2 A g^{-1} to 10 A g^{-1}; Figure S6: Galvanostatic charge and discharge curves of the δ-MnO_2 cathode at various current densities ranging from 0.2 A g^{-1} to 10 A g^{-1}; Figure S7: High-angle annular bright-field scanning TEM (HAABF-STEM) image and the corresponding elemental mappings, respectively; Figure S8: (a,b) TEM, HR-TEM and SAED pattern of W-MnO_2: (a,c) charged (1.8 V). (b,d) discharged (1 V); Figure S9: EIS of (a) the W-MnO_2 and (b) δ-MnO_2 cathodes at different stages during the discharge process. The equivalent circuit model for (c) the W-MnO_2 and d) the δ-MnO_2 electrode; Figure S10: CV tests for δ-MnO_2 cathode at various scan rates ranging from 0.3 to 3 mV/s; Figure S11: (a,b) b values of different peaks in CV curves for W-MnO_2 and δ-MnO_2 cathodes: (a) W-MnO_2. (b) δ-MnO_2; Figure S12: The proportion of capacitive and diffusion contributions at various scan rates for δ-MnO_2; Figure S13: GITT test of the δ-MnO_2 cathode; Table S1: Comparison of cycling performance with other recent studies [31–37].

Author Contributions: Conceptualization, H.T. and S.C.; methodology, H.Z. (Huanlin Zhang), Y.Z. and X.S.; software, L.L. and T.M.; validation, H.T., H.Z. (Hang Zhang) and Y.Z.; investigation, H.T. and Y.Z.; resources, X.S. and S.C.; data curation, H.T.; writing—original draft preparation, H.T.; writing—review and editing, S.C.; supervision, S.C.; funding acquisition, S.C. All authors have read and agreed to the published version of the manuscript.

Funding: This research was funded by the National Natural Science Foundation of China (Nos. 51872197 and 52271246). This study was additionally funded by Tianjin Research Innovation Project for Postgraduate Students (Project No. 2021YJSB171).

Institutional Review Board Statement: Not applicable.

Informed Consent Statement: Not applicable.

Data Availability Statement: The data presented in this study are available on request from the corresponding author.

Conflicts of Interest: The authors declare no conflict of interest.

References

1. Macrelli, A.; Olivieri, M.; Lamperti, A.; Russo, V.; Bozzini, B.; Menegazzo, M.; Bussetti, G.; Casari, C.S.; Li Bassi, A. Nanostructured $Zn_xMn_{3-x}O_4$ thin films by pulsed laser deposition: A spectroscopic and electrochemical study towards the application in aqueous Zn-ion batteries. *Electrochim. Acta* **2023**, *442*, 141909. [CrossRef]
2. Zampardi, G.; La Mantia, F. Prussian blue analogues as aqueous Zn-ion batteries electrodes: Current challenges and future perspectives. *Curr. Opin. Electrochem.* **2020**, *21*, 84–92. [CrossRef]
3. Liu, Z.; Qin, L.; Cao, X.; Zhou, J.; Pan, A.; Fang, G.; Wang, S.; Liang, S. Ion migration and defect effect of electrode materials in multivalent-ion batteries. *Prog. Mater. Sci.* **2022**, *125*, 100911. [CrossRef]
4. Tang, B.; Fang, G.; Zhou, J.; Wang, L.; Lei, Y.; Wang, C.; Lin, T.; Tang, Y.; Liang, S. Potassium vanadates with stable structure and fast ion diffusion channel as cathode for rechargeable aqueous zinc-ion batteries. *Nano Energy* **2018**, *51*, 579–587. [CrossRef]
5. Chen, D.; Lu, M.; Cai, D.; Yang, H.; Han, W. Recent advances in energy storage mechanism of aqueous zinc-ion batteries. *J. Energy Chem.* **2021**, *54*, 712–726. [CrossRef]
6. Song, M.; Tan, H.; Chao, D.; Fan, H.J. Recent advances in Zn-ion batteries. *Adv. Funct. Mater.* **2018**, *28*, 1802564. [CrossRef]
7. Fegade, U.; Jethave, G.; Khan, F.; Al Ahmed, A.; Karmouch, R.; Shariq, M.; Inamuddin; Ahmer, M.F. Recent development of aqueous zinc-ion battery cathodes and future challenges: Review. *Int. J. Energ. Res.* **2022**, *46*, 13152–13177. [CrossRef]
8. Liu, S.; Sun, Y.; Yang, J.; Zhang, Y.; Cai, Z. Highly loaded and binder-free molybdenum trioxide cathode material prepared using multi-arc ion plating for aqueous zinc ion batteries. *Materials* **2022**, *15*, 5954. [CrossRef]
9. Li, T.; Li, H.; Yuan, J.; Xia, Y.; Liu, Y.; Sun, A. Recent advance and modification strategies of transition metal dichalcogenides (TMDs) in aqueous zinc ion batteries. *Materials* **2022**, *15*, 2654. [CrossRef]
10. Karapidakis, E.; Vernardou, D. Progress on V_2O_5 cathodes for multivalent aqueous batteries. *Materials* **2021**, *14*, 2310. [CrossRef]
11. Wang, D.; Wang, L.; Liang, G.; Li, H.; Liu, Z.; Tang, Z.; Liang, J.; Zhi, C. A Superior δ-MnO_2 cathode and a self-healing Zn-δ-MnO_2 battery. *ACS Nano* **2019**, *13*, 10643–10652. [CrossRef]
12. Alfaruqi, M.H.; Islam, S.; Mathew, V.; Song, J.; Kim, S.; Tung, D.P.; Jo, J.; Kim, S.; Baboo, J.P.; Xiu, Z.; et al. Ambient redox synthesis of vanadium-doped manganese dioxide nanoparticles and their enhanced zinc storage properties. *Appl. Surf. Sci.* **2017**, *404*, 435–442. [CrossRef]
13. Chamoun, M.; Brant, W.R.; Tai, C.; Karlsson, G.; Noréus, D. Rechargeability of aqueous sulfate Zn/MnO_2 batteries enhanced by accessible Mn^{2+} ions. *Energy Storage Mater.* **2018**, *15*, 351–360. [CrossRef]
14. Zuo, Y.; Meng, T.; Tian, H.; Ling, L.; Zhang, H.; Zhang, H.; Sun, X.; Cai, S. Enhanced H^+ storage of a MnO_2 cathode via a MnO_2 nanolayer interphase transformed from manganese phosphate. *ACS Nano* **2023**, *17*, 5600–5608. [CrossRef]
15. Sun, W.; Wang, F.; Hou, S.; Yang, C.; Fan, X.; Ma, Z.; Gao, T.; Han, F.; Hu, R.; Zhu, M.; et al. Zn/MnO_2 battery chemistry with H^+ and Zn^{2+} coinsertion. *J. Am. Chem. Soc.* **2017**, *139*, 9775–9778. [CrossRef]
16. Kundu, D.; Adams, B.D.; Duffort, V.; Vajargah, S.H.; Nazar, L.F. A high-capacity and long-life aqueous rechargeable zinc battery using a metal oxide intercalation cathode. *Nat. Energy* **2016**, *1*, 16119. [CrossRef]
17. Chen, J.; Xiao, B.; Hu, C.; Chen, H.; Huang, J.; Yan, D.; Peng, S. Construction strategy of VO_2 @V_2C 1D/2D heterostructure and improvement of zinc-ion diffusion ability in VO_2 (B). *ACS Appl. Mater. Inter.* **2022**, *14*, 28760–28768. [CrossRef]
18. Dong, Y.; Jia, M.; Wang, Y.; Xu, J.; Liu, Y.; Jiao, L.; Zhang, N. Long-life zinc/vanadium pentoxide battery enabled by a concentrated aqueous $ZnSO_4$ electrolyte with proton and zinc ion co-intercalation. *ACS Appl. Energ. Mater.* **2020**, *3*, 11183–11192. [CrossRef]
19. Zhang, N.; Dong, Y.; Jia, M.; Bian, X.; Wang, Y.; Qiu, M.; Xu, J.; Liu, Y.; Jiao, L.; Cheng, F. Rechargeable aqueous Zn-V_2O_5 battery with high energy density and long cycle life. *ACS Energy Lett.* **2018**, *3*, 1366–1372. [CrossRef]
20. Chen, L.; Ruan, Y.; Zhang, G.; Wei, Q.; Jiang, Y.; Xiong, T.; He, P.; Yang, W.; Yan, M.; An, Q.; et al. Ultrastable and high-performance Zn/VO_2 battery based on a reversible single-phase reaction. *Chem. Mater.* **2019**, *31*, 699–706. [CrossRef]
21. Lu, K.; Song, B.; Zhang, J.; Ma, H. A rechargeable Na-Zn hybrid aqueous battery fabricated with nickel hexacyanoferrate and nanostructured zinc. *J. Power Sources* **2016**, *321*, 257–263. [CrossRef]
22. Yang, Q.; Mo, F.; Liu, Z.; Ma, L.; Li, X.; Fang, D.; Chen, S.; Zhang, S.; Zhi, C. Activating C-coordinated iron of iron hexacyanoferrate for Zn hybrid-ion batteries with 10,000-cycle lifespan and superior rate capability. *Adv. Mater.* **2019**, *31*, 1901521. [CrossRef] [PubMed]
23. Yi, H.; Qin, R.; Ding, S.; Wang, Y.; Li, S.; Zhao, Q.; Pan, F. Structure and properties of Prussian blue analogues in energy storage and conversion applications. *Adv. Funct. Mater.* **2021**, *31*, 2006970. [CrossRef]
24. Zhang, L.; Chen, L.; Zhou, X.; Liu, Z. Towards high-voltage aqueous metal-ion batteries beyond 1.5 V: The zinc/zinc hexacyanoferrate system. *Adv. Energy Mater.* **2015**, *5*, 1400930. [CrossRef]
25. Zhao, Y.; Zhu, Y.; Zhang, X. Challenges and perspectives for manganese-based oxides for advanced aqueous zinc-ion batteries. *Infomat* **2020**, *2*, 237–260. [CrossRef]
26. Ren, Q.; Yuan, Y.; Wang, S. Interfacial strategies for suppression of Mn dissolution in rechargeable battery cathode materials. *ACS Appl. Mater. Inter.* **2022**, *14*, 23022–23032. [CrossRef]
27. Fang, G.; Zhu, C.; Chen, M.; Zhou, J.; Tang, B.; Cao, X.; Zheng, X.; Pan, A.; Liang, S. Suppressing manganese dissolution in potassium manganate with rich oxygen defects engaged high-energy-density and durable aqueous zinc-ion battery. *Adv. Funct. Mater.* **2019**, *29*, 1808375. [CrossRef]
28. Poyraz, A.S.; Laughlin, J.; Zec, Z. Improving the cycle life of cryptomelane type manganese dioxides in aqueous rechargeable zinc ion batteries: The effect of electrolyte concentration. *Electrochim. Acta* **2019**, *305*, 423–432. [CrossRef]

29. Zhou, M.; Chen, Y.; Fang, G.; Liang, S. Electrolyte/electrode interfacial electrochemical behaviors and optimization strategies in aqueous zinc-ion batteries. *Energy Storage Mater.* **2022**, *45*, 618–646. [CrossRef]
30. Verma, V.; Kumar, S.; Manalastas, W.; Srinivasan, M. Undesired reactions in aqueous rechargeable zinc ion batteries. *ACS Energy Lett.* **2021**, *6*, 1773–1785. [CrossRef]
31. Guo, S.; Liang, S.; Zhang, B.; Fang, G.; Ma, D.; Zhou, J. Cathode interfacial layer formation via in situ electrochemically charging in aqueous zinc-ion battery. *ACS Nano* **2019**, *13*, 13456–13464. [CrossRef]
32. Luo, S.; Cao, X.; Su, Q.; Zhang, Y.; Liu, S.; Xie, X.; Liang, S.; Pan, A. Layered barium vanadate cathodes for aqueous zinc batteries: Enhancing cycling stability through inhibition of vanadium dissolution. *ACS Appl. Energ. Mater.* **2021**, *4*, 6197–6204. [CrossRef]
33. Wu, B.; Zhang, G.; Yan, M.; Xiong, T.; He, P.; He, L.; Xu, X.; Mai, L. Graphene scroll-coated α-MnO_2 nanowires as high-performance cathode materials for aqueous Zn-ion battery. *Small* **2018**, *14*, 1703850. [CrossRef]
34. Guo, J.; Ming, J.; Lei, Y.; Zhang, W.; Xia, C.; Cui, Y.; Alshareef, H.N. Artificial solid electrolyte interphase for suppressing surface reactions and cathode dissolution in aqueous zinc ion batteries. *ACS Energy Lett.* **2019**, *4*, 2776–2781. [CrossRef]
35. Li, S.; Yu, D.; Liu, L.; Yao, S.; Wang, X.; Jin, X.; Zhang, D.; Du, F. In-situ electrochemical induced artificial solid electrolyte interphase for MnO@C nanocomposite enabling long-lived aqueous zinc-ion batteries. *Chem. Eng. J.* **2022**, *430*, 132673. [CrossRef]
36. Liu, Y.; Zhi, J.; Hoang, T.K.A.; Zhou, M.; Han, M.; Wu, Y.; Shi, Q.; Xing, R.; Chen, P. Paraffin Based cathode-electrolyte interface for highly reversible aqueous zinc-ion battery. *ACS Appl. Energ. Mater.* **2022**, *5*, 4840–4849. [CrossRef]
37. Zhang, L.; Zhang, B.; Hu, J.; Liu, J.; Miao, L.; Jiang, J. An in situ artificial cathode electrolyte interphase strategy for suppressing cathode dissolution in aqueous zinc ion batteries. *Small Methods* **2021**, *5*, 2100094. [CrossRef]
38. Chen, X.; Cao, Z.; Xing, L.; Liao, Y.; Qiu, Y.; Li, W. Improved Li-storage performance with PEDOT-decorated MnO_2 nanoboxes. *Nanoscale* **2017**, *9*, 18467–18473. [CrossRef]
39. Guo, C.; Zhou, Q.; Liu, H.; Tian, S.; Chen, B.; Zhao, J.; Li, J. A case study of β- and δ-MnO_2 with different crystallographic forms on ion-storage in rechargeable aqueous zinc ion battery. *Electrochim. Acta* **2019**, *324*, 134867. [CrossRef]
40. Wu, P.; Dai, S.; Chen, G.; Zhao, S.; Xu, Z.; Fu, M.; Chen, P.; Chen, Q.; Jin, X.; Qiu, Y.; et al. Interfacial effects in hierarchically porous α-MnO_2/Mn_3O_4 heterostructures promote photocatalytic oxidation activity. *Appl. Catal. B Environ.* **2020**, *268*, 118418. [CrossRef]
41. Puttaswamy, R.; Nagaraj, R.; Kulkarni, P.; Beere, H.K.; Upadhyay, S.N.; Balakrishna, R.G.; Sanna Kotrappanavar, N.; Pakhira, S.; Ghosh, D. Constructing a high-performance aqueous rechargeable zinc-ion battery cathode with self-assembled mat-like packing of intertwined Ag(I) pre-inserted $V_3O_7{\cdot}H_2O$ microbelts with reduced graphene oxide core. *ACS Sustain. Chem. Eng.* **2021**, *9*, 3985–3995. [CrossRef]
42. Cook, J.B.; Kim, H.S.; Yan, Y.; Ko, J.S.; Robbennolt, S.; Dunn, B.; Tolbert, S.H. Mesoporous MoS_2 as a transition metal dichalcogenide exhibiting pseudocapacitive Li and Na-ion charge storage. *Adv. Energy Mater.* **2016**, *6*, 1501937. [CrossRef]
43. Deiss, E. Spurious chemical diffusion coefficients of Li^+ in electrode materials evaluated with GITT. *Electrochim. Acta* **2005**, *50*, 2927–2932. [CrossRef]
44. Chen, W.; Li, G.; Pei, A.; Li, Y.; Liao, L.; Wang, H.; Wan, J.; Liang, Z.; Chen, G.; Zhang, H.; et al. A manganese-hydrogen battery with potential for grid-scale energy storage. *Nat. Energy* **2018**, *3*, 428–435. [CrossRef]
45. Ilton, E.S.; Post, J.E.; Heaney, P.J.; Ling, F.T.; Kerisit, S.N. XPS determination of Mn oxidation states in Mn (hydr)oxides. *Appl. Surf. Sci.* **2016**, *366*, 475–485. [CrossRef]
46. Gao, X.; Wu, H.; Li, W.; Tian, Y.; Zhang, Y.; Wu, H.; Yang, L.; Zou, G.; Hou, H.; Ji, X. H^+-insertion boosted α-MnO_2 for an Aqueous Zn-ion battery. *Small* **2020**, *16*, 1905842. [CrossRef]
47. Pan, H.; Shao, Y.; Yan, P.; Cheng, Y.; Han, K.S.; Nie, Z.; Wang, C.; Yang, J.; Li, X.; Bhattacharya, P.; et al. Reversible aqueous zinc/manganese oxide energy storage from conversion reactions. *Nat. Energy* **2016**, *1*, 16039. [CrossRef]
48. Liu, N.; Wu, X.; Fan, L.; Gong, S.; Guo, Z.; Chen, A.; Zhao, C.; Mao, Y.; Zhang, N.; Sun, K. Intercalation pseudocapacitive Zn^{2+} storage with hydrated vanadium dioxide toward ultrahigh rate performance. *Adv. Mater.* **2020**, *32*, 1908420. [CrossRef]
49. Zhang, R.; Liang, P.; Yang, H.; Min, H.; Niu, M.; Jin, S.; Jiang, Y.; Pan, Z.; Yan, J.; Shen, X.; et al. Manipulating intercalation-extraction mechanisms in structurally modulated δ-MnO_2 nanowires for high-performance aqueous zinc-ion batteries. *Chem. Eng. J.* **2022**, *433*, 133687. [CrossRef]
50. Zuo, Y.; Liu, P.; Ling, L.; Tian, M.; Wang, Z.; Tian, H.; Meng, T.; Sun, X.; Cai, S. Boosted H^+ intercalation enables ultrahigh rate performance of the δ-MnO_2 cathode for aqueous zinc batteries. *ACS Appl. Mater. Inter.* **2022**, *14*, 26653–26661. [CrossRef]
51. Pang, Q.; Sun, C.; Yu, Y.; Zhao, K.; Zhang, Z.; Voyles, P.M.; Chen, G.; Wei, Y.; Wang, X. $H_2V_3O_8$ nanowire/graphene electrodes for aqueous rechargeable zinc ion batteries with high rate capability and large capacity. *Adv. Energy Mater.* **2018**, *8*, 1800144. [CrossRef]
52. Wang, X.; Xi, B.; Ma, X.; Feng, Z.; Jia, Y.; Feng, J.; Qian, Y.; Xiong, S. Boosting zinc-ion storage capability by effectively suppressing vanadium dissolution based on robust layered barium vanadate. *Nano Lett.* **2020**, *20*, 2899–2906. [CrossRef]
53. Cook, J.B.; Kim, H.S.; Lin, T.C.; Lai, C.H.; Dunn, B.; Tolbert, S.H. Pseudocapacitive charge storage in thick composite MoS_2 Nanocrystal-based electrodes. *Adv. Energy Mater.* **2017**, *7*, 1601283. [CrossRef]

materials

MDPI

Article

Highly Loaded and Binder-Free Molybdenum Trioxide Cathode Material Prepared Using Multi-Arc Ion Plating for Aqueous Zinc Ion Batteries

Sainan Liu [1,*], Yangyang Sun [1], Jing Yang [1], Yi Zhang [1,*] and Zhenyang Cai [2,*]

1 School of Minerals Processing and Bioengineering, Central South University, Changsha 410083, China
2 School of Materials Science and Engineering, Central South University, Changsha 410083, China
* Correspondence: liusainanbaby@163.com (S.L.); yee_z10@csu.edu.cn (Y.Z.); csuczy@csu.edu.cn (Z.C.)

Abstract: Aqueous zinc-ion batteries (ZIBS) are becoming more popular as the use of energy storage devices grows, owing to advantages such as safety and an abundant zinc supply. In this study, molybdenum powder was loaded directly on carbon fiber cloth (CFC) via multi-arc ion plating to obtain Mo@CFC, which was then oxidatively heated in a muffle furnace for 20 min at 600 °C to produce high mass loading α-MoO_3@CFC (α-MoO_3 of 12–15 mg cm^{-2}). The cells were assembled with α-MoO_3@CFC as the cathode and showed an outstanding Zn^{2+} storage capacity of 200.8 mAh g^{-1} at 200 mA g^{-1} current density. The capacity retention rate was 92.4 % after 100 cycles, along with an excellent cycling performance of 109.8 mAh g^{-1} following 500 cycles at 1000 mA g^{-1} current density. Subsequently, it was shown that CFC-loaded α-MoO_3 cathode material possessed significantly improved electrochemical performance when compared to a cell constructed from commercial MoO_3 using conventional slurry-based electrode methods. This work presents a novel yet simple method for preparing highly loaded and binder-free cathodic materials for aqueous ZIBs. The results suggest that the highly loaded cathode material with a high charge density may be potentially employed for future flexible device assembly and applications.

Keywords: α-MoO_3@CFC; binder-free; highly loaded; multi-arc ion plating; aqueous zinc-ion battery

Citation: Liu, S.; Sun, Y.; Yang, J.; Zhang, Y.; Cai, Z. Highly Loaded and Binder-Free Molybdenum Trioxide Cathode Material Prepared Using Multi-Arc Ion Plating for Aqueous Zinc Ion Batteries. *Materials* **2022**, *15*, 5954. https://doi.org/10.3390/ma15175954

Academic Editors: Huang Zhang and Yuan Ma

Received: 29 July 2022
Accepted: 23 August 2022
Published: 29 August 2022

1. Introduction

Lithium-ion batteries have been widely used in consumer gadgets and automobiles as a result of their excellent energy density. However, the inherent disadvantages, including lack of safety, increased cost, and environmental concerns have limited their applications in grid-scale energy storage [1–3]. Aqueous zinc-ion batteries (ZIBs) are seen as a viable alternative to lithium-ion batteries because of their benefits, such as a high theoretical capacity (820 mAh g^{-1}), cost efficiency, environmental friendliness, and safety [4,5]. To date, various cathode materials, including manganese-based [6–8], vanadium-based [9–12], and Prussian blue analogues [13–15] have been investigated and used in aqueous ZIBs. During the preparation of electrode materials at the laboratory level, for a high mass specific capacity, an active material mass (such as mentioned above) of no more than 3 mg cm^{-2} may be commonly employed. It is still difficult to meet the demand for higher mass loading to achieve superior energy density and high capacity [16,17]. That is to say, further increasing the loading of active materials remains a great challenge. In addition, with the rapid development of flexible wearable devices, the research on flexible electrodes has attracted extensive attention [18–20]. Combining the two, a more suitable preparation of high-mass-loading flexible self-supporting electrode materials is needed, which not only can improve the capacity of aqueous ZIBs, but additionally promote the practical application of flexible wearable devices.

In this regard, several studies have used the in situ growth of active materials on flexible substrates to create cathode materials for ZIBs. For example, a series of manganese-based and vanadium-based cathode materials with a high mass loading using different synthetic methods (e.g., 3D printing and the hydrothermal method) have been studied [21–24]. Among various candidate cathode materials, orthogonal molybdenum trioxide (α-MoO_3) is theoretically promising as a high mass loading electrode material for high-performance ZIBs because of its unique layered structure and remarkable thermodynamic stability [25,26]. To meet the requirements of high mass loading, multi-arc ion plating may be a feasible method, where the metal is directly evaporated on the solid cathode target by arc discharge, so as to deposit a thin film on the surface of the substrate [27,28]. To meet the requirements of flexibility, it is feasible to use carbon fiber cloth (CFC) as the deposition substrate [29,30]. In the current study, a novel method for producing an α-MoO_3@CFC cathode with high mass loading and no binder via multi-arc ion plating and simple heat treatment is presented. This approach allows for a significant quantity of active material (about 12–15 mg cm^{-2}) with a high binding strength to be placed on the CFC without the use of a conductive agent or binder, thereby improving the problems of active material shedding and the deterioration of battery performance. As a result, the prepared α-MoO_3@CFC demonstrated an areal capacity of 2.61 mAh cm^{-2} at 200 mA g^{-1} current, and a higher energy density of 200.8 Wh kg^{-1} (a cathode energy density normalized to the cathode mass). This work may not only increase the ZIB capacity, but also facilitate the practical use of flexible wearable devices.

2. Experiments

2.1. Material Synthesis

A two-step procedure was used to fabricate the α-MoO_3@CFC. First, the molybdenum target (Juno Metal Materials Co., Baoji, China) was placed in the multi-arc ion coating machine (SH007, Changsha Divine Arc Ion Coating Co., Changsha, China) along with the CFC, both of which had been sonicated with deionized water, acetone, and anhydrous ethanol. Following that, electrodeposition was initiated for 3 h, and the Mo@CFC was removed when the furnace temperature decreased to room temperature. The mass loading of Mo@CFC was about 8–10 mg cm^{-2}. The mass ratio of Mo and CFC (about 11–13 mg cm^{-2}) was about 0.62–0.91. Finally, Mo@CFC was oxidatively heated for 20 min in a muffle furnace at 600 °C to produce α-MoO_3@CFC.

2.2. Material Characterization

Characterization of the specimens was conducted using an X-ray diffractometer (XRD, Rigaku DX-2500) and Cu Kα rays (λ = 1.54178Å) in the 3° to 80° range, and the morphology and structure were examined in detail using a scanning electron microscope (Nova Nano SEM 230) and a high-resolution transmission electron microscope (HRTEM, FEI Tecnai G2 F20, 200 kv).

2.3. Electrochemical Measurements

The batteries utilized in this work for testing electrochemical performance were all coin cells, type CR2016, and all of the constructed ones were run in air. The anode was made from a zinc foil with a 12 mm diameter, whereas the manufactured α-MoO_3@CFC was cut into small square pieces of 1 cm × 1 cm for the cathode, with active material mass loading of about 12–15 mg cm^{-2}. A glass microfiber filter (GFF Whatman) was employed as a separator, and the electrolyte was a 2 M $ZnSO_4$ solution. They were then assembled layer by layer and sealed with a fully automated sealer (MSK-110D) before being kept at room temperature for 24 h for electrochemical testing.

The cathode for the commercial MoO_3 cells was fabricated by coating a slurry of commercial MoO_3 powder (purchased from Aladdin Industrial Corporation), polyvinylidene difluoride (PVDF), and acetylene black, with an 8:1:1 mass ratio on the stainless-steel mesh, which was cut in a 12 mm diameter. The cathode, electrolyte, separator, and assembly procedures are the same as described above. Moreover, for the battery testing equipment, galvanostatic charge–discharge measurements (LAND CT2001A) were made between 0.4 and 1.4 V. Cyclic voltammetry (CV) was carried out at a scan rate of 1 mV s^{-1} using an electrochemical workstation (CHI-600C, Shanghai, China). To conduct electrochemical impedance spectroscopy (EIS), a ZAHNER-IM6ex electrochemical workstation (ZAHNER Co., Kronach, Germany) was employed.

2.4. Results and Discussion

Figure 1 schematically represents a basic two-step synthesis of α-MoO_3@CFC. Mo@CFC is directly produced by multi-arc ion plating. As a molybdenum source, six pure molybdenum targets (99.99% purity) are evenly distributed on both sides of the vacuum chamber, whereas the treated CFC is fixed as a carrier on a rotating rack located at the upper end of the vacuum chamber, allowing molybdenum to be uniformly plated on. The Mo@CFC is then placed within a quartz crucible and heated for 20 min at 600 °C, allowing Mo@CFC to fully oxidize to α-MoO_3@CFC.

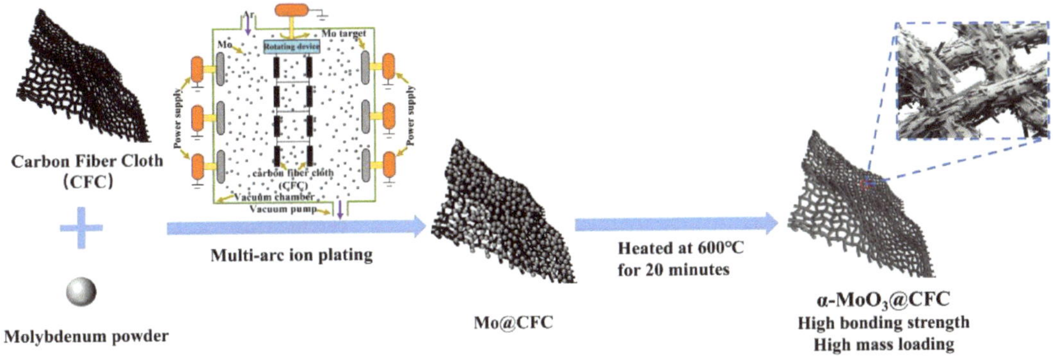

Figure 1. Schematic representation showing the synthesis of α-MoO_3@CFC.

As depicted by the Mo@CFC and α-MoO_3@CFC in Figure S1 of the Supporting Information, a layer of off-white α-MoO_3 evenly covers the surface of the black pure CFC following 20 min of heating at 600 °C, forming α-MoO_3@CFC with good flexibility.

All of the Mo@CFC diffraction peaks are in line with the standard PDF card 42-1120, as shown in Figure 2b. Figure 2a,b suggest that not only Mo is uniformly plated on the surface of CFC, but the native structures of Mo and CFC are not disrupted throughout the preparation process. All of the diffraction peaks of MoO_3 are indexed well to orthorhombic α-MoO_3 (JCPDS No.05-0508), with high crystallinity. In contrast, there is a diffraction peak attributed to Mo_4O_{11} (JCPDS No. 05-0337), indicative of the existence of low-valence-state Mo after the oxidation treatment. The EDS result (Figure 2d) also suggests that a high purity α-MoO_3@CFC was prepared.

The XRD patterns of α-MoO_3@CFC at various heating temperatures are presented in Figure S2a. When the temperature increases, the crystallinity of α-MoO_3 increases gradually. At 600°C, the characteristic peak of Mo disappears, which indicates that Mo loaded on CFC is completely oxidized to α-MoO_3 at this temperature. Figure S2b displays the XRD patterns of α-MoO_3@CFC heated for different times at 600 °C (40 min and 60 min). With the increase in heating time, the oxidation of Mo is more complete. At 60 min, the characteristic peak of Mo_4O_{11} disappears. However, with the extension of time, the CFC was destroyed by oxidation, and the mechanical properties of the samples weakened. Therefore, 20 min was determined as being the optimal preparation condition.

SEM images of pure CFC, Mo@CFC, and α-MoO_3@CFC are shown in Figure 3a–c. Because of the enormous amount of Mo powder present on the surface, the diameter of the previously smooth CFC becomes much thicker following multi-arc ion plating. After oxidation, α-MoO_3@CFC with a one-dimensional rod-like shape are obtained, as presented in Figure 3c. Furthermore, the SEM image of commercial MoO_3 is shown in Figure S3, presenting an irregular morphology. The TEM image further confirms the typical rod-like structure, with a uniform width of about 25 nm and a length of up to about 2 μm. As presented in Figure 3e, the HRTEM image displays clear lattice fringes with a d-spacing of 0.3814 nm, in good agreement with the (200) crystalline planes of the orthorhombic MoO_3. Moreover, the rod-like structure demonstrates the monocrystalline structure, as indicated by the SAED pattern (Figure 3f). The energy dispersive spectroscopy (EDS) elemental mapping images show the compositions of Mo, O, and C (Figure 3g,g1–g3). It is worth mentioning that the distribution of C is relatively unapparent, due to the high loading on CFC.

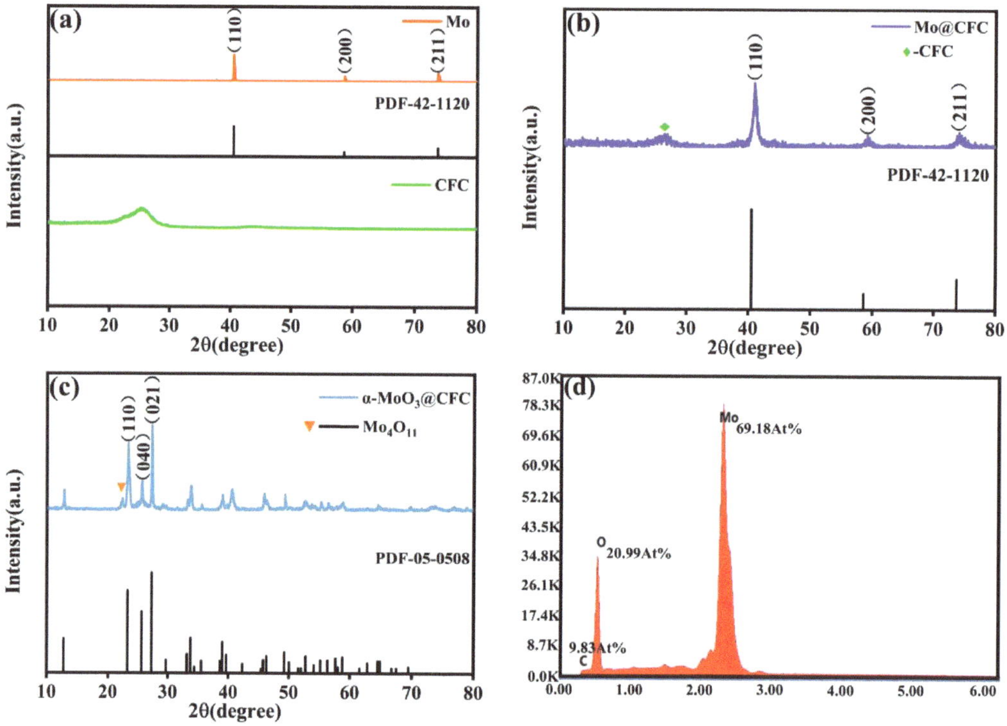

Figure 2. The patterns of X-ray diffraction for (**a**) Mo powder and pure CFC, and (**b**) Mo@CFC; (**c**,**d**) X-ray diffraction and EDS of α-MoO_3@CFC.

The electrochemical performances of ZIBs containing α-MoO_3@CFC and commercial MoO_3 as active cathode materials are shown in Figure 4. Figures 4a and S5 display the classic CV curves for α-MoO_3@CFC and commercial MoO_3 in the potential window of 0.4–1.4 V, at a 1 mV s^{-1} scan rate. Figure 4a represents a group of redox peaks located at around 0.83 V and 0.48 V, indicating typical battery behavior, with the oxidation peak situated around 0.83 V arising from the extraction of Zn^{2+} from α-MoO_3@CFC; and the reduction peak located near 0.48 V can be assigned to the insertion of Zn^{2+} into MoO_3 [25]. A comparison of the cycling performance of α-MoO_3@CFC and commercial MoO_3 powder at a current density of 200 mA g^{-1} is shown in Figure 4b. The initial commercial MoO_3 has a specific capacity of just 125.6 mAh g^{-1} and decays to 55.8 mAh g^{-1} following 100 cycles, with a capacity retention rate amounting to 44.4%. In contrast, the specific capacity of α-MoO_3@CFC stabilizes at around 200 mAh g^{-1} after 100 cycles, with a capacity retention rate of 92.4%. Additionally, the calculated areal capacity of this Zn//α-MoO_3@CFC battery reaches 2.61 mAh cm^{-2} (200.8 mAh g^{-1} based on the average mass of 13 mg cm^{-2}), with an excellent energy density of 200.8 Wh kg^{-1}.

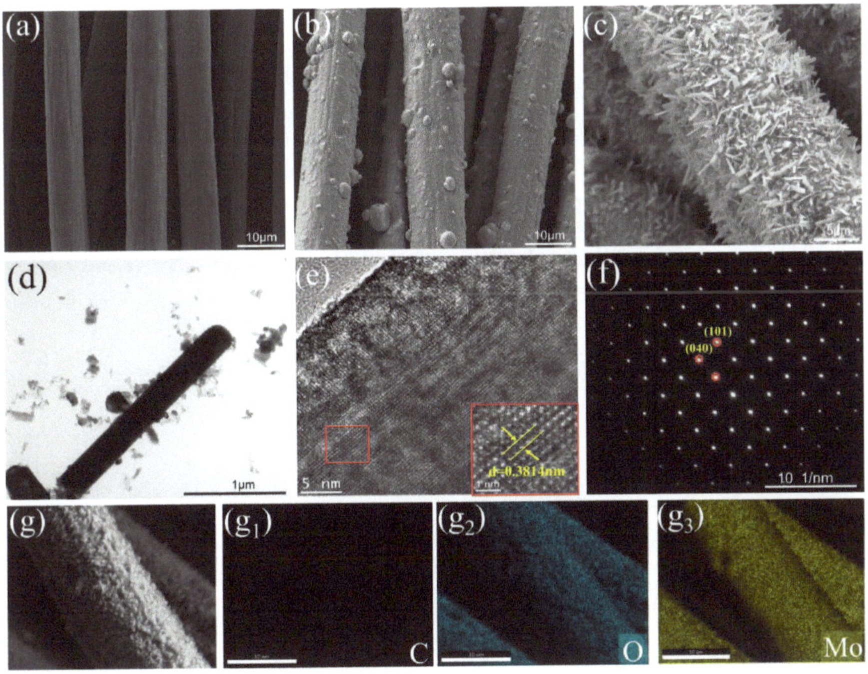

Figure 3. SEM image of (**a**) CFC, (**b**) Mo@CFC, and (**c**) α-MoO_3@CFC heated at 600 °C for 20 min; (**d–g**) TEM image, HRTEM image, SAED image, and EDS mapping of C (**g1**), O (**g2**), and Mo (**g3**) of the α-MoO_3@CFC.

Figure 4c shows the curves of the α-MoO_3@CFC-containing cathode at the 1st, 2nd, 5th, 50th, and 100th cycles, charged and discharged at 200 mA g^{-1}. The initial discharge and charge capacities were equal to 409.3 and 288 mA h g^{-1}, respectively. These values correspond to a Coulombic efficiency of 70.4%. The series of charge/discharge plateaus agree well with the CV curves shown in Figure 4a. More notably, the rate performance of α-MoO_3@CFC is significantly improved in comparison to that of commercial MoO_3, as is evident in Figure 4d, with specific capacities of 483.4, 196.2, 138.8, 114.6, and 95.4 mAh g^{-1} at current densities ranging from 100 mA g^{-1} to 2000 mA g^{-1}, respectively. The specific capacity increased to 308.2 mAh g^{-1} when the current density was returned to 100 mA g^{-1}, and it remained steady in future cycles. The long-term cycling performance of α-MoO_3@CFC at a current density of 1000 mA g^{-1} is shown in Figure 4e, where the specific capacity of α-MoO_3@CFC remained at 109.8 mAh g^{-1} following 500 cycles, whereas the commercial MoO_3 had poor performance, with a rapid decline of the capacity. In addition, the ex situ SEM images of α-MoO_3@CFC and commercial MoO_3 after 20 cycles at a current density of 1000 mA g^{-1} are shown in Figure S4. Obviously, the mechanical stability of α-MoO_3@CFC is better.

In order to better comprehend the excellent electrochemical performance of the α-MoO_3@CFC electrode, electrochemical impedance spectroscopy (EIS) measurements are shown in Figure 5. The Nyquist plots of both α-MoO_3@CFC and commercial MoO_3 consist of semicircles at high frequencies and a straight line at lower frequencies. These regions correspond to the resistance of the electrode surface (R_f) and a charge-transfer resistance (R_{ct}), respectively [26,31,32]. Fitting the EIS data obtained for the α-MoO_3@CFC and commercial MoO_3 electrodes before cycling showed that R_{ct} values of the cathodes based on the α-MoO_3@CFC and commercial MoO_3 materials are equal to 43.73 and 82.09 Ω, respectively. These results confirmed that α-MoO_3@CFC possessed better conductivity, as well as stronger ion and electron migration ability, than commercial MoO_3. From a comparison of this work with some reported studies (high mass loading), in some respects, α-MoO_3@CFC-based electrodes for ZIBs have possible application prospects (see Table 1). The main reason for the improved performance may be the strategy of a binder-free electrode.

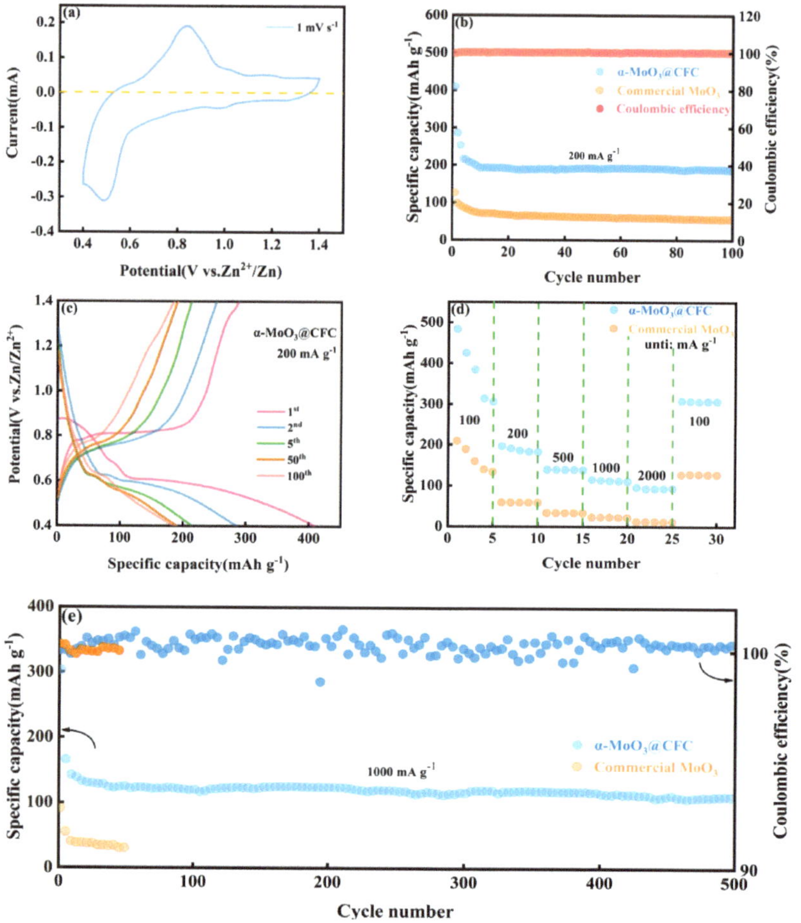

Figure 4. Electrochemical performance of α-MoO_3@CFC and commercial MoO_3. (**a**) Presented are 1 mV s^{-1} CV curves of α-MoO_3@CFC, (**b**) cycle performance of α-MoO_3@CFC and commercial MoO_3 powder at 200 mA g^{-1} density, (**c**) galvanostatic charge–discharge curves of α-MoO_3@CFC, (**d**) rate performance of α-MoO_3@CFC and commercial MoO_3 powders. (**e**) Long-term cycling performance of α-MoO_3@CFC and commercial MoO_3 powders at a current density of 1000 mA g^{-1}.

Table 1. Comparison of electrochemical properties of α-MoO_3@CFC and reported aqueous zinc-ion cathode materials.

Type of Material	Cycle Number, Cycling Capacity (mAh g^{-1}) (Current Density (mA g^{-1}))	Mass Loading (mg cm^{-2})	Ref.
α-MoO_3@CFC	100th, 200.8 (200) 500th, 109.8 (1000)	12–15	This Work
MoO_3 nanowires	400th, 171.14 (400)	11.1	[25]
MoO_3 nanobelt	100th, 254.77 (100)	4.4–4.5	[33]

Table 1. *Cont.*

Type of Material	Cycle Number, Cycling Capacity (mAh g^{-1}) (Current Density (mA g^{-1}))	Mass Loading (mg cm^{-2})	Ref.
$V_3O_7{\cdot}H_2O$ nanoarray	50th, 323 (100) 800th, 155 (2000)	5	[21]
$H_{11}Al_2V_6O_{23.2}$@graphene	400th, 131.7 (2000)	15.7	[24]
3D printed FeVO/rHGO	650th, 126.4 (2000)	12.4	[23]
MnO_2 composite electrode	300th, 184 (200)	9.5	[16]

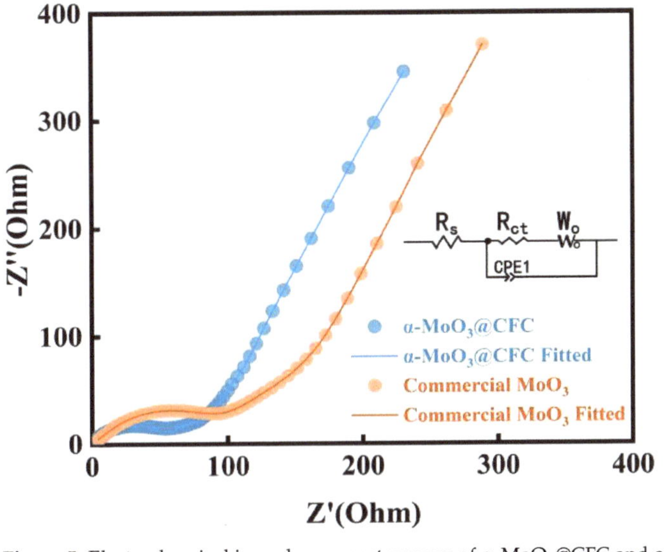

Figure 5. Electrochemical impedance spectroscopy of α-MoO_3@CFC and commercial MoO_3.

3. Conclusions

In summary, ZIBs are fabricated with a high mass loading (12–15 mg cm^{-2}) and binder-free cathode material, α-MoO_3@CFC, prepared using multi-arc ion plating and simple heating. When operated at a current density of 200 mA g^{-1}, the α-MoO_3@CFC electrode has an electrochemical specific capacity of 200.8 mAh g^{-1} after 100 cycles. After 500 cycles at a current density of 1000 mA g^{-1}, it still has a capacity of 109.8 mAh g^{-1}. The exceptional mechanical characteristics and high areal capacity of 2.61 mAh cm^{-2} with an excellent energy density of 200.8 Wh kg^{-1} reveal considerable promise for flexible wearable and grid energy storage applications.

Supplementary Materials: The following supporting information can be downloaded at: https://www.mdpi.com/article/10.3390/ma15175954/s1, Figure S1: Macroscopic morphology of (a) Mo@CFC, (b) α-MoO_3@CFC. Figure S2: XRD patterns of α-MoO_3@CFC at (a) different heating temperatures, (b) different heating times at 600°C. Figure S3: SEM image of commercial MoO_3. Figure S4: The ex-situ SEM images of α-MoO_3@CFC (a) and commercial MoO_3 (b) after 20 cycles at a current density of 1000 mA g^{-1}. Figure S5: 1 mV s^{-1} CV curves of commercial MoO_3.

Author Contributions: S.L.: conceptualization, investigation, supervision, writing—review and editing, funding acquisition. Y.S.: conceptualization, methodology, investigation, data curation, software, visualization, writing—original draft. J.Y.: data curation, formal analysis, software. Y.Z. and Z.C.: supervision, writing—review and editing. All authors have read and agreed to the published version of the manuscript.

Funding: This research was funded by the National Natural Science Foundation of China under Grant No. 52104285.

Institutional Review Board Statement: Not applicable.

Informed Consent Statement: Not applicable.

Data Availability Statement: The data presented in this study are available on reasonable request from the corresponding author.

Conflicts of Interest: The authors declare that they have no conflict of interest.

References

1. Tarascon, J.M.; Armand, M. Issues and challenges facing rechargeable lithium batteries. *Nature* **2001**, *414*, 359–367. [CrossRef] [PubMed]
2. Lu, L.; Han, X.; Li, J.; Hua, J.; Ouyang, M. A review on the key issues for lithium-ion battery management in electric vehicles. *J. Power Sources* **2013**, *226*, 272–288. [CrossRef]
3. Kim, T.; Song, W.; Son, D.-Y.; Ono, L.K.; Qi, Y. Lithium-ion batteries: Outlook on present, future, and hybridized technologies. *J. Mater. Chem. A* **2019**, *7*, 2942–2964. [CrossRef]
4. Shi, Y.; Chen, Y.; Shi, L.; Wang, K.; Wang, B.; Li, L.; Ma, Y.; Li, Y.; Sun, Z.; Ali, W.; et al. An Overview and Future Perspectives of Rechargeable Zinc Batteries. *Small* **2020**, *16*, e2000730. [CrossRef]
5. Chen, H.; Qin, H.; Chen, L.; Wu, J.; Yang, Z. V_2O_5@CNTs as cathode of aqueous zinc ion battery with high rate and high stability. *J. Alloy. Compd.* **2020**, *842*, 155912. [CrossRef]
6. Huang, J.; Wang, Z.; Hou, M.; Dong, X.; Liu, Y.; Wang, Y.; Xia, Y. Polyaniline-intercalated manganese dioxide nanolayers as a high-performance cathode material for an aqueous zinc-ion battery. *Nat. Commun.* **2018**, *9*, 2906. [CrossRef]
7. Wu, J.; Chi, X.W.; Liu, Y.Z.; Yang, J.H.; Liu, Y. Electrochemical characterization of hollow urchin-like MnO_2 as high-performance cathode for aqueous zinc ion batteries. *J. Electroanal. Chem.* **2020**, *871*, 114242. [CrossRef]
8. Xu, J.W.; Gao, Q.L.; Xia, Y.M.; Lin, X.S.; Liu, W.L.; Ren, M.M.; Kong, F.G.; Wang, S.J.; Lin, C. High-performance reversible aqueous zinc-ion battery based on iron-doped alpha-manganese dioxide coated by polypyrrole. *J. Colloid Interface Sci.* **2021**, *598*, 419–429. [CrossRef]
9. Tang, B.; Zhou, J.; Fang, G.; Guo, S.; Guo, X.; Shan, L.; Tang, Y.; Liang, S. Structural Modification of V_2O_5 as High-Performance Aqueous Zinc-Ion Battery Cathode. *J. Electrochem. Soc.* **2019**, *166*, A480–A486. [CrossRef]
10. Wei, T.; Li, Q.; Yang, G.; Wang, C. Highly reversible and long-life cycling aqueous zinc-ion battery based on ultrathin $(NH_4)_2V_{10}O_{25}\cdot 8H_2O$ nanobelts. *J. Mater. Chem. A* **2018**, *6*, 20402–20410. [CrossRef]
11. Qin, M.L.; Liu, W.M.; Shan, L.T.; Fang, G.Z.; Cao, X.X.; Liang, S.Q.; Zhou, J. Construction of V_2O_5/NaV_6O_{15} biphase composites as aqueous zinc-ion battery cathode. *J. Electroanal. Chem.* **2019**, *847*, 113246. [CrossRef]
12. Mao, F.F.; Li, Y.W.; Zou, Z.G.; Huang, B.; Yang, J.W.; Yao, J.H. Zn^{2+} storage performance and structural change of orthorhombic V_2O_5 nanowires as the cathode material for rechargeable aqueous zinc-ion batteries. *Electrochim. Acta* **2021**, *397*, 139255. [CrossRef]
13. Ni, G.; Xu, X.; Hao, Z.; Wang, W.; Li, C.; Yang, Y.; Zhou, C.; Qin, L.; Chen, W.; Yao, X.; et al. Tuning the Electrochemical Stability of Zinc Hexacyanoferrate through Manganese Substitution for Aqueous Zinc-Ion Batteries. *ACS Appl. Energy Mater.* **2020**, *4*, 602–610. [CrossRef]
14. Lu, K.; Song, B.; Zhang, Y.; Ma, H.; Zhang, J. Encapsulation of zinc hexacyanoferrate nanocubes with manganese oxide nanosheets for high-performance rechargeable zinc ion batteries. *J. Mater. Chem. A* **2017**, *5*, 23628–23633. [CrossRef]
15. Li, Z.; Liu, T.; Meng, R.; Gao, L.; Zou, Y.; Peng, P.; Shao, Y.; Liang, X. Insights into the Structure Stability of Prussian Blue for Aqueous Zinc Ion Batteries. *Energy Environ. Mater.* **2020**, *4*, 111–116. [CrossRef]
16. Song, Y.; Duan, S.; Yang, D.; Dong, R.; Guo, D.; Sun, X.; Liu, X.-X. 3D Exfoliated Carbon Paper toward Highly Loaded Aqueous Energy Storage Applications. *Energy Technol.* **2019**, *7*, 1900892. [CrossRef]
17. Jiao, T.; Yang, Q.; Wu, S.; Wang, Z.; Chen, D.; Shen, D.; Liu, B.; Cheng, J.; Li, H.; Ma, L.; et al. Binder-free hierarchical VS_2 electrodes for high-performance aqueous Zn ion batteries towards commercial level mass loading. *J. Mater. Chem. A* **2019**, *7*, 16330–16338. [CrossRef]
18. Li, Y.; Fu, J.; Zhong, C.; Wu, T.; Chen, Z.; Hu, W.; Amine, K.; Lu, J. Recent Advances in Flexible Zinc-Based Rechargeable Batteries. *Adv. Energy Mater.* **2018**, *9*, 1802605. [CrossRef]
19. Yu, P.; Zeng, Y.; Zhang, H.; Yu, M.; Tong, Y.; Lu, X. Flexible Zn-Ion Batteries: Recent Progresses and Challenges. *Small* **2019**, *15*, e1804760. [CrossRef]
20. Yao, M.; Yuan, Z.; Li, S.; He, T.; Wang, R.; Yuan, M.; Niu, Z. Scalable Assembly of Flexible Ultrathin All-in-One Zinc-Ion Batteries with Highly Stretchable, Editable, and Customizable Functions. *Adv. Mater.* **2021**, *33*, e2008140. [CrossRef]
21. Chen, D.; Lu, M.; Wang, B.; Cheng, H.; Yang, H.; Cai, D.; Han, W.; Fan, H.J. High-mass loading $V_3O_7\cdot H_2O$ nanoarray for Zn-ion battery: New synthesis and two-stage ion intercalation chemistry. *Nano Energy* **2021**, *83*, 106157. [CrossRef]
22. Lei, J.; Yao, Y.; Wang, Z.; Lu, Y.-C. Towards high-areal-capacity aqueous zinc–manganese batteries: Promoting MnO_2 dissolution by redox mediators. *Energy Environ. Sci.* **2021**, *14*, 4418–4426. [CrossRef]

23. Ma, H.; Tian, X.; Wang, T.; Tang, K.; Liu, Z.; Hou, S.; Jin, H.; Cao, G. Tailoring Pore Structures of 3D Printed Cellular High-Loading Cathodes for Advanced Rechargeable Zinc-Ion Batteries. *Small* **2021**, *17*, e2100746. [CrossRef] [PubMed]
24. Zhang, W.; Liang, S.; Fang, G.; Yang, Y.; Zhou, J. Ultra-High Mass-Loading Cathode for Aqueous Zinc-Ion Battery Based on Graphene-Wrapped Aluminum Vanadate Nanobelts. *Nanomicro Lett.* **2019**, *11*, 69. [CrossRef]
25. He, X.; Zhang, H.; Zhao, X.; Zhang, P.; Chen, M.; Zheng, Z.; Han, Z.; Zhu, T.; Tong, Y.; Lu, X. Stabilized Molybdenum Trioxide Nanowires as Novel Ultrahigh-Capacity Cathode for Rechargeable Zinc Ion Battery. *Adv. Sci.* **2019**, *6*, 1900151. [CrossRef]
26. Liu, Y.; Wang, J.; Zeng, Y.; Liu, J.; Liu, X.; Lu, X. Interfacial Engineering Coupled Valence Tuning of MoO_3 Cathode for High-Capacity and High-Rate Fiber-Shaped Zinc-Ion Batteries. *Small* **2020**, *16*, e1907458. [CrossRef]
27. Wang, Z.; Tian, C.; Tolstogouzov, A.; Liang, F.; Zou, C.; Li, S.; Gusev, S.I.; Yousaf, M.I.; Pelenovich, V.; Zuo, W.; et al. Microstructure and Rutherford Backscattering Spectrometry of Hard/Lubricant Mo-Ti-Al-N Multilayered Coatings Prepared by Multi-Arc Ion Plating at Low Substrate Rotation. *Coatings* **2020**, *10*, 101. [CrossRef]
28. Wang, H.-D.; Xu, B.-S.; Liu, J.-J.; Zhuang, D.-M. Molybdenum disulfide coating deposited by hybrid treatment and its friction-reduction performance. *Surf. Coat. Technol.* **2007**, *201*, 6719–6722. [CrossRef]
29. Xu, N.; Yan, C.; He, W.; Xu, L.; Jiang, Z.; Zheng, A.; Wu, H.; Chen, M.; Diao, G. Flexible electrode material of V_2O_5 carbon fiber cloth for enhanced zinc ion storage performance in flexible zinc-ion battery. *J. Power Sources* **2022**, *533*, 231358. [CrossRef]
30. Qian, Y.; Meng, C.; He, J.; Dong, X. A lightweight 3D Zn@Cu nanosheets@activated carbon cloth as long-life anode with large capacity for flexible zinc ion batteries. *J. Power Sources* **2020**, *480*, 228871. [CrossRef]
31. Tang, B.; Tian, N.; Jiang, J.; Li, Y.; Yang, J.; Zhu, Q. Investigation of zinc storage capacity of WS_2 nanosheets for rechargeable aqueous Zn-ion batteries. *J. Alloy. Compd.* **2022**, *894*, 162391. [CrossRef]
32. Wu, T.H.; Liang, W.Y. Reduced Intercalation Energy Barrier by Rich Structural Water in Spinel $ZnMn_2O_4$ for High-Rate Zinc-Ion Batteries. *ACS Appl. Mater. Interfaces* **2021**, *13*, 23822–23832. [CrossRef] [PubMed]
33. Wang, L.; Yan, S.; Quilty, C.D.; Kuang, J.; Dunkin, M.R.; Ehrlich, S.N.; Ma, L.; Takeuchi, K.J.; Takeuchi, E.S.; Marschilok, A.C. Achieving Stable Molybdenum Oxide Cathodes for Aqueous Zinc-Ion Batteries in Water-in-Salt Electrolyte. *Adv. Mater. Interfaces* **2021**, *8*, 2002080. [CrossRef]

nanomaterials

MDPI

Article

Synthesis and Electrochemical Performance of the Orthorhombic $V_2O_5{\cdot}nH_2O$ Nanorods as Cathodes for Aqueous Zinc Batteries

Xiaoping Tan [1,2,3], Gaoli Guo [1,2], Kaidi Wang [1,2] and Huang Zhang [1,2,3,*]

1 Research & Development Institute of Northwestern Polytechnical University in Shenzhen, Shenzhen 518057, China; iamxptan@mail.nwpu.edu.cn (X.T.); iamglguo@mail.nwpu.edu.cn (G.G.); iamkdwang@mail.nwpu.edu.cn (K.W.)
2 Institute of Flexible Electronics, Northwestern Polytechnical University, Xi'an 710072, China
3 Ningbo Research Institute of Northwestern Polytechnical University, Ningbo 315103, China
* Correspondence: iamhzhang@nwpu.edu.cn

Abstract: Aqueous zinc-ion batteries offer the greatest promise as an alternative technology for low-cost and high-safety energy storage. However, the development of high-performance cathode materials and their compatibility with aqueous electrolytes are major obstacles to their practical applications. Herein, we report the synthesis of orthorhombic $V_2O_5{\cdot}nH_2O$ nanorods as cathodes for aqueous zinc batteries. As a result, the electrode delivers a reversible capacity as high as 320 mAh g^{-1} at 1.0 A g^{-1} and long-term cycling stability in a wide window of 0.2 to 1.8 V using a mild $ZnSO_4$ aqueous electrolyte. The superior performance can be attributed to the improved stability of materials, inhibited electrolyte decomposition and facilitated charge transfer kinetics of such materials for aqueous zinc storage. Furthermore, a full cell using microsized Zn powder as an anode within capacity-balancing design exhibits high capacity and stable cycling performance, proving the feasibility of these materials for practical application.

Keywords: zinc batteries; $V_2O_5{\cdot}nH_2O$ nanorods; cathode materials; Zn anode; aqueous batteries

Citation: Tan, X.; Guo, G.; Wang, K.; Zhang, H. Synthesis and Electrochemical Performance of the Orthorhombic $V_2O_5{\cdot}nH_2O$ Nanorods as Cathodes for Aqueous Zinc Batteries. *Nanomaterials* **2022**, *12*, 2530. https://doi.org/10.3390/nano12152530

Academic Editor: Carlos Miguel Costa

Received: 30 June 2022
Accepted: 21 July 2022
Published: 23 July 2022

1. Introduction

With the fast development of renewable energy sources, there is an urgent need to develop low-cost, safe and sustainable energy storage devices to bridge the gap between intermittent energy generation and the distribution of power consumption [1,2]. Currently, lithium-ion batteries (LIBs) are dominating the market for powering portable electronics and electric vehicles, which is taking advantage of their high energy/power densities and efficiency [3]. However, volatile and flammable organic solvents in the electrolytes bring safety concerns [4]. Aqueous rechargeable batteries have emerged as a promising alternative technology, which benefit from the use of water-based electrolytes and similar working mechanisms making them especially competitive for large-scale energy storage [5,6].

Aqueous zinc-ion batteries (AZIBs) are one of the most promising systems due to their use of zinc metal as an anode, which has proper redox potential (−0.76 V vs. standard hydrogen electrode) for aqueous electrolytes, and high theoretical capacity (819 mAh g^{-1}; 5851 mAh mL^{-1}) [7–11]. However, the cathode materials capable of efficient Zn storage present the most critical challenge to achieving high-performance AZIBs. So far, only few cathode candidates have been reported for reversible Zn insertion/extraction, including manganese oxides, Refs. [12–14] Prussian blue analogues, Refs. [15,16] vanadium oxides, Refs. [17–19] and organic compounds [20–22].

Orthorhombic vanadium pentoxide (V_2O_5) holds great potential as a cathode material for AZIBs, owing to its high theoretical Zn-storage capacity of 589 mAh g^{-1} (based on the two-electron redox reaction of V as redox center) [23–25]. Its typically layered structure enables the efficient (de-)intercalation of metal ions, providing high capacity and rate capability [26–28]. However, its long-term cyclability is still unsatisfactory, due to the weak interactions of intercalated ions of Zn^{2+} with interlayers. Moreover, the dissolution of V-based compounds appears to be another issue in aqueous batteries, which affects cycling stability, especially in electrolytes with a low-pH value. To improve the redox kinetics and prolong the cycle life, various strategies have been reported. On one hand, by chemically pre-intercalated metal ions or foreign molecules, the layered vanadium oxides showed well-improved reversibility and cycling performance. In this respect, both alkali/alkaline-earth metal ions (Li^+, [29] Na^+, [30] Mg^{2+}, [31] Ca^{2+}, [32] etc.) and polar molecules (water, [33] pyridine, [34] etc.) have been adopted to pre-intercalate into the spaces between two bilayers, thus resulting in improved reaction kinetics and cycling stability of vanadium oxides. On the other hand, novel nanostructure design could be another approach to enabling efficient Zn-insertion in V_2O_5 for high-performance rechargeable aqueous Zn batteries [35–39].

Herein, we report the synthesis and electrochemical properties of orthorhombic $V_2O_5{\cdot}nH_2O$ nanorods as cathode materials for AZIBs. In a mild 3 m ("m": mol kg^{-1}) $ZnSO_4$ aqueous electrolyte, the electrochemical performance of a $Zn/V_2O_5{\cdot}nH_2O$ cell was investigated in a wide voltage window from 0.2 to 1.8 V. Furthermore, full-cell configuration using microsized Zn powders as the anode is also reported, providing a realistic design of the Zn/V_2O_5 system for practical energy storage applications.

2. Materials and Methods

2.1. Material Synthesis

The $V_2O_5{\cdot}nH_2O$ nanorods were synthesized using a typical hydrothermal method. Specifically, 0.6 g commercial V_2O_5 (98%, Alfa, China) was dispersed in 60 mL deionized water. After stirring for 30 min, 8 mL H_2O_2 (31wt%, Adamas, China) was added dropwise and vigorous stirring continued at 40 °C for 1 h. Then, the yellowish solution was transferred into a 100 mL Teflon-lined autoclave and kept at 200 °C for 48 h. After natural cooling, the powders were collected by centrifugation after being washed with DI water and ethanol. The final product was annealed at 250 °C for 2 h with a heating rate of 2 °C min^{-1} in air. Commercial V_2O_5 powders without any treatment and V_2O_5 nanorods annealed at 500 °C were used as a control.

2.2. Material Characterizations

The crystal phase of the product was characterized by X-ray diffraction (XRD) on a Bruker D8 Advance diffractometer (Cu Kα, Bruker, Karlsruhe, Germany). The morphology and microstructure were characterized by scanning electron microscopy (SEM, Gemini 300, ZEISS, Oberkochen, Germany) and transmission electron microscopy (TEM, FEI Tecnai G2 F30, 300 kV, Hillsboro, OR, USA). X-ray photoelectron spectrometer (XPS, ULVAC PHI Quantera, Chigasaki, Japan) was carried out to detect the surface chemical states of both pristine powders and cycled electrodes. The thermogravimetric analysis (TGA) was performed by a thermogravimetric analyzer (Netzsch, TGA-209F, Selb, Germany) at a ramping of 10 °C min^{-1}.

2.3. Electrochemical Measurements

The electrochemical performance was investigated in CR2032-type coin cells. The cathode and Zn powder (99.99%, 600 mesh, Aladdin, Shanghai, China) electrodes were fabricated by doctor-blade casting slurries composed of 80 wt% active materials, 10 wt% carbon black (Super C65, Imerys) and 10 wt% polyvinylidene fluoride as binder (PVDF, Solef 6020, Solvay, Brussels, Belgium) in *N*-methyl-2-pyrrolidone (NMP, anhydrous, Innochem, Beijing, China) on stainless steel foil (type-316, thickness: 0.02 mm). After drying at 80 °C in air, electrodes of a diameter of 12 mm were punched and further dried at 12 °C under vacuum. The areal mass loading of V_2O_5 was about 1.5 mg cm^{-2}. The areal mass loading of Zn powders was about 0.8–1.0 mg cm^{-2}. The electrolyte was prepared by dissolving

3 mol $ZnSO_4 \cdot 7H_2O$ (99.5%, ACROS, Geel, Belgium) in 1 kg Millipore water (including the crystal water in the salt). Metallic zinc foil discs (0.03 mm thickness, Ø 16 mm, 99.95%, Goodfellow, Huntingdon, UK) and glass fiber discs (GF/D, Whatman, Ø 18 mm, Maidstone, UK) were punched as anodes and separators. Galvanostatic cycling was performed on an Automated Battery Test System (CT-4000, NEWARE, Shenzhen, China) in a voltage range from 0.2 to 1.8 V. The electrochemical workstation (CHI 660E, Shanghai, China) was used for the cyclic voltammetry (CV) and electrochemical impedance spectroscopy (EIS) tests. The EIS spectra were recorded in a frequency range from 100 kHz to 0.01 Hz with an AC amplitude of 5 mV at OCV states. The GITT test was performed to study the kinetics of cation diffusion in the V_2O_5 nanorod electrode using the following equation: $D_{Zn^{2+}} = (4/\pi\tau)*[n_M V_M/S]^2[\Delta E_S/\Delta E_t]^2$, in which the τ represents the relaxation time (s), n_M and V_M are the moles (mol) and molar volume (cm^{-3} mol^{-1}) of V_2O_5, respectively, S is the electrode−electrolyte interface area (cm^2), ΔE_S and ΔE_t are the steady-state voltage change and overall cell voltage after the application of a current pulse in a single step, respectively [35]. All the measurements were performed at room temperature (20 °C).

2.4. Differential Electrochemical Mass Spectrometry Measurement

The differential electrochemical mass spectrometry (DEMS) measurement was performed in an ECC-Air cell (EL-CELL, Hamburg, Germany). The cell was fabricated with a gas inlet and outlet and was connected to a bench-top gas mass spectrometer (Heiden HPR-20 QIC, Warrington, UK). The electrochemical measurement was controlled by a LAND CT2001A system (LAND, Wuhan, China). Upon the cell cycling, a constant stream of Ar (flow rate: 0.5 mL min^{-1}) was maintained to take the gas products from the DEMS cell to the mass spectrometer.

3. Results

The $V_2O_5 \cdot nH_2O$ nanorods were prepared using a simple hydrothermal method followed by a moderate temperature annealing treatment. The illustration of the orthorhombic $V_2O_5 \cdot nH_2O$ nanorods is presented in Figure 1a, which visually demonstrates the designed nanorod structure. The crystal structure of orthorhombic $V_2O_5 \cdot nH_2O$ is built out of edge/corner-sharing VO5 tetragonal pyramids that form the layered structure [23]. The interlayer spacing can be intercalated by cations and molecules. The crystal phase of the nanorods was identified by XRD. As shown in Figure 1b, the powder XRD pattern of the as-synthesized $V_2O_5 \cdot nH_2O$ nanorods sample clearly exhibits characteristic peaks of the standard orthorhombic V_2O_5 phase (JCPDS no. 41-1426) with Pmmn space group, in which the peaks at 2θ 15.4°, 20.3°, 21.7°, 25.6°, 26.1°, 31.0°, 32.4° and 33.3°, correspond to the (200), (001), (101), (201), (110), (400), (301) and (011) planes, respectively [40]. The sharp and high-intensity diffraction peaks indicate the good crystallinity of the obtained materials.

To detect the surface chemical states on the nanorod materials, XPS analysis was performed. The high-resolution V 2p XPS spectra are shown in Figure 1c. The binding energies of the XPS spectra were calibrated using C 1s at 284.6 eV as a reference. The mainly deconvoluted V $2p_{3/2}$ and V $2p_{1/2}$ peaks at 517.5 and 525.0 eV can be attributed to the existence of V^{5+} in the orthorhombic $V_2O_5 \cdot nH_2O$ [40,41]. Small peaks were also detected at lower binding energies, indicating the existence of tetravalent vanadium (V^{4+}) species [42,43]. This implies that the vanadium in the $V_2O_5 \cdot nH_2O$ nanorod surface layer was not completely oxidized when annealed at 250 °C, in accordance with the literature [37]. The O 1s spectrum (Figure S1, Supplementary Materials) exhibits distinct peaks at around 532.6 and 531.2 eV, which can be attributed to the OH^- and pre-intercalated water molecules, while the main peak at 530.2 eV can be related to the vanadium oxide [44]. The existence of such a V-hydroxide-based interfacial structure is believed to enable the enhancement of interfacial behavior with facilitated electrochemical kinetics [45,46]. Thermogravimetric analysis was used to investigate the stability of the materials. As shown in Figure S2, the result shows a weight loss from room temperature to 370 °C under N_2 flowing, which is related to the removal of pre-intercalated water and the decomposition of hydroxides. However, the sample exhibited a lower weight loss in the same temperature range under O_2 flowing, and a slight weight increase was observed after 400 °C originating from the oxidization of vanadium to a higher valance state [37,47]. Figure 1d,e and Figure S3a display the SEM images of the $V_2O_5{\cdot}nH_2O$ nanorods. It can be seen that the synthesized $V_2O_5{\cdot}nH_2O$ materials are composed of nanorods with a diameter of 30–300 nm and with various lengths. The commercial V_2O_5 powders are flake-shaped with an average size of ~1 um (Figure S4). The microstructure of the $V_2O_5{\cdot}nH_2O$ nanorods is further confirmed by transmission electron microscopy (TEM) images (Figure S3b and Figure 1f). The results clearly demonstrate the typically one-dimensional structure of the materials. The fast Fourier transform (FFT) profile (Figure 1f inset) and the high-resolution TEM image (Figure 1g) clearly exhibit lattice fringes of 0.34 nm, which refer to the (110) planes of orthorhombic $V_2O_5{\cdot}nH_2O$. The EDX results indicate the presence of V and O elements in the synthesized $V_2O_5{\cdot}nH_2O$ nanorods with an atomic ratio of 0.41, which is close to the theoretical value of V_2O_5 of 0.40 (Figure S3c,d).

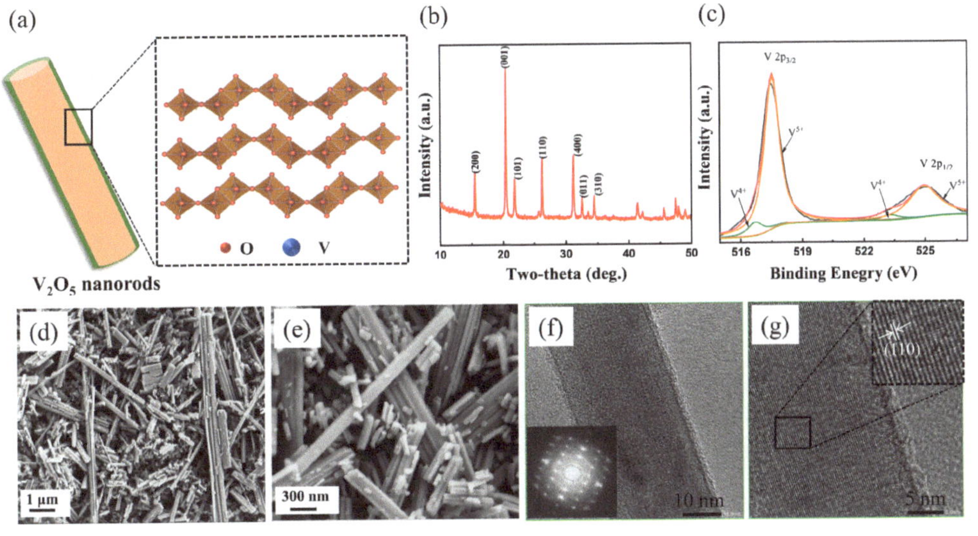

Figure 1. (**a**) Schematic illustration of the orthorhombic V_2O_5 nanorod structure. XRD pattern (**b**), V 2p XPS spectrum (**c**), SEM images (**d**,**e**) and TEM images (**f**,**g**) of the synthesized V_2O_5 materials.

To investigate the Zn-storage properties of the $V_2O_5{\cdot}nH_2O$ nanorods, coin cells were fabricated using zinc foil as the anode in 3 m $ZnSO_4$ aqueous electrolytes. Figure 2a shows the cyclic voltammetry (CV) plots at a scan rate of 3 mV s^{-1} in a voltage window of 0.2–1.8 V. Upon the initial cathodic scan, there was a broad reduction peak at ~0.9 V, and a weak peak at ~0.6 V, indicating the stepwise insertion of Zn^{2+} into the $V_2O_5{\cdot}nH_2O$ nanorods [23]. During the anodic scan, two oxidation peaks at 1.16 and 1.28 V were observed due to the extraction of Zn from the host. In the deep cycles, all the peaks did not shift, however, the currents increased slightly, which can be attributed to the structural activation of the materials due to a gradual insertion of Zn^{2+}. Figure S5 shows the CV curves of the $V_2O_5{\cdot}nH_2O$ nanorod electrode at various scan rates from 0.5 to 5 mV s^{-1}. As the scan rate increases, the curves retain a similar shape, indicating the fast redox reaction kinetics of these nanorod materials for Zn batteries. It should be mentioned that the working voltage window applied for CV measurement was larger than that of most reported vanadium-based cathodes in either Zn $(CF_3SO_3)_2$ or $ZnSO_4$ electrolytes [19,23,44], there was no obvious oxygen evolution reaction detected upon the anodic scan at 1.8 V, which shows the unique electrochemical stability of such materials in aqueous $ZnSO_4$ electrolyte.

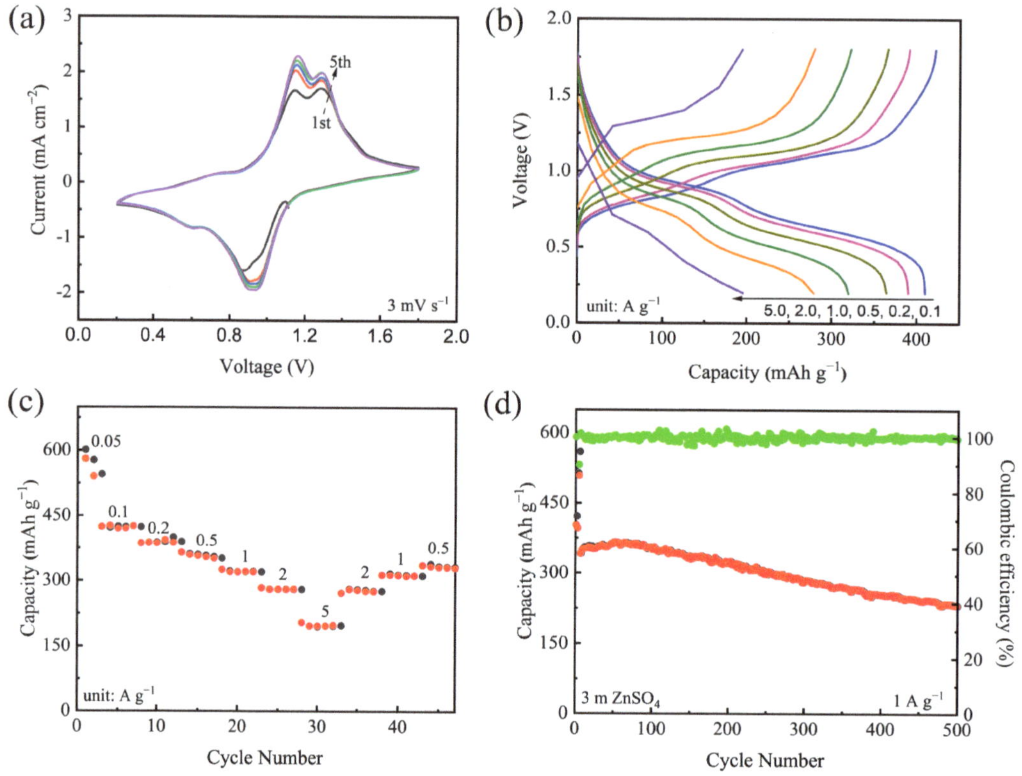

Figure 2. (**a**) CV curves of the $V_2O_5{\cdot}nH_2O$ nanorod electrode in a 3 m $ZnSO_4$ aqueous electrolyte using Zn foil as a counter electrode at a scan rate of 3 mV s^{-1}. (**b**) Galvanostatic discharge–charge (GCD) voltage profiles of the $V_2O_5{\cdot}nH_2O$ nanorod electrode at current densities from 0.1 to 5.0 A g^{-1}. (**c**) Specific capacities of the $V_2O_5{\cdot}nH_2O$ nanorod electrode at various current densities for 5 cycles. (**d**) Long-term cycling performance and corresponding Coulombic efficiencies of the $V_2O_5{\cdot}nH_2O$ nanorod electrode at 1.0 A g^{-1}.

The galvanostatic charge–discharge (GCD) test was performed under different current densities from 0.1 to 5.0 A g^{-1} in a voltage range of 0.2–1.8 V (Figure 2b). The voltage profiles of the $V_2O_5{\cdot}nH_2O$ nanorods at a low-current density of 0.1 A g^{-1} exhibit distinct plateaus at around 0.9 V and 0.6 V, in agreement with the CV results. Even when increasing the current density to 2.0 A g^{-1} the plateaus still remained, indicating a good kinetic reaction of the electrode materials. The specific capacities of the electrode at various current densities from 0.05 A g^{-1} to 5.0 A g^{-1} for five cycles are presented in Figure 2c. The electrode can deliver reversible capacities of 540.4, 410.2, 390.2, 364.5, 320.1, 297.3 and 197.7 mAh g^{-1} at 0.05, 0.1, 0.2, 0.5, 1.0, 2.0 and 5.0 A g^{-1}, respectively. When the current returns to lower values, the capacities can almost be recovered. It should be noted that capacity fading and low Coulombic efficiency were commonly observed at low current density, which can be attributed to the inevitable side reaction of the electrode materials with the aqueous electrolyte [48]. A long-term cycling test was also performed at 1.0 A g^{-1} with three initial cycles at low current density as an activation step. As seen in Figure 2d, the capacity can maintain at 230.3 mAh g^{-1} after 500 cycles with an average Coulombic efficiency of ~99.7%. The selected GCD profiles can be found in Figure S6. For comparison, the performance of commercial V_2O_5 (Figure S7) was also investigated, and the electrode only delivered a capacity of 130 mAh g^{-1} at 1.0 A g^{-1}, which is much lower than that of the $V_2O_5{\cdot}nH_2O$ nanorod. Moreover, the V_2O_5 nanorod sample annealed at 500 °C, which removed the intercalated water, showed poor cycling stability with low specific capacity, as displayed in Figure S8. Note that the capacities of commercial V_2O_5 and V_2O_5 nanorods annealed at a higher temperature increased in the initial cycles, which can be attributed to the transformation of V_2O_5 into hydrated V_2O_5 [49]. However, the nanorod structure annealed at a moderate temperature (250 °C) enabled superior high-rate capability of orthorhombic V_2O_5 and long-term cycling stability in aqueous Zn batteries.

On the basis of the CV data at various scan rates, the electrochemical kinetic process can be determined by the following power–law relationship:

$$i = \mathrm{a}v^{\mathrm{b}}$$

where i and v denote the response current and scan rate, respectively, and a and ^{b} are both adjustable parameters. Figure 3a shows the b values of peaks 1–4 to be 0.53, 0.66, 0.72, and 0.62. A b value of 0.5 indicates that the current is diffusion-controlled. The achieved results imply that the corresponding redox reactions of the V_2O_5 nanorod electrode at the peak regions are mainly limited by ion diffusion within the scan rates ranging from 0.5 to 5 mV s^{-1}. To separate the capacitive contribution, the equation $i(v) = \mathrm{k}_1 v + \mathrm{k}_2 v^{1/2}$ was used to distinguish the ion intercalation part ($\mathrm{k}_1 v$) and the capacitance part ($\mathrm{k}_2 v^{1/2}$), as shown in Figure 3b. At 5 mV s^{-1}, 41.7% of the total current (capacity) is capacitive, indicating that the diffusion contribution holds the dominant position in the total capacity, and the corresponding reaction is mainly limited by the Zn^{2+} ion diffusion rate.

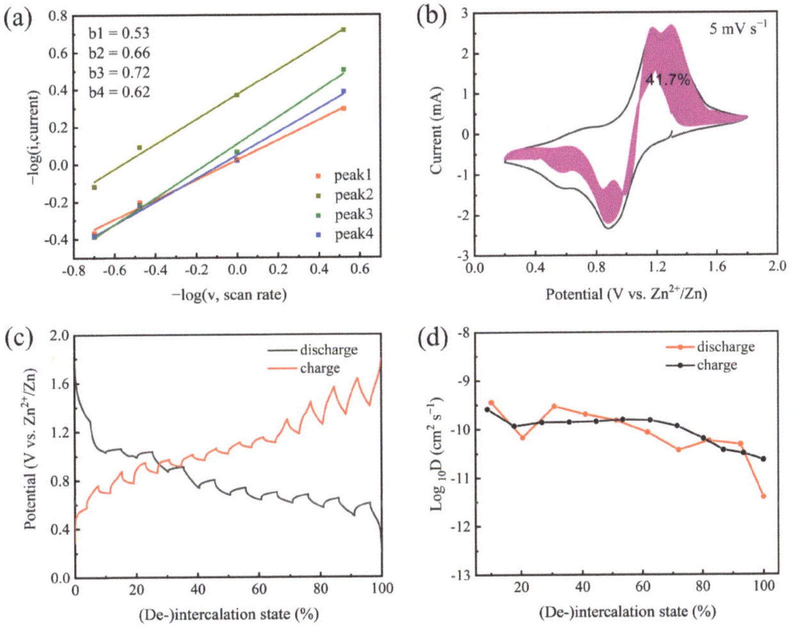

Figure 3. (**a**) The plots of–log (i) versus–log (v) of cathodic and anodic peaks. (**b**) Capacitive contribution (pink part) and diffusion-controlled contribution (void part) at 5 mV s^{-1}. (**c**) Discharge–charge curves of $V_2O_5{\cdot}nH_2O$ nanorods in GITT measurement. (**d**) The diffusion coefficients of Zn^{2+} upon the 2nd discharge and charge progresses of the $V_2O_5{\cdot}nH_2O$ nanorods.

In order to obtain the Zn^{2+} ion diffusion coefficient of V_2O_5 nanorod electrodes, GITT tests were carried out during the second cycle (Figure 3c). The corresponding diffusion coefficient of Zn^{2+} (D_{Zn}) during the second cycle was calculated to estimate the kinetics of $V_2O_5{\cdot}nH_2O$ nanorods, which was in the range of ca. 10^{-9}–10^{-11} cm^2 s^{-1} (Figure 3d) and was higher than the that of bulk V_2O_5 [39]. The Zn^{2+} ion diffusivity on discharging at the last stage was lower than the average value due to the Zn^{2+} ion enriched in the cathode. Overall, the $V_2O_5{\cdot}nH_2O$ nanorods showed a high diffusion coefficient, demonstrating the rapid diffusion kinetics of Zn^{2+} in the materials.

To understand the structural evolution and surface chemical change of the electrode material, XRD and XPS measurements were applied to the electrodes at different charge and discharge states. Figure 4a shows the typical charge–discharge curve of the $V_2O_5{\cdot}nH_2O$ nanorod electrode, indicating the selected states for the measurements. The XRD profiles at open-circuit voltage (OCV), different depth of discharge and state of charge states are presented in Figure 4b. The diffraction peak of (001) plane at 26.3° shifts to higher angles during the discharge process from OCV to 0.8 V and a new peak at 13.6° appears, indicating that the Zn intercalation leads to the formation of a new phase with larger interlayer spacing [36,50]. Upon deep discharge to 0.2 V, the new peaks at 24.1° and 28.2° were clearly observed, which can be attributed to the ZnV_3O_8 (JCPDS no. 24-1481) and $Zn_2V_2O_7$ (JCPDS no. 52-1893) phases. This phenomenon with new phases resulting from Zn insertion in vanadium-based oxides cathodes is commonly reported in the literature [18,33,48]. As reported, the Zn-based precipitation, such as the $ZnSO_4[Zn(OH)_2]_3{\cdot}xH_2O$ (ZHS) phase, is widely detected when using $ZnSO_4$ aqueous electrolyte, however, the signal of the ZHS phase was not observed in our case possibly due to the fact that the concentration of OH^- is not high enough in the electrolyte to generate large amounts of ZHS [51,52]. When the electrode was charged to 1.1 V, the diffraction peak of (001) plane moves back to its original position, indicating the materials' stability after Zn extraction. At the fully charged state, the main peaks of the orthorhombic V_2O_5 phase can be fully recovered. Thus, from the XRD results, we observed highly reversible Zn insertion with induced zinc vanadium oxides in the $V_2O_5{\cdot}nH_2O$ nanorod materials alongside new phase changes.

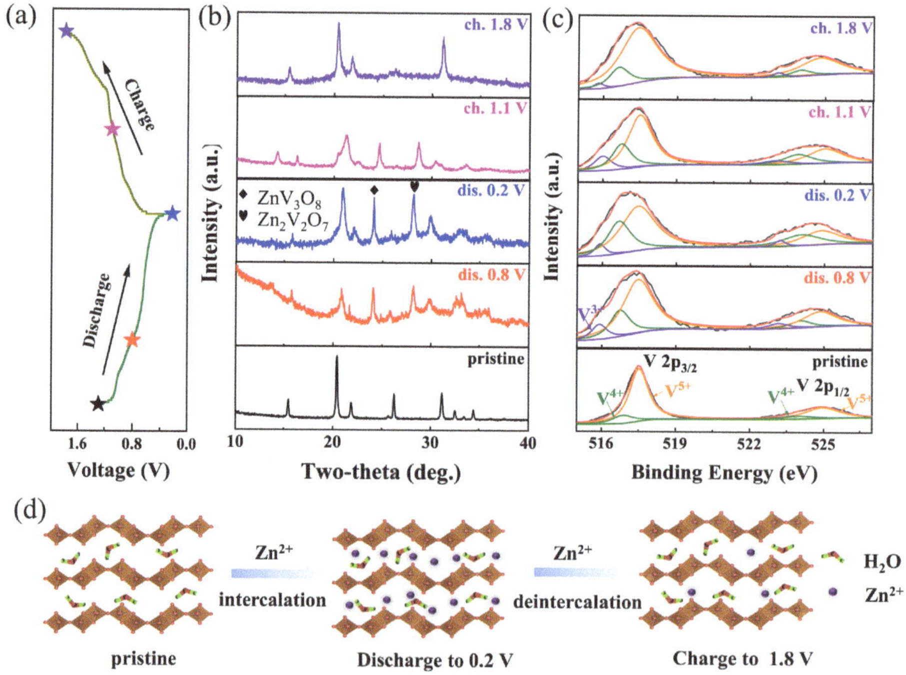

Figure 4. (**a**) Typical galvanostatic discharge–charge (GCD) profiles of the V_2O_5 nanorods in aqueous Zn batteries. (**b**) XRD patterns and (**c**) V 2p XPS spectra of the V_2O_5 nanorod materials at different discharge and charge states. (**d**) The schematic of ion storage mechanism upon discharging and charging.

Focusing on the surface valence states of vanadium during the Zn^{2+} intercalation process, the V 2p XPS spectra of $V_2O_5{\cdot}nH_2O$ nanorods at different electrochemical stages are shown in Figure 4c and Figure S9. Similar to the pristine material, the fresh electrode exhibits two V $2p_{3/2}$ peaks at 517.5 eV and 516.8 eV, which can be assigned to the V^{5+} and V^{4+} species. The analysis of the discharged cathode revealed a third contribution at 515.8 eV that corresponds to V^{3+} species, along with an increase in V^{4+} and a decrease in V^{5+} as a result of Zn^{2+} intercalation and the consequent reduction of V_2O_5 framework. During the charging process, the V is oxidized from V^{3+} and V^{4+} to V^{5+} due to the simultaneous Zn extraction. It is worth mentioning that there is still trace amount of V^{3+} remaining in the fully charged electrode, which can be attributed to the incomplete extraction of Zn^{2+}. On the basis of the above electrochemical results, we propose a Zn-storage mechanism in Figure 4d, showing the typical (de-)intercalation behaviors upon the discharge–charge processes of the $V_2O_5{\cdot}nH_2O$ materials.

SEM was conducted on the electrode after 200 cycles to investigate the morphology changes after the long-term Zn (de-)intercalation process. The SEM images (Figure S10 and 5a) demonstrate that there is no obvious morphology change and the nanorod structure almost remains. Such a good material stability ensures the cycling stability of the $V_2O_5{\cdot}nH_2O$ nanorod material for aqueous Zn-storage. To evaluate the interfacial behaviors, Nyquist plots, consisting of a semicircle associated to the charge transfer at electrode–electrolyte interface and a slop line reflecting the ion diffusion in the bulk electrode, were collected on the commercial V_2O_5 powder and $V_2O_5{\cdot}nH_2O$ nanorod electrodes before and after 200 cycles (Figure 5b) [35]. The commercial V_2O_5 exhibits a significant increase in the charge-transfer resistance after cycling, while the $V_2O_5{\cdot}nH_2O$ nanorod only possesses an increase from ~40 Ω to ~140 Ω, suggesting that the designed nanostructure with a hydroxide-based interfacial structure can efficiently stabilize the electrode materials and facilitate charge transfer at the interface, consequently prolonging the cycling stability and reversibility upon Zn insertion.

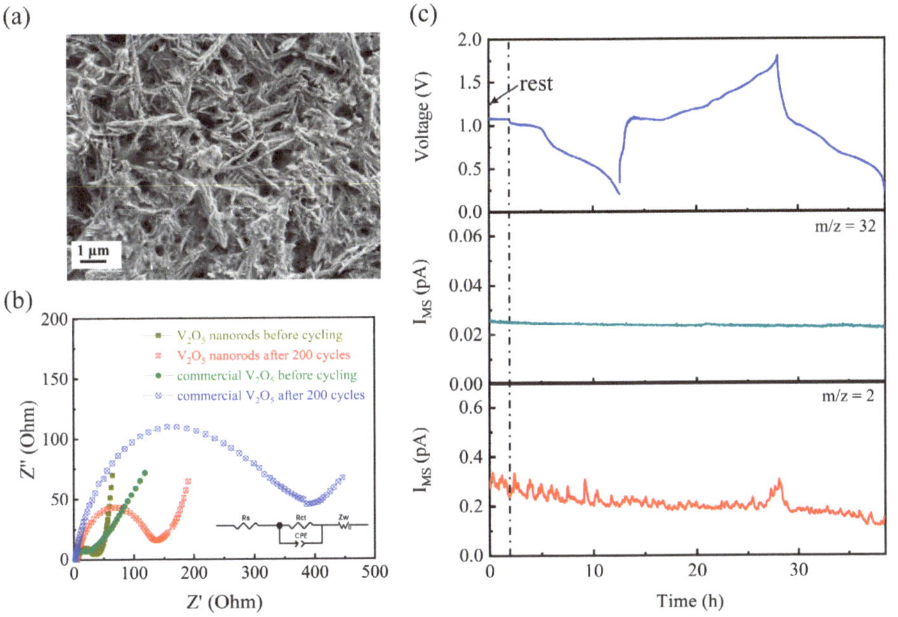

Figure 5. (**a**) SEM image of the cycled electrode after 200 cycles. (**b**) Nyquist plots of the commercial V_2O_5 powder and $V_2O_5{\cdot}nH_2O$ nanorod electrodes before and after cycling. (**c**) Online DEMS data for Zn/ $V_2O_5{\cdot}nH_2O$ cell in 3 m $ZnSO_4$ aqueous electrolyte in the voltage window of 0.2–1.8 V.

Although cycling in a wide window can result in higher energy, the electrolyte decomposition generated from the water electrolysis gets more stress. In situ analysis of electrolyte compatibility with electrodes (both Zn and V_2O_5) was conducted by DEMS, to detect the gas evolution of the cell upon cycling. The mass spectrometric current traces for O_2 (m/z = 32) and H_2 (m/z = 2) are presented in Figure 5c. There was no O_2 evolution detected upon the initial one and a half cycles. The electrochemically induced H_2 peak only occurred when charged to 1.8 V, while the current intensity was rather low (only ~0.35 pA) [53].

Up to now, most of the research on aqueous Zn batteries has directly used Zn foil as the anode, which is not feasible for real industrial application due to its unmatching to current electrode manufacturing technologies [54]. Therefore, a Zn/$V_2O_5 \cdot nH_2O$ battery using microsized Zn powders as the anode was also fabricated. Such a technology promises capacity balance design in a full battery. Figure 6 shows the electrochemical performance of the Zn (powder)/$V_2O_5 \cdot nH_2O$ cell in the voltage range from 0.2 to 1.8 V. This cell is cathode-limited in a cathode to anode capacity ratio of 1:1.2 (according to the initial capacity of V_2O_5 and theoretical capacity of Zn). As shown in Figure 6a, the cell can deliver a reversible capacity of ~500 mAh g^{-1} at 0.05 A g^{-1}, and maintain at 330 mAh g^{-1} at 1.0 A g^{-1}, demonstrating similar capacities to the cells using Zn foil as the anode. Based on the active materials in both electrodes, the energy density of such cells reaches 255 Wh Kg^{-1}. The rate performance is presented in Figure 6b. As seen, the cell can deliver capacities of 415.6, 379.7, 350.0, 316.3, 280.3 and 184.9 mAh g^{-1} at 0.1, 0.2, 0.5, 1.0, 2.0 and 5.0 A g^{-1}, respectively, and the capacity can recover to 270.0 mAh g^{-1} when the current decreases to 2.0 A g^{-1}. It should be noted that the capacity decay was faster than that of a half-cell and the capacities at higher current densities were slightly lower than those of cells using zinc foils as anodes, which may result from vanadium dissolution and the severe side reaction of zinc powders in aqueous electrolytes and can be further improved by structural modification and electrolyte optimization [54–57]. The cell was also cycled at 1.0 A g^{-1} for 100 cycles (Figure 6c). The capacity can retain at 157 mAh g^{-1} after cycling with high CEs of ~98.9%, The electrochemical performance of reported orthorhombic V_2O_5 materials in aqueous Zn batteries are summarized in Table S1, apparently proving the superior cycle life, high efficiency and high-capacity utilization not only in cells using metallic Zn foil as the anode but also in realistic configuration.

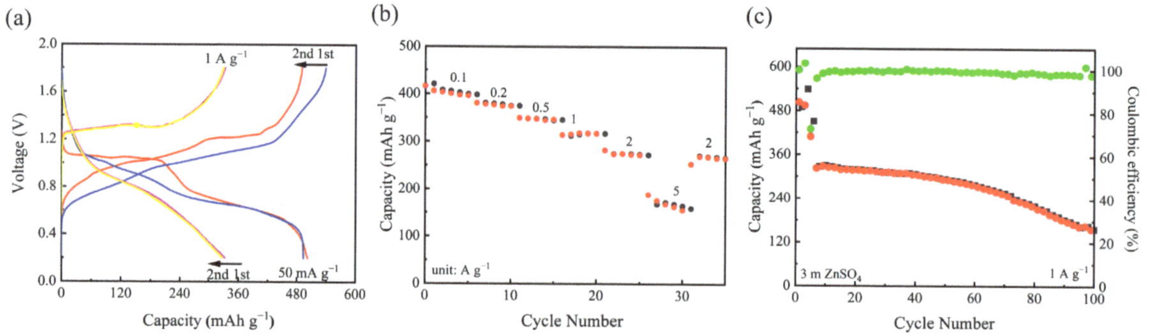

Figure 6. (**a**) GCD profiles of the Zn/$V_2O_5 \cdot nH_2O$ cell using Zn powders as the anode at 50 mA g^{-1} and 1 A g^{-1} for the 1st and 2nd cycles. (**b**) Rate performance of the Zn/$V_2O_5 \cdot nH_2O$ cell at current densities from 0.1 to 5.0 A g^{-1}. (**c**) Cycling performance of the cell at 1.0 A g^{-1}. The cell was initially cycled at 0.05 A g^{-1} for three cycles as an activation step.

4. Conclusions

In summary, high-performance orthorhombic $V_2O_5 \cdot nH_2O$ nanorod materials have been synthesized with mixed vanadium valences, a hydroxide surface layer and pre-intercalated water characteristics. As cathodes for aqueous zinc batteries in a 3 m $ZnSO_4$

aqueous electrolyte, the obtained materials show superior electrochemical performance in terms of high specific capacity, power capability and cycling stability. By using both structural analysis and electrochemical characterizations, it was proven that the unique nanostructure and interfacial behavior significantly improved the material stability of orthorhombic V_2O_5 in aqueous media, suppressed the electrolyte decomposition at high operating voltage, and facilitated charge transfer kinetics. Furthermore, a realistic full cell using Zn powder as the anode with capacity balance design was also reported to deliver excellent capacity utilization and cycling stability. This work provides a strategy for enabling orthorhombic V_2O_5 materials to be efficient in aqueous Zn batteries and proves the feasibility of such materials for real industrial applications.

Supplementary Materials: The following supporting information can be downloaded at: https://www.mdpi.com/article/10.3390/nano12152530/s1, Figure S1: O 1s XPS spectrum; Figure S2: TG curves; Figure S3: SEM, TEM images and EDX results; Figure S4: V 2p XPS spectrum and SEM image of commercial V_2O_5; Figure S5: CV curves; Figure S6: GCD profiles at 1.0 A g^{-1}; Figure S7: Electrochemical performance of commercial V_2O_5; Figure S8: Electrochemical performance of V_2O_5 nanorods annealed at 500 °C; Figure S9: ex-situ XPS spectra of $V_2O_5{\cdot}nH_2O$ nanorods; Figure S10: SEM image of $V_2O_5{\cdot}nH_2O$ nanorods after 200 cycles; Table S1: Electrochemical performance of the reported orthorhombic V_2O_5 materials in aqueous Zn batteries. References [23,24,35–37,58,59] are cited in the supplementary materials.

Author Contributions: Conceptualization: H.Z.; methodology: X.T.; software: X.T. and G.G.; validation: X.T., G.G. and H.Z.; formal analysis: X.T. and K.W.; investigation: X.T.; resources: H.Z.; data curation: X.T. and H.Z.; writing—original draft preparation: X.T.; writing—review and editing: H.Z.; visualization: X.T. and K.W.; supervision: H.Z.; project administration: H.Z.; funding acquisition: H.Z. All authors have read and agreed to the published version of the manuscript.

Funding: This work was financially supported by the Guangdong Basic and Applied Basic Research Foundation (2020A1515110587), the Ningbo Natural Science Foundation (202003N4055), the Natural Science Basic Research Program of Shaanxi (2021JQ-110), the Natural Science Foundation of Chongqing (cstc2020jcyj-msxmX0667), and the Fundamental Research Funds for the Central Universities.

Institutional Review Board Statement: Not applicable.

Informed Consent Statement: Not applicable.

Data Availability Statement: Not applicable.

Conflicts of Interest: The authors declare no conflict of interest.

References

1. Chu, S.; Majumdar, A. Opportunities and challenges for a sustainable energy future. *Nature* **2012**, *488*, 294. [CrossRef] [PubMed]
2. Kittner, N.; Lill, F.; Kammen, D.M. Energy storage deployment and innovation for the clean energy transition. *Nat. Energy* **2017**, *2*, 17125. [CrossRef]
3. Goodenough, J.B.; Kim, Y. Challenges for Rechargeable Li Batteries. *Chem. Mater.* **2010**, *22*, 587–603. [CrossRef]
4. Liu, K.; Liu, Y.; Lin, D.; Pei, A.; Cui, Y. Materials for lithium-ion battery safety. *Sci. Adv.* **2018**, *4*, eaas9820. [CrossRef]
5. Chao, D.; Zhou, W.; Xie, F.; Ye, C.; Li, H.; Jaroniec, M.; Qiao, S.-Z. Roadmap for advanced aqueous batteries: From design of materials to applications. *Sci. Adv.* **2020**, *6*, eaba4098. [CrossRef]
6. Zhang, H.; Liu, X.; Li, H.; Hasa, I.; Passerini, S. Challenges and Strategies for High-Energy Aqueous Electrolyte Rechargeable Batteries. *Angew. Chem. Int. Ed.* **2021**, *60*, 598–616. [CrossRef]
7. Xu, W.; Wang, Y. Recent Progress on Zinc-Ion Rechargeable Batteries. *Nano-Micro Lett.* **2019**, *11*, 90. [CrossRef]
8. Zeng, X.; Hao, J.; Wang, Z.; Mao, J.; Guo, Z. Recent progress and perspectives on aqueous Zn-based rechargeable batteries with mild aqueous electrolytes. *Energy Storage Mater.* **2019**, *20*, 410–437. [CrossRef]
9. Zhao, Y.; Zhu, Y.; Zhang, X. Challenges and perspectives for manganese-based oxides for advanced aqueous zinc-ion batteries. *InfoMat* **2020**, *2*, 237–260. [CrossRef]
10. Gao, J.; Xie, X.; Liang, S.; Lu, B.; Zhou, J. Inorganic Colloidal Electrolyte for Highly Robust Zinc-Ion Batteries. *Nano-Micro Lett.* **2021**, *13*, 69. [CrossRef]
11. Daskalakis, S.; Wang, M.; Carmalt, C.J.; Vernardou, D. Electrochemical Investigation of Phenethylammonium Bismuth Iodide as Anode in Aqueous Zn^{2+} Electrolytes. *Nanomaterials* **2021**, *11*, 656. [CrossRef]

12. Chao, D.; Ye, C.; Xie, F.; Zhou, W.; Zhang, Q.; Gu, Q.; Davey, K.; Gu, L.; Qiao, S.-Z. Atomic Engineering Catalyzed MnO_2 Electrolysis Kinetics for a Hybrid Aqueous Battery with High Power and Energy Density. *Adv. Mater.* **2020**, *32*, 2001894. [CrossRef] [PubMed]
13. Sun, W.; Wang, F.; Hou, S.; Yang, C.; Fan, X.; Ma, Z.; Gao, T.; Han, F.; Hu, R.; Zhu, M.; et al. Zn/MnO_2 Battery Chemistry With H^+ and Zn^{2+} Coinsertion. *J. Am. Chem. Soc.* **2017**, *139*, 9775–9778. [CrossRef] [PubMed]
14. Zhong, C.; Liu, B.; Ding, J.; Liu, X.; Zhong, Y.; Li, Y.; Sun, C.; Han, X.; Deng, Y.; Zhao, N.; et al. Decoupling electrolytes towards stable and high-energy rechargeable aqueous zinc–manganese dioxide batteries. *Nat. Energy* **2020**, *5*, 440–449. [CrossRef]
15. Zampardi, G.; La Mantia, F. Prussian blue analogues as aqueous Zn-ion batteries electrodes: Current challenges and future perspectives. *Curr. Opin. Electrochem.* **2020**, *21*, 84–92. [CrossRef]
16. Li, Z.; Liu, T.; Meng, R.; Gao, L.; Zou, Y.; Peng, P.; Shao, Y.; Liang, X. Insights into the Structure Stability of Prussian Blue for Aqueous Zinc Ion Batteries. *Energy Environ. Mater.* **2021**, *4*, 111–116. [CrossRef]
17. Soundharrajan, V.; Sambandam, B.; Kim, S.; Alfaruqi, M.H.; Putro, D.Y.; Jo, J.; Kim, S.; Mathew, V.; Sun, Y.-K.; Kim, J. $Na_2V_6O_{16}{\cdot}3H_2O$ Barnesite Nanorod: An Open Door to Display a Stable and High Energy for Aqueous Rechargeable Zn-Ion Batteries as Cathodes. *Nano Lett.* **2018**, *18*, 2402–2410. [CrossRef]
18. Liu, X.; Zhang, H.; Geiger, D.; Han, J.; Varzi, A.; Kaiser, U.; Moretti, A.; Passerini, S. Calcium vanadate sub-microfibers as highly reversible host cathode material for aqueous zinc-ion batteries. *Chem. Commun.* **2019**, *55*, 2265–2268. [CrossRef]
19. Kundu, D.; Adams, B.D.; Duffort, V.; Vajargah, S.H.; Nazar, L.F. A high-capacity and long-life aqueous rechargeable zinc battery using a metal oxide intercalation cathode. *Nat. Energy* **2016**, *1*, 16119. [CrossRef]
20. Zhao, Q.; Huang, W.; Luo, Z.; Liu, L.; Lu, Y.; Li, Y.; Li, L.; Hu, J.; Ma, H.; Chen, J. High-capacity aqueous zinc batteries using sustainable quinone electrodes. *Sci. Adv.* **2018**, *4*, eaao1761. [CrossRef]
21. Wan, F.; Zhang, L.; Wang, X.; Bi, S.; Niu, Z.; Chen, J. An Aqueous Rechargeable Zinc-Organic Battery with Hybrid Mechanism. *Adv. Funct. Mater.* **2018**, *28*, 1804975. [CrossRef]
22. Zhang, S.; Long, S.; Li, H.; Xu, Q. A high-capacity organic cathode based on active N atoms for aqueous zinc-ion batteries. *Chem. Eng. J.* **2020**, *400*, 125898. [CrossRef]
23. Zhang, N.; Dong, Y.; Jia, M.; Bian, X.; Wang, Y.; Qiu, M.; Xu, J.; Liu, Y.; Jiao, L.; Cheng, F. Rechargeable Aqueous Zn–V_2O_5 Battery with High Energy Density and Long Cycle Life. *ACS Energy Lett.* **2018**, *3*, 1366–1372. [CrossRef]
24. Zhou, J.; Shan, L.; Wu, Z.; Guo, X.; Fang, G.; Liang, S. Investigation of V_2O_5 as a low-cost rechargeable aqueous zinc ion battery cathode. *Chem. Commun.* **2018**, *54*, 4457–4460. [CrossRef] [PubMed]
25. Chernova, N.A.; Roppolo, M.; Dillon, A.C.; Whittingham, M.S. Layered vanadium and molybdenum oxides: Batteries and electrochromics. *J. Mater. Chem.* **2009**, *19*, 2526–2552. [CrossRef]
26. Ding, J.; Zheng, H.; Gao, H.; Liu, Q.; Hu, Z.; Han, L.; Wang, S.; Wu, S.; Fang, S.; Chou, S. In Situ Lattice Tunnel Distortion of Vanadium Trioxide for Enhancing Zinc Ion Storage. *Adv. Energy Mater.* **2021**, *11*, 2100973. [CrossRef]
27. Ding, J.; Gao, H.; Ji, D.; Zhao, K.; Wang, S.; Cheng, F. Vanadium-based cathodes for aqueous zinc-ion batteries: From crystal structures, diffusion channels to storage mechanisms. *J. Mater. Chem. A* **2021**, *9*, 5258–5275. [CrossRef]
28. Ding, J.; Gao, H.; Liu, W.; Wang, S.; Wu, S.; Fang, S.; Cheng, F. Operando constructing vanadium tetrasulfide-based heterostructures enabled by extrinsic adsorbed oxygen for enhanced zinc ion storage. *J. Mater. Chem. A* **2021**, *9*, 11433–11441. [CrossRef]
29. Yang, Y.; Tang, Y.; Fang, G.; Shan, L.; Guo, J.; Zhang, W.; Wang, C.; Wang, L.; Zhou, J.; Liang, S. Li^+ intercalated $V_2O_5{\cdot}nH_2O$ with enlarged layer spacing and fast ion diffusion as an aqueous zinc-ion battery cathode. *Energy Environ. Sci.* **2018**, *11*, 3157–3162. [CrossRef]
30. He, P.; Zhang, G.; Liao, X.; Yan, M.; Xu, X.; An, Q.; Liu, J.; Mai, L. Sodium Ion Stabilized Vanadium Oxide Nanowire Cathode for High-Performance Zinc-Ion Batteries. *Adv. Energy Mater.* **2018**, *8*, 1702463. [CrossRef]
31. Ming, F.; Liang, H.; Lei, Y.; Kandambeth, S.; Eddaoudi, M.; Alshareef, H.N. Layered $Mg_xV_2O_5{\cdot}nH_2O$ as Cathode Material for High-Performance Aqueous Zinc Ion Batteries. *ACS Energy Lett.* **2018**, *3*, 2602–2609. [CrossRef]
32. Xia, C.; Guo, J.; Li, P.; Zhang, X.; Alshareef, H.N. Highly Stable Aqueous Zinc-ion Storage Using Layered Calcium Vanadium Oxide Bronze Cathode. *Angew. Chem. Int. Ed.* **2018**, *57*, 3943–3948. [CrossRef] [PubMed]
33. Mengyu, Y.; Pan, H.; Ying, C.; Shanyu, W.; Qiulong, W.; Kangning, Z.; Xu, X.; Qinyou, A.; Yi, S.; Yuyan, S.; et al. Water-Lubricated Intercalation in $V_2O_5{\cdot}nH_2O$ for High-Capacity and High-Rate Aqueous Rechargeable Zinc Batteries. *Adv. Mater.* **2018**, *30*, 1703725. [CrossRef]
34. Dong, J.; Jiang, Y.; Wei, Q.; Tan, S.; Xu, Y.; Zhang, G.; Liao, X.; Yang, W.; Li, Q.; An, Q.; et al. Strongly Coupled Pyridine-$V_2O_5{\cdot}nH_2O$ Nanowires with Intercalation Pseudocapacitance and Stabilized Layer for High Energy Sodium Ion Capacitors. *Small* **2019**, *15*, 1900379. [CrossRef]
35. Qin, H.; Chen, L.; Wang, L.; Chen, X.; Yang, Z. V_2O_5 hollow spheres as high rate and long life cathode for aqueous rechargeable zinc ion batteries. *Electrochim. Acta* **2019**, *306*, 307–316. [CrossRef]
36. Li, Y.; Huang, Z.; Kalambate, P.K.; Zhong, Y.; Huang, Z.; Xie, M.; Shen, Y.; Huang, Y. V_2O_5 nanopaper as a cathode material with high capacity and long cycle life for rechargeable aqueous zinc-ion battery. *Nano Energy* **2019**, *60*, 752–759. [CrossRef]
37. Liu, F.; Chen, Z.; Fang, G.; Wang, Z.; Cai, Y.; Tang, B.; Zhou, J.; Liang, S. V_2O_5 Nanospheres with Mixed Vanadium Valences as High Electrochemically Active Aqueous Zinc-Ion Battery Cathode. *Nano-Micro Lett.* **2019**, *11*, 25. [CrossRef]

38. Lu, Y.; Zhu, T.; van den Bergh, W.; Stefik, M.; Huang, K. A High Performing Zn-Ion Battery Cathode Enabled by In Situ Transformation of V_2O_5 Atomic Layers. *Angew. Chem. Int. Ed.* **2020**, *59*, 17004–17011. [CrossRef]
39. Wang, X.; Ma, L.; Zhang, P.; Wang, H.; Li, S.; Ji, S.; Wen, Z.; Sun, J. Vanadium pentoxide nanosheets as cathodes for aqueous zinc-ion batteries with high rate capability and long durability. *Appl. Surf. Sci.* **2020**, *502*, 144207. [CrossRef]
40. Cao, A.-M.; Hu, J.-S.; Liang, H.-P.; Wan, L.-J. Self-Assembled Vanadium Pentoxide (V_2O_5) Hollow Microspheres from Nanorods and Their Application in Lithium-Ion Batteries. *Angew. Chem. Int. Ed.* **2005**, *44*, 4391–4395. [CrossRef]
41. Sun, W.; Gao, G.; Zhang, K.; Liu, Y.; Wu, G. Self-assembled 3D N-CNFs/V_2O_5 aerogels with core/shell nanostructures through vacancies control and seeds growth as an outstanding supercapacitor electrode material. *Carbon* **2018**, *132*, 667–677. [CrossRef]
42. Liu, Y.; Zhang, C.; Liu, C.; Song, H.; Nan, X.; Cao, G. The effect of nitrogen annealing on lithium ion intercalation in nickel-doped lithium trivanadate. *Sci. Bull.* **2016**, *61*, 587–593. [CrossRef]
43. Zanarini, S.; Di Lupo, F.; Bedini, A.; Vankova, S.; Garino, N.; Francia, C.; Bodoardo, S. Three-colored electrochromic lithiated vanadium oxides: The role of surface superoxides in the electro-generation of the red state. *J. Mater. Chem. C* **2014**, *2*, 8854–8857. [CrossRef]
44. Zong, Q.; Du, W.; Liu, C.; Yang, H.; Zhang, Q.; Zhou, Z.; Atif, M.; Alsalhi, M.; Cao, G. Enhanced Reversible Zinc Ion Intercalation in Deficient Ammonium Vanadate for High-Performance Aqueous Zinc-Ion Battery. *Nano-Micro Lett.* **2021**, *13*, 116. [CrossRef] [PubMed]
45. Zhu, Y.; Ren, J.; Yang, X.; Chang, G.; Bu, Y.; Wei, G.; Han, W.; Yang, D. Interface engineering of 3D $BiVO_4$/Fe-based layered double hydroxide core/shell nanostructures for boosting photoelectrochemical water oxidation. *J. Mater. Chem. A* **2017**, *5*, 9952–9959. [CrossRef]
46. Li, X.; Kou, Z.; Xi, S.; Zang, W.; Yang, T.; Zhang, L.; Wang, J. Porous $NiCo_2S_4$/FeOOH nanowire arrays with rich sulfide/hydroxide interfaces enable high OER activity. *Nano Energy* **2020**, *78*, 105230. [CrossRef]
47. Yan, B.; Li, X.; Bai, Z.; Lin, L.; Chen, G.; Song, X.; Xiong, D.; Li, D.; Sun, X. Superior sodium storage of novel VO_2 nano-microspheres encapsulated into crumpled reduced graphene oxide. *J. Mater. Chem. A* **2017**, *5*, 4850–4860. [CrossRef]
48. Zhang, H.; Liu, X.; Li, H.; Qin, B.; Passerini, S. High-Voltage Operation of a V_2O_5 Cathode in a Concentrated Gel Polymer Electrolyte for High-Energy Aqueous Zinc Batteries. *ACS Appl. Mater. Interfaces* **2020**, *12*, 15305–15312. [CrossRef]
49. Zhu, K.; Wu, T.; Huang, K. Understanding the Dissolution and Phase Transformation Mechanisms in Aqueous Zn/α-V_2O_5 Batteries. *Chem. Mater.* **2021**, *33*, 4089–4098. [CrossRef]
50. Senguttuvan, P.; Han, S.-D.; Kim, S.; Lipson, A.L.; Tepavcevic, S.; Fister, T.T.; Bloom, I.D.; Burrell, A.K.; Johnson, C.S. A High Power Rechargeable Nonaqueous Multivalent Zn/V_2O_5 Battery. *Adv. Energy Mater.* **2016**, *6*, 1600826. [CrossRef]
51. Dong, Y.; Jia, M.; Wang, Y.; Xu, J.; Liu, Y.; Jiao, L.; Zhang, N. Long-Life Zinc/Vanadium Pentoxide Battery Enabled by a Concentrated Aqueous $ZnSO_4$ Electrolyte with Proton and Zinc Ion Co-Intercalation. *ACS Appl. Energy Mater.* **2020**, 3, 11183–11192. [CrossRef]
52. Xu, X.; Chen, Y.; Zheng, D.; Ruan, P.; Cai, Y.; Dai, X.; Niu, X.; Pei, C.; Shi, W.; Liu, W.; et al. Ultra-Fast and Scalable Saline Immersion Strategy Enabling Uniform Zn Nucleation and Deposition for High-Performance Zn-Ion Batteries. *Small* **2021**, *17*, 2101901. [CrossRef] [PubMed]
53. Zhao, J.; Zhang, J.; Yang, W.; Chen, B.; Zhao, Z.; Qiu, H.; Dong, S.; Zhou, X.; Cui, G.; Chen, L. "Water-in-deep eutectic solvent" electrolytes enable zinc metal anodes for rechargeable aqueous batteries. *Nano Energy* **2019**, *57*, 625–634. [CrossRef]
54. Li, Q.; Wang, Y.; Mo, F.; Wang, D.; Liang, G.; Zhao, Y.; Yang, Q.; Huang, Z.; Zhi, C. Calendar Life of Zn Batteries Based on Zn Anode with Zn Powder/Current Collector Structure. *Adv. Energy Mater.* **2021**, *11*, 2003931. [CrossRef]
55. Du, W.; Huang, S.; Zhang, Y.; Ye, M.; Li, C.C. Enable commercial Zinc powders for dendrite-free Zinc anode with improved utilization rate by pristine graphene hybridization. *Energy Storage Mater.* **2022**, *45*, 465–473. [CrossRef]
56. Wang, F.; Borodin, O.; Gao, T.; Fan, X.; Sun, W.; Han, F.; Faraone, A.; Dura, J.A.; Xu, K.; Wang, C. Highly reversible zinc metal anode for aqueous batteries. *Nat. Mater.* **2018**, *17*, 543–549. [CrossRef]
57. Hao, Y.; Zhang, S.; Tao, P.; Shen, T.; Huang, Z.; Yan, J.; Chen, Y. Pillaring Effect of K Ion Anchoring for Stable V_2O_5-Based Zinc-Ion Battery Cathodes. *ChemNanoMat* **2020**, *6*, 797–805. [CrossRef]
58. Chen, H.; Qin, H.; Chen, L.; Wu, J.; Yang, Z. V_2O_5@CNTs as cathode of aqueous zinc ion battery with high rate and high stability. *J. Alloys Compd.* **2020**, *842*, 155912. [CrossRef]
59. Chen, X.; Wang, L.; Li, H.; Cheng, F.; Chen, J. Porous V_2O_5 nanofibers as cathode materials for rechargeable aqueous zinc-ion batteries. *J. Energy Chem.* **2019**, *38*, 20–25. [CrossRef]

MDPI AG
Grosspeteranlage 5
4052 Basel
Switzerland
Tel.: +41 61 683 77 34

MDPI Books Editorial Office
E-mail: books@mdpi.com
www.mdpi.com/books

www.ingramcontent.com/pod-product-compliance
Lightning Source LLC
Chambersburg PA
CBHW041418210926
51850CB00013B/3307